Lecture Notes in Computer Science 6964

Commenced Publication in 1973
Founding and Former Series Editors:
Gerhard Goos, Juris Hartmanis, and Jan van Leeuwen

Carlos Delgado Kloos Denis Gillet
Raquel M. Crespo García
Fridolin Wild Martin Wolpers (Eds.)

Towards Ubiquitous Learning

6th European Conference
on Technology Enhanced Learning, EC-TEL 2011
Palermo, Italy, September 20-23, 2011
Proceedings

Springer

Volume Editors

Carlos Delgado Kloos
Universidad Carlos III de Madrid
Avenida Universidad 30, 28911 Leganés (Madrid), Spain
E-mail: cdk@it.uc3m.es

Denis Gillet
École Polytechnique Fédérale de Lausanne
1015 Lausanne, Switzerland
E-mail: denis.gillet@epfl.ch

Raquel M. Crespo García
Universidad Carlos III de Madrid
Avenida Universidad 30, 28911 Leganés (Madrid), Spain
E-mail: rcrespo@it.uc3m.es

Fridolin Wild
The Open University
Walton Hall, Milton Keynes, MK7 6AA, UK
E-mail: f.wild@open.ac.uk

Martin Wolpers
Fraunhofer Institute for Applied Information Technology (FIT)
Schloss Birlinghoven, 53754 Sankt Augustin, Germany
E-mail: martin.wolpers@fit.fraunhofer.de

ISSN 0302-9743 e-ISSN 1611-3349
ISBN 978-3-642-23984-7 ISBN 978-3-642-23985-4 (eBook)
DOI 10.1007/978-3-642-23985-4
Springer Heidelberg Dordrecht London New York

Library of Congress Control Number: 2011935864

CR Subject Classification (1998): H.4, H.3, I.2, C.2, H.5, J.1

LNCS Sublibrary: SL 2 – Programming and Software Engineering

Typesetting: Camera-ready by author, data conversion by Scientific Publishing Services, Chennai, India

Printed on acid-free paper

Springer is part of Springer Science+Business Media (www.springer.com)

Preface

These are the proceedings of the sixth edition of the European Conference on Technology-Enhanced Learning (EC-TEL). The conference was held in the beautiful bay of Mondello close to Palermo (Italy) during September 20–23, 2011. EC-TEL is now an established conference and the main reference in technology-enhanced learning in Europe and worldwide. Previous editions of the conference have been held in Crete, Greece (2006 and 2007), Maastricht, The Netherlands (2008), Nice, France (2009), and Barcelona, Spain (2010).

This year's edition was devoted to *ubiquitous learning* understood from the widest point of view. EC-TEL 2011 pushed further the ubiquitous learning paradigm by not only tackling the challenges of exploiting new trendy devices in various contexts, but also by investigating ways to meet and support formally and informally the learners in their learning playgrounds and social environments thanks to innovative scenarios.

EC-TEL has been as always very competitive. For the 2011 edition, 158 paper submissions were received from which 30 were selected as full papers. This makes an acceptance rate of 19%. In addition, 12 papers were presented as short papers and there were 8 posters. There were many interesting papers on topics such as Web 2.0 and social media, recommender systems, learning analytics, collaborative learning, interoperability of tools, etc.

Two first-class keynote speakers were present at the conference and their ideas are also reflected in these proceedings. Chris Dede of the Graduate School of Education of Harvard University gave a talk to present the perspective from the US with the title "Emerging Technologies, Ubiquitous Learning, and Educational Transformation." From Europe, Carlo Perrotta of Futurelab, UK, gave a talk about "Ubiquitous Learning vs. the Value of Boundaries: Reflections on Five Years of 'Innovation in Education'." It is now tradition to have the European Commission present the expectations of forthcoming calls of the Framework Programme. This year, as well, Javier Hernández Ros, new Head of Unit of Cultural Heritage and Technology-Enhanced Learning at the European Commission, gave a talk about the Framework Programme and its forthcoming eighth call.

A number of thought-provoking workshops were organized prior to the main conference. Twenty-five workshop proposals were received, from which ten were selected. An additional workshop about "Technology-Enhanced Learning for Mathematics and Science" was held in the memory of Erica Melis of DFKI, who recently passed away, as a tribute to her active engagement in the field.

There were many other highlights at the conference:

- The Doctoral Consortium, which plays a key role in the education of future PhDs in the field

– The Industry Track, where representatives of industry presented their vision of the field in two interesting panel sessions: "Learning in the Cloud" and "Learning and Social Networks."

This conference was made possible thanks to the help of many enthusiastic people. On the one hand, all the researchers from academia and industry who submitted papers, posters, and demos to the conference. Without them, there would not be any conference. We hope that those papers did not find their way into the conference, because of its incredibly high standards, will have the opportunity to see the light elsewhere. Thanks are due also to the Program Committee, who did an excellent job in the paper selection process and provided valuable feedback for improving the papers. Special thanks go to Mario Allegra and his team at the CNR Institute for Educational Technology in Palermo for the excellent local organization. Also many thanks to the Chairs of the different committees that formed the conference organization for an excellent and smooth collaboration.

Partial support was received from the STELLAR Network of Excellence, the Spanish Ministry of Science and Innovation, Desire2Learn, the UNIR University and the Universidad Carlos III de Madrid. Additionally, thanks are also due to many other supporting institutions, namely: the European Association of Technology-Enhanced Learning EATEL, the Fraunhofer Institut FIT, the EPFL, the eMadrid research project of the region of Madrid, the ROLE project of the European Commission, and the Spanish Association TELSpain. For handling the organization of submitted papers, the handy system EasyChair was used.

Above all we would like to thank you, the reader of these proceedings and possibly also participant of the conference, because it is you who can harness the advances reported in these proceedings and take them to the next level.

September 2011

Carlos Delgado Kloos
Denis Gillet
Raquel M. Crespo García
Fridolin Wild
Martin Wolpers

Organization

Executive Committee

General Chair

Martin Wolpers	Fraunhofer FIT, Germany

Program Chairs

Carlos Delgado Kloos	Universidad Carlos III de Madrid/eMadrid/TelSpain, Spain
Denis Gillet	École Polytechnique Fédérale de Lausanne, Switzerland

Local Organization Chair

Mario Allegra	Istituto per le Tecnologie Didattiche, Italy

Workshop Chair

Stefanie Lindstaedt	Know-Center Graz, Austria

Poster and Demo Chair

Fridolin Wild	The Open University, UK

Dissemination Chair

Abelardo Pardo	Universidad Carlos III de Madrid/eMadrid/TelSpain, Spain

Industrial Session Chair

Daniel Burgos	ATOS/International University of La Rioja/TELspain, Spain

Doctoral Consortium Chairs

Katherine Maillet	Télécom SudParis, France
Mike Sharples	University of Nottingham, UK

Publication Chair

Raquel M. Crespo García	Universidad Carlos III de Madrid, Spain

Program Committee

Heidrun Allert	Christian-Albrechts-Universität zu Kiel, Germany
Charoula Angeli	University of Cyprus, Cyprus
Luis Anido Rifón	Universidade de Vigo, Spain
Sylvester Arnab	The Serious Games Institute, UK
Inmaculada Arnedillo-Sánchez	Trinity College Dublin, Ireland
Juan I. Asensio-Pérez	University of Valladolid, Spain
Nicolas Balacheff	CNRS, France
Francesco Bellotti	University of Genoa, Italy
Adriana Berlanga	Open Universiteit, The Netherlands
Carl Blesius	Harvard Medical School, USA
Katrin Borcea-Pfitzmann	Technische Universität Dresden, Germany
Bert Bredeweg	University of Amsterdam, The Netherlands
Francis Brouns	Open Universiteit, The Netherlands
Peter Brusilovsky	University of Pittsburgh, USA
Daniel Burgos	ATOS Origin Research & Innovation, Spain
Rafael A. Calvo	University of Sydney, Austria
Lorenzo Cantoni	Università della Svizzera italiana, Italy
Linda Castañeda	University of Murcia, Spain
Manuel Castro	UNED, Spain
Mohamed Amine Chatti	Informatik 5 - RWTH Aachen, Germany
Giuseppe Chiazzese	CNR, Italy
Agnieszka Chrzaszcz	A.G.H. University of Science and Technology, Poland
Raquel M. Crespo García	Universidad Carlos III de Madrid, Spain
Ulrike Cress	Knowledge Media Research Center, Germany
Alexandra Cristea	University of Warwick, UK
Laura Czerniewicz	Centre for Educational Technology, South Africa
Giorgio Dabormida	ELGI, Italy
Paul De Bra	Eindhoven University of Technology, The Netherlands
Carlos Delgado Kloos	Universidad Carlos III de Madrid, Spain
Christian Depover	Université de Mons-Hainaut, France
Michael Derntl	RWTH Aachen University, Germany
Philippe Dessus	LSE, Grenoble, France
Darina Dicheva	Winston-Salem State University, USA
Stefan Dietze	Knowledge Media Institute, Open University, UK
Pierre Dillenbourg	EPFL, Switzerland
Yannis Dimitriadis	University of Valladolid, Spain
Vania Dimitrova	University of Leeds, UK

Peter Dolog	Aalborg University, Denmark
Hendrik Drachsler	Open Universiteit, The Netherlands
Jon Dron	Athabasca University, Canada
Benedict Du Boulay	University of Sussex, UK
Erik Duval	K.U.Leuven, Belgium
Martin Ebner	University of Graz, Austria
Marta Enrech Larrea	Universitat Oberta de Catalunya, Spain
Al Essa	Desire2Learn, Canada
Dieter Euler	Universität St. Gallen, Switzerland
Baltasar Fernández-Manjón	Universidad Complutense de Madrid, Spain
Carmen Fernández-Panadero	Universidad Carlos III de Madrid, Spain
Christine Ferraris	Université de Savoie - Equipe Scénarios, France
Dragan Gasevic	Athabasca University, Canada
Denis Gillet	Swiss Federal Institute of Technology in Lausanne (EPFL), Switzerland
Fabrizio Giorgini	Giunti Labs, Italy
Christian Glahn	Open Universiteit, The Netherlands
Monique Grandbastien	LORIA, UHP Nancy1, France
David Griffiths	University of Bolton, UK
Begona Gros	Universitat Oberta de Catalunya, Spain
Christian Guetl	Graz University of Technology, Austria
Joerg Haake	FernUniversität in Hagen, Germany
Andreas Harrer	Catholic University Eichstätt-Ingolstadt, Germany
Anne Helsdingen	CELSTEC OUNL, The Nertherlands
Davinia Hernández-Leo	Universitat Pompeu Fabra, Spain
Knut Hinkelmann	University of Applied Sciences Northwestern Switzerland FHNW
Ulrich Hoppe	University Duisburg-Essen, Germany
Emmanuel Jamin	ATOS, Spain
Patrick Jermann	Ecole Polytechnique Fédérale de Lausanne, Switzerland
Sanna Järvelä	University of Oulu, Finland
Marco Kalz	Open Universiteit, The Netherlands
Nikos Karacapilidis	University of Patras, Greece
Michael Kickmeier-Rust	University of Graz, Austria
Barbara Kieslinger	Centre for Social Innovation, Austria
Paul Kirschner	Open Universiteit, The Netherlands
Ralf Klamma	RWTH Aachen University, Germany
Joris Klerkx	Katholieke Universiteit Leuven, Belgium
Tomaž Klobučar	Jozef Stefan Institute, Slovenia
Rob Koper	Open Universiteit, The Netherlands
Nicole Kraemer	University Duisburg-Essen, Germany
Milos Kravcik	RWTH Aachen, Germany
Barbara Kump	TU Graz, Austria

Pierre Tchounikine	University of Grenoble, France
Stefaan Ternier	Open Universiteit, The Netherlands
Stefan Trausan-Matu	"Politehnica" University of Bucharest, Romania
Martin Valcke	Ghent University, Belgium
Jan Van Bruggen	Open Universiteit, The Netherlands
Christine Vanoirbeek	Ecole Polytechnique Federale Lausanne, Switzerland
Katrien Verbert	K.U.Leuven, Belgium
Aisha Walker	University of Leeds, UK
Gerhard Weber	University of Education Freiburg, Germany
Wim Westera	Open Universiteit, The Netherlands
Fridolin Wild	The Open University, UK
Katrin Wodzicki	Knowledge Media Research Center, Germany
Martin Wolpers	Fraunhofer Institute of Applied Information Technology, Germany
Volker Zimmermann	IMC AG, Germany

Additional Reviewers

Börner, Dirk
Dascalu, Mihai
Di Serio, Angela
Fazeli, Soude
Hsiao, I-Han
Kleanthous, Styliani
Lorenz, Anja
Rebaque, Pablo
Rebedea, Traian
Rega, Isabella
Schirru, Rafael
Sie, Rory
Tardini, Stefano
Ziebarth, Sabrina

Table of Contents

Invited Papers

Full Papers

Short Papers

Poster Papers

Emerging Technologies, Ubiquitous Learning, and Educational Transformation

Chris Dede

Harvard Graduate School of Education, 323 Longfellow Hall, 13 Appian Way, Cambridge, Massachusetts 02138, United States of America
Chris_Dede@Harvard.edu

Abstract. Emerging technologies are enabling ubiquitous learning. This can empower a structural change away from classrooms as the primary place of learning, the school day as the primary educational time, and the teacher as the primary source of information. Mobile devices can allow teachers to link to tutors, coaches, and mentors outside of school in a seamless web of support for each student. My colleagues and I are conducting research on sophisticated analytics to mine rich datastreams collected on students' devices, using each learner's interactions to help in developing personalized educational experiences. We also are studying "augmented realities" that infuse virtual data and authentic, simulated experiences into real world settings, facilitating transfer of learning from classrooms to life. However, to realize the full power of ubiquitous learning for educational transformation, educators must overcome numerous challenges related to devices and infrastructure, safety and privacy, digital assets and assessments, and human capital.

Keywords: mobile devices, ubiquitous learning, broadband, educational transformation, augmented reality, emerging technologies.

1 Introduction

We live in a time when the industrial era school system is on the verge of collapse. Our society can no longer afford a labor-intensive model of education that uses expensive human resources inefficiently. In the United States, we now see student-teacher ratios in some urban settings climbing to unworkable levels of 40 plus, even 60 pupils per class (Dolan, 2011; Dillon, 2011). This is not a temporary financial dislocation due to an economic downturn, but a permanent sea-change that has already happened in every other service sector of developed countries' economies.

Further, in K-12 schooling, our stellar illustrations of success are based on personal heroism, educators who make sacrifices in every other part of their lives in order to help their students. These are wonderful stories of saint-like dedication, but such a model for educational improvement is unscalable to typical teachers. We have not found a way to be effective and affordable at scale, and our financial resources are inexorably dwindling.

C. Delgado Kloos et al. (Eds.): EC-TEL 2011, LNCS 6964, pp. 1–8, 2011.

All other professions are successfully transforming to affordable models that use technology to empower typical practitioners to be effective. The United States Department of Education's 2010 National Educational Technology Plan (NETP) presents a transformational vision for 21st century education that builds on insights about modern interactive media gained from other sectors of society, but also depicts new processes and structures that recognize the unique challenges of helping students learn life-long and life-wide. This paper centers on the ways emerging technologies enable ubiquitous learning anyplace and anytime, to aid in achieving a structural change away from the industrial-era model of classrooms as the primary place of learning, the school day as the primary educational time, and the teacher as the primary source of information.

2 A Framework for 21st Century Education

The NETP (U.S. Department of Education, 2010) is a rich document that readers can interpret from a variety of perspectives. In this paper, the lens I will use is that of redesigning industrial-era schooling. I believe that our society should transform its current educational system into a different model better suited to prepare students for the opportunities and challenges of an emerging global, knowledge-based civilization (Dede, 2007). What elements in the Plan are suggestive about foundations for this redesign (Dede, 2010)?

Learning

- Learning can no longer be confined to the years we spend in school or the hours we spend in the classroom: It must be life-long, life-wide, and available on demand. (page 9)
- Technology provides access to a much wider and more flexible set of learning resources than is available in classrooms and connections to a wider and more flexible set of "educators," including teachers, parents, experts, and mentors outside the classroom. (pp. 11-12)
- Engaging and effective learning experiences can be individualized or differentiated for particular learners (either paced or tailored to fit their learning needs) or personalized, which combines paced and tailored learning with flexibility in content or theme designed to fit the interests and prior experience of each learner. (page 12)

Assessment

- Through multimedia, interactivity, and connectivity it is possible to assess competencies that we believe are important and that are aspects of thinking highlighted in cognitive research. It also is possible to directly assess problem-solving skills; make visible sequences of actions taken by learners in simulated environments; model complex reasoning tasks; and do it all within the contexts of relevant societal issues and problems that people care about in everyday life. (page 27)
- When students are learning online, there are multiple opportunities to exploit the power of technology for formative assessment. The same technology that supports learning activities gathers data in the course of learning that can be

used for assessment… As students work, the system can capture their inputs and collect evidence of their problem-solving sequences, knowledge, and strategy use, as reflected by the information each student selects or inputs, the number of attempts they make, the number of hints and feedback given, and the time allocation across parts of the problem. (pp. 29-30)

Teaching

- Connected teaching offers a vast array of opportunities to personalize learning. Many simulations and models for use in science, history, and other subject areas are now available online, including immersive virtual and augmented reality environments that encourage students to explore and make meaning in complex simulated situations (Dede 2009). To deeply engage their students, educators need to know about their students' goals and interests and have knowledge of learning resources and systems that can help students plan sets of learning experiences that are personally meaningful… Although using technology to personalize learning is a boost to effective teaching, teaching is fundamentally a social and emotional enterprise. The most effective educators connect to young people's developing social and emotional core (Ladson-Billings, 2009; Villegas & Lucas, 2002) by offering opportunities for creativity and self-expression. Technology provides an assist here as well… Digital authoring tools for creating multimedia projects and online communities for sharing them with the world offer students outlets for social and emotional connections with educators, peers, communities, and the world at large. Educators can encourage students to do this within the context of learning activities, gaining further insights into what motivates and engages students—information they can use to encourage students to stay in school (pp. 41-42).
- All institutions involved in preparing educators should provide technology-supported learning experiences that promote and enable the use of technology to improve learning, assessment, and instructional practices. This will require teacher educators to draw from advances in learning science and technology to change what and how they teach, keeping in mind that everything we now know about how people learn applies to new teachers as well. The same imperatives for teacher preparation apply to ongoing professional learning. Professional learning should support and develop educators' identities as fluent users of advanced technology, creative and collaborative problem solvers, and adaptive, socially aware experts throughout their careers. (page 44)

Productivity

- One of the most basic assumptions in our education system is time-based or "seat-time" measures of educational attainment…Time-based measures were appropriate in their day, but they are not now when we know more about how people learn and we have access to technology that can help us accommodate different styles and paces of learning. As we move to online learning and learning that combines classroom and online learning, time-based measures will increasingly frustrate our attempts to provide learning experiences that lead to achievement and the pursuit of postsecondary education that our modern world requires. (page 68)

- Another basic assumption is the inflexible way we organize students into age-determined groups, structure separate academic disciplines, organize learning into classes of roughly equal size with all the students in a particular class receiving the same content at the same pace, and keep these groups in place all year. The last decade has seen the emergence of some radically redesigned schools, demonstrating the range of possibilities for structuring education. For example, organizing education around the demonstration of competence rather than seat time opens up a wide range of possibilities. The first school district to win the Baldridge Quality Award, Chugach School District in Alaska, achieved remarkable gains in student outcomes after mobilizing its community to identify the competencies it wanted to see in high school graduates and shifting to a performance-based system in which diplomas were awarded on the basis of performance on the district's assessment of those competencies. (pp. 68-69)
- As we seek ways to extend learning time, in addition to considering the amount of time students spend in school, we should also look at whether we can provide engaging and powerful learning experiences through other means. For example, we know that students' lives outside school are filled with technology that gives them 24/7 mobile access to information and resources and allows them to participate in online social networks and communities where people from all over the world share ideas, collaborate, and learn new things. Our education system should leverage students' interest in technology and the time they currently spend learning informally outside the regular school hours to extend learning time in a way that motivates them even more. (pp. 70-71)

I believe these NETP ideas about learning, assessment, teaching, and productivity are most powerful when implemented via a "distributed" model of formal education, in which parents trained and licensed as tutors, informal educators (e.g., museum staff, librarians) certified as coaches, and community members prepared and licensed as mentors are also paid "professional educators" (Dede, 2010). In such a 21st century educational system, schools of education would prepare, license, and provide professional support for teachers, tutors, coaches, and mentors who were trained to orchestrate their coordinated activities through the use of a sophisticated technology infrastructure.

3 Mobile Broadband Devices and Ubiquitous Learning

What technology infrastructure can enable this ambitious vision of distributed formal education? Mobile broadband devices now have "six senses" (Dede & Bjerede, 2011): (1) knowing where you are, (2) interacting with networks, (3) sensing local content and services, (4) discovering relevant things, (5) enhancing your surroundings with information and simulation, and (6) learning your interests as well as how and with whom you like to learn. This new capacity for learning infused throughout the world is a powerful way of moving beyond the traditional model of learning isolated from the world in classroom settings.

Mobile broadband devices can enable customized learning life-long and life-wide, as well as providing tools and applications that allow teachers, tutors, coaches, and mentors to orchestrate their efforts into a seamless web of support. Our research on virtual performance assessments (http://vpa.gse.harvard.edu) is developing insights about how sophisticated analytics can mine rich datastreams collected from learners' digital interactions to help with diagnosing what types of personalized educational experiences are good next steps (Clarke-Midura, Dede, & Norton, 2011). In conjunction with this work, we are designing and studying engaging *augmented realities* that infuse virtual data and authentic, simulated experiences into real world settings, facilitating transfer of learning from classrooms to life situations.

3.1 EcoMOBILE as an Example of Research on Augmented Reality

My colleagues and I are just beginning a Qualcomm-funded research project, EcoMobile, to determine the strengths and limits of augmented reality as a complement to classroom learning in science. Knowledge about ecosystems and populations is an important strand of the life science content standards, and the processes underlying ecosystems exemplify sophisticated causal mechanisms (e.g., systems dynamics) foundational for advanced science and mathematics. However, even after instruction, students often hold inaccurate interpretations about ecosystems' structural patterns and inherent causal complexity (Grotzer & Basca, 2003). Teachers struggle to convey in hands-on, engaging ways difficult concepts with causality such as time delays, spatial distance, non-obvious causes, and population-level effects.

To meet this shortfall in current, largely textbook-based curricula, with U.S. Department of Education Institute of Education Sciences' funding we have developed and are studying EcoMUVE: a multi-user virtual environment (MUVE)-based curriculum that addresses grades 6 through 8 life science National Science Education Standards (www.ecomuve.org). EcoMUVE is an inquiry-based, four-week curriculum that incorporates student experiences in immersive, simulated virtual ecosystems to enhance student understanding of ecosystems science, the inquiry process, and the complex causality inherent in ecosystems dynamics. Our research findings show promising results on its perceived value, usability, and implementation feasibility, along with gains in student learning and motivation (Metcalf et al, 2011).

We hypothesize that student engagement, understanding, and self-efficacy in science would be enhanced if students experiencing EcoMUVE could also use mobile broadband devices (MBDs) to explore the real world ecosystems in their own locality. To study these hypotheses, we have begun work on EcoMOBILE: a complementary set of learning experiences using MBDs to access virtual information and simulated experiences while immersed in real world ecosystems. MBDs allow students to collect and share data using probeware, cameras, and microphones; access on-site information about ecosystem components; and visit geo-referenced locations to observe critical components of the ecosystem and to experience virtual simulations related to causality.

Some aspects of ecosystem ecology are difficult to translate into a virtual environment, such as virtually representing the true diversity in forms and behaviors of organisms in a real environment. Also, the feel and smell of an ecosystem after a recent rain cannot be digitally replicated. Further, real world experiences that mirror aspects of the virtual world may aid in transfer and generalization of learning. Students have a need to connect the abstract ideas they are learning in science class to experiences they have in the real world. This process can be mediated by the affordances offered by mobile broadband devices. The EcoMOBILE project is studying the extent to which current technologies can accomplish this goal in an affordable, practical manner, as well as what implementation challenges are involved in going to scale.

4 Enabling Transformation

The transition from industrial-era schooling to 21st century distributed education can involve evolutionary, revolutionary, or disruptive transformation (Dede & Bjerede, 2011). *Evolutionary* transformation centers on using mobile devices within and outside of classrooms to enable 1:1 access to computing, digital textbooks, and facile collaboration among teams of students. *Revolutionary* transformation, in contrast, focuses on using mobile broadband to expand human support for learning beyond the classroom and school day to invent new structures for formal education, such as the "distributed learning" model sketched above. As an alternative to these types of internal, deliberate change for current institutions, *disruptive* transformation builds on Harvard Business School Professor Clay Christensen's concept of new forms of educational institutions, such as virtual schools and online universities, using mobile devices as part of a strategy for externally undercutting and eliminating inflexible, traditional models of education (Christensen, Horn, & Johnson, 2008). The emergence of 21st century learning will likely involve various groups implementing all three of these transformative approaches.

Any of these transformative strategies faces many challenges. The most difficult problems are "knots" because they required coordination across stakeholders to resolve. As an example of such a knot, educators and parents do not want to invest in mobile infrastructure unless they are certain high quality educational experiences are available – yet content providers and technology vendors do not want to invest in developing these experiences unless they are sure a strong market for those services exists. Investment by public and private stakeholders is necessary to untie this knot.

Four key areas that must be "unknotted" to realize the power of mobile broadband for ubiquitous learning are:

- *Devices and Infrastructure:* How can we best balance educational investments between the classic infrastructure of wired computers and the emerging infrastructure of wireless mobile devices?
- *Safety and Privacy:* How can we use Internet access and digital student data to enhance education, while preventing various forms of abuse?
- *Digital Assets and Assessments:* How can we drive innovation in digital learning materials and services when the education market is notoriously fragmented and slow to adopt, and when the strengths and limits of mobile devices for learning are not well understood?
- *Human Capital:* How can we empower educators and other stakeholders to realize the potential of anytime, anyplace mobile learning through evolutionary, revolutionary, and disruptive transformations that move beyond the model of industrial-era schooling?

Complicating the challenge is that barriers in each area create difficulties for progress in the others.

5 Conclusion

A ubiquitous technology infrastructure that supports anytime, anyplace learning as central to 21st century education. I believe that every student and educator should have a mobile broadband device, with training and support for its optimal usage to empower learning. Policy makers should systematically explore mechanisms to fund such an infrastructure for every district, school, and student, regardless of economic status.

The opportunity of mobile learning is best set within a larger framework of education empowered by technology, as illustrated by the 2010 U.S. National Educational Technology Plan. Recommendations from the Plan that particularly speak to the issues discussed in this paper include:

- *1.3* States, districts, and others should develop and implement learning resources that exploit the flexibility and power of technology to reach all learners anytime and anywhere.
- *2.3* Conduct research and development that explores how embedded assessment technologies, such as simulations, collaboration environments, virtual worlds, games, and cognitive tutors, can be used to engage and motivate learners while assessing complex skills.
- *3.2* Leverage social networking technologies and platforms to create communities of practice that provide career-long personal learning opportunities for educators within and across schools, preservice preparation and in-service education institutions, and professional organizations.
- *4.2* Ensure that every student and educator has at least one Internet access device and appropriate software and resources for research, communication, multimedia content creation, and collaboration for use in and out of school.

Working to advance these objectives is a path forward to achieve the potential of emerging technologies to enable transformation from industrial-era schooling to 21st century education.

Acknowledgments. Research on the EcoMUVE and Virtual Performance Assessment projects is funded by the U.S. Department of Education's Institute of Education Sciences. The EcoMOBILE project is funded by Qualcomm's Wireless Reach Initiative. All conclusions in this paper are those of the author and do not necessarily reflect the views of the funders.

References

1. Clarke-Midura, J., Dede, C., Norton, J.: Next Generation Assessments for Measuring Complex Learning in Science. In: The Road Ahead for State Assessments, pp. 27–40. Rennie Center for Education and Public Policy, Cambridge (2011), `http://renniecenter.issuelab.org/research`

2. Christensen, C.M., Horn, M.B., Johnson, C.W.: Disrupting Class: How Disruptive Innovation will Change the Way the World Learns. McGraw-Hill, New York (2008)
3. Dede, C., Bjerede, M.: Mobile Learning for the 21st Century: Insights from the 2010 Wireless EdTech Conference. Qualcomm, San Diego (2011), http://wirelessedtech.com
4. Dede, C.: Reflections on the Draft National Educational Technology Plan 2010: Foundations for Transformation. Educational Technology 50(6), 18–22 (2010)
5. Dede, C.: Immersive Interfaces for Engagement and Learning. Science 323(5910), 66–69 (2009)
6. Dede, C.: Reinventing the Role of Information and Communications Technologies in Education. In: Smolin, L., Lawless, K., Burbules, N. (eds.) Information and Communication Technologies: Considerations of Current Practice for Teachers and Teacher Educators (NSSE Yearbook 2007), vol. 106(2), pp. 11–38. Blackwell Publishing, Malden (2007)
7. Dillon, S.: Tight budgets Mean Squeeze in Classrooms. New York Times, March 6 (2011), http://www.nytimes.com/2011/03/07/education/07classrooms.html?_r=1&pagewanted=print
8. Dolan, M.: Detroit Schools Cuts Plan Approved. Wall Street Journal, February 22 (2011), http://online.wsj.com/article/SB10001424052748703610604576158783513445212.html
9. Grotzer, T.A., Basca, B.B.: How Does Grasping the Underlying Causal Structures of Ecosystems Impact Students' Understanding? Journal of Biological Education 38(1), 16–29 (2003)
10. Ladson-Billings, G.: The dreamkeepers: Successful teachers of African American children. Wiley, San Francisco (2009)
11. Metcalf, S., Kamarainen, A., Tutwiler, M.S., Grotzer, T., Dede, C.: Ecosystem Science Learning via Multi-User Virtual Environments. International Journal of Gaming and Computer-Mediated Simulations 3(1), 86–90 (2011)
12. U.S. Department of Education: Transforming American Education: Learning Powered by Technology (2010), http://www.ed.gov/technology/netp-2010
13. Villegas, A.M., Lucas, T.: Preparing culturally responsive teachers. Journal of Teacher Education 53(1), 20–32 (2002)

Ubiquitous Learning vs. the Value of Boundaries: Reflections on Five Years of 'Innovation in Education'

Carlo Perrotta

Institute of Education, University of London, UK; Futurelab Education, London, UK
c.perrotta@ioe.ac.uk

Abstract. This paper offers a reflective account of the relationship between innovation and ubiquitous learning in education. The paper argues that the tendency towards removing all boundaries which lies at heart of 'ubiquitous learning' certainly offers visionary promises, but it might also lead to a condition characterised by the absence of all structuring elements. At the level of the individual learner, this condition might result in a sense of fragmentation and to a focus on safety and emotional well-being. At an institutional level it might legitimise a drive towards very reductive forms of accountability, in which technology is used mainly for surveillance and behaviour control. The paper proposes that the boundaries of educational technology should be acknowledged and further investigated, in order to identify practical strategies that strengthen, rather than dilute, the agency of learners in ubiquitous learning landscapes, by tracing clear distinctions and demarcations which can structure their experiences.

Keywords: Ubiquitous, Innovation, Boundaries, Interdisciplinary.

1 Introduction

Futurelab is a UK based charity devoted to 'transforming' education through innovative technology and practice. Its story so far has been one of optimism and euphoria about the empowering potential of technology in education, coupled with access to funding and opportunities to create networks across the worlds of education, industry and academia. In the UK, Futurelab has been one of the strongest proponents of a technology-based approach in formal schooling, based on the development of '21st century skills' like enquiry based learning or digital literacy[1]. Implicit in much of Futurelab's work is a set of assumptions about how to overcome the limits and the barriers that have been holding back the modernisation of education, not only in the UK but at a global level. This set of assumptions can be summarised by the word 'blurring': the blurring of formal and informal spaces where young people interact and learn, the blurring of boundaries between curricular subjects, which emphasises breadth, eclecticism and process skills over depth and 'content', the blurring of roles in the pedagogic relationship between adults and young people. In this respect, it

[1] http://www.futurelab.org.uk/our-work

C. Delgado Kloos et al. (Eds.): EC-TEL 2011, LNCS 6964, pp. 9–14, 2011.

could be argued that the ubiquitous potential of technology-enhanced learning - the potential to occur anywhere, across physical and social boundaries – is the inspiration behind Futurelab's idea of 'innovation in education'. For someone who has been involved in several discussions and projects based on such idea, this conference is a welcome opportunity for 'taking stock', by casting a more critical look on the aforementioned assumptions, in the hope that the resulting reflection will have something useful to offer to those who are interested in 'ubiquitous learning': possibly the most visionary and innovative notion currently available in educational technology.

2 Ubiquity: The Perfect Innovation

In 1991 Mark Weiser introduced the concept of 'Ubiquitous Computing'. He predicted that technological innovation would lead to the pervasiveness of 'smart' devices in all spheres of life. Ubiquitous and persistent access to technology led many to envisage a concurrent futuristic scenario for technology-enhanced learning. In this scenario, learners would have opportunities to work with other learners and to collaborate locally, nationally and internationally, overcoming language barriers, time barriers and establishing learning networks anywhere and at any time of day. Fast and adaptive links between learners and resources would become widespread, established through a range of connected devices and maintained across a range of locations. 'Intelligent' technology would augment the learning experience; for example, networks of sensors and RFID (Radio-Frequency Identification) objects would gather data, seamlessly and unbeknownst to learners, providing immediate feedback in more effective ways.

This scenario is already a reality in some respects, while in others it is still a compelling- or frightening - vision of the future. As it is often the case with technology-inspired visions of the future, the more challenging aspects tend to be either submerged under layers of unwarranted enthusiasm, or diluted in overly negative caricatures of a 'dystopian' world in which we are dominated by machines. No doubt this is a fraught debate which polarises opinions, but nonetheless is a useful one, especially in the field of educational technology where ambivalence and contradiction are often avoided due to the deep-rooted belief that technology is a 'transformational' force for good (Selwyn, 2010).

There are however some critical points, raised mostly in economics and sociology, which provide us with insights in the broader dynamics of innovation that underpin ubiquitous computing and its most closely related companion: ubiquitous commerce. In this paper, I would like to argue that this broader focus can be a good starting point to begin an exploration of the challenges and contradictions of ubiquitous learning as well.

From a socio-economic perspective, the primary purpose of innovation is to remove barriers to economic development by altering the parameters of existing markets or creating entirely new ones. In order to accomplish this purpose, innovation processes have become heavily reliant on technology, or rather on the techno-utopian assumption that all possible limits (natural, social, geographical) can be circumvented or transcended with the right mix of ingenuity and invention (Harvey, 2005; 2010). In this respect, it could be argued that the emergence of ubiquitous computing is underpinned by the visionary aim to create a 'perfect' innovation landscape, in which all boundaries and limits are virtually removed, and where technology allows information to flow seamlessly and uninterruptedly around individuals, immersing them in an eco-system of connected devices which are constantly collecting data, constantly adapting, constantly offering opportunities to consume. In other words, there appears to be a strong relationship between innovation-driven economics and the trend towards ubiquitous computing, and this might go some way to explaining why practical applications were initially envisaged in commerce, rather than learning. The study of these applications led to the identification of some challenges and to calls for further, more critically minded research (Bonn et al, 2005).

For instance, empirical research in 'U-commerce' (ubiquitous commerce) has suggested that faced with uninterrupted feedback loops of data collection and adaptation, which would allow 'smart products' to be unpredictably and uncannily 'personal', individuals might react angrily and wish for a return to more predictable and less disorienting consumer patterns (Sheng et al, 2008). In another study, the extent to which people viewed their privacy protection as critical had an influence on how much they were willing to accept RFID-based information services. In the same way, expected emotional reactions to shopping in an 'augmented' supermarket were negatively related to how much people valued their privacy (Rothensee & Spiekermann, 2008).

Now, these insights shed some light on the challenges associated with blurring or removing altogether the boundaries between private and public information in U-commerce environments, but can they help our understanding of 'U-Learning'? I believe they can, as they convey an awareness of problematic implications-mainly in relation to privacy- which surely have a broad relevance beyond the world of commerce. However, I would also like to argue that in ubiquitous learning the problem of private vs. public is only one part of a more complex and nuanced dynamic, characterised by the existence of other 'valuable' boundaries which help structure the experience of learners, and which should be acknowledged and negotiated, rather than ignored. I will explain this point in the next section.

3 The Value of Boundaries

There are authors who, independently and from radically different disciplinary backgrounds, have made similar suggestions: there are certain boundaries which should be acknowledged and possibly used as a framework to develop a more realistic, more socially just and even more empirically sound understanding of education and learning. For instance, the cultural boundaries between areas of knowledge (Young and Muller, 2010); the boundaries that stem from subjective experiences of identity and belonging (Wortham, 2006); not least the bounded nature of human cognition, that is, the cognitive architecture that supports, and constrains, deep understanding and expertise development (Mayer, 2003). Moreover, recent contributions to the literature on the 'effects' of ICT for learning suggest that the most robust evidence of ICT use in education focused on specific uses of ICT (Cox & Marshall, 2007). Where the aim of the research has been to investigate ICT in a 'boundless' fashion, without clearly identifying the range and type of ICT use, then unclear or contradictory results were obtained. This suggestion draws on evidence that shows how technology environments take many forms and require radically different types of learning, hence an investigation of energy consumption could be carried out using the spreadsheet application Excel, or a software modelling environment like Model Builder.

"In the case of Excel, the learner needs to understand the relationship between mathematical equations and tabular means of presenting and inserting these. In the case of Model Builder, the learner needs to learn new modelling syntax based on natural language and how this can be used in conjunction with icons and images on the screen" (Cox and Marshall, 2007 p62-63).

Different technologies not only invoke different social practices and different trajectories of expertise and learning, but also different, distinct domains of knowledge. Failure to acknowledge this intrinsic 'differentiatedness' might lead to creeping and covert forms of inequalities, in which the more disadvantaged learners are the worst damaged, as they lack the resources needed to reconfigure and rediscover the boundaries which have been artificially submerged. In other words, the tendency towards removing all boundaries which lies at heart of 'ubiquitous learning' certainly offers visionary promises, but it might also lead to a condition characterised by the absence of all structuring elements, and by the conflation of different times and spaces, and different types of knowledge, in one constant undifferentiated 'learning flow'.

"Educational boundaries are social but also real, not arbitrary, that is, they cannot be dissolved, at least in the short term, without serious consequences for most if not all learners. What such de-differentiating mechanisms are most likely to achieve is not to dissolve the boundaries, but to render them invisible—an invisibility that is exaggerated for the more disadvantaged" (Young and Muller, 2010 pp. 18/19).

This condition would very much resemble what some sociologists call a 'crisis of representation': a state of confusion and uncertainty as to how to represent the world and make sense of it, in which fragmentation and defensive forms individualism tend to thrive (Harvey, 1990). Similarly, drawing on insights from the psychology of 'self-regulation', we could hypothesise that a situation in which learning opportunities are totally distributed and pervasive might cause in many learners, most likely those already lagging behind or coming from disadvantaged backgrounds, a state of constant stress in which they would rather focus on maintaining their emotional well-being within clearly defined bounds, rather than embarking on a boundless and daunting path of growth and self-development (Boekaerts & Niemivierta, 2000).

More broadly, we could also hypothesise a 'worst case scenario' occurring in institutional contexts like schools and universities, which are unlikely to disappear in the near ubiquitous future we are facing. In these complex settings, frustration at the boundless nature of 'ubiquitous learning' might lead to a paradoxical outcome whereby pervasive technology becomes a tool in very conservative and strict regimes of individual and collective accountability. I will explain this point further.

There is some evidence suggesting that a strong and explicit emphasis on innovation is sometimes accompanied in education by a more implicit trend towards 'traditional' pedagogic models, which can be more easily standardised to increase performance. This evidence is provided by Lubienski's account of the charter school experiment in the US (Lubienski, 2003). According to Lubienski, a mix of conditions drove the model of market-based innovation that underpinned the charter school experiment - i.e. consumer choice encouraging competition and hence innovation - to a rather contradictory outcome: the homogenisation of such schools around conservative forms of curricular organisation and assessment.

I think we could reframe Lubienski's analysis, or parts of it, as an account of the failure to acknowledge the boundaries I have described thus far. In the midst of a 'crisis of representation' caused by the inability to grasp the differentiatedness and complexity of learning processes, the American Charter Schools resorted to using the rhetoric of innovation as a marketing stratagem, a form of window dressing, to "shape their intake as much as possible through image presentation (...)" (Lubienski, 2003, p.424), while actual pedagogic practices were becoming ossified, due the intrinsic 'disciplining' pressure of a performance based accountability framework that rewarded measurable and standardised performance (test scores). Still according to Lubienski, "Employing standardized practices and strategies based on 'appearances' may be a more effective and less costly option for an organization than experimenting with new approaches or mediating between conflicting goals (DiMaggio and Powell 1983)" (ibid).

In such a scenario, it is not totally unrealistic to hypothesise that the 'rhetoric' of ubiquitous computing will be, at best, part of marketing and 'branding' strategies to shape student intake, with little or no relevance with actual learning and only superficially implemented in that respect. At worst, ubiquitous computing might become easily subjected to the demands of ever tighter and stricter accountability agendas, for instance by being used mostly for constant assessment, surveillance and behaviour control. I would like to emphasise that this is by no means a prediction, but only a suggestion: a 'heightened sensitivity' around these risks might help the identification of more critical and more socially relevant research foci in ubiquitous learning. In this type of research, the exploration of which boundaries should be recognised and valued might help us develop a more realistic understanding of what can – and what cannot - be achieved through 'ubiquitous learning', and more broadly through technology-driven innovation in education. This understanding would hopefully underpin the identification of practical strategies that strengthen, rather than dilute, the agency of learners in ubiquitous learning landscapes, by tracing clear distinctions and demarcations which can structure their experiences and can counter the spreading of damaging crises of representation.

4 Concluding Remarks

In this short paper, I have drawn on my own professional experience with ubiquitous learning, to articulate a critical reflection around the use of technology to remove boundaries and limits in education. My main contention is that an unchecked and uncritical enthusiasm in the potential of 'ubiquity' might lead to the trivialisation or obfuscation of important boundaries. At the level of the individual learner, this condition might result in a 'crisis of representation' in which confusion and uncertainty lead to a sense of fragmentation and to a focus on safety and emotional well-being, rather than self-development and independence. At an institutional level, this condition might in a similar way legitimise a drive towards very conservative and reductive forms of accountability, in which a most paradoxical outcome might be that ubiquitous technology becomes part of a 'totalitarian' agenda that aims to exercise "increasingly sophisticated, increasingly dubious forms of influence and control" (Fielding, 2006, p. 366).

This paper has drawn on insights from economics, sociology and psychology. As such, it also represents a tentative suggestion to ground current and future research efforts in ubiquitous learning in a truly 'socio-cultural' and interdisciplinary perspective. A perspective, that is, that attempts to bridge the incommunicability between disciplines that describe and explain similar objects of study using different languages, but when brought together can contribute to more robust and more socially relevant forms of scholarship.

References

1. Boekaerts, M., Niemivirta, M.: Self-regulated learning: Finding a balance between learning goals and ego-protective goals. In: Boekaerts, M., Pintrich, P.R., Zeidner, M. (eds.) Handbook of Self-Regulation, pp. 417–451. Academic Press, San Diego (2000)
2. Bohn, J., Coroama, V., Langheinrich, M., Mattern, F., Rohs, M.: Social, Economic, and Ethical Implications of Ambient Intelligence and Ubiquitous Computing. In: Weber, W., Rabaey, J., Aarts, E.H.L. (eds.) Ambient intelligence. Springer, Heidelberg (2005)
3. Cox, M., Marshall, G.: Effects of ICT: Do we know what we should know? Education and Information Technologies 12(1), 59–70 (2007)
4. DiMaggio, P.J., Powell, W.W.: The iron cage revisited: Institutional isomorphism and collective rationality in organizational fields. American Sociological Review 48(2), 147–160 (1983)
5. Fielding, M.: Leadership, personalisation and high performance schooling: naming the new totalitarianism. School Leadership & Management 26(4), 347–369 (2006)
6. Harvey, D.: The Enigma of Capital and the Crises of Capitalism. Oxford University Press, Oxford (2010)
7. Harvey, D.: A Brief History of Neoliberalism. Oxford University Press, Oxford (2005)
8. Harvey, D.: The Condition of Postmodernity: An Enquiry into the Origins of Cultural Change. Blackwell, Malden (1990)
9. Lubienski, C.: Innovation in Education Markets: Theory and Evidence on the Impact of Competition and Choice in Charter Schools. American Educational Research Journal 40(2), 395–443 (2003)
10. Mayer, R.: Should there be a three-strikes rule against pure discovery learning? The case for guided methods of instruction. American Psychologist 59(1), 14–19 (2004)
11. Rothensee, M., Spiekermann, S.: Between Extreme Rejection and Cautious Acceptance Consumers, Reactions to RFID-Based IS in Retail. Social Science Computer Review 26(1), 75–86 (2008)
12. Selwyn, N.: Schools and Schooling in the Digital Age: a Critical Analysis. Routledge, London (2011)
13. Sheng, H., Fui-Hoon, F., Keng, S.: An Experimental Study on Ubiquitous commerce Adoption: Impact of Personalization and Privacy Concerns. Journal of the Association for Information Systems 9(6), article 15 (2008)
14. Weiser, M.: The Computer for the 21st Century. Scientific American, 94–104 (September 1991)
15. Wortham, S.: Learning identity: The joint emergence of social identification and academic learning. Cambridge University Press, Cambridge (2006)
16. Young, M., Muller, J.: Three Educational Scenarios for the Future: lessons from the sociology of knowledge. European Journal of Education 45, 11–27 (2010)

The MIROR Project

Anna Rita Addessi[1] and Gualtiero Volpe[2]

[1] University of Bolgna, Italy
annarita.addessi@unibo.it
[2] University of Genoa, Italy
gualtiero.volpe@unige.it

Abstract. The MIROR project is an European Union funded project under the 7th Framework programme (Technology-enhanced learning). The project is in its first year and so the aim of this paper is to explain its background and rationale, to describe its aims and structure. The MIROR Project deals with the development of an innovative adaptive system for music learning and teaching based on the reflexive interaction paradigm. The reflexive interaction paradigm is based on the idea of letting users manipulate virtual copies of themselves, through specifically designed machine-learning software referred to as interactive reflexive musical systems (IRMS). MIROR aims at extending the IRMS paradigm with the analysis and synthesis of multisensory expressive gesture to increase its impact on the musical pedagogy of young children. The MIROR platform will be developed in the context of early childhood music education. It will act as an advanced cognitive tutor, designed to promote specific cognitive abilities in the field of music improvisation, composition and body performance, both in formal learning contexts (kindergartens, primary schools, music schools) and informal ones (at home, kinder centers, etc.).

Keywords: young children music education, MIROR platform, reflexive interaction, expressive gesture analysis.

1 Introduction

The MIROR project is an European Union funded project under the 7th Framework programme (*FP7/2007-2013, Technology-enhanced learning, grant agreement n° 258338*). The objective of the MIROR project is to develop and validate a novel software platform for music learning and teaching based on the notion of *reflexive interaction*. The MIROR platform will address improvisation, composition, and body performance contexts both in “formal” situations (music schools, individual and collective lessons) and “informal” ones (e.g. at home). The technical design of this educational software will integrate from the beginning pedagogical constraints stemming from psychological and pedagogical experiments conducted throughout the project. The project builds upon a cross-disciplinary research team and exploits the synergies between learning and cognition in humans and machines. The Consortium is composed by the University of Bologna (coordinator), Dr Anna Rita Addessi; SONY-Computer Science Laboratory, Paris, Dr François Pachet; University of Genoa,

C. Delgado Kloos et al. (Eds.): EC-TEL 2011, LNCS 6964, pp. 15–28, 2011.

Dr Gualtiero Volpe; University of Exeter, Dr Susan Young; University of Gothenburg, Prof. Bengt Olsson; University of Athens, Dr Christina Anagnostopoulou, and Compedia, Israel, Dr Shai Newman.

1.1 The Problem of Musical Learning in an Interactive Scenario

How does musical learning in children take place? How is it affected by old and new technologies? These questions are directly relevant for setting the background to the MIROR project.

Recent studies dealing with musical invention in very young children (2-4 years) have suggested that the origin of new musical ideas is structurally anchored in the development of sympathetic interaction established between the adult and the child while playing with musical instruments [51][31]. Furthermore, according to some developmental theories, the adult/infant relationship has an important role in the affective and cognitive development of the child [22][29][46][42][26]. The question therefore arises as to which models of cognitive development and learning are produced when these relationships are established not between two human subjects, but between a child and a machine. New technologies can be considered not only as "tools" for didactic support, but also as languages and "brainframe" [18] that affect, form, and profoundly shape the processes of music learning and the musicality of children. In a review of music technology in education, Webster [48] concludes that there is a scarcity of research on using music technology with young children. The reasons for this may lie in ideologies and established traditions of early childhood music education practice. From birth children are immersed in everyday musical worlds mediated increasingly by digital technologies. They arrive in pre-school education equipped with a range of competences and concepts about music and musical process derived from these experiences. The issue is not whether digitised technologies should be part of early childhood music education (for if early childhood education purports to 'start with the child', then they are already present in experiences children bring with them), but:

- how pedagogical approaches need to transform in order to best serve the competences children have, [23][52].
- the lack of collaboration between the system design and educational communities.
- the lack of music education in schools and pre-schools for childhood and early childhood: music education is still often absent in primary school, and even more so in nurseries and kindergartens, in spite of the important role the musical experience and expression plays in children's daily life. [3][26][28].

2 The MIROR Project: Music Interaction Relying on Reflexion

In this context, the MIROR project is grounded on the concepts and technologies known as *Reflexive Interaction*, initially elaborated at the SONY FRANCE-Computer Science Laboratory in Paris, which represent a new generation of computationally augmented musical environments.

The first prototype of interactive reflexive musical systems, the Continuator [33] was elaborated for adult musicians. Then, it was experimented with the children.

The effectiveness of the promising pedagogical concept has been largely demonstrated through previous experiments carried out since 2003. The experiments (the pilot protocol, didactic experiences in the kindergarten, experiments in teaching improvisation classes) have shown the extraordinary potential of these new generation of software for educational purposes not only in the specific field of the music education but also in the wider field of learning strategies [1][2][19][6]. The scientific results of these experiments have been recognised by several scholars who are experts in music education and technologies, and who have highlighted both the technological and research methodological innovations of these studies: [49][9][8][20][21][30][53].

Building on the promising results of these studies, the MIROR project aims at developing the potential of IRMS and turn it into a novel form of pedagogical software and associated pedagogy. The main output of the project will be the MIROR platform, and will be specifically designed for musical education in young children, at the school and at home, and in more general way in the kinder centres, children hospital departments, centres for mother and child, etc.

The possibility of making and sharing music allows individuals to recognize themselves in a community, but often the formal/informal music is not accessible to everyone for some reasons. Thanks to its technological and pedagogical characteristics, the MIROR platform will exploit the social potential of collaborative music making, and represents not only a revolutionary appliance for learning/teaching music education with young children, but also a powerful tool of aggregation and (music) education in situation of multicultural conflicts, desertion of school, and "cultural poverty".

2.1 Objectives

The MIROR project aims primarily at developing the potential of IRMS for the benefit of music *education*. This includes the design, implementation and validation of concrete pedagogical scenarios in which these IRMS organize and stimulate the learning / teaching processes in the domains of music improvisation, composition and body performance. More precisely, MIROR's primary goals are the following:

1. Developing an innovative adaptive and intuitive system for music education, based on the "reflexive interaction" paradigm: the MIROR Platform. The platform is developed in the context of early childhood music education, as a new learning appliances and advanced tutors, able to promote specific cognitive processing and abilities in the field of music exploration, improvisation and composition. The MIROR Platform addresses music improvisation (MIROR-Improvisation), composition (MIROR-Composition), and exploration with/of body gesture (MIROR-Body gesture). The core architecture of the MIROR platform is displayed in Figure 1.
2. Developing a detailed analysis aiming at assessing the impact of the reflexive interaction paradigm for both music learning and general cognitive/learning processes.
3. Promoting an active approach to musical culture, based on the "music-making" concept, rather than on "music-consuming"; promoting the use of

the MIROR Platform in children and adults, enabling a wider access to music by experts as well as non experts, and enhance the diffusion of music culture in EU society through experiments with the MIROR Platform in several European countries, producing an User Guide and Teachers' Guide.

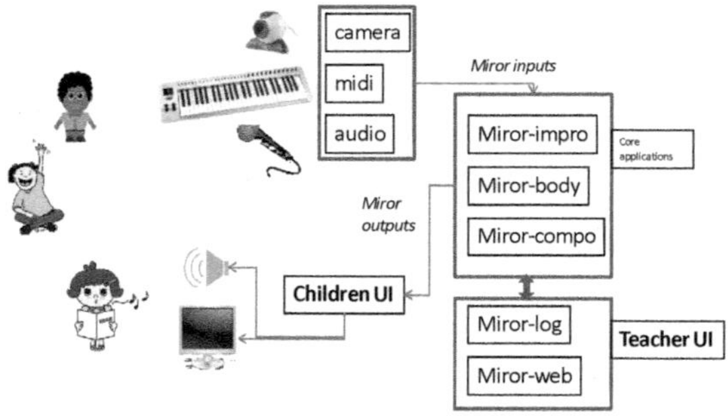

Fig. 1. The overall architecture of MIROR, with the various inputs, outputs, core modules, and user interfaces

3 The Paradigm of Reflexive Interaction

The reflexive interaction paradigm is based on the idea of letting users manipulate virtual copies of themselves, through specifically designed machine-learning software referred to as interactive reflexive musical systems (IRMS). Reflexive Interaction software are essentially intelligent mirrors, that continuously attempt to learn and reproduce the musical behavior of the users.

3.1 Interactive Reflexive Musical Systems

Interactive Reflexive Musical Systems, IRMS in short, were originally conceived at the SONY FRANCE Computer Science Laboratory in Paris [34]. The notion of IRMS emerged from experiments in novel forms of man-machine interactions, in which users essentially manipulates an "image" of themselves. Traditional approaches in man-machine interactions consist of designing algorithms and interfaces that help the user to solve a given, predefined task. Departing from these approaches IRMS are designed without a specific task in mind, but rather as intelligent "mirrors". Interactions with the users are analysed by IRMS to build progressively a model of this user in a given domain (such as musical performance). The output of an IRMS is a *mimetic response* to a user interaction. Target objects (e.g., melodies) are eventually created as a *side-effect* of this interaction, rather than as direct products of a co-design by the user. This idea took the form of a concrete project dealing with musical improvisation, The Continuator. The Continuator is able to interactively learn and reproduce music of "the same style" as a human playing a keyboard, and it is perceived as a stylistic musical mirror: the musical phrases generated by the system are similar but different from those played by the users. Technically, the Continuator is based on the integration of a machine-learning component specialized in learning and producing musical streams, in an interactive system. Many algorithms have been proposed to model musical style, from the early days of information theory of the 50s, to the works of David Cope [14], who showed that a computer could generate new music "in the style" of virtually any known composer. However, these mimetic performances were the result of off-line processes, often involving human intervention, in particular for structuring musical pieces. The Continuator was the first system to propose a musical style learning algorithm in a purely interactive, real-time context [33]. In a typical session with the Continuator, a user freely plays musical phrases with a (Midi) keyboard, and the system produces an immediate answer, increasingly close to its musical style. As the session develops, a dialogue takes place between the user and the machine, in which the user tries to "teach" the machine his/her musical language. As an example, Figure 2 shows how a simple melody is continued by the Continuator.

Fig. 2. A simple melody (top staff) is continued by the Continuator in the same style [33]

3.2 IRMS and Expressive Gesture Analysis

The basic concept in the paradigm of Reflexive Interaction is to establish a dialogue between the user and the machine, in which the user tries to "teach" the machine his/her musical language. The effectiveness of such a dialogue can be greatly increased by introducing the dimensions of expressivity and emotion. The MIROR project includes the design of an expressive interface for real-time analysis of non-verbal expressive communication through multimodal, expressive features to further develop the children capacity for improvisation, composition and creative performance. Expressivity and emotion have been demonstrated to be a key factor in human-machine interaction (e.g., see research on Affective Computing [38]). It has also been recognized that embodiment, in terms of movement and gesture plays a relevant role in musical learning of children [1][31][27][7[[16]. An innovative aspect of the MIROR project is to endow Interactive Reflexive Musical Systems with the capability of exploiting mechanisms of expressive emotional communication. In particular, this novel generation of IRMS (*emotional IRMS*) integrate modules able to analyze and process in real-time the *expressive content* conveyed by users through their *full-body movement* and *gesture*. Concretely, techniques are being developed for extracting expressive motion features characterizing the behaviour of the users (e.g., their motoric activation, contraction/expansion, hesitation, impulsivity, fluency) and to map such features onto real-time control of acoustic parameters. Inputs for such techniques are mainly images from one or more video-cameras or from novel sensors such as the Kinect.. Further analysis techniques (e.g., based on machine learning) are applied to classify motion with respect to motion categories and/or to compute trajectories in high-level expressive spaces derived from dimensional theories of emotion [41,44]. As a result, such techniques enable using expressive gesture for improvisation (e.g., users by means of their expressive gesture can mould and modify in real-time the sound and music content they are creating), for composition (e.g., the choice of the mapping between expressive features and acoustic parameters is a compositional issue), for creation (e.g., the availability of high-level gesture features provides novel possibilities and novel languages for music creation) [11][12][13].

Research on emotional IRMS is grounded on the EyesWeb XMI platform [12], www.eyesweb.org (see Fig. 3). EyesWeb XMI (for eXtended Multimodal Interaction) is an open platform supporting on the one hand multimodal processing both in its conceptual and technical aspects, and allowing on the other hand fast development of robust application prototypes for use in artistic performances, in education, in interactive multimedia installations and applications.

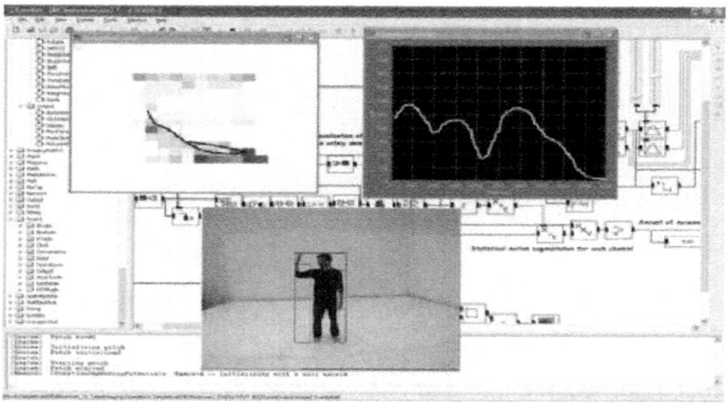

Fig. 3. A running EyesWeb application for expressive gesture processing. Expressive gesture qualities such as energy and fluidity are extracted and mapped in a 2D space (trajectory in the upper left window). Occupation rates for regions in such a space are then computed. In the case displayed in the figure a higher amount of quick and fluid gestures is detected.

4 The IRMS and the Children

4.1 Previous Studies

An extensive series of exploratory studies was carried out with 3-5 year-old children since 2003 [1][2][4][5][6][19][35]. These studies showed that the IRMS paradigm not only "works with children" but that it yields a novel and efficient approach to learning / teaching, as it develops fascinating and spectacular child/machine interaction. This pioneering work was followed by many others, for instance [52]. One innovative feature of the Continuator is the creation of a natural, organic dialogue with the child. This dialogue is based on turn-taking rules, and on the mechanism of repetition and variation, which are both natural mechanisms observed in infant/mother interactions [42][25][45]. Between the Continuator and the child a circular interaction is set up, in which the child's musical style influences the system, which answers by borrowing the child's musical fragments, in a continuous dialoguing improvisation. These studies have produced a number of results that suggest that IRMS have a potential not fully exploited.

Nature of the interaction. The results would suggest that the Continuator is able to develop interesting child/computer interaction, very similar to that between humans. This phenomenon seems to have its origins in the ability of the system to replicate the musical style of the children. The interaction based on repetition/variation allows the children to organise their musical discourse, passing from exploration to genuine musical invention.

The system creates a state of *well-being* very similar to the one described in the Flow theory of Csikzsentmihalyi [15]. The presence of the state of Flow in children playing with (B) and without (A) the Continuator over three sessions was measured. The results show that when a child plays alone with the Continuator (task B), he/she reaches a constant high level of Flow. Considering the two tasks, it can be observed that the percentage of the Flow presence is higher in task B than in A (B= 54%, A= 42%) [2].

Attention Span: By attention span is understood as the subjects' tendency to persist in their contact with objects or activities, irrespective of any underlying aim. The attention span of the children was measured for each task (A=the child playing the keyboard alone B=the same child playing with an other child, C=the same child playing the keyboard with the Continuator launched; and D=the same child playing with an other child and the Continuator)) . The results showed that tasks B and D (i.e. those involving the system) produced the longest mean times of attention [2].

Exploration and music improvisation: The children explored the keyboard and means of making sound in a myriad of different ways. The analysis of the improvisations revealed rhythmic and melodic patterns, synchronization on the same pulse, forms of song and accompaniment, individual improvisation styles, brief formal constructions based on imitation, repetition, alternation, and contrast. Both in the exploration and in the improvisations, personalized styles could be observed in their approach to producing sounds, or handling of the instrument.

Thinking in sound:: an important observation made was that the main channel of children / machine interaction is listening, encouraging the children to *think in sound.*

Teaching and collaborative playing: The interaction with the Continuator also developed autonomy and intrinsic motivation, enhanced collaborative playing and particular learning models as self-learning, self-regulation and self-practising.

4.2 Educational Theories Implemented by IRMS

Thanks to its capacities to replicate the musical behavior and evolve in an organic fashion with the user, IRMS translate into technological design several theoretical concepts of learning development and the theory of creativity. Mirror behavior: The capacity to replicate the behavior of others grounds on one part on non-conscious processing known as the "chameleon effect" [Chartrand, 1999]. Neuroscientific studies root these non-conscious mechanisms in the mirror neuron system (MNS), a network of neurons, which become active during the execution and observation of actions [39]. A similar structure based on repetition and variation has been observed by Stern [42] in the vocal relationship between mother and infant, and by Imberty [26] in the field of musical development. Similar interactions have also been observed recently in young children and adults while they play [51][32]. This kind of interpersonal dimension has been recognized as a potential source of musical creativity for young children. In the field of pedagogical theories, the theory of variation (Pramling et Al. 2009) shall allow further studies about this issue

In particular, IRMS promote especially the "optimal experience" described by the Flow Theory introduced by Csikszentmiahlyi [15]. The Continuator mentioned above has been shown to be a type of "Flow machine" in the sense that it produces, by definition, a response corresponding to the skill level of the user [34]. Most importantly, Flow theory introduces a series of concrete indicators, which can be measured during recorded experiments [5], This strong potential of IRMS can be used to enhance the state of "well-being" and creativity in children and adults over the interaction with the MIROR Platform. IRMS also exploit the Vygotskian concept of zone of proximal development (ZPD). In this way, IRMS establish an interaction between pairs, where the mirroring reflection creates a balance between challenges and skills, a basis to create Flow experiences and creative processes. This characteristic enables the MIROR Platform to enhance self-regulation, self-initiated activities, and the learner-centered approach. Furthermore, IRMS support children in mixing old musical skills with new ones, in an original and autotelic way, according to the "fiction cognitive" perspective [24] where the innovative technology enables the subject to see and listen in a more original way, bringing out previous childhood experiences. IRMS generate very complex reactions, where the children are expected to form differentiated judgments about "self" and "others". In literature, these forms of awareness are considered crucial for the building of the child's identity. IRMS, by means of its mirror effect, help towards the construction of a "musical self", or, in the words of [51], a "Second self", where not only the machine seems to think, but also think like the user. Finally, the MIROR project owes to the Laban Movement Analysis (LMA), elaborated by the Hungarian dance artist and theorist Rudolf Laban (1879-1958)[28]. LMA has been widely used in the field of dance education and was applied also to music and movement education. This analytical approach is the basis of the *expressive gesture analysis* implemented by the EyesWeb technology. The MIROR project exploits this theoretical framework coming from the dance context to develop musical abilities of children.

4.3 Related Projects in Music Education and New Technology and Steps Forward

The issue of teaching how to improvise. Unlike most of existing projects based on teaching to perform the musical repertoire, the MIROR Project will focus on the *improvisation, composition* and *creative performance*.

The body issue in music education. The MIROR Project will explicitly integrate multi-modal interface gesture analysis in the learning/cognition processes activated during child/machine interaction. It will lead to new forms of musical expression, based on the integration of music and dance in particular.

The issue of fixed musical objectives. The IRMS emphasizes a particular kind of interaction where the aims are not established by neither the machine nor by the user, but by the actual interaction between child and machine.

The issue of adaptation to learners. the centre of attention in the reflexive interaction process is not the end-product (the music), but the subject engaged in the interaction. Reflexive interactions naturally produce a learner-centred approach.

The issue of attractiveness. The experiments realised so far with IRMS have showed a high level of excitement, attention span and intrinsic motivation in children playing with the Continuator where the only interface is the keyboard. These findings show that the attractiveness of IRMS is based on the conceptual and technical features of the software rather than on external or nicely designed GUIs.

The issue of formal, non formal and informal music education. Thanks to the support of the member of the Advisory and Liaison Board, the MIROR Project will be concerned not only with formal music education, but also with basic music education in compulsory schooling, in informal education and in different social contexts and organisational learning, particular social contexts, special needs groups and therapeutic settings.

Music therapy and new technology. The "reflexive interaction" could be a versatile device to enhance and stimulate expressive behaviour and communication in therapeutic setting. Furthermore, the optimal experience generated by the interaction with IRMS creates states of "well being", having a strong therapeutic potential. We aim at exploiting this potential of IRMS in a systematic way, in collaboration with specialists in special needs education, psychiatry and music therapy.

Methodological issue: Psychological and pedagogical evaluation. The MIROR Project will test the system's prototype by means of both psychological experiments into learning/cognition processes activated during the child/IRMS interaction, and pedagogical validation in practical contexts and music classroom settings.

5 Pedagogical Issues

Experiments conducted so far with IRMS have shown the extraordinary potential of this paradigm for building exciting, autotelic interactive environments, for boosting creativity and for addressing productively several pedagogical goals. In particular, the analysis of videos of children's interactions revealed a certain number of problems, but also gave rise to new ideas, notably concerning interaction modes.

The previous experiments were based on a constraint-free system, the Continuator, with no explicit pedagogical constraint implemented. A substantial part of the technical program of the project consists of extending this component in substantial ways, by integrating explicitly pedagogical constraints for using it in classroom settings. In a pedagogical context, other interaction modes are being developed, to take into account explicitly the role of the teacher, constraints coming from pedagogical goals, the possibility of making music with other children and the possibility of interacting through gesture.

The pedagogical issues are mostly related to the difficulty of organizing a pedagogical setting and material around the basic IRMS type of interaction. Pedagogical programs have to include scenarios where specific dimensions of music must be emphasized. The issue here is to add pedagogical constraints in the basic IRMS interaction so as to allow the development of such scenarios easily. Furthermore, the organizational learning contexts have to be tested and validated, on the basis of new studies.

Another important extension of IRMS into the MIROR Platform is the pedagogic exploitation of the possibility of communicating with the machine through body movements. These issues are addressed by introducing the expressive gesture analysis, implemented in the MIROR-Body Gesture prototype. The developed techniques are integrated in prototypes of emotional IRMS, addressing live music and dance performance and oriented to pedagogical application in music and dance education.

A specific work package is dedicated to the implementation of the children and teacher *interfaces*.

Finally, didactic and pedagogical materials will be elaborated and grouped in the User's Guide, notably for children, teachers and parents.

6 The Consortium and the Work Plan

The MIROR Consortium is composed of partners with complementary expertise, in order to create a synergic team able to successfully cover all the competences necessary to carry out the project. The three MIROR prototypes (Impro, Compo and Body Gesture) are developed in collaboration with the psycho-pedagogical partners indicating the psycho and pedagogical constraints of the platform on the basis of the previous studies and on the basis of the experiments that are elaborated and carried out over the project. This spiral model of design is fundamental in the case of implementation of didactical software, but it is more fundamental in the field of early child music education technology. This field demands a specific expertise on the cognitive, affective and sensory-motor child development, infant musicality and music learning processes. Their expertise in different methods of music education research will lead to a cross-disciplinary study based on experimental protocols, observation, case-study, action research, empirical questionnaire. The presence of 2 different technological partners enables the Consortium to put together 2 different interactive technologies: music interaction and multimodal gesture analysis. The first is the pioneer of the idea of reflexive musical interaction. The second will enhance the reflexive concept by mean the results of the prior project on gesture analysis.

Expert in music analyis: Music analysis of improvisation and production of the children interacting with the IRMS will lead to an analysis of the cognitive processes involved in musical learning and development.

Final users, mainly schools, have the key task of testing and evaluating the realised tools and will also participate to the dissemination activities. The Advisory and Liaison Board is created for this kind of contribution.

The SME is foreseen for marketing analysis and industrial exploitation plan.

Fig. 4. Children playing with the MIROR-Improvisation prototype (Protocol no. 1, Bologna-Italy, May 2011)

The integration of researches of cross-disciplinary teams allows the creation of an effective spiral characterised by the concept of inter-reliability. The results of one team, notably the technological team, influence and are influenced by the results of another team (psycho and pedagogical team). Different research methods and approaches are used (basic research, experimental, observation, action research): these factors enable the MIROR Project to study different aspects of the problem and to validate the results by means of cross-control.

MIROR is now reaching the end of its first Year. Ongoing activities and initial results include the first version of the MIROR prototypes and an initial series of experiments on the MIROR-Improvisation prototype, based on the MIROR Platform.

Web site of the MIROR project: www.mirorproject.eu

References

1. Addessi, A.R., Pachet, F.: Experiments with a musical machine: musical style replication in 3/5 year old children. British Journal of Music Education 22(1), 21–46 (2005)
2. Addessi, A.R., Pachet, F.: Young children confronting the Continuator, an interactive reflective musical system. Musicae Scientiae, Special Issue, 13–39 (2006)
3. Addessi, A.R., Young, S. (eds.): Proceedings of MERYC 2009. Bononia University Press, Bologna (2009)

4. Addessi, A.R., Ferrari, L., Pachet, F.: Without touch, without seeing. Children playing with a cybernetic musician. In: ECME Proceedings, Chinese Cultural University, Taipei, Taiwan (2006)
5. Addessi, A.R., Ferrari, L., Carlotti, S., Pachet, F.: Young children musical experience with a flow machine. In: Proceedings of ICMPC9 and 6th ESCOM Conference. Bononia University Press, Bologna (2006)
6. Benghi, D., Addessi, A.R.: Teaching to improvise with the Continuator. In: 28th ISME Proceedings, Bologna, pp. 882–883 (2008)
7. Boone, R.T., Cunningham, J.G.: Children's decoding of emotion in expressive body movement: the development of cue attunement. Developmental Psychology 34, 1007–1016 (1998)
8. Bryan-Kinns, N.: Computer in support of musical expression, University of London, UK (2004)
9. Burnard, P.: Routes to understanding musical creativity. In: Bresler, L. (ed.) International Handbook of Research in Arts Education, vol. 2. Springer, The Netherlands (2007)
10. Conklin, D., Anagnostopoulou, C.: Discovery of segmental patterns in music. INFORMS Journal on Computing 18(3), 285–293 (2006)
11. Camurri, A., Volpe, G. (eds.): GW 2003. LNCS (LNAI), vol. 2915. Springer, Heidelberg (2004)
12. Camurri, A., De Poli, G., Leman, M., Volpe, G.: Toward communicating expressiveness and affect in multimodal interactive systems for performing art and cultural applications. IEEE Multimedia Magazine 12(1), 43–53 (2005)
13. Camurri, A., Canepa, C., Volpe, G.: Active listening to a virtual orchestra through an expressive gestural interface: the Orchestra Explorer. In: Proceedings of NIME-2007, New York University, NY, USA (2007)
14. Cope, D.: Experiments in Musical Intelligence. A-R Editions, Winsconsin, Madison (1996)
15. Csikzsentmihalyi, M.: Creativity: flow and the psychology of discovery and invention. Harper Collins Edition, New York (1996)
16. Davidson, J.W.: What type of information is conveyed in the body movements of solo musician performers. Journal of Human Movement Studies 6, 279–301 (1994)
17. Delalande, F.: Le condotte musicali. Clueb, Bologna (1993)
18. De Kerckhove, D.: Brainframes: technology, mind and business. Bosch & Keuning, Belgium (1991)
19. Ferrari, L., Addessi, A.R., Pachet, F.: New technologies for new music education: the Continuator in a classroom setting. In: Proceedings of ICMPC9 and 6th ESCOM Conference, Bononia University Press, Bologna (2006); (revised version in print on the International Journal of Music Education)
20. Fober, D.: Miroirs technologiques pour la pratique instrimentale. In: Rencontres Musicales Pluridisciplinaires, GRAME-Centre national de creation musicale, Lyon, pp. 87–92 (2006)
21. Fober, D., Letz, S., Orlarey, Y.: VEMUS- Une école de musique eropéenne visrtuelle. In: Actes des Journées d'Informatique Musicale JIM 2007, GRAME-Centre national de creation musicale, Lyon (2007)
22. Fogel, A.: Oltre gli individui: un approccio storico-relazionale alla teoria e alla ricerca sulla comunicazione. In: Il rapporto madre-bambino, Carracci, Roma, pp. 123–157 (2000)
23. Fölkestad, G., Hargreaves, D.J., Lindström, B.: Compositional strategies in computer-based music-making. British Journal of Music Education 15(1), 83–97 (1998)
24. Guerra, L. (ed.): Educazione e tecnologie. I nuovi strumenti della mediazione didattica. Junior Edizioni, Azzano San Paolo (2010)

25. Ilari, B., Gluschankoff, C.(eds.): Music in the early years: research, theory and practice. Special Issue of Early Child Development and Care 179(6) (2009)
26. Imberty, M.: La musique creuse le temps. De Wagner à Boulez: Musique, psychologie, psycanalyse. L'Harmattan, Paris (2005)
27. Imberty, M., Gratier, M. (eds.): Temps, geste et musicalité. L'Harmattan, Paris (2007)
28. Laban, R., Lawrence, F.C.: Effort. Macdonald & Evans Ltd, Plymouth (1947)
29. Malloch, S.: Mothers and infants and communicative musicality. Musicae Scientiae Special Issue, 29–58 (2000)
30. McGregor, C.: The practicing and processing of music rudiments: a comparison of a traditional versus online music appreciation class. Report, Southwestern College (2007)
31. McPherson, G. (ed.): The child as musician: a handbook of musical development. Oxford University Press, Oxford (2006)
32. Olsson, B.: Social issues on music education. In: Bresler, L. (ed.) The international handbook of research on arts education. Springer, The Netherlands (2007)
33. Pachet, F.: Musical interaction with style. Journal of Music Research 32(3), 333–341 (2003)
34. Pachet, F.: Enhancing Individual Creativity with Interactive Reflexive Musical Systems. In: Deliège, I., Wiggins, G. (eds.) Musical Creativity: current research in theory and practice. Psychology Press, Hove (2006)
35. Pachet, F., Addessi, A.R.: When children reflect on their playing style: The Continuator. ACM Computers in Entertainment 2(2) (2004)
36. Pachet, F., Roy, P.: Markov Constraints: Steerable Generation of Markov Sequences. Constraints 16(2), 148–172 (2011)
37. Pachet, F., Roy, P., Barbieri, G.: Finite-Length Markov Processes with Constraints. In: Proc. of IJCAI 2011, Barcelona (July 2011)
38. Picard, R.: Affective Computing. MIT Press, Cambridge (2011)
39. Rizzolatti, G., Fogassi, L., Gallese, V.: Mirrors of the mind. Scientific American 295, 54–61 (2006)
40. Rowe, R.: Interactive Music Systems. The MIT Press, Cambridge (1993)
41. Russell, J.A.: A circumplex model of affect. Journal of Personality and Social Psychology 39, 1161–1178 (1980)
42. Stern, D.: The present moment in psychotherapy and in everyday life. Norton, New York (2004)
43. Sawyer, K.R.: Improvised conversations: music, collaboration, and development. Psychology of Music 27, 192–216 (1999)
44. Tellegen, A., Watson, D., Clark, L.A.: On the dimensional and hierarchical structure of affect. Psychological Science 10(4), 297–303 (1999)
45. Trehub, C.: Musical predisposition in infancy: an update. In: Perez, I., Zatorre, R. (eds.) The cognitive neuroscience of music, pp. 3–20. Oxford University Press, New York (2009)
46. Trevarthen, C.: Musicality and the intrinsic motive pulse: evidence from human psychobiology and infant communication. Musicae Scientiae, Special Issue, 155–215 (2000)
47. Turkle, S.: The second self: computers and the human spirit. Simon and Schuster, New York (1984)
48. Webster, P.R.: Computer-based technology and music teaching and learning. In: Colwell, R., Richardson, C. (eds.) The new handbook of research on music teaching and learning, pp. 416–439. Oxford University Press, Oxford (2002)

49. Webster, P.R.: Computer-based technology and music. In: Bresler, L. (ed.) International handbook of research in arts education, vol. 2, pp. 1311–1328. Springer, The Netherlands (2007)
50. Wallbott, H.G.: Bodily expression of emotion. European Journal of Social Psychology 28, 879–896 (1998)
51. Young, S.: The interpersonal dimension: a potential source of musical creativity for children. Musicae Scientiae, Special Issue 2003-2004, 175–191 (2004)
52. Young, S.: Interactive music technology in early childhood music education. In: Proceedings of the 9th ICMPC and the 6th ESCOM Conference. Bologna University Press, Bologna (2006)
53. Young, S.: Music 3-5. Routledge, London (2008)

CROKODIL - A Platform for Collaborative Resource-Based Learning

Mojisola Anjorin[1], Christoph Rensing[1], Kerstin Bischoff[2], Christian Bogner[3], Lasse Lehmann[1], Anna Lenka Reger[6], Nils Faltin[4], Achim Steinacker[5], Andy Lüdemann[6], and Renato Domínguez García[1]

[1] Technische Universität Darmstadt, Darmstadt, Germany
[2] L3S Research Center, Hannover, Germany
[3] Technische Universität Kaiserslautern, Kaiserslautern, Germany
[4] imc AG, Saarbrücken, Germany
[5] intelligent views GmbH, Darmstadt, Germany
[6] Siemens AG, München, Germany
{Firstname.Lastname}@crokodil.de

Abstract. On-the-job learning is primarily a personal knowledge acquisition process accomplished increasingly based on resources found on the Web. These days, collaboratively learning from and with others on the Web is taking on a very prominent position in this learning process. CROKODIL aims to provide support for collaborative learning based on web resources. In this paper, we introduce our learning scenario and an evaluation of our target group. We describe our pedagogical concepts, and present the results of an evaluation of these concepts. CROKODIL supports the semantic tagging of resources as well as the collaborative use of these resources and their information. Social networking functionalities are integrated in the platform to encourage and support collaborative learning. We also present some extensions to the base functionality of the platform, such as resource recommendations and interfaces for the integration in existing learning management systems.

Keywords: Resource-based learning, Collaborative learning, Social networks, Web resources, Recommender systems.

1 Introduction

Knowledge acquisition is a lifelong process which has to be continuously performed due to rapidly evolving knowledge bases and the fast development of new technologies. In the last ten years, searching for information via the Internet has become a very important way to keep oneself up-to-date. On-the-job, there is the need to constantly find out more about certain topics or to look for solutions to current problems which someone else might have already solved. In many cases, there are no completely didactically worked out learning objects or learning courses available which could equip the learner or worker with the knowledge needed to complete a said task. A shift is needed from the pure form of instructional learning to the more flexible resource-based learning [1], which focuses more on problem solving and

C. Delgado Kloos et al. (Eds.): EC-TEL 2011, LNCS 6964, pp. 29–42, 2011.

critical thinking. As a result, knowledge acquisition is increasingly accomplished by utilizing resources found on the Web such as open learning content from an educational institution (like iTunes U[1]) or user-generated content like on YouTube[2] or Slideshare[3] as well as collaboratively constructed resources such as wikis and blogs [2]. Besides using and generating content, learners collaborate with other learners using different applications like social networks, discussion boards, wikis or forums. They ask questions, discuss answers, share documents, or recommend resources on certain subjects, which were found to be helpful, to other learners. The advantages of communication and interaction in communities and social networks can no longer be ignored as a very important part of the learning process [1].

This paper is organized as follows. Section 2 explains advantages and challenges of collaborative resource-based learning. Section 3 summarizes existing approaches to address these challenges. In Section 4, we introduce four different learning scenarios which we analyze to determine the requirements for our platform. The results of an exemplary study of two of the target groups are presented. The pedagogical concept and its evaluation are discussed in Section 5. In Section 6, we present the CROKODIL platform, describing its different elements and extensions. This paper concludes with an outlook on future work.

2 Advantages and Challenges in Resource-Based Learning

As mentioned in the introduction, resource-based learning offers many advantages when compared to traditional instructional learning. Rakes [2] shows that learners prefer to learn through interactions with various learning resources rather than through instructional teaching. Studies performed in [3] have shown that learning with web resources highly depends on the personal situation, preferences and previous knowledge of the learner as opposed to the learner's type. Especially for users who actively aim to organize their learning process, the usage of web resources and services from the so called Web2.0, in addition to or in place of traditional learning approaches is helpful.

However, resource-based learning poses challenges for learners both from a technical as well as a pedagogical perspective: there often is no teacher who structures the learning process and prepares the learning material. Additionally, resources on the Web are rarely created by the authors with the intent of creating learning material. The learners, by themselves, have to assess the trustworthiness of the resources and select a variety of these resources which are relevant to their educational goals. Hence, learners have to perform the entire learning process in a self-directed manner. This competence to learn autonomously on the Web cannot be taken for granted. "Learning to learn" is one of the eight key competences listed in the European Commission Reference Framework for lifelong learning [4]. It means the ability to pursue and persist in learning, to organize one's own learning, including the effective management of time and information, both individually and in groups. Therefore, people have to be trained in this kind of learning in educational institutions. In such

[1] http://www.apple.com/education/itunes-u/, Online 2011-03-29
[2] http://www.youtube.com/, Online 2011-03-29
[3] http://www.slideshare.net/, Online 2011-03-29

educational institutions, which are predominantly characterized by instructional learning, collaborative resource-based learning should be integrated in the form of episodes. Such episodes could be small, like a learning task in which learners have to research online in order to answer a specific question, or large when learners perform an extensive research, like an independent preparation of a complex thesis.

Furthermore, learners have to manage different tasks in the overall process of resource-based learning in addition to the actual reading and learning process. Amongst other challenges, learners have to phrase search terms, select relevant web pages from search results and organize and structure web pages they have found for later use [5]. The technological challenge therefore, is to support the user in all the processes he has to perform in addition to the actual learning of the content of these web resources. Our overall goal in CROKODIL is to support all these tasks in one platform.

3 Related Work

Traditional systems for technology enhanced learning mainly focus on the support of the organization of institutional learning scenarios as in different kinds of Learning Management Systems (LMS), or of collaborative learning as in various Computer Supported Collaborative Learning applications (CSCL) or in the management of different tools used in personal learning as in Personal Learning Environments (PLE). For resource management in resource-based learning learners usually use different tools. A survey we performed [6] has shown that learners use search engines, word processing tools, memos or browser bookmarks. So far, social bookmarking platforms like Delicious[4] or reference management programs for research articles like Citavi[5] or Mendeley[6] are rarely used. The authors in [7] give an extensive review of nine different social semantic bookmarking systems: BibSonomy[7], SOBOLEO[8], Fuzzzy[9], GroupMe![10], Twine[11], ZigTag[12], Faviki[13], gnirz[14] and Annotea[15]. These systems allow the user to add annotations to resources by tagging, rating or commenting. Each system, however, implements these features differently and a different emphasis is given to the degree of automated support offered to the user. Two examples of tagging and annotation systems applied specifically to the learning scenario are OATS (Open Annotation and Tagging System) [8] and FolksAnnotation [9]. OATS is an open-source tool supporting learners in collaboratively creating and sharing

[4] http://www.delicious.com/, Online 2011-03-29
[5] http://www.citavi.com/, Online 2011-03-29
[6] http://www.mendeley.com/, Online 2011-03-29
[7] http://www.bibsonomy.org/, Online 2011-03-29
[8] http://tool.soboleo.com/, Online 2011-03-29
[9] http://www.fuzzzy.com/, Online 2011-03-29
[10] http://groupme.org/GroupMe/, Online 2011-03-29
[11] http://www.evri.com/, Online 2011-03-29
[12] http://www.zigtag.com/, Online 2011-03-29
[13] http://www.faviki.com/pages/welcome/, Online 2011-03-29
[14] http://code.google.com/p/gnizr/, Online 2011-29-03
[15] http://www.annotea.org/, Online 2011-03-29

knowledge using highlights, tags and notes in HTML based learning content. FolksAnnotation is a semantic metadata tool for annotating learning resources using folksonomies and domain ontologies. These tools however lack pedagogic support for the learning process and only support isolated steps of resource-based learning.

In the CROKODIL project[16], we aim to provide a holistic solution. We therefore develop pedagogical and technological concepts to support collaborative resource-based learning. We integrate these concepts in a novel learning platform and are thus providing support for all the different tasks which are part of collaborative resource-based learning.

4 Target Group Scenarios and Evaluation

Resource-based learning does not only take place in a personal setting outside of educational institutions. It is rather necessary to introduce this form of learning even within educational institutions and provide the students with the skills needed to learn autonomously with resources from the Web. Resource-based learning is thus integrated into the overall instructional setting and thereby gains a new and different scope. Resource-based learning can be implemented as a short episode like answering a question posed by the lecturer or as a longer episode, for example during the preparation of an elaboration on a specific topic. To expedite the use of resource-based learning in educational institutions, we couple the CROKODIL platform with a learning management system. This allows institutions to integrate resource-based learning in the traditional instructional learning process, as well as providing access to content and resources from an LMS in resource-based learning. In the CROKODIL project educational institutions are involved in the design and the evaluation of the new platform: IBB (Institut für Berufliche Bildung) and Siemens Professional Education. Both institutions implement two different scenarios each which are described briefly in the following. For two of these scenarios, we report on an initial target group survey.

4.1 Target Group Scenarios

Scenario 1: Re-education in Information Technology (IBB)
The first scenario is the provision of further education in information technology to people out-of-work. These learners are generally between 25 and 46 years of age and already have some work experience. They receive one-day tasks from the lecturer, which they accomplish independently by using textbooks and resources found on the Internet. CROKODIL will be integrated in this daily self-directed learning process.

Scenario 2: Education Program for School Dropouts (IBB)
The second scenario encompasses an educational program for school dropouts to support them in finding apprenticeship vacancies. Therefore, in addition to this education program, the trainees partake in internships and classroom trainings. The classroom training day is usually held in an instruction-oriented manner where teachers hold teacher-centered lessons combined with exercises and tests. In this

[16] http://www.crokodil.de, Online 2011-03-29

scenario, a continuous use of the CROKODIL platform is hardly reconcilable with the existing rigid learning structure as described above. Thus an episode-based approach to resource-based learning is planned. Short and concise phases of resource-based learning will be integrated into the existing learning procedure. A possible CROKODIL episode in this case could encompass trainees researching in a group on a certain topic and presenting their findings later on to the class later on.

Scenario 3: Bachelor of Arts in Business Administration + Industrial Clerk (Siemens Professional Education)
The participants in the third scenario are pursuing a Bachelor's degree in Business Administration in combination with a commercial vocational qualification (as an industrial clerk). As part of their vocational training, they have to prepare papers and presentations on various business subjects. The preparation of these papers is mostly self-directed. The title and the content structure, however, have to be agreed upon with their trainer. The target group is divided into two subgroups, each consisting of 10-15 students. CROKODIL will be used in this exploratory learning process to access and manage various internal company resources (e.g. Siemens libraries, intranet resources, Web Based Trainings (WBTs), e-Books) as well as publicly accessible web resources.

Scenario 4: Bachelor of Engineering + Electronics Technician for Automation Technology (Siemens Professional Education)
In the last scenario, the participants are pursuing a Bachelor's degree in Engineering in combination with a vocational qualification (as an electronics technician for Automation Technology). At present, the learning content is communicated via lectures, tutorials and Computer Based Trainings (CBTs). The degree of self-directed learning could be enhanced by the use of the CROKODIL platform. The integration of additional publicly accessible resources could also be advantageous.

4.2 Target Group Assessment for Scenario 1 and Scenario 2

In order to get a better picture of the target group of the platform and to judge the background of the users regarding resource-based learning, self-regulated learning and their experience with social communities and Web 2.0 tools, we conducted an initial target group survey with the trainees from scenarios 1 and 2. From Scenario 1, we evaluated 15 participants: 14 male, 1 female, with an average age of 32.8 (SD 7). For Scenario 2, we evaluated 11 female and 7 male participants, having an average age of 18.7 (SD 1.6). Most of the survey items were rated using a 5-point Likert scale ranging from 0 (never) to 4 (almost always). Table 1 shows some selected results.

All participants of the survey are intense web users, going online once or several times a day and feeling quite confident handling this medium. Both groups predominantly use the Internet at home. For the learners in Scenario 1, the most popular internet activities are private surfing, email, online shopping and education related activities. However, most participants as well oftentimes use the Internet during re-education measures (avg. 3.4, SD 0.63) as the primary means for searching information with respect to re-training tasks (avg. 3.7, SD 0.46). For the school dropouts of Scenario 2, internet based communication services like email, instant messaging and social networks are a lot more prominent.

Table 1. Selected results of the target group survey

		Scenario 1		*Scenario 2*	
		Mean	SD	Mean	SD
Which tools do you use for researching information?	Search engine(s)	3.9	0.258	3.7	0.594
	Wikipedia	3.1	0.516	2.2	0.981
	Portal(s)	1.2	0.802	0.9	1.125
	Specialized databases	2.0	1.177	0.2	0.775
Which browser functionalities do you use?	Bookmarking	2.2	1.568	1.5	1.506
	Browser History	2.1	1.387	1.7	1.047
	Add-ons	2.4	1.500	1.5	1.356
Which services do you use for communication with colleagues or co-learners?	Email	1.7	1.345	1.3	1.073
	Phone	2.0	1.309	2.9	0.957
	SMS / MMS	1.0	1.195	2.9	1.023
	Instant Messaging	0.7	0.816	2.6	1.057
	Social networks	0.8	0.775	2.9	1.023

The participants from Scenario 1 show a heavy reliance on search engines and Wikipedia when browsing the Web to find relevant information. Portals and specialized databases are however rarely visited. Some participants also mentioned community related bulletin boards/newsgroups as valuable sources they sometimes refer to. The participants in Scenario 2 almost exclusively rely on search engines and to some extent Wikipedia as well. Expert groups or specialized databases are not used at all. With respect to organizing and re-finding relevant web resources, some users regularly make use of bookmarking and the browser history. For some users, we observe an interesting pattern between these two browser functionalities. Users, who frequently bookmark resources, less often resort to their browser history and vice versa. Tagging is however not a very common practice. Half of the users in Scenario 1 do not know the term tagging and only a few infrequently tag pictures, videos or web resources. In Scenario 2, only 30% are familiar with tagging. Those who are, use tagging frequently, in particular for pictures.

From our evaluation, YouTube, Facebook[17] and its German variants MeinVZ[18]/StudiVZ[19] are the most popular social networks. However, only a small number of users actively use these platforms on a regular basis. For those who do, the former two social networks are mainly used for communication with close friends and acquaintances. There are some significant differences between the two target groups with respect to social networking. In Scenario 1, on the one hand, communication with friends or colleagues/co-learners within such social platforms is still rare compared to more traditional media like telephone or email. YouTube, however, serves as a source for finding interesting content. In Scenario 2, on the other hand, Facebook and its German counterpart MeinVZ are used by almost all respondents; more than half of them frequently visit these sites to communicate with friends, acquaintances and co-learners, to publish content, to stay up-to-date and to give recommendations. This difference in the usage of social networks can be most

[17] http://www.facebook.com/, Online 2011-03-29
[18] http://www.meinvz.net/, Online 2011-03-29
[19] http://www.studivz.net/, Online 2011-03-29

probably explained by considering the age difference between the two groups, where the participants in Scenario 2 have a lower average age.

5 Pedagogical Concepts and Evaluation

As mentioned in Section 2, there is a pedagogical challenge regarding collaborative resource-based learning - competences in self-regulated and resource-based learning have to be imparted to learners. The aim of the CROKODIL platform is to mediate these competences to the learners. To complement this, a tutor will help with task formulation and tracking attendance in an institutional setting.

5.1 Pedagogical Concept

The central pedagogical elements in CROKODIL are *activities*. The definition of activities should be the starting point for each learning process within the platform. Additionally, the recurring inspection of an activity is the main means for reflecting on the learning process. Activities are basically used to organize tasks often set by a tutor, or to structure goals users wish to accomplish within their learning episode. Activities have the following features and characteristics [10]:

- Activities generally possess a name and a description.
- Activities can be structured hierarchically, i.e. an activity can have sub-activities and super-activities. This helps the learner in identifying the sub-steps belonging to a task, as well as structuring it accordingly before starting to work on the task itself.
- An arbitrary number of resources can be allocated to an activity.
- Users can describe their experiences, ask for help or discuss various aspects with other users in form of a dedicated discussion thread.
- An activity can be worked on collaboratively e.g. by adding other users or groups to the activity.
- Activities often result in documents containing the outcome of the work on the activity.

5.2 Evaluation of the Pedagogical Concept

We organized a one-day workshop with the learners from Scenario 1 (see Section 4) to evaluate the pedagogical concept before implementing it in the first prototype of the CROKODIL platform [10].

From the 32 participants of the re-education program, two working groups were formed: an experimental group of students who were briefed on our pedagogical concept and would apply this concept during the workshop day, and the second group who were not briefed and would be using the usual form of learning. Both groups received a task description from the tutor and were instructed to solve this task on their own by applying prior knowledge and researching for resources on the Internet. The experimental group was instructed to follow the pedagogical concept, i.e. to define and structure activities which are needed to solve the given task. In addition, they were told to document which web resources they used for the activities and which experiences were made while working on the activities. Half-way through the day the members of the experimental group presented their intermediary results in order to receive corrections and guidance regarding the correct implementation of the pedagogical concept. At the end of the day both groups presented their final results.

The evaluation of the observed process and the results [10], show that the experimental group assesses working with the pedagogical concept as a very positive and valuable experience. Especially the reflection process and the formulation of the relevant activities were found to be very important. From our observations, in the short time-frame of the evaluation, the experimental group was able to understand and apply the main didactic components of the pedagogical concept. A weak point, however, was the structuring of the activities in the different sub-activities. We gathered that more support functionality or explicit instructions are needed from either the system or from the tutor to help the users work more effectively and efficiently. These observations confirm the suggestions made by the European Commission in [4], regarding the need for learners to acquire competences for self-directed learning. Our didactic concept already caters for this by allowing the tutor to define appropriate guidelines even before the session starts. The tutor can create activities in the system which the learners can then extend with their own sub-activities.

6 The CROKODIL Platform

6.1 Overview

As mentioned in Section 2, learners have to manage different tasks in the overall process of resource-based learning in addition to the actual reading and learning process. The CROKODIL platform supports the learner in these tasks. Of crucial importance in resource-based learning are the resources from the Web, which are used by the learner during the learning process. When searching for resources on the Internet, the learner receives support in storing, describing and sharing resources which seem interesting for the present activity. In addition to searching on the Internet, the learner can also search through the resources on the CROKODIL platform which have already been stored by other learners in the learning community. Potentially relevant resources on the platform are recommended to the learner.

Resources in the platform are easy to store, annotate and describe as learners can simply tag these resources. CROKODIL enables users to add a concept type to each tag, which allows them to explicitly depict the concept they have in mind. This concept of semantic tagging is described in [11]. Based on an analysis of possible tag types used in resource-based learning, a type set that contains topic, person, activity, event, location, type, and a plain tag (i.e. a tag with no specific type attached) was implemented. The tags and resources are represented in the data model as a node, whereas the assignment of a tag to the resource is represented as an edge between them. A relationship is created between a user and a node (which could be a resource or tag). A complex knowledge network is thereby formed. We use K-Infinity[20] as a

[20] Intelligent Views: http://www.i-views.de, Online 2011-03-29

management platform which provides all components needed for building semantic networks. We access and query the semantic network via a web based front-end and we have integrated external data sources. K-Infinity supports different knowledge processing steps ranging from the simple definition of complex knowledge models, to the central and local maintenance of knowledge structures, as well as the use and distribution of available knowledge.

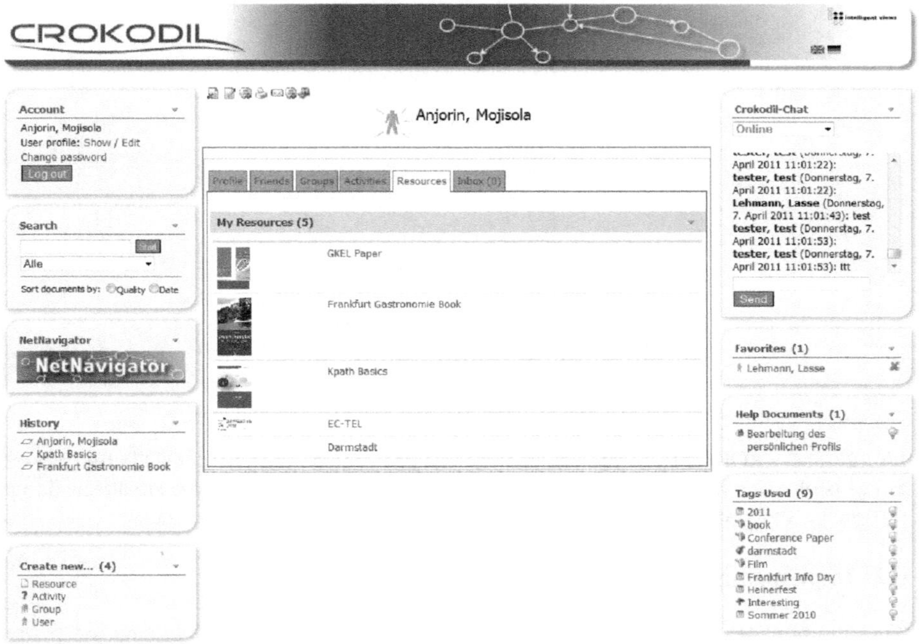

Fig. 1. The CROKODIL Web Portal showing the tab *Resources*, the other tabs show the user's profile, friends, groups, activities and message inbox. The user can also take part in the online chat. The tags used are listed on the right-hand side for easy and quick access.

For the collection of resources and tags and for copying parts of a web resource (called snippets) as a description of the resource, we developed an add-on for the Web browser Firefox, as shown in Fig. 2. Alternatively, the user can use a custom web application, the CROKODIL web portal, to add resources and information about the resources, as shown in Fig. 1. In the portal, the user can also search for tags or tagged resources, and can view all stored resources.

The portal not only offers support for the management of resources, it also allows the creation of a social network of learners. Each learner has a profile and can create friendship relationships with other users. Users of the platform can also form study groups. These social relationships extend the knowledge network. To support the collaboration among the students, capabilities for synchronous (text chat) and asynchronous (personal messages or discussion threads) communication are integrated into the platform. In addition to these basic functions more features are implemented in the platform. These are now described in more detail in the following sections.

Fig. 2. The CROKODIL Firefox add-on is integrated as a sidebar in the Firefox browser. Snippets of a web page are dragged and dropped on to the sidebar and these are attached to the activated activity. In addition, similar resources are recommended to the learner.

6.2 Implementation of the Pedagogical Concept

The components suggested in the pedagogical concept presented in Section 5 have been implemented in CROKODIL [10]:

- A new tag type "activity" has been defined. Resources found during a learning activity can be tagged with this activity. These resources are thus assigned to this activity.
- Activities can be described by various attributes. Of crucial importance is the attribute "result document". This allows the learners to store their own documentation of the learning outcome of the activity.
- Activities can be organized hierarchically by the relation "is part of activity" and its (inverse) relation "is parent activity to". This allows the structuring of complex activities.
- Each activity provides a text box to document the learner's experiences. Comments to these experiences can also be added. All contributions are shown as a list of text sections showing the user's name, the date and the time of creation.

6.3 Recommender Services in CROKODIL

The relations between tags, users and resources as described in [12] are analysed and based on the information gained from the structure of the semantic network, recommender services are given to the learner. Recommendations of resources based on semantic similarity between textual resources are also possible. The algorithms applied are presented in [13]. Complementing structural and content-based approaches for personalized recommendations, we will also experiment with knowledge propagation along the social network of CROKODIL users. Social connections are very relevant as they convey additional user information about interests, tastes or preferences. Moreover, studies indicate that especially in taste-driven domains like music or books, familiarity with the person generating the recommendation is appreciated by users and may lead to more precise recommendations [14,15]. However, sociological findings imply that not all members of a network are equally useful, but the different kinds of ties hold different potential, especially regarding the novelty and reliability of information [16].

A first step is to assess the strength of ties in our learner networks. Interactions between CROKODIL learners in form of direct communication or learning group membership will be combined with indirect evidence such as bookmarking, viewing or copying another user's resources. Together with preference information about tags, resources and activities as well as structural information about the local social graph, different kinds of ties (like support, influence) and their strength will be identified. Those ties will be exploited to propagate relevant information and to recommend experts.

6.4 Integration in Social Networks and Learning Management Systems

In order to enrich user profiles within CROKODIL, as well as to increase external visibility of our platform, we plan to integrate CROKODIL with social networks which are popular to our target group according to our questionnaire (see Section 4). We will import relevant user profile attributes, user related metadata like tags and resources; as well as social connections and group memberships. Such external information spares the user from filling in profiles multiple times and, more importantly, to model our users and the ties among them more precisely, e.g. by determining on how many other platforms pairs of friends interact. Facebook offers, for example, FacebookConnect as a way to integrate user information into one's own website. It can also be used to publish interesting content on a user's Facebook newsfeed, so friends become aware of activities within CROKODIL.

Besides, Web 2.0 tagging platforms like Delicious are established bookmarking systems that already have a lot of user-generated tags for web resources. Thus, to overcome the problem of data sparseness in the early stage, we are fetching popular tags for URLs bookmarked as well in CROKODIL. External tags can be displayed or only be used internally within similarity computations for recommendations.

The CROKODIL platform is currently being integrated with the CLIX[21] Learning Management System. This allows the integration of episodes of resource-based learning within traditional courses and simplifies the transfer of resources between both systems. For example, learners can individually compile a list of resources in CROKODIL, transfer them to the LMS, and then discuss them online within the LMS e.g. which resources to use for a joint presentation. Similarly, resources can be

[21] http://clix.de, Online 2011-03-29

provided by the instructor in the LMS and then transferred by the learners into the CROKODIL platform. The technical integration is realized in two steps. The LMS CLIX is extended to support OpenSocial[22] gadgets. This includes a proposed extension of the OpenSocial specification allowing the upload of documents to the hosting container. A CROKODIL gadget will then be developed that allows a learner to browse his resources in the CROKODIL system. Resources can be displayed as topic lists or in a graphical network view. The learner selects resources from the list and triggers their upload into the LMS. A resource can be a link to an internet page or a document file. In both cases, metadata from CROKODIL, like tags and notes is also transferred. The OpenSocial specification is already supported by many social networks and it can be expected that soon the major LMS systems will support it, too. The CROKODIL gadget can then be simply plugged into these LMSs to integrate them with CROKODIL.

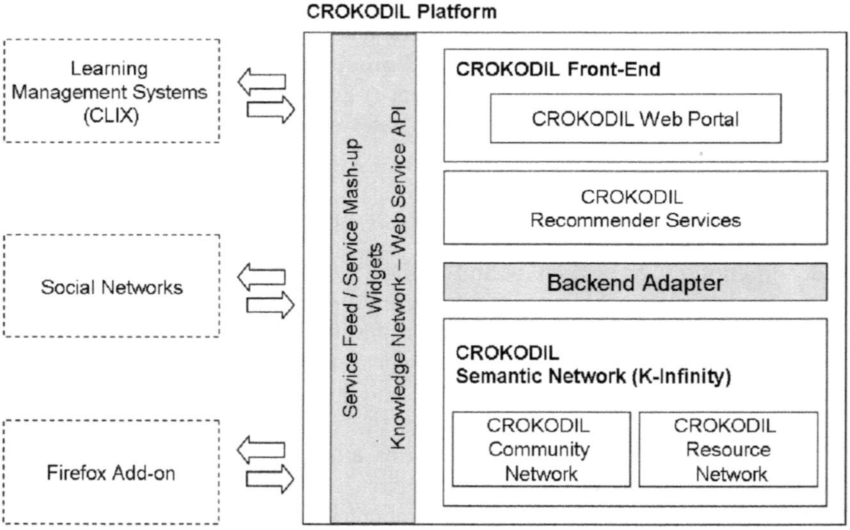

Fig. 3. An overview of the CROKODIL platform architecture showing the most important parts as described in the sections above

7 Future Steps and Challenges

In the future, we plan to implement different recommender systems to inform learners about similar resources from other learners. In addition, tags will be suggested which may be more appropriate for describing the resources. We will investigate and implement relevance ranking measures to improve the searching for and retrieval of relevant resources on the platform. We presently have released the first prototype of CROKODIL and plan to have further releases including these additional technical features. We also plan to conduct an extensive evaluation of the first prototype of the

[22] http://www.opensocial.org/, Online 2011-03-29

CROKODIL platform. The scenarios provided by our project partners IBB and Siemens AG will be considered for this evaluation. A comparative evaluation of the different recommender algorithms proposed will show the strengths and weaknesses of each approach; possibly leading to hybrid approaches considering structure, content, metadata, and social behaviour amongst users on the platform.

Acknowledgments. This work is supported by funds from the German Federal Ministry of Education and Research under the mark 01 PF 08015 A and from the European Social Fund of the European Union (ESF). The responsibility for the contents of this publication lies with the authors.

References

1. Rakes, G.: Using the Internet as a Tool in a Resource- Based Learning Environment. Technology 36(5), 52–56 (1996)
2. Hannafin, M.J., Hill, J.: Resource-based learning. In: Spector, M., Merrill, M.D., van Merrienboer, J., Driscoll, M.P. (eds.) Handbook of research on educational communications and technology, 3rd edn., pp. 525–536. Erlbaum, Mahwah (2007)
3. Oberhuemer, P.: Didactic Models and Graphical Learning Modeller. PROLIX Project, EU (2007), http://www.prolix.eu
4. European Commission: Key Competences for Lifelong Learning – A European Framework (2007), available http://ec.europa.eu/dgs/education_culture/publ/pdf/ll-learning/keycomp_en.pdf
5. Scholl, P., Mann, D., Rensing, C., Steinmetz, R.: Support of Acquisition and Organization of Knowledge Artifacts in Informal Learning Contexts. In: EDEN Annual Conference 2007, Naples, p. 16 (2007)
6. Böhnstedt, D., Scholl, P., Benz, B., Rensing, C., Steinmetz, R., Schmitz, B.: Einsatz persönlicher Wissensnetze im Ressourcen-basierten Lernen. In: Seehusen, S., Lucke, U., Fischer, S. (eds.) DeLFI 2008: 6. e-Learning Fachtagung Informatik, pp. 113–124 (2008)
7. Braun, S., Schora, C., Zacharias, V.: Semantics to the Bookmarks: A Review of Social Semantic Bookmarking Systems. In: Proc. of the 5th I-SEMANTICS, pp. 445–454 (2009)
8. Bateman, S., Farzan, R., Brusilovsky, P., McCalla, G.: OATS: the open annotation and taggind system. In: Proceedings of I2LOR 2006, Montreal, Canada (2006), proceedings, http://www.lornet.org/Default.aspx?tabid=64
9. Al-Khalifa, H.S., Davis, H.C.: Folksannotation: A semantic metadata tool for annotating learning resources using folksonomies and domain ontologies. Innovations in Information Technology (2006)
10. Rensing, C., Bogner, C., Prescher, T., Domínguez García, R., Anjorin, M.: Aufgabenprototypen zur Unterstützung der Selbststeuerung im Ressourcen-basierten Lernen. To be published in proceedings from DeLFI 2011: 9. e-Learning Fachtagung Informatik (2011)
11. Böhnstedt, D., Scholl, P., Rensing, C., Steinmetz, R.: Collaborative Semantic Tagging of Web Resources on the Basis of Individual Knowledge Networks. In: Houben, G.-J., McCalla, G., Pianesi, F., Zancanaro, M. (eds.) UMAP 2009. LNCS, vol. 5535, pp. 379–384. Springer, Heidelberg (2009)

12. Rensing, C., Scholl, P., Böhnstedt, D., Steinmetz, R.: Recommending and FindingMultimedia Resources in Knowledge Acquisition Based on Web Resources. In: Proceedings of 19th International Conference on Computer Communications and Networks, pp. 1–6. IEEE eXpressConference Publishing, Los Alamitos (2010)
13. Scholl, P., Böhnstedt, D., Domínguez García, R., Rensing, C., Steinmetz, R.: Extended Explicit Semantic Analysis for Calculating Semantic Relatedness of Web Resources. In: Wolpers, M., Kirschner, P.A., Scheffel, M., Lindstaedt, S., Dimitrova, V. (eds.) EC-TEL 2010. LNCS, vol. 6383, pp. 324–339. Springer, Heidelberg (2010)
14. Sinha, R.R., Swearingen, K.: Comparing Recommendations Made by Online Systems and Friends. In: Proc. DELOS Workshop: Personalisation and Recommender Systems in Digital Libraries (2001)
15. Bonhard, P., Sasse, M.A., Harries, C.: The devil you know knows best: how online recommendations can benefit from social networking. In: Proc. BCS-HCI, pp. 77–86 (2007)
16. Granovetter, M.: The Strength of Weak Ties: a Network Theory Revisited. Sociological Theory 1, 201–233 (1983)

Remote Collaborative Multi-user Informal Learning Experiences: Design and Evaluation

Ernesto Arroyo, Valeria Righi, Roger Tarrago, Patricia Santos,
Davinia Hernández-Leo, and Josep Blat

GTI, ICT Department, Universitat Pompeu Fabra, Roc Boronat 138,
08018 Barcelona, Spain
{ernesto.arroyo,valeria.righi,roger.tarrago,patricia.santos,
davinia.hernandez,josep.blat}@upf.edu

Abstract. This paper presents a customizable system used to develop a collaborative multi-user problem solving game. It addresses the increasing demand for appealing informal learning experiences in museum-like settings. The system facilitates remote collaboration by allowing groups of learners to communicate through a videoconferencing system and by allowing them to simultaneously interact through a shared multi-touch interactive surface. A user study with 20 user groups indicates that the game facilitates collaboration between local and remote groups of learners. The videoconference and multi-touch surface acted as communication channels, attracted students' interest, facilitated engagement, and promoted inter- and intra-group collaboration—favoring intra-group collaboration. Our findings suggest that augmenting videoconferencing systems with a shared multi-touch space offers new possibilities and scenarios for remote collaborative environments and collaborative learning.

Keywords: Computer supported collaborative learning, ubiquitous learning, informal learning, serious games, social awareness.

1 Introduction

This paper presents a customizable system designed to support collaborative multi-user problem solving in remote environments. It addresses the need for collaborative informal learning experiences in museum-like environments. The system combines multi-touch interaction and videoconferencing to support remote multi-user collaborative learning experiences. The system has been used to develop a serious game where groups of remote museum visitors interact with each other. During the game students put in practice their collaborative, communication and negotiation skills. They also share their knowledge acquired in the museum with their game-mates. The interactive experience is capable of engaging students in collaborative problem solving activities. It takes advantage of students' curiosity and attractiveness to new technologies to meet the increasing demand for appealing informal learning experiences [19] for use in museum-like settings.

C. Delgado Kloos et al. (Eds.): EC-TEL 2011, LNCS 6964, pp. 43–56, 2011.

The system has been evaluated with 39 university first-year students in a controlled environment. The results indicate that the system promotes collaboration and communication among participants. The appeal of the multi-touch surface attracted students, facilitating engagement, and inter- and intra-group collaboration. Participants were more inclined to collaborate with co-located partners and provided feedback related to the screen setup for facilitating the ergonomics of the communication through videoconference.

2 Background

Moore (1989) proposed three main types of interaction that can be supported by synchronous interactive technology: the instructor–learner interaction, learner–learner interaction and learner–content interaction [15]. Several online and desktop-based systems and applications have been developed to support the three types of interaction, allowing simultaneous collaboration between multiple learners (remote or co-located). These systems have proved useful in facilitating collaboration and provision of instant formative feedback and improving the cognitive proximity and identity decoupling of students [20]. Some of these collaborative systems use readily available commercial hardware, such as the Microsoft Multipoint[1] project, which integrates up to 25 mice and allows students in a classroom to simultaneously interact and work together on a single desktop computer. Each student is equipped with a computer mouse and Multipoint software assigns a unique mouse pointer to each student, giving him or her equal opportunity to participate in coordinated, collaborative, or cooperative activities through single display groupware [18]. Other collaborative systems include online tools to allow learners to participate using a personal computer. For instance, GroupScribbles[2] (GS) is an application for realizing collaborative learning activities especially designed for primary school students [20]. Students and teachers interact with the system from their own computers and use several artifacts (such as adhesive notes, bulletin boards, whiteboards, stickers) to share their knowledge. They can create their own artifacts in their private space and share them in the public space. As [13] identified, by using GS students have the opportunity of playing active roles in their learning, they can interact with their peers and teachers. GS has been designed to allow multiple users interacting with the system at the same time; however it does not support a group of students interacting with the same computer at the same time.

Touch and Tangible User Interfaces (TUI) have already proved useful in supporting interaction between multiple learners. In touch and TUI learners interact and manipulate digital information through touch and physical objects [5]. Interacting with tabletops, smartboards, multi-touch screens and tangible artifacts, facilitates simultaneous interaction and provides new opportunities to envisage innovative

[1] http://www.microsoft.com/windows/multipoint/
[2] http://groupscribbles.sri.com

learning scenarios: interactive tables might support participation balance in f2f collaborative learning [3], multi-touch whiteboards might facilitate group learning [11], and tangible artifacts might be used to solve design problems [9].

Videoconferencing systems can support collaborative learning activities at a distance6]. They offer a direct communication channel, allowing students to visualize each other and to freely voice their ideas and opinions, thus facilitating an effective remote collaboration experience [12]. Online collaborative systems incorporate videoconferencing and also provide the means for sharing a virtual desktop or for exchanging digital information, such as pictures, presentations, documents, etc. Examples of existing commercial products include flashmeeting[3], wimba[4] and Adobe Connect[5]. Although video conferencing systems offer several advantages for remote learning, their use seldom goes beyond video calls in informal educational settings involving small to medium sized groups [21] [2]. This is not only due to their excessive installation costs and complex setup (for specialized equipment), but mainly to their lack of support for collaboration on complex scenarios that go beyond document and desktop sharing. For instance, when dealing with mixed settings with co-located and remote groups of learners, neither video-conferencing nor online collaboration systems support interaction within the local group. On one hand videoconferencing provides f2f (or group2group) communication, but restricts simultaneous participation and interaction. Furthermore, typical videoconferencing systems often force users to take-turns when working in groups by signaling their intent to participate and waiting until allowed to intervene. On the other hand online collaborative applications and single display-display groupware support simultaneous interaction over a shared space, but lack (f2f) natural communication between remote or local learners.

Augmenting videoconferencing systems with multi-touch interaction allows multiple learners to interact simultaneously over a tangible shared space, to maintain verbal communication, and to share non-verbal cues through an additional visual link [17]. This offers new possibilities and scenarios for remote collaborative environments and collaborative learning. For instance, in galleries or museum-like environments, visitors may want to collaborate in order to share their views, knowledge and interpretation of the elements exhibited. The social interactions can lead to learning outcomes, and together with the multi-touch experience engage the participants that might be strangers otherwise [26]. New interactive learning experiences can draw museum visitors and motivate students to further explore an exhibition during a school visit. Moreover, museums in different locations can be complementary to each other and the learning experienced by visitors in one museum could enrich the learning of visitors in the other.

---

[3] Flashmeeting Website, http://flashmeeting.e2bn.net

[4] Wimba Website, http://www.wimba.com/products/wimba_classroom

[5] Adobe Connect Website, http://www.adobe.com/products/adobeconnect.html

3 A Remote Collaborative Multi-user Learning Experience

This section describes the design criteria used to fulfill the main *collaboration* and *communication* goals. We illustrate the design criteria by the development of a concrete experience involving two remote museums sharing a common installation. The goal of this experience is to facilitate ***collaborative problem solving*** while working in remote environments. The project also seeks to support one-to-one and many-to-many ***communication*** to allow more than one learner to participate and ***interact*** in informal learning experiences between remote and co-located learners. We describe how the system features and design considerations attempt to accomplish the communication, collaboration and interaction goals.

3.1 Use Case and Scenario

This project was developed within the context of two Spanish museums that exhibited an art collection simultaneously in Figueres and Barcelona cities. The common exhibition was titled "La col·lecció de col·leccions de Guy Selz" (The collection of collections of Guy Selz). It included random and hectic selection of objects from around the world that were collected by Guy Selz through his life travels [10]. Some objects included anecdotes and descriptions about their origin, uses, and importance. Each museum hosted an un-categorized sample from the collection with the intention to allow visitors to come up with their own categories. The unique characteristics of this use-case directed the application to become an additional channel for museum visitors to explore the exhibition and increase the sense of connectedness between two spaces. All the while, allowing museum visitors to interact and collaborate.

With respect to the use of technology all stakeholders agreed that incorporating multi-touch surfaces and videoconferencing could enhance the visitor and learner experience, however they did not agree on a single scenario. Envisioned scenarios included treasure hunting activities, collaborative puzzles, interactive videoconferencing and performances, collaborative music composition, collaborative tagging, expertise sharing, collaboratively exploring high definitions images, among others. Thus, it was necessary to develop an agile flexible system to provide the means for enacting different types of interactive learning games and experiences with different user flows and interface behaviors. The Toy Museum of Catalonia requested for the experience to be a game (one major design constrain) as they receive frequent guided school visits and students had expressed their interest in playing while at the museum (expected from a toy museum). Thus, during a coordinated activity in both museums, groups of students visiting the Arts Santa Monica museum in Barcelona and the Toy Museum of Catalonia in Figueres are invited to play a game with other groups of students also visiting a museum in a different city. The game is related to objects they have observed and experienced during their museum visit.

3.2 Educational Games

Part of the design criteria has been based on the motivational generalizations and design principles proposed by [19]. First of all the game has been designed to provide *formative feedback* to help students to acquire the necessary expertise that they need to learn about the contents of the museum. Feedback with information about the pieces selected during the game and hints to find the correct solution are provided to the students. In addition the tasks have been designed in order that *students feel confident and competent* to solve them without many difficulties. Students interact with different types of artifacts and the game provides three levels in which students perform different tasks. All the tasks have been designed to *stimulate students' motivation* but also to provide *learning content that can be meaningful and interesting* for them.

3.3 Collaboration

Inter- and Intra-Group Collaboration

The effectiveness of communication and collaboration can be enhanced if group's activities are coordinated [6]. Thus the primary goal is to facilitate collaborative activities within a group of co-located students, while also facilitating remote collaboration and coordination between groups of students in different locations. The experience has been designed to facilitate **intra-group** and **inter-group** collaboration by incorporating *focused collaborative problem-solving tasks* that require collaborating with other participants and at the same time, they provide opportunities to *attain social and academic goals* [4][19].

Intra-group collaboration refers to discussion and interaction elicited to solve a problem task, in which co-located members are involved. At the other hand, inter-group collaboration refers to problem solving activities that involve members belonging to remote groups. In order to facilitate intra- and inter- group collaboration, the system provides students with a shared workspace where simultaneous and multi-user interactions are allowed for both co-located and remote participants. Table 1 describes the elements that have been specifically incorporated into the game to reinforce collaboration and discussion among remote groups: 1) students must complete problem-solving tasks and perform operations together with the remote partners in order to progress; 2) students can simultaneously work in different parts of the task both individually or collaboratively; 3) students must focus and provide help on a task that should be completed by the remote group.

Physical and Verbal collaboration

In collaborative environments individual participation can take place both through verbal communication and through physical actions [8]. Our system supports both **physical** and **verbal** intra- and inter-group communication and collaboration. It provides a videoconference channel for verbal communication and a shared virtual space for physical communication and collaboration.

Physical **inter-group and intra-group** collaboration is supported by simultaneous interactions with a multi-touch surface that allows more than one users to interact at the same time. Students can interact at the same time with the game interface elements among co-located (**intra-group**) and remote (**inter-group**) learners over a shared virtual space. On the shared virtual space deictic gestures are possible thanks to visual elements that provide awareness of the remote actions. Multi-touch screen on both ends facilitate **inter- and intra-group** students to have the same opportunities to collaborate and interact with the artifacts in order to solve problems together. A videoconference displayed on a vertical screen supports verbal inter-group collaboration, allowing remote participants to talk and discuss about game tasks.

3.4 Equity of Collaboration

Simultaneous interaction can prevent *turns-taking* behavior among participants and at the same time can promote equitable interaction [22]; however it can also result in parallel working and distribution of tasks among users [22]. In a remote environment parallel working and taking-turn behavior might constrain reciprocity and communication, therefore limiting collaboration among remote members. We adopted several design criteria to avoid limiting collaboration and promote an equal participation among remote partners. A multi-touch table allows for direct manipulation of digital content while at the same time allows for multi-user interaction. Several studies about coordination in tabletops settings have demonstrated that distributed tabletop interaction might be influenced by workspace configuration, leading to territoriality behavior and unequal participation among users [24]. We designed the virtual space with the objective of allowing one or more remote learners to interact simultaneously over a shared collaborative workspace. The system uses a "*what you see is what I see*" interaction metaphor [23] in which all participants share the same visual perspective of the shared workspace (same-side configuration). This configuration allows both remote groups to have *equal opportunities for interaction*, allowing them to make contributions for solving the problem and avoiding territoriality division of the surface [24]. Moreover, in order to facilitate an equal participation among co-located users, items of the interface were equally distributed on the whole surface.

3.5 Communication and Awareness

An important issue when designing remote collaborative environments is to provide a common channel for communication. This channel should facilitate emerging conversations and allow users to be aware of the actions of their remote partner [6]. The system includes a continuous videoconference link that promotes verbal communication and awareness between remote groups of students. The aim is that groups will use the videoconference to coordinate their actions verbally over the videoconference link. The system also supports remote awareness by using several visual cues and elements on the shared workspace that help learners understand the actions of the remote partners, providing common ground for deictic gestures that support communication. The system uses visual pointers of the finger in order to reduce interference within the space produced by multiple participants interacting simultaneously [25]. When user touches the screen a colored circle is "painted" under his finger (each location is represented by specific color). In this way, the actions of each student are immediately visualized on the screen of the remote location: when students move an object in one location, remote students will also see the same object moving marked with the color of the remote group.

4 Prototype Description

Two stations placed in two remote locations compose the system. Each station includes two 32' screens: one horizontal multi-touch screen, and a vertical screen (Fig. 1). The horizontal screen displays content related to the exhibition that can be seen and accessed from both stations. The vertical screen displays a video feed from each location and complementary information related to the learning experience (instructions, status, hints, and support material).

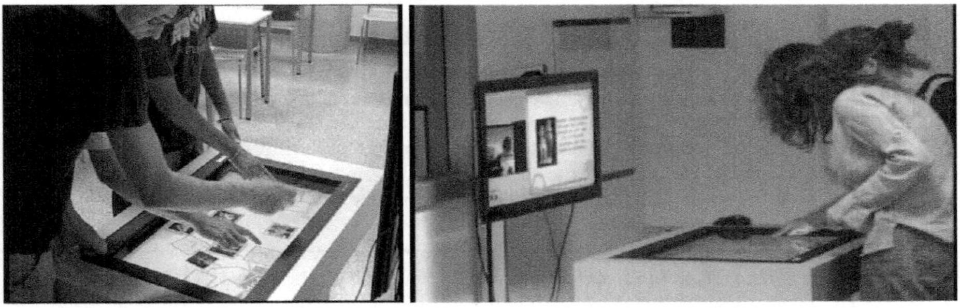

Fig. 1. The system includes two remotely located stations. *Right:* Participants interact on a horizontal surface while sharing the same workspace with remote partners. A vertical screen shows a video feed facilitating communication between the two locations. *Left*: Simultaneous co-located participants categorizing objects by country of origin and by functionality.

4.1 Game Levels

Collaborative tasks are introduced in stages in order to guide and facilitate discovering and exploring the educational material. The game is divided into three sequential stages and requires at least one participant on each remote location. Each stage starts with a brief description of the rules on the vertical screen. Coherent interaction throughout the different stages of the game facilitates transitioning between stages. In stage zero, or at the start of the game, users select the number of participants on each remote location-- from one, up to three players and cannot be changed during the game. The system uses the number of participants to control the numbers of fingers required to collaboratively select pieces in the following stages. The first stage presents information about the elements in the game and provides useful information for completing the remainder stages. This stage promotes negotiation and collaborative decision-making by requiring participants to coordinate their interaction and cooperatively select an item (cooperative manipulation [14]). The other two stages require participants to group items with respect to geographical or functional categories. In this phase of the experience all participants can simultaneously interact on the surface moving and placing objects on proper placeholders. These stages also include "special items" that have been designed to encourage cooperation and information sharing between remote participants. Table 1 summarizes the main features of the game and their relationship to the type of collaborative behavior elicited.

Table 1. Game stages and their relationship to collaboration objectives

Stage	Game related goals	Type of collaboration elicited	Type of Interaction that support the elicited collaboration	Elements of the game that support the elicited collaboration
1	*Exploration*	*Inter-group collaboration* (negotiation) through physical and verbal participation	Selection of objects through *cooperative gesture*	Shared workspace Remote Awareness Videoconference
	Participants explore a collection of ten images distributed on the multi-touch horizontal surface. Players on both locations must agree and choose an item in order to receive further information. All members from both groups must simultaneously select the same item by placing their finger on it (cooperative gesture). Once "all" players simultaneously select an item, they are presented with additional imagery and a textual description related to the selected item. The description includes information about the object's origin, functionality, creation date, and how it became part of the museum collection.			
2	*Categorization* of images based on place of origin	*Inter- and intra-group collaboration* through physical and verbal participation.	*Simultaneous remote and co-located interaction*	Shared workspace Special piece Remote awareness Videoconference
	Participants must organize the objects according to their country of origin. The surface table includes a world map with place holders. Students can organize objects simply by dragging and placing each object onto the appropriate geographical location. Visual feedback is given for correct answers, while visual and auditory hints are displayed on the vertical screens for incorrect answers. Each of ten objects can be controlled independently by holding it or by dragging it, however members from either group can take control over it only after the item is released. There are two special items, one for each team. When selected, all others pieces are grayed out and cannot be moved, while a hint appears on the vertical screens and an alert signal is played signaling players to focus on the hint. Once the piece is correctly located, the others items are released.			
3	*Categorization* of images based on function classes	*Inter- and intra-group collaboration* through physical and verbal participation.	*Simultaneous remote and co-located interaction*	Shared workspace Special piece Remote awareness Videoconference
	This level is conceptually identical to the second one. However, on this stage players must organize items based on their functionality (religious, popular, or entertainment). The type of interaction and collaboration allowed are identical to the second level. Two special items are also presented and behave as described above.			

5 Evaluation

We conducted a user study to explore the efficiency of the design criteria in promoting collaboration and discussion among participants. The user study provided quantitative and qualitative data to assess how the collaboration features in game lead to different collaborative behaviors, and to assess user's perception and satisfaction with the game.

Table 2. User study's units of analysis and coding scheme for physical and verbal intra-group and inter-group collaboration

Goal	Unit of analysis	INTRA-GROUP definition	INTER-GROUP definition
Quantitative analysis through video coding			
Collaboration	Time in seconds	*Physical*: helping moving a piece, or touching the surface to indicate the place where to move the piece *Verbal:* suggesting, verbally, the place where to move the piece	*Physical*: touching the surface to indicate the place where to move the piece; (in Level 1, touching the surface to indicate the item they want to select). *Verbal:* suggesting, verbally through videoconference, the place where to move the piece; (in Level 1, suggesting through videoconference the item to select).
Equity of Collaboration	Gini coefficient	---	Takes the collaboration rate for each group (physical & verbal), and computes the Gini coefficient between two groups.
Qualitative analysis through questionnaire			
Perceived Communication	Likert scale	2 questions assessing how much participants perceived having communicated with their co-located partners	2 questions assessing how much participants perceived to have communicated with their remote partners
Perceived Collaboration	Likert scale	Idem perceived communication	2 questions assessing how much participants perceived to have collaborated with their co-locate partners
Overall satisfaction	Open question	List 3 aspects of the experience that you liked/enjoyed List 3 aspects of the experience that you didn't like.	

Collaboration and equity

We are interested in exploring three aspects of collaboration: 1) to what extent the game promoted inter- and intra-group collaboration among the different stages of the game, 2) the existence of a preferred type of collaboration, i.e. a tendency of users in collaborating more with the co-located members (intra-group collaboration) than with the remote ones (inter-group collaboration), and 3) to what extent the game promoted equal inter-group participation, i.e. if both groups equally collaborated with the other team or if the process of collaboration was mostly initiated by one of the two teams.

We measured both physical and verbal *intra-group* and *inter-group collaboration* by identifying the main patterns of participants' interactions and representing them through scores computed as described in table 2. The unit of analysis for group collaboration was defined as the sum of individual scores coming from the members of the same group. We used the *Gini Coefficient* to express the equitable nature of collaboration among remote teams as reported in [7]. It uses the contribution of each group to compute the deviation with respect of equal participation between the two groups. The group contributions for calculating the Gini coefficient have been computed as described in table 2.

Overall Satisfaction
We collected information about overall user satisfaction using a follow-up questionnaire divided into four main categories based on the following criteria: *perceived inter- and intra-group communication, perceived inter- and intra-group collaboration, overall satisfaction of the experience*. Overall satisfaction was assessed using close and open qualitative questions to elicit positive and negative aspects of four main categories related to aspects of the *game*, the *videoconference*, the social *nature* of the game (e.g. collaboration), the *interaction*, and *issues* regarding the installation.

5.1 Method

The evaluation consisted of 10 one-shot experimental sessions in which 20 groups (with 2 to 3 persons per group) of 39 first-year university students (14 women, 25 men) played all game levels (lasting on avg. 5.4 min.) in a controlled environment. The stations were set up in two remote rooms at a university campus. Participants in each room were given few minutes to familiarize with the multi-touch table and to introduce themselves to the remote team through videoconference while one researcher explained the main goal of the game to both teams. The researchers remained in the room throughout the session, but did not interact with participants unless in response to specific problems related to technical difficulties. All sessions were video recorded and the actions of each participant were analyzed via video coding using the scheme represented in table 2.

5.2 Results

Collaboration
Analysis of *collaborative patterns* in users' activities showed that collaborative engagement of participants decreased as the game progressed: on average, participants collaborated for 23.3%, 16.7%, and 13.5% of the total time for level one, two and three respectively. Further analysis on second and third levels, in which both inter and intra-group collaboration are elicited, showed a significant difference ($t=9.65$, $p< 0.001$) in the time spent collaborating in the two modalities (intra-group: 77.8%, Vs. inter-group: 22.2%). This indicates that participants were more engaged in talking and interacting with the members of the same team than in discussing with the remote team.

During the entire game, users preferred verbal communication as channel for collaboration rather than physical interaction (e.g. deictic gestures) on the surface ($p < .001$). This is particularly true for the inter-group collaboration in level 2 and 3; instead in level 1 (in which cooperative interaction is required; see Table 1), participants used more deictic gestures to communicate with the remote partners with respect to the other levels; however statistical analysis did not disclose any significant difference between levels. This supports the fact that in cooperative manipulation, where the actions of remote users must be strictly coordinated, is particular crucial to support the interaction with elements that provide remote awareness. Additionally, we observed a learning curve regarding simultaneous interactions; on the second stage participants exhibited a turn-taking behavior, while on the 3rd level they interacted with the objects simultaneously and independently. Participants went from an average of 5.2 times giving turns on the second level to only 2.3 times on the third level.

Table 3. Averages of time passed by each group in intra and inter group collaboration

Level	**Intra-group Collaboration****			**Inter-group Collaboration****			**Gini Coefficient**
	Total	*Physical*	*Verbal*	*Total*	*Physical*	*Verbal*	
1	-	-	-	100	45.2	54.8	.52
2	74	35.1	64.9	26	13.9	86.1	.25
3	81.9	34.7	65.3	18.1	8.5	91.5	.28

** *The values represent percentage respect to the total time spent collaborating.*

Regarding *perceived communication and collaboration*, participants scored significantly higher on the *perceived intra-group* communication (mean (M.): 3.1) than the inter-group (M. 2.1), ($t = 3.99$, $p < .001$). Also on the level of collaboration, participants perceived to have collaborating more with the members of the same team (M. 3.13), rather than with the other team (M. 2.4) ($t = 3.42$, $p < 0.001$), confirming the results of the quantitative analysis.

Equity of collaboration

Analysis of Gini coefficient revealed differences regarding equity of inter-group collaboration. In the first level one group mostly predominated the task deciding verbally which item to select, while on the others levels both groups initiated collaboration with the remote team, almost equally (.52 vs. .25 and .28).

Overall Satisfaction, Positive and Negative Aspects

The follow-up questionnaire showed that the overall satisfaction of the experience was high (M. 3.8, SD. 0.66, in a scale from 1 to 5). Responses about the positive aspects indicate that participants appreciated the collaborative nature of the game and the type of interaction allowed by the multi-touch interface. Whereas aspects related with the videoconference and general features of the game (such as its intuitiveness, the educative role etc.) scored low values (Fig. 2). Analysis of the negative aspects of the game revealed several problems, such as the lack of clarity in the goal of the game, the lack of competition, and the game subject. Many negative complains addressed the audio quality of the videoconference that caused difficulty in the communication with the remote team.

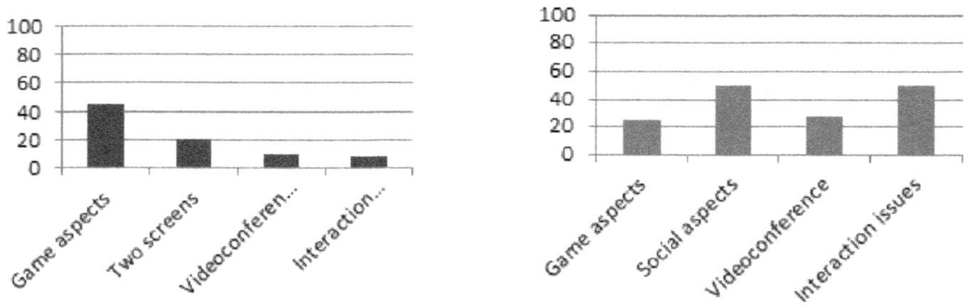

Fig. 2. Positive (*left*) and negative (*right*) aspects according to users' opinions: participants appreciated the social collaborative nature of the game and the multi-touch interface, found the split-screen setup uncomfortable and suggested several game improvements.

5.3 Discussion

The results indicate that the different stages of the game promoted inter- and intra-group collaboration as participants collaborated up to 23% of the time. There was a tendency of users in collaborating intra-group as participants spent around 80% of the time collaborating with their peers. This is partially due to the bad choice of using a two-screen setup; one vertical and one horizontal, as many users did no looked at the vertical videoconferencing screen while interacting on the horizontal screen. This is also supported by the responses about negative aspects of the experience, as many users complained for the lack of affordance of the videoconference while interacting. The results also indicate that both groups collaborated equally in collaborative tasks (levels 2 and 3), while cooperative tasks (level 1) elicited a predominated behavior by one of the two teams. In general, remote collaboration was higher in level 1 as the type of interaction required cooperation between participants, whereas in the latter stages participants freely engaged in intra-group interaction.

Despite problems reported regarding the videoconference video quality, the audio link provided a natural way for learners to communicate with remote participants. Participants preferred verbal communication as channel for collaboration rather than physical interaction on the surface. This was especially true for tasks not requiring cooperative actions, as in this case users needed visual cues of remote actions to coordinate the movement with the remote partners. Although our system supports simple cooperative manipulation, i.e. selection of items, the findings suggest that this kind of interaction promotes a higher level of collaboration, but might lead to unequal participation among remote partners.

Participants reported several deficiencies in the game design that should be addressed in future iterations: the lack of a clear goal, competition, and intuitiveness. Another important drawback is that the game can only be used when there is at least one participant on each remote location. This drawback can be easily addressed by offering single and collaborative modes of interaction in future iterations. Overall, the results indicate that participants appreciated the collaborative nature of the game and the type of interaction allowed by the multi-touch interface.

6 Conclusions and Future Work

This paper has introduced a customizable system used to develop collaborative informal learning experiences in remote museum-like environments. An interactive learning experience combines multi-touch interaction and videoconferencing to engage students in multi-user collaborative problem solving activities. The system facilitates remote collaboration allowing participants to communicate through videoconferencing, in a vertical screen, and through a shared interactive workspace displayed on a horizontal screen.

The findings assessing the game features show that the system promoted different collaborative behaviors and enabled inter- and intra-group collaboration and cooperation. Participants collaborated up to 23% of the time with a tendency for intra-group collaboration. Moreover, remote groups collaborated equally in collaborative tasks while collaboration was mostly initiated by one of the two teams for the cooperative tasks. Finally, user's perception and satisfaction was positive as participants appreciated the collaborative nature of the game, the type of interaction allowed by the multi-touch interface, and the videoconference communication channel. To conclude, our findings lend support to the view that augmenting videoconferencing systems with a tangible shared space offers new possibilities and scenarios for remote collaborative environments and collaborative learning.

Future work includes addressing the design problems encountered; improving the intuitiveness of the interface, adding competition aspects to the game, conducting new experiences in other educational spaces and devices such as tablets. Moreover, we are currently working on integrating videoconferencing within a 3D collaborative-shared space to support a more ergonomic interaction between remote participants, allowing for eye contact while interacting with the surface at the same time [1].

Acknowledgments. This work has been partially funded by the LEARN 3 project (TIN2008-05163/TSI) and by the European Fund for Regional Development as part of the Catalonian FEDER 2007-2013 project.

References

1. Arroyo, E., Ardaiz, O., Righi, V., Blat, J.: Distributed Multi-touch Virtual Collaborative Environments. In: Proceedings of CTS 2010, Illinois, USA, pp. 635–636 (2010)
2. Anastasiades, P.S., Vitalaki, E., Gertzakis, N.: Collaborative learning activities at a distance via interactive videoconferencing in elementary schools: Parents' attitudes. Computers & Education 50, 1527–1539 (2008)
3. Bachour, K., Kaplan, F., Dillenbourg, P.: An interactive Table for Supporting Participation Balance in Face-to-Face Collaborative Learning. IEEE Transactions on Learning Technologies 3(3), 203–213 (2010)
4. Baker, M., Lund, K.: Promoting reflective interactions in a CSCL environment. Journal of Computer Assisted Learning 13(3) (2003)
5. Brereton, M., McGarry, B.: An observational study of how objects support engineering design thinking and communication: implications for the design of tangible media. In: Proceedings of SIGCHI 2000, pp. 217–224. ACM Press, New York (2000)

6. Ellis, C.A., Gibbs, J.B., Rein, G.: Groupware: some issues and experiences. Commun. ACM 34(1), 39–58 (1991)
7. Fitze, M.: Discourse and participation in ESL face-to-face and written electronic conferences. International Journal of Corpus Linguistics 11(4), 511–512 (2006)
8. Harris, A., Rick, J., Bonnett, V., Yuill, N., Fleck, R., Marshall, P., Rogers, Y.: Around the table: Are multiple-touch surfaces better than single-touch for children's collaborative interactions? In: Proceedings of CSCL 2009, pp. 335–344. ISLS (2009)
9. Jermann, P., Zufferey, G., Dillenbourg, P.: Tinkering or Sketching: Apprentices' Use of Tangibles and Drawings to Solve Design Problems. In: Dillenbourg, P., Specht, M. (eds.) EC-TEL 2008. LNCS, vol. 5192, pp. 167–178. Springer, Heidelberg (2008)
10. Joan I Rosa, J.: La col·lecció de col·leccions de Guy Selz. ACTAR, Arts Santa Monica, Barcelona (2010)
11. Kemp, R., Kemp, E., Mohanarajah, T.: Supporting Group Learning Using a Digital Whiteboard. In: Masoodian, M., Jones, S., Rogers, B. (eds.) APCHI 2004. LNCS, vol. 3101, pp. 594–598. Springer, Heidelberg (2004)
12. Lewis, G.J., Hasan, S.M., Alexandrov, et al: Collaborative Virtual Environment for Advanced Computing. Computers and Artificial Intelligence 24(1) (2005)
13. Looi, C.-K., Chen, W., Ng, F.-K.: Collaborative activities enabled by GroupScribbles: An exploratory study of learning effectiveness. Comp. & Ed. 54(1), 14–26 (2010)
14. Margery, D.M., Arnaldi, B., Plouzeau, N.: A general framework for cooperative manipulation in virtual environments. In: Proceedings of Virtual Environments 1999, pp. 169–178. Springer, New York (1999)
15. Moore, M.: Three types of interaction. The American J. of Distance Ed. 3, 1–7 (1989)
16. Anastasiades, P.S., Vitalaki, E., Gertzakis, N.: Collaborative learning activities at a distance via interactive videoconferencing in elementary schools: Parents' attitudes. Computers & Education 50, 1527–1539 (2008)
17. Pauchet, A., Coldefy, F., Lefebvre, L., et al.: Tabletops: Worthwhile Experience of Collocated and Remote Collaboration. In: Proc. TABLETOP 2007, pp. 27–34 (2007)
18. Pawar, U., Pal, J., Toyoma, K.: Multiple Mice for Computers in Education in Developing Countries. In: ICTD 2006, pp. 64–71 (2006)
19. Pintrich, P.R.: A motivational science perspective on the role of student motivation in learning and teaching contexts. J. of Educational. Psychology 95(4), 667–686 (2003)
20. Roschelle, J.: GroupScribbles: A tool for highly interactive Tablet PC classrooms. In: HP Webinar on Teaching, Learning, & Technology in Higher Ed. (2006)
21. Schenker, J.L.: Economic woes boost videoconferencing. Bloomberg Businessweek (2008), `http://www.businessweek.com/` (retrieved)
22. Stanton, D., Neale, H.R.: The effects of multiple mice on children's talk and interaction. Journal of Computer Assisted Learning 19(2), 220–228 (2003)
23. Stefik, M., Bobrow, D.G., et al.: WYSIWIS revised: early experiences with multi-users interfaces. ACM Trans. Inf. Syst. 5(2), 147–167 (1987)
24. Tang, A., Neustaedter, C., Greenberg, S.: VideoArms: Embodiments for Mixed Presence Groupware. In: BCS-HCI (2006)
25. Tuddenham, P., Robinson, P.: Distributed Tabletops: Supporting Remote and Mixed-Presence Tabletop Collaboration. In: TABLETOP 2007, pp. 19–26 (2007)
26. Vom Lehn, D., Heath, C., et al.: Exhibiting Interaction: Conduct and Collaboration in Museums and Galleries. Symbolic Interaction J. 24(2), 189–216 (2001)

Automatic Identification of Tag Types in a Resource-Based Learning Scenario

Doreen Böhnstedt, Lasse Lehmann, Christoph Rensing, and Ralf Steinmetz

Multimedia Communications Lab (KOM), Technische Universität Darmstadt,
Rundeturmstr. 10,
64283 Darmstadt, Germany
{Doreen.Boehnstedt,Lasse.Lehmann,Christoph.Rensing,
Ralf.Steinmetz}@kom.tu-darmstadt.de

Abstract. When users use tags they often have a rich semantic structure in mind, which can not be fully explicated using existing tagging systems. However, a tagging system needs to be simple in order to be successful, otherwise it will not be accepted by users. In our ELWMS.KOM system for the support of self-regulated Resource-Based Learning users can assign specific semantic types to the tags they use in order to manage their web-based learning resources. However studies have shown that most users would appreciate an automatic identification of tag types. In this paper we present a knowledge-based approach for the automatic identification of the tag types used in the ELWMS.KOM system. Evaluations conducted on different corpora show that the algorithm works with an overall accuracy of up to 84%.

Keywords: Tagging, Tag Type Identification, Semantic Tagging.

1 Introduction

In an evolving world with fast changing circumstances and requirements learners are often required to learn self-directed in a demand-oriented manner. To close their knowledge gaps learners more and more use various resources from the WWW. This learning paradigm is called Resource-Based Learning (RBL) [8]. In order to be able to learn efficiently with resources gathered from the web, learners need to be supported accordingly: The overhead generated by tasks like search, organization and storage of the resources needs to be reduced to a minimum, and challenges like cognitive overload and disorientation caused by the vast amount of web resources and their hyperlinked structure need to be addressed. We have designed, implemented and evaluated ELWMS.KOM [2], a system for the support of self-directed Resource-Based Learning. The main goal of the ELWMS.KOM system is to give learners an accessible tool to organize their (web-)resources according to their cognitive model. To achieve this, we need to provide a means for the learners to make this model explicit, which usually is done by having the learners provide additional information about the web-resources. In recent years, researchers have perceived a mismatch regarding the creation of metadata. On the one hand, users generally avoid creating formalized metadata and do not want to fill out standardized forms to provide

C. Delgado Kloos et al. (Eds.): EC-TEL 2011, LNCS 6964, pp. 57–70, 2011.

structured information about their documents [4]. On the other hand with the advent of the so-called Web 2.0, tagging systems have become popular, having users voluntarily generate vast amounts of tags (and thus metadata). One reason for the success of tagging in comparison to creation of formalized metadata is the fact that tagging generally is simple and does not follow formalized rules. Users basically can tag how and whatever they want.

Thus, we decided to choose tagging as a way for the users to make the model they have in mind explicit. In addition, we give them the possibility to optionally assign a semantic type (like event, person or location) to a tag – thus performing *semantic tagging*. For example, an attendee of the EC-TEL 2010 conference in Barcelona who heard a talk about the paper "*Extended Explicit Semantic Analysis for Calculating Semantic Relatedness of Web Resources*" by Philipp Scholl will very unlikely remember the full title of the paper when she wants to revisit the paper half a year later. However the probability that she remembers the event ("EC-TEL 2010"), the location ("Barcelona"), the person ("Philipp Scholl") or the topics ("TEL", "Semantic Relatedness") is very high. Thus if she or other users organized the named resource accordingly, she could access the paper later on very efficiently. The type of a tag, i.e. whether it is an event, a location, a person, a topic etc., constitutes important semantic information about a tag, which can be used to better structure and keep track of the information. An overview of related (semantic) tagging application is given in [3].

Our goal is to make the users explicate their semantic models in a more structured manner without loosing the simplicity and accessibility (and thus the success) of tagging systems. Although in our ELWMS.KOM system users have the possibility to specify the type of a given tag in a very simple and intuitive way, our final goal is to automatically identify the type of a given tag. The knowledge-based algorithm for the automatic identification we propose (Section 4) as well as its evaluation (Section 5) represent the main contributions of this paper. Further, we analyze related work in the area (Section 3) and – in the following section – shortly describe the ELWMS.KOM system, introduce the semantic tag types used and present studies that confirm our selection of types.

2 Semantic Tagging in a Resource-Based Learning Scenario

ELWMS.KOM is a system for the support of Resource-Based Learning consisting of an add-on for the browser Firefox to efficiently insert web resources into the semantic network by tagging and a web-based platform for search and retrieval [2]. User-based evaluations of the research and learning environment ELWMS.KOM have shown that it has the potential to support the resource management in RBL scenarios [10]. Learners are more active and plan their learning process better than without such a support. They are satisfied with their outcome while using ELWMS.KOM and its semantic tagging concept is broadly accepted.

The ELWMS.KOM prototype is the basis for the CROKODIL project[1] which focuses on the community aspects of Resource-Based Learning and provides a web-based community platform for the support of RBL. The tag types relevant for

[1] http://www.crokodil.de, Retrieved online: 2011-03-31

ELWMS.KOM (and CROKODIL) are detailed and enhanced with usage examples in the following, while Figure 1 shows an exemplary semantic network:

- Person / organization: e.g. author or referrer of a resource, a person the resource is connected with or is about
- Location: e.g. a location where a resource was found or is connected with
- Type: e.g. genre or mime type of a resource (e.g. blog, wiki)
- Event: e.g. a conference where a resource was presented
- Topic: what is a resource about
- Goal / Task: the goal (e.g. task or knowledge demand) a resource was searched for

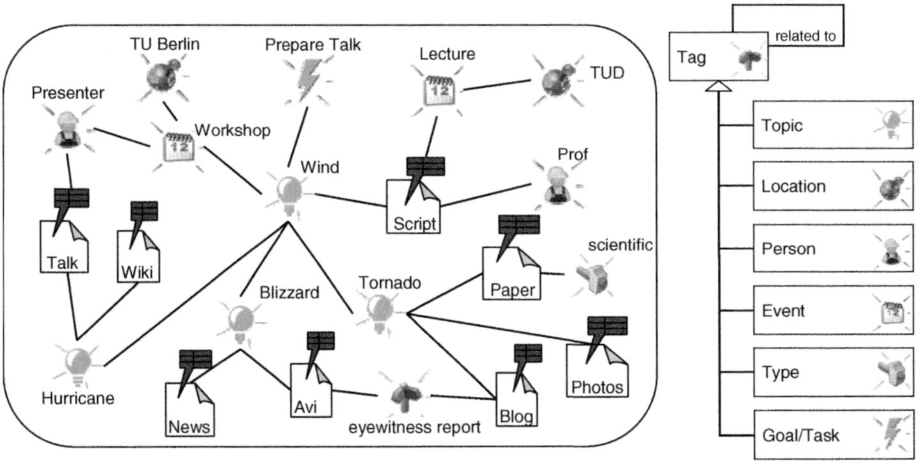

Fig. 1. ELWMS.KOM – exemplary semantic network

This list of tag types was confirmed in two ways in user studies we have conducted. On the one hand, the users were asked whether the given types would satisfy their needs. This was approved by a vast majority of test persons. On the other hand, the tags that remained unassigned to one of the given tag types were inspected manually to judge whether an additional type could be identified, which was not the case. The goal type in the above list of tag types takes a special position because it is meant as an instrument for the users to organize their self-directed learning processes [10]. This goal type will always be set manually by the users and is thus excluded from the automatic identification. While the feature of tagging with tags of different types was generally well accepted in evaluations conducted so far, nearly all users stated that an *automatic identification* of the type of a tag would be great. Additionally, in the evaluations there was a significant amount of tags that remained without a type, even if it could have been assigned to one of the given tags. This issue could be addressed by an automatic identification approach, too.

Further, an analysis of tags and resources regarding the languages used has shown, that the ELWMS.KOM / CROKODIL system is used across languages. Even though most users in the evaluation have German as their native language, 75% of the resources and 30% of the tags were in English (named entities not counted).

Our goal is to provide a method to automatically detect the type of a tag in ELWMS.KOM and CROKODIL. Given this application scenario, the requirements for such an approach are as follows:

- The algorithm needs to classify a given tag into the types given above (plus a further category for tags that do not fall into one of the given categories).
- Since in our application scenario users use tags and resources across languages, the algorithm needs to work across at least German and English.

3 Related Work

In the following, we describe related approaches for the categorization of tag types and their automatic identification. Table 1 gives an overview of the most important approaches in the given context. The first three approaches also perform an automatic identification of the proposed tag types.

Table 1. Overview of tag categorization approaches

Wartena	Bischoff	TagExplorer	Golder	ELMWS
Topic	Topic	Thing	Topic	Topic
Attribute	Type		Type	Type
Author	Author/Viewer	Person/Group	Author/Owner	Person
Opinion	Rating		Rating	
	Usage Context		Task	Goal
Self reference	Self reference		Self reference	
	Location	Location		Location
			Refinement	
	Time	Time		
		Activity/Event		Event

All five approaches share *topics* and *persons* as tag types. Four of them share *events or dates* and the *type* of a given resource. Three of the systems identify *opinions* or *ratings* as well as *goals / usage context* and *location* as additional tag types.

Wartena [14] classifies tags used for books in the literature tagging system LibraryThing [13]. He defines two main categories *work* and *user*. The *topic* for the content of a book, *attribute* for genre and *usage context* of a book, and the *author* being assigned to the *work* main category, while *opinion* reflecting the personal opinion of a reader and *self reference*, which reflects the current status of a book (damaged, lent) assigned to the *user* main category. Wartena uses a machine learning approach for classification. Features used include eccentricity of tags (user-specific and for the whole corpus), frequency of tags, book ratings, author names and the containment of specific strings like "author", "reading", "great", "prize" etc. (see [14] for a complete list). For evaluation, 565 tags of the LibraryThing system were extracted and labelled manually with the categories on the two given levels (see Section 3.1). On the categorization task using the two top level categories the approach has a classification accuracy of 92% (with a baseline quality – which is according to the distribution of categories in the ground truth – of 71%). Using all categories, the accuracy of the method reaches 75% (baseline: 27%).

Bischoff et al. [1] analyzed tags in Flickr, LastFM and Delicious and identified different categories of tags in these systems (see Table 1). In the following, we describe the characteristics of the Delicious tags, since this categorization is the most closely related to our application scenario and has the most overlap with our approach. Tag types encompassing *opinions* or *ratings* are not needed in our system because the system provides other means for this. The automatic classification method proposed by Bischoff et al. relies on a twofold approach. The five tag types *time*, *self reference*, *location*, *author* and *type* are detected using a mix of known methods for named entity recognition, tag type specific heuristics as well as manually created or automatically extracted lists. Tags which can not be assigned to one of these five types by these methods are assigned to one of the three remaining types using a machine learning approach. Features used include the number of terms, number of characters, tag frequency, word type and the WordNet category a tag can be mapped to. The approach is evaluated on three corpora, extracted from the systems Delicious, LastFM and Flickr. All extracted tags have been labelled manually to create a ground truth. The corpus is dominated by topic tags, which make up ~67%. The approach achieves on the Delicious corpus, which is the most relevant in the given context, an average F-Measure of 69.2%.

In the **TagExplorer** system [12], Flickr tags are categorized into *location* (where the picture was taken), *subject* (corresponds to topic in other systems), *names* (corresponds to persons), *activity* and *time*. TagExplorer detects the tag type of Flickr tags using WordNet. Each tag type is mapped to specific WordNet categories. The tag label to classify is searched in WordNet and the first noun in the result list is used to determine the category. The authors assume that the WordNet categories are correct and the first noun search result represents the most probable type of that tag. Since for a direct mapping of tags to WordNet categories only a coverage of 52% of the tags can be achieved [11], so the ClassTag system [7] is used in addition. Here, information from Wikipedia is used as external source to provide a better coverage of WordNet categories. According to an extrapolation done by the authors, 69% of Flickr tags can be classified using this approach, while the mapping of tags to WordNet using the ClassTag system works with a precision of 72%.

Golder and Huberman [5], have analyzed Delicious tags and identified seven different tag types of which only the type "*Refining Category*" is not already present in other approaches. This category provides a means to further extend the functionality of Delicious by using tags to group tags and resources according to the alphabet or numbers.

The related work on tag categorization shows that while there is a common subset of categories used by most approaches, the exact combination of categories is always dependent on the specific use case and usage scenario. This also applies to the interpretation of a category itself: Where e.g. in TagExplorer a *location* states where a Flickr picture was taken, in other approaches the usage of this category is more free. On the other hand, the tag type *goal* is very relevant for a system supporting self-regulated RBL while it is not that important for tagging images in Flickr or music in LastFM. Since each approach uses different tag types and different corpora, a comparison or competitive evaluation of the different approaches is not feasible. However, the results give an idea on how well the approaches perform.

4 Automatic Identification of Tag Types

In our approach for automatic identification of tag types, we strongly rely on external corpora. In the following we shortly describe the external databases we use followed by a description of the algorithm in its entirety.

4.1 External Databases Used

Our algorithm consists of classifiers for each tag type we want to detect and most of the classifiers use one or more external databases and corpora for classification. In the following, we shortly describe the databases our algorithm makes use of.

Freebase[2] is a freely available database containing general knowledge. It contains 20 million entries for topics or objects like persons, events and others. The Freebase community maintains the data and aggregates them from other sources. Entries in Freebase possess structured information described by schemata. The main language is English although entries in different languages exist. Freebase ranks the results of a query according to its relevance using Lucene[3]. A freebase query returns ranked search results which are allocated to different Freebase categories. These categories can be mapped to the desired types.

GeoNames[4] is a location database that contains 10 million geographic names, representing 7.5 million topographic objects grouped in different main classes (e.g. state, waters, city…) and subclasses (airport, government building etc.). Each location is stored in the language of the country the object is located in as well as different other languages. English, however, is the language mostly used in GeoNames.

DBpedia[5] is a community project which aims to extract and provide structured information from Wikipedia. Partly the information is automatically extracted from Wikipedia info boxes contained in specific article types (like e.g. cities, states).

WordNet[6] is an English thesaurus containing verbs, nouns, adjectives and adverbs, grouped in so-called synsets (groups of synonyms). These synsets are allocated to categories (e.g. location, event, food …). Further, synsets possess different types of relations to other synsets, e.g. hyponymic, antonymic or holonymic relations.

Other databases used include a list of holidays[7], a database of first and last names [6], a list of web genres[8] as well as the dictionary BEOLINGUS[9]. The latter is used to

[2] http://www.freebase.com/, Retrieved online: 2011-06-20
[3] http://lucene.apache.org/, Retrieved online: 2011-06-20
[4] http://www.geonames.org/, Retrieved online: 2011-06-20
[5] http://dbpedia.org/, Retrieved online: 2011-06-20
[6] http://wordnet.princeton.edu/, Retrieved online: 2011-06-20
[7] http://de.wikipedia.org/wiki/Feiertage_in_Deutschland, Retrieved online: 2011-03-31
[8] http://www.webgenrewiki.org/index.php5/Genre_Classes_List, Retrieved online: 2011-03-31
[9] http://dict.tu-chemnitz.de/, Retrieved online: 2011-03-31

determine the lexical category of terms contained in a tag label. All databases and sources are freely available and none of them have been adapted or extended in the course of this work.

4.2 Tag Type Identification Algorithm

There are two different challenges in the automatic identification of a tag type: On the one hand we need to identify the set of types that are valid for a given tag, e.g. the tag "Merkel" can be a *person* meaning the German chancellor "Angela Merkel" or a *location* meaning Merkel, Texas. We determine the valid types by querying type specific databases like GeoNames for locations. On the other hand we need to decide which one of the valid tag types is plausible for a given tag. Thus in cases where different types are possible for a tag, we need a means to put the tag types in an order. This is what the generic Freebase database is used for. The main assumption we make is that the ranking of search results in Freebase reflects the probability for the conventionality of the semantic manifestation of the queried label represented by the search result. **Figure 2** shows the classification process. The only input to the system is the tag that needs to be classified, while the output is the type proposed by the algorithm, or optionally depending on the use case, a ranked list of types. The single steps of the process are described in detail in the following.

The **Preprocessing** step (1) involves filtering of special characters and detection of tag borders. The latter is necessary for the preprocessing of tags from the Delicious corpus only (cf. Sect. 5.1). Delicious tags are automatically divided into separate tags when a space is entered by a user, i.e. if a user gives two terms separated by a space these terms will always be treated as single tags. In order to have multi-term tags, users use various other delimiters instead. Often characters like '_' or '-' or a special notation like “camelCase” is used. Thus the preprocessing step tries to divide these tag compositions into single terms. Due to the fact that in the given scenario, many tags as well as resources are in German as well as in English but most of the entries in external databases are in English a **translation** is necessary for some of the tag types – specifically for topics. This is done using interlanguage links in Wikipedia. If an article matching the tag is present in the German Wikipedia and the corresponding article contains an interlanguage link to the English version of the article, the label of the English version is used as translation, otherwise no translation is used. We use this means for translation to avoid noise caused by translation ambiguity. The resulting tag and - if available - its translation are used as search terms in the Freebase processing step.

The original tag and - if available - its translation is used to conduct a **Freebase Query** (2). This query returns a list of Freebase entries as result. The tag types used in our scenario can be mapped to Freebase categories which are used to structure entries of the Freebase database. The Freebase entries that result from the query are matched with the tag used in the query. A Freebase entry matches, if the case-folded tag exactly matches the case-folded label of the Freebase entry, except for the tag type *person*, where a containment of the tag in the entry label is counted as matching as well. The top ten matching results are divided and each result is forwarded to the respective classifier together with its rank within the top ten. The ranks of translated tags are only forwarded to the topic, type and event classifiers. Since the person classifier doesn’t use translations at all and the location classifier uses other sources for translation.

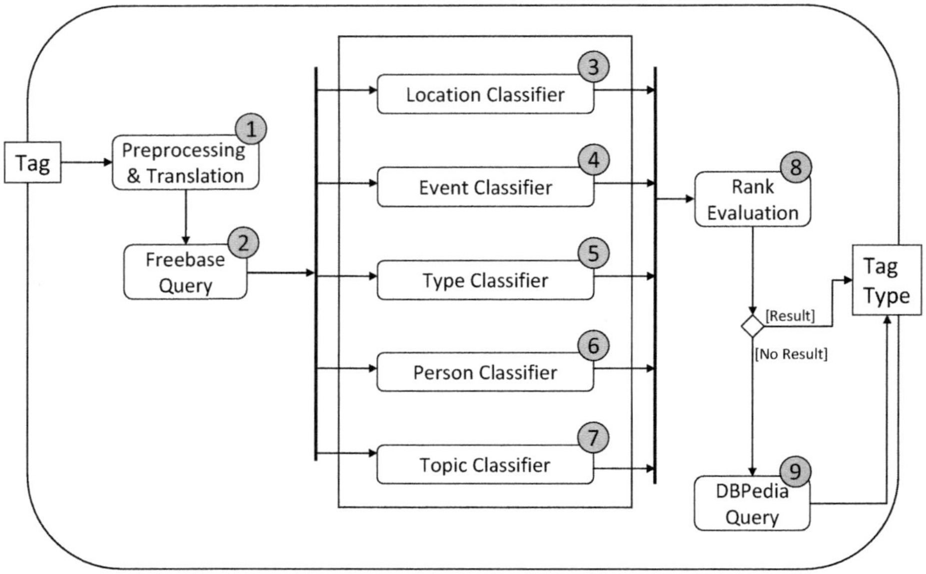

Fig. 2. Overview of tag type identification approach

The **Location Classifier** (3) in the first instance queries the location database GeoNames with the original tag. When a tag label equals one of the German or English aliases of a result this result including all German and English aliases is stored in a list. Matching in that case means that the case-folded tag matches the likewise case-folded alias exactly, thus usually there is only one GeoNames result for which several different aliases are stored. The list resulting from GeoNames is then used to query Freebase and the minimal rank for all different aliases is returned by the classifier. Since a translation is already provided by GeoNames in the form of aliases, the Wikipedia translation of the original tag is not used by this classifier.

In addition to the ranked entries provided by Freebase, the **Event Classifier** (4) uses heuristics that give a bias towards the *event* type if they apply. If the tag label contains a four digit number (like 2011) or a combination of numbers between 1 and 31 and the name of a month, the classifier returns a rank of zero (i.e. the highest possible rank). If these heuristics do not apply, a database containing holidays is queried. If the tag label is contained in this database, again zero is returned. If none of these methods produce results the minimal Freebase rank of the original tag and its translation is returned.

The **Type Classifier** (5) first uses a list of web genres which was extracted, translated using Wikipedia and stored in a database. If the tag matches one of the entries in the database, rank zero is returned. Otherwise the classifier returns the minimum Freebase rank of the original tag and its translation.

The **Person Classifier** (6) uses a string-based heuristic to determine whether the tag is comprised of patterns like "B Gates" or "Bill G". If this heuristic applies rank zero is returned. As stated above, for classifying *persons*, the Freebase query in step (2) is done slightly different than for other tag types. On the one hand, no translation is used since persons are named entities and on the other hand, the matching between tag label and result does not have to be exact, but the containment of the tag in the result label is sufficient. Thus, the tag "Merkel" will match "Angela Merkel". Since Freebase only contains known personalities like politicians, musicians or writers, we need a means to identify names for unknown persons. Therefore an additional database containing German and English first and last names is used. The more frequently a name matching the tag label is, the higher the probability that the tag is a name and the lower the rank that is returned by the classifier. The minimal rank returned by this heuristic is one.

The **Topic Classifier** (7) classifies all Freebase entries as topic which can not be mapped to one of the other tag types and returns the respective rank. If the tag label is not found in Freebase but a matching Wikipedia entry is found, a rank of 11 (the maximal Freebase rank plus one) is returned. If this does not apply as well, WordNet and BEOLINGUS are used to determine the word types of the tag label. If most terms of the tag label are nouns, it is classified according to [1] as topic with a rank of 12.

During the **Rank Evaluation** (8) step, the ranks returned by the different classifiers are compared and the tag type with the lowest integer value rank is returned as the proposed tag type. It is possible that two types are assigned the same rank, due to the fact that tags are translated before querying the external data sources. In that case, the rank of the tag type determined by the original tag will always be prioritized over the type received with a translated tag.

The **DBpedia Query** (9) is an optional step and a fallback solution that is used in the unlikely case that the process so far did not lead to any result, i.e. none of the classifiers could identify the tag as the respective type. The tag is used to query DBpedia and the first result in the list that can be mapped to one of the tag types using the DBpedia category system is returned as the proposed type of the tag.

Due to the fact that the approach is solely based on heuristics and external sources (which can be cached locally), the algorithm can identify the type of a tag in near real-time, i.e. the average decision time is 416ms.

5 Evaluation

We have evaluated the proposed approach using two different corpora. The first one is comprised of tags taken from Delicious while the other corpus is constructed from usage data of the RBL system ELWMS.KOM described above.

5.1 Setting and Methodology

For both corpora used, the approach has to solve a categorization problem. Thus standard information retrieval measures like accuracy, precision, recall and F-measure can be used for evaluation.

The **Delicious corpus** consists of tags collected from the social bookmarking service Delicious[10]. To make sure that the corpus contains tags of all desired tag types like *locations*, *persons* and *events*, Delicious was searched for specific keywords (like "vacation" for getting *location* tags or "war" for getting *event* tags). 225 different

[10] http://www.delicious.com/, Retrieved online: 2011-06-20

resources from 143 users were retrieved. The corpus was then built out of all tags of resources found this way, excluding the tags initially searched for. All tags were then manually categorized by three human raters in order to construct a ground truth for the identification of tag types. The raters were able to inspect the tagged web resource before making their decision. In addition to the ELWMS.KOM tag types, the raters could assign a tag to an additional "not classified" category for tags that could not clearly be assigned to one of the named types. **Table 2** shows details for the Delicious corpus. The inter-rater agreement calculated using the free-marginal multi-rater variant of Fleiss Kappa [9] for this corpus was 0.66. Although there is considerable agreement among the raters, this also shows that even for human raters this classification task is not easy and the decision on the type of a tag can be subjective (see Sect. 5.2). Therefore mainly two subsets of the corpus are used for the evaluation: the subset of tags where the raters completely agreed on the type and the set where two of the three raters agreed.

Table 2. Details on evaluation corpora

	Delicious Corpus	**ELWMS Corpus**
Resources considered	225	422
Different users	143	21
Overall tags	1272	1161
Tags uniformly rated by all three raters	620 (100%)	493 (100%)
... classified as person	106 (17.1%)	168 (34.1%)
... classified as location	144 (23.2%)	16 (3.2%)
... classified as event	16 (2.6%)	56 (11.4%)
... classified as type	27 (4.4%)	19 (3.9%)
... classified as topic	288 (46.5%)	234 (47.5%)
... not classified	39 (6.3%)	0 (0%)
Tags with a 2/3 majority of raters agreeing	1018 (100%)	948 (100%)
... classified as person	141 (13.9%)	217 (22.9%)
... classified as location	159 (15.6%)	19 (2.0%)
... classified as event	25 (2.5%)	73 (7.7%)
... classified as type	58 (5.7%)	64 (6.8%)
... classified as topic	519 (51.0%)	567 (59.8%)
... not classified	116 (11.4%)	8 (0.8%)

The **ELWMS** corpus was generated during one of the user studies conducted for the ELWMS.KOM system. During the usage of the system, users searched for web-resources in order to complete tasks or to do research on specific topics. The users used the ELWMS.KOM Firefox add-on to tag the resources they found with tags of the given types (see Sect. 2). To construct the corpus, 1161 tags from 21 different users of the system were collected. The obtained tags were again manually categorized by two additional human raters to confirm the tag types and smooth out the subjectivity, resulting in three ratings per tag in total. On this corpus the agreement of the raters was 0.67. Again, two agreement subsets of this corpus were used for evaluation.

5.2 Results

Table 3 shows the confusion matrix when running our algorithm on the uniformly rated Delicious corpus. The fact that all three raters classified the tag types uniformly means that a classification of this subset is comparably straightforward. In this configuration, the approach reaches an overall accuracy – i.e. ratio of correctly classified tag types – and F-measure of 81.3%, showing that our approach works considerably well.

Table 3. Results for uniformly rated tags on Delicious corpus

a	b	c	d	e	f	← classified as
91	2	0	0	6	7	a = Person
2	136	0	0	4	2	b = Location
0	0	15	0	0	1	c = Event
1	0	1	8	9	8	d = Type
15	17	10	1	230	15	e = Topic
4	0	0	0	11	24	f = Not categorized
0.81	0.88	0.58	0.89	0.88	0.42	Precision
0.86	0.94	0.94	0.30	0.81	0.62	Recall
0.83	0.91	0.71	0.44	0.84	0.5	F-measure
					81.3%	F-measure (Ø)

Table 4. Results for tags with 2/3 majority on Delicious corpus

a	b	c	d	e	f	← classified as
108	4	0	0	17	12	a = Person
3	150	0	0	4	2	b = Location
0	0	20	0	3	2	c = Event
2	0	2	19	26	9	d = Type
43	26	14	11	389	36	e = Topic
22	4	5	1	40	44	f = Not categorized
0.61	0.82	0.49	0.61	0.81	0.42	Precision
0.77	0.94	0.80	0.33	0.75	0.38	Recall
0.68	0.87	0.61	0.43	0.78	0.40	F-measure
					71.2%	F-measure (Ø)

The confusion matrix shows that there is generally a high confusion between the *topic* type and the other tag types. This is due to the fact that the word type based heuristic in the topic classifier marks noun based tags automatically as topics even if they have not been identified as topics by the raters. Another reason is the freedom of raters to classify a *person* tag as topic, e.g. if a resource is *about* a person and thus the person can be seen as topic of the resource. This distinction is not made by our approach, which always classifies such a tag as *person*.

When taking into account tags where two of the three raters agreed on the type, the task becomes significantly harder, which is obvious since the human raters were not able to agree on a distinctive tag type as well. In such a scenario (shown in Table 4) the accuracy drops to 71.7% (the F-measure to 71.2%). However, when taking into account the rating of the third rater as a valid option, an F-measure of 78.2% is reached by our approach. Again the main source of errors is the classification of topics, specifically the distinction between topics and uncategorized tags.

When running the approach on the ELWMS corpus, it generally shows similar results. Again the overall accuracy and F-measure are significantly better on the "easy" classification tasks (where all raters agreed on the tag type). Here the approach reaches almost 84% (79.1% accuracy – see Table 5).

Table 5. Results for uniformly rated tags on ELWMS corpus

a	b	c	d	e	f	← classified as
140	2	0	0	2	24	a = Person
0	15	0	0	1	0	b = Location
0	0	53	0	3	0	c = Event
0	5	0	4	10	0	d = Type
15	2	1	0	178	38	e = Topic
0	0	0	0	0	0	f = Not categorized
0.90	0.63	0.98	1.0	0.92	0.0	Precision
0.83	0.94	0.95	0.21	0.76	0.0	Recall
0.87	0.75	0.96	0.35	0.83	0.0	F-measure
					83.7%	F-measure (Ø)

Table 6. Results for tags with 2/3 majority on ELWMS corpus

a	b	c	d	e	f	← classified as
169	5	1	0	13	29	a = Person
1	16	0	0	2	0	b = Location
6	0	62	0	4	1	c = Event
0	5	0	6	53	0	d = Type
33	7	11	4	415	97	e = Topic
0	0	0	0	3	5	f = Not categorized
0.81	0.48	0.84	0.60	0.85	0.04	Precision
0.78	0.84	0.85	0.09	0.73	0.63	Recall
0.79	0.62	0.84	0.16	0.79	0.07	F-measure
					74.0%	F-measure (Ø)

Again, the algorithm shows slightly weaker results on the harder version of the corpus. However, with 74%, the F-measure is still quite good. Again the major reason for errors is the confusion between *topic* tags and uncategorized tags.

In an additional evaluation we used a corpus solely containing English tags and named entities for both corpora to show the (possibly) negative influence of translations in the approach. In almost all cases the algorithm performed better with English tags (and named entities) only. This is an expected result due to the fact that nearly all external sources we used are mostly in English. On this version of the Delicious corpus the algorithm achieved an F-measure of 86.6% and 70.2% respectively and on the ELWMS corpus 85.4% and 73.6%. However, the difference is still small enough to conclude that the cross-language capabilities of our approach work well.

Overall, the results of our approach are comparable (if not better) to other approaches in the area. As said before, due to the differences in tag categories and usage scenarios, a competitive evaluation of approaches is not feasible. Although we have evaluated the approach only for corpora containing a mix of German and English tags and resources, theoretically the approach is applicable to other languages as well, as long as there is a Wikipedia language version that is suited to serve as translation corpus.

6 Conclusions

The semantic tagging approach based on tag types, shown in this paper, enables users of the Resource-Based Learning systems ELWMS.KOM and CROKODIL to easily build a semantic network for organizing web-resources according to their cognitive model. Furthermore, these tag types can be used to provide different visualizations, e.g. geographic maps with location tags and calendar views with event tags. In this paper, we have proposed a knowledge-based real-time-capable approach for the context-free identification of these tag types. On the tested corpora, our approach shows satisfying results that would allow for an automatic classification of tags without user interaction and with a high reliability. Thus, the algorithm reduces the manual effort for building the semantic network.

At the moment, the algorithm is context-free, i.e. it does not take into account the context of a given tag (e.g. the resource tagged, the user who tags or other tags for the same or similar resources) and makes a decision solely based on external knowledge. The evaluations have shown that this approach works very well. However, our approach will always return the most plausible type for a tag. That means that if for instance a user really means "Merkel, Texas" instead of "Angela Merkel" the algorithm will fail. The algorithm can theoretically work perfectly only if it takes into account the context of a tag, for example the content of the tagged resource. Thus our next step will be to match the content of the tagged resource with the content of the Freebase articles returned by the algorithm and adapt the ranking accordingly. One other possibility for further work is to utilize the ranked list of tag types to make other meanings of a tag visible to a user and thus broaden his horizon. Thus, the algorithm could also provide direct "learning support".

Acknowledgments. This work was supported by funds from the German Federal Ministry of Education and Research under the mark 01 PF 08015 A and from the European Social Fund of the European Union (ESF). The responsibility for the contents of this publication lies with the authors.

References

1. Bischoff, K., Firan, C.S., Kadar, C., Nejdl, W., Paiu, R.: Automatically identifying tag types. In: Huang, R., Yang, Q., Pei, J., Gama, J., Meng, X., Li, X. (eds.) ADMA 2009. LNCS, vol. 5678, pp. 31–42. Springer, Heidelberg (2009)
2. Böhnstedt, D., Scholl, P., Benz, B., Rensing, C., Steinmetz, R., Schmitz, B.: Einsatz persönlicher Wissensnetze im Ressourcen-basierten Lernen. In: Fischer, S., Seehusen, S., Lucke, U. (eds.) DeLFI 2008, Köllen, Bonn, GI, LNI, vol. P-132, pp. 113–124 (2008)
3. Böhnstedt, D., Scholl, P., Rensing, C., Steinmetz, R.: Collaborative semantic tagging of web resources on the basis of individual knowledge networks. In: Houben, G.-J., McCalla, G., Pianesi, F., Zancanaro, M. (eds.) UMAP 2009. LNCS, vol. 5535, pp. 379–384. Springer, Heidelberg (2009)
4. Duval, E., Hodgins, W.: A LOM research agenda. In: Proceedings of the twelfth international conference on World Wide Web, pp. 1–9 (2003)
5. Golder, S.A., Huberman, B.A.: Usage patterns of collaborative tagging systems. Journal of Information Science 32(2), 198–208 (2006)
6. Lokaiczyk, R.: Clustering und Klassifikation von Personennamen. Diploma thesis, Universität Leipzig (2006)
7. Overell, S., Sigurbjörnsson, B., Zwol, R.: Classifying tags using open content resources. In: Proceedings of the Second ACM International Conference on Web Search and Data Mining, New York, USA, pp. 64–73 (2009)
8. Rakes, G.C.: Using the internet as a tool in a resource-based learning environment. Educational Technology 36(5), 52–56 (1996)
9. Randolph, J.J.: Free-marginal multirater kappa: An alternative to fleiss' fixed-marginal multirater kappa. In: Joensuu Learning and Instruction Symposium (2005)
10. Scholl, P., Benz, B., Böhnstedt, D., Rensing, C., Steinmetz, R., Schmitz, B.: Einsatz und Evaluation eines Zielmanagement-Werkzeugs bei der selbstregulierten Internet-Recherche. In: Fischer, S., Seehusen, S., Lucke, U. (eds.) DeLFI 2008, Köllen, Bonn, GI, LNI, vol. P-132, pp. 125–136 (2008)
11. Sigurbjörnsson, B., Zwol, R.: Flickr tag recommendation based on collective knowledge. In: Proceedings of the 17th international conference on World Wide Web, New York, USA, pp. 327–336 (2008)
12. Sigurbjörnsson, B., Zwol, R.: Tagexplorer: Faceted browsing of flickr photos. Technical Report, Yahoo! Research (2010)
13. Voss, J.: Librarything - Web 2.0 für Literaturfreunde und Bibliotheken. In: Mitteilungsblatt der Bibliotheken in Niedersachsen und Sachsen-Anhalt, pp. 12–13 (2007)
14. Wartena, C.: Automatic classification of social tags. In: Lalmas, M., Jose, J., Rauber, A., Sebastiani, F., Frommholz, I. (eds.) ECDL 2010. LNCS, vol. 6273, pp. 176–183. Springer, Heidelberg (2010)

QuizMap: Open Social Student Modeling and Adaptive Navigation Support with TreeMaps

Peter Brusilovsky[1], I-Han Hsiao[1], and Yetunde Folajimi[2]

[1] School of Information Sciences, University of Pittsburgh, USA
[2] University of Ibadan, NIGERIA
{peterb,ihh4}@pitt.edu, yetunde_folajimi@yahoo.com

Abstract. In this paper, we present a novel approach to integrate social adaptive navigation support for self-assessment questions with an open student model using QuizMap, a TreeMap-based interface. By exposing student model in contrast to student peers and the whole class, QuizMap attempts to provide social guidance and increase student performance. The paper explains the nature of the QuizMap approach and its implementation in the context of self-assessment questions for Java programming. It also presents the design of a semester-long classroom study that we ran to evaluate QuizMap and reports the evaluation results.

Keywords: TreeMap, Open Student Model, Open User Model, Information Visualization, Self-Assessment, Social Guidance.

1 Introduction

Open student modeling and adaptive navigation support are amongst the most popular technologies in modern personalized E-Learning. Open student models (also referred to as open learner models) allow students to observe their progress and reflect on their successes and failures [1-2]. Open models are especially beneficial when presented in a visual form, providing students with an easy-to-grasp and holistic view on their learning [3-5]. Adaptive navigation support [6] guides students to the most appropriate education content. This technology is known for its ability to help students acquire knowledge faster [7], improve learning outcomes [8-10] and reduce navigational overhead [7].

In our recent work on topic-based adaptive navigation support [11-12], we attempted to combine the benefits of open student models and adaptive navigation support. We structured educational hyperspace into topics and annotated topic links with adaptive icons showing both the progress of student knowledge of the topic and relevance of each topic to the current learning goal. In this way, a set of link annotations offered a visualization of student topic-level progress, while each of these links provided an immediate access to learning content for this topic. As our studies in three different domains demonstrated [11, 13], the combined approach allowed to significantly increase the quality of student learning and student motivation to work with non-mandatory learning content. Yet, we believed that the combined approach

C. Delgado Kloos et al. (Eds.): EC-TEL 2011, LNCS 6964, pp. 71–82, 2011.

has not reached its full potential since, by the nature of topic-based navigation support, it was provided on the topic level, i.e. the system was able to recommend the whole topic as most appropriate to a specific student at a given time, but was not able to distinguish individual questions within this topic as more or less relevant.

This paper reports our attempt to explore a richer integration of open student modeling and adaptive navigation support. The key idea of our new approach was to enhance the original topic-based navigation support with social navigation [14]. Social navigation guides users to relevant information by showing the traces of past users' work. In our past work, we demonstrated that social navigation in education context could successfully guide students to relevant content down to the level of an individual item [15]. Main challenge in adopting classic social navigation approach to our target content, a set of parameterized self-assessment questions for Java, is that the work of past users with questions can be reflected by two parameters (instead of usual one): the number of attempts to solve a specific question (a parameterized question can be attempted several times!) and the success rate (the fraction of correct answers) for the question. We believed that both kinds of information could be important to guide future users to most appropriate questions. To address this challenge, we explored the use of TreeMaps for student progress visualization. TreeMap is an expressive tile-based visualization approach that allows presenting two parameters at the same time using the size and the color of individual tiles, while easily integrating the progress of many individual users in a single map.

The focus of the paper is *QuizMap*, a TreeMap-based visual interface for accessing self-assessment questions in Java, which provides using a combination of open social student modeling and social navigation support. In addition to the unique integration of social navigation, open student modeling, and navigation support, QuizMap extends the benefits of visualizing the student models from the cognitive aspects to the social aspects. By exposing student model in contrast to his/her peers, QuizMap is expected to provide social guidance and increase student performance. Following a brief review of similar work, we introduce the target domain and explain how a QuizMap for this domain is organized. Next we explain the design of a semester-long classroom study that we ran to evaluate QuizMap and report the evaluation results. Finally, we summarize this work and discuss future research plans.

2 A Summary of Related Work

2.1 Open Student Models

A range of benefits have been reported on opening the student models to the learners, such as increasing the learner's awareness of the developing knowledge, difficulties and the learning process, students' engagement, motivation, and knowledge reflection [3-5]. Dimitrova et al. [16] explored interactive open learner modeling by engaging learners to negotiate with the system during the diagnosis process. Chen et al. [17] investigated the active open learner models in order to motivate learners to improve their academic performance. Both individual and group open learner models were studied and demonstrated the increase of reflection and helpful interactions among teammates. In [6], Bull & Kay described a framework to apply open user models in adaptive learning environment and provided many in-depth examples.

2.2 Social Visualization in E-learning

While visualization as an approach becomes more and more popular in E-Learning context, we can find only a handful of work that explores the value of social visualization, i.e., presenting students some information about the whole class or their peers. Vassileva and Sun [18] investigated the community visualization in the online communities. They summarized that social visualization allows peer-recognition and provides students the opportunity to build trust in others and in the group. Bull & Britland [19] used OLMlets to research the problem of facilitating group collaboration and competition. The results showed that optionally releasing the models to their peers increases the discussion among the students and encourages them to start working sooner. CourseVis [20] is another example, yet one of the few open user model systems that provide graphical visualization for multiple users of groups to teachers and learners. It helps instructors to identify problems early and prevent some of the problems with distance learning. In our own work on self-assessment quizzes with IntrospectiveViews interface [21], we embedded open user models into parallel IntrospectiveViews visualization and demonstrated that knowledge-based and social guidance combined are more effective in guiding the students to appropriate questions that they are ready to handle than knowledge-based guidance alone.

2.3 TreeMaps

A TreeMap is a space-filling visualization method for representing hierarchical information [22]. By dividing the display area into a nested sequence of rectangles whose areas are associated to attributes of the data set, it effectively illustrates the structural information with slices and dices. TreeMaps have been applied to a wide variety of domains ranging from financial analysis [23], petroleum engineering [24] to network security analysis [25]. Some studies have focused on specialized techniques to visualize large number items on a TreeMap without aggregation [26]. The innovative idea to use TreeMaps to visualize a model of individual learner knowledge was first suggested in [27].

3 QuizMap – A TreeMap-Based Open Social Student Model

QuizMap is a TreeMap representing the work of a user group (such as a class) with self-assessment questions. We customized the TreeMap by using the size and color of the rectangles to display the performance of the student. To adapt the TreeMap approach to the context of self-assessment questions, we structured system's TreeMap into 4 levels. Each level of the TreeMap clusters different level of information in detail. The top level consists of 1 root node, which represents the summary information of the entire class, including the overall attempts, successful rate and average statistics. The second level consists of 21 nodes corresponding to topics covered within the class. Under each topic node, next level is formed by the parameterized self-assessment questions belonged to the topic. The bottom level of the TreeMap shows performance of each individual student in a group for each question. The QuizMap structure is presented in Figure 1.

The sizes of the rectangle for each node represent the amount of work done. The color indicates the amount of knowledge gained (credited with each successful answer). The student's own performance is colored in orange and to contrast with the rest of the class, colored in blue. The darker the color, the higher success it presents and vice versa. Both reddish yellow and bluish color tints can be decomposed into 10 different "shades" (Figure 2). All the absolute values of the performance are displayed when user hover over the rectangle. These two different color schemes are meant to make it easier for the student to compare his or her performance with the performance of individual peers and the whole class.

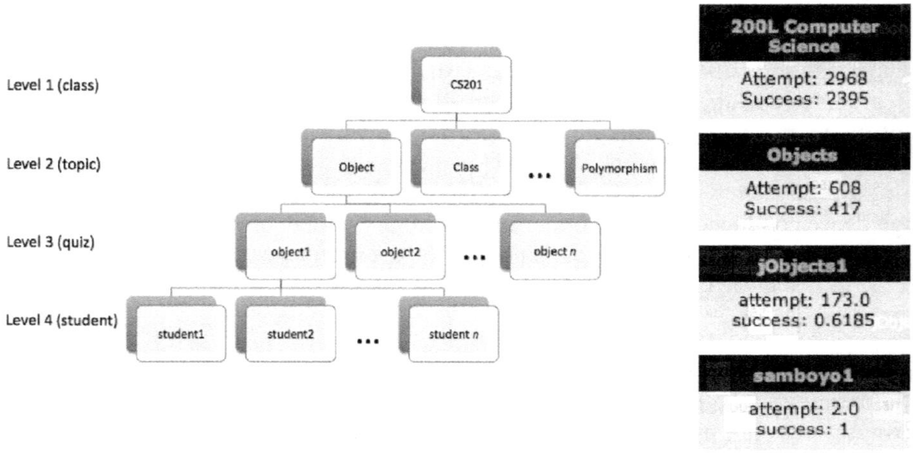

Fig. 1. QuizMap structure

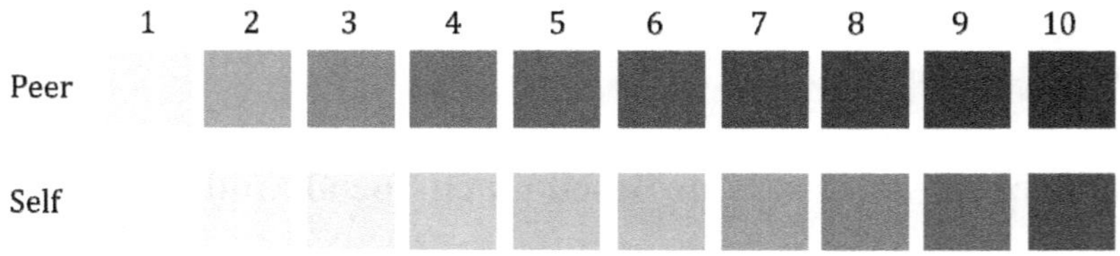

Fig. 2. QuizMap rectangle color shades indication

To illustrate the use of the TreeMap in the context of self-assessment quizzes, Figure 3 represents an overview of QuizMap. To answer a quiz, a student has to select the question from each topic in the QuizMap. Upon the selection, QuizMap will pop a separate window to display the question (Figure 4). Each question asks the student to predict the results of execution of a specific Java program (i.e., mentally execute the program and enter the final value of some variable of the text to be printed by the program.) All questions are parameterized, i.e., include a random parameter, which the system instantiates when the question is delivered to a student. As a result, the student can attempt to answer the same question multiple times with different values of the

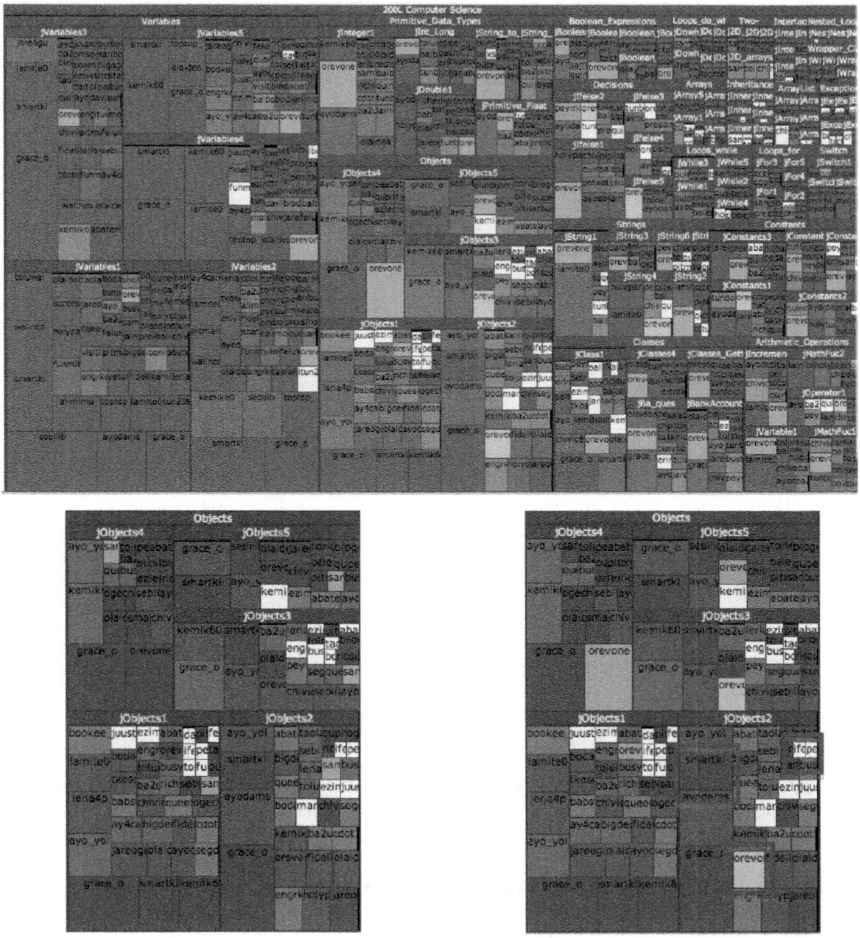

Fig. 3. An overview of QuizMap (top); A zoom in view on topic *Objects* of two students, student A (bottom-left) & student B (bottom-right)

parameter, which helps to achieve the mastery level. The implementation and functionalities of parameterized self-assessment quizzes were described in detail in [12].

The bottom part of Figure 3 shows two zoom-in views of the same topic, Objects, for two students. It demonstrated, that the amount of work done by the student A (Figure 3 bottom left) was relatively the same amount of questions on this topic. The color indicates a roughly 70% successful rate across all questions that s/he attempted. It suggested that this student had been consistent on performing different complexity levels of questions. Such way of evaluation can also be found throughout the class on his/her model. However, the other zoom-in view of QuizMap by student B (Figure 3 bottom right) displayed a different scenario. The student focused on working on certain questions, especially the *jObject4* question, which reached relatively high attempts. Throughout the class, he also followed the similar pattern of work. He had more attempts on a particular set of questions repeatedly and achieved the 50~70% successful rate. It suggested that this student might have troubles in those topics. Therefore, he kept trying again and again on the same questions to improve himself. In this open social student model TreeMap, students are expected to identify the strengths and weaknesses of themselves and their peers. For example, in the example of lower QuizMap, the student was struggling (low successful rate) with the question *jObject2* under the topic *Objects*. QuizMap provides opportunities for him to discover stronger peers by recognizing dark blue rectangles and vice versa. This student should also realize that who have less success on this specific question by recognizing the lighter blue rectangles. Those students may have lower chances to help him achieve a better understanding in this question.

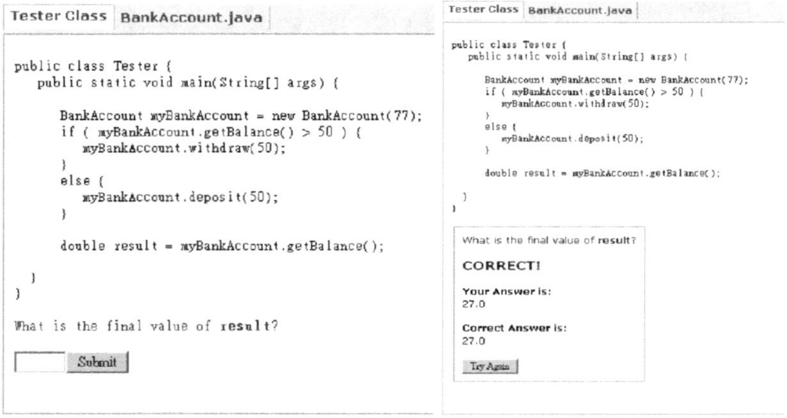

Fig. 4. A parameterized self-assessment question (left); a correctly answered question with evaluation results (right)

4 Classroom Study and Evaluation Results

We conducted a classroom study in the Programming and Algorithms course offered by University of Ibadan. The students were second year Computer Science majors. There are 86 students in the class – 52 male and 34 female. Out of them, 77 students were taking the course first time while 9 were repeating the course. The essence of the course is to build on the foundation they already have and teach Algorithm concept using Java and C++, thus enabling them build complete working program from the algorithm. Lectures were conducted through face-to-face interaction with the students. Assignments were submitted online by email attachment. Students already had introductory knowledge of Java in the first semester. Therefore, the QuizMap was introduced to the class as a supplemental tool. Students were encouraged to use QuizMap after being acknowledged that QuizMap quizzes will appear in the exam up to 10% of the marks. A major problem encountered by the students during the semester was the internet access issue. Access to internet in the school lab was only available for very limited hours which did not fit properly into the students' schedule most of the time. Sometimes, electricity was also a problem. As such, students could not use the computers in the laboratories at those times.

4.1 System Usage and Learning Gain

We analyzed the log data on students' interaction with the social visualization on the self-assessment quizzes (QuizMap) and compared the usage with the data from a comparable Object-Oriented Programming class at the University if Pittsburgh where students accessed the self-assessment quizzes using a traditional course portal with no visualization (QuizJET). Table 1 shows the basic statistics on both systems. There were 65 students who used the QuizMap. They made 2961 attempts to the questions, on average 45.55 questions per student. Students achieved 79.30% on average successful rate in answering the self-assessment questions. On average, students tried 4.55 distinct topics, 17.07 distinct questions and had 4.29 visits on the QuizMap. Comparing to QuizJET, the students who worked with QuizMap made less attempts and explored fewer topics (it could be related to the computer and internet access problems). Despite that, they achieved almost a much higher success rate. This level of success rate is typical for question access mediated by adaptive navigation support [11-13]. This provides some evidence that social navigation support is comparable to classic adaptive navigation support by its effectiveness. To obtain more reliable evidence a study should be repeated with more comparable groups.

Table 1. Summary of the overall usage on QuizJET and QuizMap

		QuizJET	QuizMap
All	**Users**	16	65
	Total Attempts	1293	2961
	Attempts per user	80.81 ± 22.06	45.55 ± 6.67
	Success rate	42.63% ± 1.99%	79.30% ± 1.94%
	Distinct Topics	7.81 ± 1.64	4.55 ± 0.59
Average	**Distinct Questions**	33.37 ± 6.50	17.07 ± 2.78
	Sessions	3.75 ± 0.53	4.29 ± 0.54
	Pre-test Scores	9.56 ± 1.29	7.55 ± 0.49
	Post-test Scores	17.12 ± 0.86	13.25 ± 0.60

To examine the connection between the amount of student work with QuizMap and their learning, we categorized students into three groups: passive, mild, and active users. Passive users are defined as ones who made less than 10 attempts; Mild users are ones who use the system moderately (10~45 attempts); Active users are defined as those who use the system above the average (45 and more attempts, see the mean in Table 1). Table 2 summarizes usage data for each group. We found that active users explored 10.13 distinct topics and 38.13 distinct questions on average. It is not surprising that the numbers were significantly higher than the usage of passive users. However, they were also significantly higher than the data of mild users. The results demonstrate a connection between student learning and system use. The more they worked on the system, the more likely they discovered more topics. In addition, we found that the more active the students were, the higher learning gain they achieved.

Table 2. Detail QuizMap usage by student activity (only consider the users who took both pre- and post- tests)

Parameters	Passive (n=13)	Mild (n=22)	Active (n=14)	*p*
Attempts	8.36	40.68	76.13	<.01
Success Rate	93.1%	73.9%	77.41%	
Distinct Topics	1.00	2.82	10.13	<.01
Distinct Questions	2.35	10.36	38.133	<.01
Average Sessions	1.71	2.77	7.00	
Learning Gain	3.71	3.77	6.4	<.05
Pre-test score	6.64	7.32	8.73	
Post –test score	10.36	11.09	15.13	

4.2 Effects on Social Guidance

What is the mechanism of social guidance? How this approach based on the "collective wisdom" of a student community can guide students to the right questions as successfully as classic knowledge-based guidance? In our past work, we found evidence that in social guidance systems stronger students with better understanding of the subject lead the way discovering most relevant resources and creating guidance trails for weaker students. In order to investigate the social guidance effect in QuizMap, we categorized students into two groups, strong and weak, based on their pre-test scores (ranging from a minimum of 0 to a maximum of 20). Strong students scored 10 or higher points in the pre-test, and weak students scored less than 10 points. In Figure 5, we plotted all attempts over the course period. X-axis denoted as course period; Y-axis denoted as the topic complexities sorted from easy to complex. Blue data points represent strong students and orange points are the weak ones. We found that both strong and weak students started simultaneously on the easy topics. However, over time stronger students tended to run ahead of weaker ones. Weaker students approached new topics after the stronger ones had already explored it. Such behavior is more noticeable for more complex topics. The pattern indicated that stronger students, indeed, guided the weaker ones to the proper questions. This allowed weaker students to achieve success rate and post-test scores that are close to those of stronger students. At the end of the course, they narrowed the knowledge gap and achieved higher learning gain that stronger ones (Table 3).

Table 3. QuizMap usage by strong/weak student

Parameters	Weak (n=29)	Strong (n=22)
Attempts	33.17 ± 6.89	54.18 ± 13.40
Success Rate	77.91% ± 3.30%	83.29% ± 2.70%
Distinct Topics	3.93 ± 0.83	5.18 ± 1.06
Distinct Questions	13.37 ± 3.64	20.23 ± 4.99
Average Sessions	3.52 ± 0.51	4.00 ± 0.67
Learning Gain (post-pre)	7.55 ± 0.89	3.22 ± 1.12
Pre score	4.86 ± 0.53	11.1 ± 0.35
Post score	12.41 ± 0.96	14.32 ± 0.98

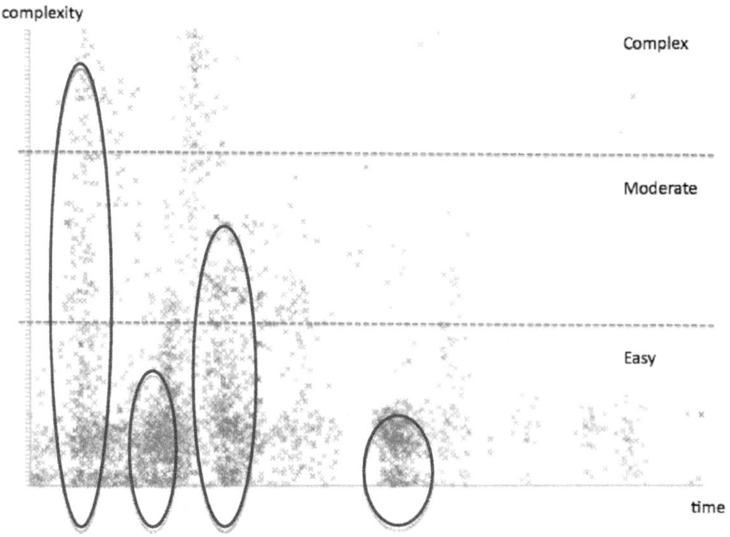

Fig. 5. Strong students guided weak students to explore the topics overtime. Blue data point represents strong students' attempt (circle selected chunk); orange data points represents the weak ones (not selected chunk).

4.3 Subjective Evaluation

At the end of the semester, the students were asked to evaluate the system by filling out a questionnaire. We received a lot of praise and criticism. 78.8% of the students considered that the QuizMap visualization motivated them to solve more quizzes. We can further hypothesize that the color contrasts on the QuizMap helps to motivate student to solve more quizzes while the rest of the student peers have already progressed on such. About 72~78% of the students *strongly agreed* or *agreed* that the QuizMap helped them to explore more topics and questions. We expected an even higher percentage of satisfaction on exploring the quizzes and questions. However, due to the large size of the class and students' active work on the system, QuizMap generated 1105 cells in total. Some students complained that there were too many cells making it easy to lose focus. Over 80% of the students strongly agreed or agreed that they benefited from the self-assessment quizzes as well as the QuizMap visualization. Figure 6 displays the detail percentage for each survey question.

5 Summary

This paper described a novel approach to integrate social navigation for self-assessment questions with open user model in a TreeMap interface. The hierarchical representation of TreeMap was implemented to help students visualize both, the state of their knowledge and the progress of the whole class. Color contrasts between personal progress and the progress of others students was used to provide social guidance. The classroom demonstrated that QuizMap visualization provided effective social guidance allowing students to achieve high quality of learning. The effect was comparable with the impact of traditional knowledge-based guidance. The potential key to the success of the social guidance is the trailblazing behavior of stronger students who explored the topics and left the trace for weaker students to follow. In general, student satisfaction with QuizMap was high. However, there is also an evidence that the QuizMap approach may not be optimal for larger classes that generate too many cells on the TreeMap, causing it to become too crowded.

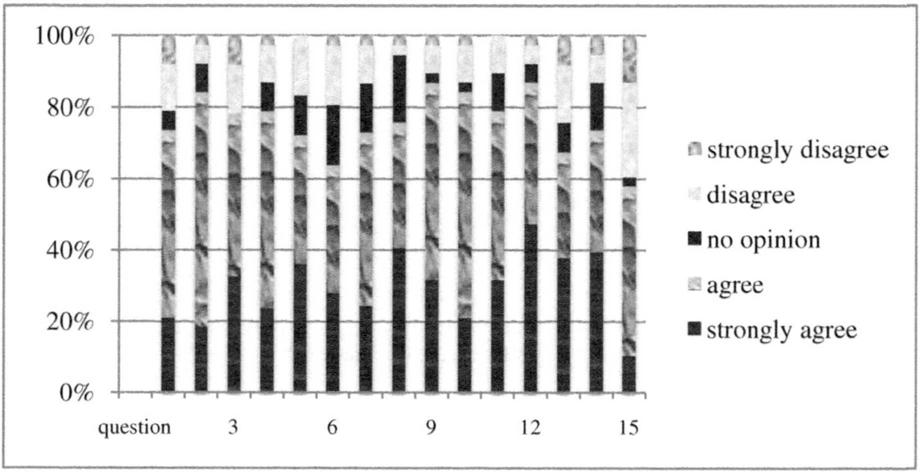

1. Online self-assessment quizzes helped to understand difficult concepts.
2. Online self-assessment quizzes helped to prepare for exams.
3. Online self-assessment quizzes contributed to my learning in this course.
4. The QuizMap visualization motivated me to solve more quizzes.
5. The QuizMap visualization helped to explore more topics (quizzes).
6. The QuizMap visualization helped to explore more questions.
7. The QuizMap visualization helped me to choose appropriate quizzes.
8. The QuizMap visualization showed the group's performance(blue color grids) and mine(orange ones) are clear.
9. I benefited from the self-assessment quizzes.
10. I also benefited from the QuizMap visualization.
11. Online self-assessment quizzes helped to discover my weak points.
12. Online self-assessment quizzes brought back forgotten concepts to my memory.
13. I am motivated to continue using online self-assessment quizzes after completing this course.
14. I think I will also benefit from self-assessment quizzes and QuizMap visualization in other courses.
15. The online self-assessment quizzes and QuizMap provided for easy navigation and I have minimal problems using them.

Fig. 6. Subjective Evaluation: Questions and Results

Acknowledgements. This research is partially supported by the National Science Foundation under Grant No. 0447083.

References

1. Bull, S., Brna, P., Pain, H.: Extending the scope of the student model. User Modeling and User-Adapted Interaction 6(1), 45–65 (1995)
2. Bull, S.: Supporting learning with open learner models. Paper presented at the 4th Hellenic Conference on Information and Communication Technologies in Education, Athens, Greece, September 29 - October 3 (2004)
3. Zapata-Rivera, J.D., Greer, J.E.: Visualizing and inspecting Bayesian belief models. International Journal of Artificial Intelligence in Education IJCAI 14, 1–37 (2004)
4. Mitrovic, A., Martin, B.: Evaluating the Effect of Open Student Models on Self-Assessment. Int. J. of Artificial Intelligence in Education 17(2), 121–144 (2007)
5. Kay, J.: Lifelong Learner Modeling for Lifelong Personalized Pervasive Learning. IEEE Transactions on Learning Technologies 1(4), 215–228 (2008)
6. Brusilovsky, P.: Adaptive navigation support in educational hypermedia: the role of student knowledge level and the case for meta-adaptation. British Journal of Educational Technology 34(4), 487–497 (2003)
7. Brusilovsky, P., Pesin, L.: Adaptive navigation support in educational hypermedia: An evaluation of the ISIS-Tutor. Journal of Computing and Information Technology 6(1), 27–38 (1998)
8. Davidovic, A., Warren, J., Trichina, E.: Learning benefits of structural example-based adaptive tutoring systems. IEEE Transactions on Education 46(2), 241–251 (2003)
9. Specht, M.: Empirical evaluation of adaptive annotation in hypermedia. Paper presented at the ED-MEDIA/ED-TELECOM 1998 - 10th World Conference on Educational Multimedia and Hypermedia and World Conference on Educational Telecommunications, Freiburg, Germany, June 20-25 (1998)
10. Kavcic, A.: Fuzzy User Modeling for Adaptation in Educational Hypermedia. IEEE Transactions on Systems, Man, and Cybernetics 34(4), 439–449 (2004)
11. Brusilovsky, P., Sosnovsky, S., Yudelson, M.: Addictive links: The motivational value of adaptive link annotation. New Review of Hypermedia and Multimedia 15(1), 97–118 (2009)
12. Hsiao, I., Sosnovsky, S., Brusilvsky, P.: Guiding Students to the Right Questions: Adaptive Navigation Support in an E-learning System for Java Programming. Journal of Computer Assisted Learning 26(4), 270–283 (2010)
13. Brusilovsky, P., Sosnovsky, S.: Engaging students to work with self-assessment questions: A study of two approaches. Paper presented at the 10th Annual Conference on Innovation and Technology in Computer Science Education, ITiCSE 2005, Monte de Caparica, Portugal (2005)
14. Dieberger, A., Dourish, P., Höök, K., Resnick, P., Wexelblat, A.: Social navigation: Techniques for building more usable systems. Interactions 7(6), 36–45 (2000)
15. Farzan, R., Brusilovsky, P.: AnnotatEd: A social navigation and annotation service for web-based educational resources. New Review in Hypermedia and Multimedia 14(1), 3–32 (2008)
16. Dimitrova, V., Self, J.A., Brna, P.: Applying Interactive Open Learner Models to Learning Technical Terminology. In: Bauer, M., Gmytrasiewicz, P.J., Vassileva, J. (eds.) UM 2001. LNCS (LNAI), vol. 2109, pp. 148–157. Springer, Heidelberg (2001)

17. Chen, Z.H., Chou, C.Y., Deng, Y.C., Chan, T.W.: Active Open Learner Models as Animal Companions: Motivating Children to Learn through Interaction with My-Pet and Our-Pet. International Journal of Artificial Intelligence in Education 17(3), 217–226 (2007)
18. Bull, S., Kay, J.: Student Models that Invite the Learner In: The SMILI Open Learner Modelling Framework. International Journal of Artificial Intelligence in Education 17(2), 89–120 (2007)
19. Vassileva, J., Sun, L.: Using Community Visualization to Stimulate Participation in Online Communities. e-Service Journal 6(1), 3–40 (2007)
20. Bull, S., Britland, M.: Group Interaction Prompted by a Simple Assessed Open Learner Model that can be Optionally Released to Peers. In: Brusilovsky, P., Grigoriadou, M., Papanikolaou, K. (eds.) Proceedings of Workshop on Personalisation in E-Learning Environments at Individual and Group Level, User Modeling (2007)
21. Hsiao, I.-H., Bakalov, F., Brusilovsky, P., König-Ries, B.: Open Social Student Modeling: Visualizing Student Models with Parallel IntrospectiveViews. In: Konstan, J.A., Conejo, R., Marzo, J.L., Oliver, N. (eds.) UMAP 2011. LNCS, vol. 6787, pp. 171–182. Springer, Heidelberg (2011)
22. Shneiderman, B.: Treemaps for Space Constrained Visualization of Hierarchies: an historical summary of Treemap research and applications (2004), `http://www.cs.umd.edu/hcil/treemaps/`
23. Wattenberg, M.: Visualizing the Stock Market. In: Proceedings of Extended Abstracts of Human Factors in Computing Systems (CHI 1999), pp. 188–189. ACM Press, New York (1999)
24. Plaisant, C., Chintalapani, G., Lukehart, C., Schiro, D., Ryan, J.: Using Visualization Tools to Gain Insight Into Your Data. In: Proc. of Society of Petroleum Engineering Annual Technical Conference, Denver, CO (2003)
25. Mansmann, F., Keim, N.: Visual support for analyzing network traffic and intrusion detection events using TreeMap and graph representations. In: Proceedings of the Symposium on Computer Human Interaction for the Management of Information Technology, Maryland, USA (2009)
26. Fekete, J., Plaisant, C.: Interactive Information Visualization of a Million Items. In: Proc. of IEEE Conference on Information Visualization, Boston, September 2002, pp. 117–124 (2002)
27. Lindstaedt, S.N., Beham, G., Kump, B., Ley, T.: Getting to know your user – unobtrusive user model maintenance within work-integrated learning environments. In: Cress, U., Dimitrova, V., Specht, M. (eds.) EC-TEL 2009. LNCS, vol. 5794, pp. 73–87. Springer, Heidelberg (2009)

Unsupervised Auto-tagging for Learning Object Enrichment

Ernesto Diaz-Aviles, Marco Fisichella, Ricardo Kawase,
Wolfgang Nejdl, and Avaré Stewart

L3S Research Center, Leibniz University Hannover, Germany
{diaz,fisichella,kawase,nejdl,stewart}@L3S.de

Abstract. An online presence is gradually becoming an essential part of every learning institute. As such, a large portion of learning material is becoming available online. Incongruently, it is still a challenge for authors and publishers to guarantee accessibility, support effective retrieval and the consumption of learning objects. One reason for this is that non-annotated learning objects pose a major problem with respect to their accessibility. Non-annotated objects not only prevent learners from finding new information; but also hinder a system's ability to recommend useful resources. To address this problem, commonly known as the cold-start problem, we automatically annotate specific learning resources using a state-of-the-art automatic tag annotation method: *α-TaggingLDA*, which is based on the Latent Dirichlet Allocation probabilistic topic model. We performed a user evaluation with 115 participants to measure the usability and effectiveness of *α-TaggingLDA* in a collaborative learning environment. The results show that automatically generated tags were preferred 35% more than the original authors' annotations. Further, they were 17.7% more relevant in terms of recall for users. The implications of these results is that automatic tagging can facilitate effective information access to relevant learning objects.

Keywords: Metadata Generation, User Study, LDA, Cold-Start, Recommender Systems.

1 Introduction

Learning strategies have shifted from a solitary activity to a collaborative web-based one [2]. In collaborative learning systems, digital collections of educational materials or Learning Objects (LOs), such as, lecture videos, notes and presentations, are made available in online repositories. Online learners are not only able to browse or search for LOs, but also enrich this content with value-added metadata.

Learning object enrichment is crucial within a collaborative setting. For example, consider a scenario in a collaborative environment where a user wants to retrieve specific documents related to their interests and uses tags to navigate to the associated resources. Ideally, if the system can effectively provide good tag coverage over the resources, the user can better navigate through document objects and be steered to the relevant resources in the system. On the contrary, if tags are either unclear, not specific for the resource, noisy, or ambiguous, then users cannot retrieve or easily locate resources. Unfortunately, the latter situation is all too common. Since users typically only

C. Delgado Kloos et al. (Eds.): EC-TEL 2011, LNCS 6964, pp. 83–96, 2011.

tag a small fraction of the documents, most of the other documents have no associated metadata. Furthermore, newly added resources, which have not yet been tagged, are hard to be located or associated with related objects. This dilemma is well known and referred to as the *new item* cold-start problem.

As outlined in the aforementioned scenario, an important prerequisite for realizing the benefit of tags in a collaborative learning setting, is that a LO actually has to have at least a minimum number of tags associated with it. When a learning resource has no associated tags, the collaborative learning system cannot provide a recommendation, nor does any descriptor exist in the tag cloud to help support navigation to the orphan resource.

One way to address the cold-start problem in collaborative systems is by using automatic tagging, which associates tags with untagged resources. State-of-the-art work in this area relies upon latent data models to make explicit, some hidden, underlying "context" (i.e., set of keywords or tags). The untagged resources are treated as a new resource, for which inferencing can be performed to bring the new resource into a known context where it can inherit an appropriate set of tags. Little has been done in the area of latent model based automatic tagging for learning objects repositories. Moreover, the usability and effectiveness of such automatic tagging has not been assessed by the learners themselves.

In this work, we propose to automatically associate tag annotations to untagged LOs by exploiting the content from different, but similar resources, found outside the boundaries of a single content repository and using a state-of-the-art method, *α-TaggingLDA*, which is based on the probabilistic topic model Latent Dirichlet Allocation [4]. In doing so, we will address the following research questions:

Q1: To what extent do the LO's annotations assigned by the authors agree with the ones assigned automatically?
Q2: From the user perspective, how relevant are the automatic generated tags comparing to the ones provided by experts?
Q3: In a social tagging recommendation scenario for LOs, are the automatic annotations better candidate terms for assisting users in the tagging process than the keywords assigned by the LO's author?

The contributions of this work are:

- an automatic tagging approach to efficiently address the cold-start problem in collaborative learning environments by relying on content from resources in an auxiliary domain, i.e., one that lies outside of the content repository of the untagged LO.
- an evaluation of our approach through a user study involving 115 participants in a online setting, with real-world data.

The rest of this paper is organized as follows. In Section 2 we will first present related work on the areas of automatic tagging and learning objects enrichment. In Section 3 we describe our approach to automatically tag learning objects. In Section 4 we present a three part evaluation to measure the effectiveness of our method and in Section 5 we describe the evaluation results, which are discussed in Section 6. Finally, in Section 7 we conclude and discuss the directions for future work.

2 Related Work

Cold-start is a common problem in many user-centric systems that seek to improve information access. Specifically for the collaborative learning environment setting, a new LO is introduced into the domain, but it has no (or incomplete) associated user-defined metadata or annotations. Automatic enhancement of LO metadata is an alternative to address this problem. In this section we present related work in the areas of automatic tagging and learning objects enrichment.

2.1 Learning Object Enrichment

Lohmann et al. [13] demonstrate the importance of additional metadata to learning resources visibility and reusability and suggest design guidelines for automatic tagging approaches. The authors suggest (i) the use of a stable set of tags for agreed description of resources, (ii) to guide the tagging process (e.g., with tag recommendations), (iii) to use text extracted from resources for starting set of tags, and (iv) the use of a small set of selectable tags for tags convergence. The two user evaluations we present in this work align with these guidelines.

Another recent system, namely, ReMashed [7,6] takes advantage of tagged and rated data of combined Web 2.0 sources, integrating the metadata from decentralized sources of content. Their work addresses the *new user* cold-start scenario and shows that a recommender system that exploits already tagged resources can mitigate the lack of user information. Our work, on the other hand, focuses on the *new item* cold-start problem, and aims to annotate untagged LOs. Once objects are tagged, it is possible to improve the performance of a recommender system [19].

Abel et al. [1] introduce the LearnWeb 2.0 environment,which supports sharing, discovering, and managing learning resources, which are spread across different Web 2.0 platforms, between learners and educators. LearnWeb 2.0 aggregates resources and enhances their metadata using functionalities from ten different Web 2.0 services. Furthermore, in order to support collaborative searching, the authors are provided with an automatic resource annotation service. Once a search result is displayed in the environment, it is automatically tagged with the corresponding query terms. This mechanism assumes that the system has enough information to retrieve the item as relevant for a given query. Furthermore, it requires an initial user interaction, i.e., search, in order to be able to annotate the resource. This is not necessarily the case for resources with sparse text, or multimedia resources with little or no metadata available. Our approach is content based and does not require any user interaction to automatically annotate the LOs.

2.2 Automatic Tagging

Numerous forms of enrichment can be used to annotate a LO such as ratings, comments and tagging, to name but a few. Tagging, in particular, has proven to be an intuitive and flexible mechanism for improving the access to information or personalizing the users' online experience. Tags are capable of facilitating search [3,10] and improving recommendations [19,18]. Tags have also been used for personalization: improving information access in collaborative tag recommendations [16] and facilitating personalized information access across disparate media types [19].

State-of-the-art methods for automatic annotation and tag recommendation rely on dimensionality reduction and are based on tensor factorization [17,16,20] or on probabilistic topic models, in particular on Latent Dirichlet Allocation (LDA), for example, Krestel et al. in [11,12] exploited resources annotated by many users, thus having a relatively stable and complete tag set, to overcome the cold-start problem. They build an LDA model from tags that have been previously assigned by users. In this way, a resource in the system is represented with tags from topics discovered by LDA. For a new resource with few or no annotations, they expand the latent topic representation with the top tags of each latent topic.

In these auto tagging systems, the performance of the aforementioned approaches highly relies on the assumption of a dense set of data upon which the model can be built. To overcome this issue, Diaz-Aviles et al. [5] introduced *α-TaggingLDA*, a method for automatic tagging resources with sparse and short textual content. In the presence of a new resource, an *ad hoc* corpus of related resources is created, and then the method applies LDA to elicit latent topics for the resource and the associated corpus. This is done in order to automatically tag resources based on the most likely tags derived from the latent topics identified. In our work this method was chosen to annotate LOs in a cold-start scenario.

In contrast to previous work, we aim to evaluate the usability and effectiveness of this automatic tagging approach, and address its potential to generate metadata for novel resources in the context of a collaborative learning environment.

3 Automatically Enhancing Learning Objects with Tags

In order to enhance the learning object with tags we use *α-TaggingLDA*, a state-of-the-art approach for automatic tagging introduced by Diaz-Aviles et al. [5]. *α-TaggingLDA* is designed to mitigate new item cold-start problems by exploiting content of resources, without relying on collaborative interactions.

The approach is based on the probabilistic topic model: Latent Dirichlet Allocation (LDA) [4], which is a generative probabilistic model for collections of discrete data such as text corpora. The basic idea of LDA is that documents are represented as random mixtures over latent topics, where each topic is characterized by a distribution over terms.

For example, an LDA model might have topics that can be labeled as EDUCATION and ENTERTAINMENT[1]. Furthermore, a topic has probabilities of generating various words such as *school*, *students*, and *teacher*, which can be classified and interpreted as EDUCATION. Naturally, the word *education* itself will have high probability given this topic. The ENTERTAINMENT topic likewise has high probability of generating words such as *film*, *music*, and *theater*.

[1] Please note that these labels are arbitrary. The algorithm does not automatically assign any particular label to the latent topics.

More formally, assume that a text collection consists of a set of documents D. Furthermore, consider the set of topics Z, the distribution $P(z \mid d)$ over topics $z \in Z$ in a particular document $d \in D$, and the probability distribution $P(t \mid z)$ over terms $t \in T$ given topic $z \in Z$, where T is the set of terms. Each term $t_i \in T$ in a document (where the index refers to the ith term token) is generated by first sampling a topic from the topic distribution, then choosing a term from the topic-term distribution. We write $P(z_i = j)$ as the probability that the jth topic was sampled for the ith term token and $P(t_i \mid z_i = j)$ as the probability of term t_i under topic j. The model specifies the following distribution over terms within a document:

$$P(t_i \mid d) = \sum_{z} P(t_i \mid z_i = j) P(z_i = j \mid d) \quad (1)$$

where $|Z|$ is the number of topics.

In LDA, the goal is to estimate the distribution topic-term $P(t \mid z)$ and the document-topic distribution $P(z \mid d)$. These distributions are sampled from Dirichlet distributions (e.g., using Gibbs sampling [8]) and indicate which terms are important for which topic and which topics are important for a particular document, respectively.

α-TaggingLDA

An overview of the *α-TaggingLDA* method is shown in Figure 1. In order to illustrate the method with an example, consider a novel LO entitled *Knowledge Technologies in Context*, this resource is new to the collaborative learning system and does not have any tag annotations assigned. The absence of tags makes it difficult for the system to consider it as candidate for recommendations, for instance.

α-TaggingLDA first extracts relevant LO's *textual content*, such as the title, description or metadata (e.g., author) and creates a document denoted as d_{LO}. Then, the LO is associated to a set of 'similar' documents, which we refer to as an *ad hoc corpus* for the LO, represented as $corpus_{LO}$.

Note that the *α-TaggingLDA* method does not impose any restriction on the similarity measure used to associate the corpus with the LO. The similarity measure could be specified based on the nature of the resources, (e.g., text documents, multimedia items) and the textual content or metadata available. For example, a particular implementation might rely upon a computationally inexpensive similarity measure or on a more complex clustering algorithm.

In our particular example, the title of the LO is used to query an Internet search engine in order to retrieve the title and snippets of the n relevant results ($n = 4$, in this case). This subset corresponds to $corpus_{LO}$.

The LO's textual content is extracted and the subset of the top n results constitute the text collection $D = \{d_{LO}\} \cup corpus_{LO}$, which is input to LDA, together with the number of topics required. In this example, the number of topics is set to two, i.e., $|Z| = 2$. The set of tags to be used to annotate the LO is denoted as $TopN_{tags}(LO)$, and its size is set to six for this particular case, i.e., $|TopN_{tags}(LO)| = 6$.

Table 1 presents an example of the output produced by LDA according to the setting described above. Topics are ordered based on the document-topic distribution $P(z \mid d)$, and within each topic, terms are ranked based on the topic-term $P(t \mid z)$ distribution.

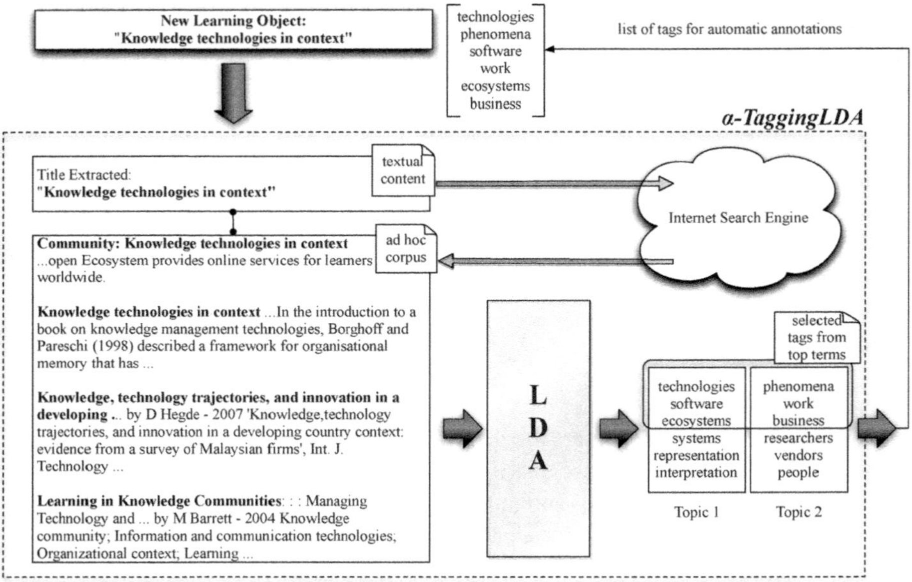

Fig. 1. *α-TaggingLDA* is applied to annotate a new LO, *Knowledge Technologies in Context*, with a list of six tags: $TopN_{tags}(LO)$ = { *technologies, phenomena, software, work, ecosystems, business* }, based on two LDA topics

For the construction of the final set of tags $TopN_{tags}(LO)$, *α-TaggingLDA* selects the first candidate tag from $Topic_1$'s top terms, the second tag from $Topic_2$'s top terms, the third tag, again from $Topic_1$'s top terms, and so forth. The final list of tag annotations for the LO in our example corresponds to $TopN_{tags}(LO)$ = { *technologies, phenomena, software, work, ecosystems, business* }. For the details of this strategy, we refer the reader to the work done by Diaz, et.al. [5].

Table 1. Example of two topics output by LDA. Topics are ordered based on the document-topic distribution $P(z \mid d)$, and within each topic, terms are ranked based on the topic-term $P(t \mid z)$ distribution.

$Topic_1$		$Topic_2$	
$P(z = 1 \mid d_{LO}) = 0.70$		$P(z = 2 \mid d_{LO}) = 0.30$	
Term t	$P(t \mid z = 1)$	Term t	$P(t \mid z = 2)$
technologies	0.45	phenomena	0.33
software	0.25	work	0.28
ecosystems	0.16	business	0.19
systems	0.11	researchers	0.15
representation	0.03	vendors	0.04
interpretation	0.01	people	0.01

4 Evaluation

In this section we measure the benefits of automatic tag annotations for a collaborative learning environment. To answer the research questions presented in Section 1, we conducted three distinct evaluations, first, an experimental evaluation followed by two user studies. The rest of the section describes each evaluation settings.

4.1 Experimental Setting

Dataset. We based our experiments on a dataset sampled from the OpenScout project collection [15]. The project gathers metadata information from learning resources located at different learning content repositories. For our evaluation, we selected learning objects whose language is English and have at least five keywords added by their author. In total, 563 learning objects, 1692 unique keywords and 3150 keywords assignment were considered for the experiments.

Metadata Enrichment. The method used for automatic metadata enrichment in our experiments, *α-TaggingLDA*, is implemented in Java. The corpus builder is based on the search results obtained by querying Yahoo!'s open search web services platform (BOSS)[2]. The titles and short text summaries (snippets) of the ten most relevant results returned are used to create ten different textual documents. The final *ad hoc* corpus for the learning object consists of these and the textual content of the resource. Then, by applying LDA on this corpus we extract the desired number of latent topics, and from them, the needed tags are inferred. The default number of topics considered was two, according to the optimal setting specified in [5].

We use the LDA with Gibbs sampling implementation provided by the Machine Learning for Language Toolkit (MALLET) [14].

4.2 Evaluation I: Author's Keywords and Automatic Tags

Our first evaluation consists of an offline study that measures the agreement between the keywords assigned to the LOs by its author and the tag annotations provided by the method. In order to quantify such agreement, we consider a recommender system setting, where the author's keyword assignments constitute our test set. The task of the collaborative learning environment is to recommend $TopN_{tags}$ relevant tags for a given LO.

We use recall, precision and f1, three widely used metrics in recommender systems [9], to assess the performance. The metrics are defined in Equations 2 and 3.

- Recall for a given author u and a learning object i is defined as:

$$recall(u,i) = \frac{|Keywords(u,i) \cap TopN_{tags}(i)|}{|Keywords(u,i)|} \quad , \; and \tag{2}$$

[2] http://developer.yahoo.com/search/boss/

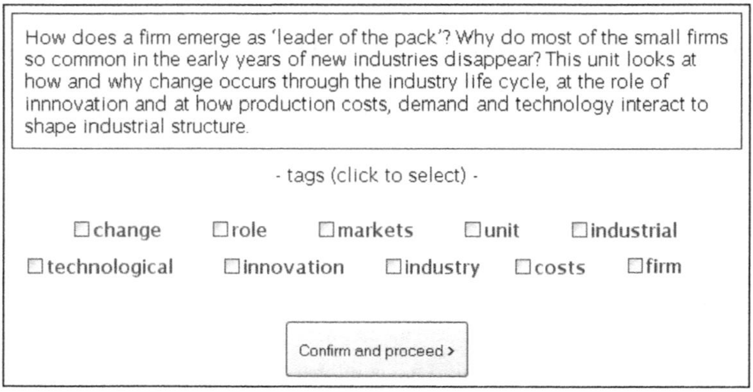

Fig. 2. Evaluation II: Guided Choice User Study Interface. Each participant was instructed to choose at least three tags from the set of ten suggested tags. Five tags were originally added by the expert/author of the content, while the remaining five were automatically generated. The tags were presented in a random order and their origin was not disclosed to the participants.

– Precision for a given author u and a learning object i is defined as:

$$precision(u, i) = \frac{|Keywords(u, i) \cap TopN_{tags}(i)|}{|TopN_{tags}(i)|} \quad , \tag{3}$$

where $Keywords(u, i)$ is the set of keywords assigned by the author u to the learning object i and $TopN_{tags}(i)$ is the set of size N corresponding to the tags automatically assigned by *α-TaggingLDA* to the given LO. In this experiment we set $N = 10$.

For the dataset, we averaged these values over all the authors. The aggregated values of recall and precision are then used to compute their harmonic mean or f1 measure as defined according to Equation 4.

$$f1 = 2 \cdot \frac{recall \cdot precision}{recall + precision} \tag{4}$$

4.3 Evaluation II: Guided Choice User Study

The goal of this experiment was to compare the automatically generated tags against the ones provided by experts.

This evaluation is a user study in which each participant was presented with basic information regarding a learning object, namely, the title and an abstract that varies from 20 up to 200 words (see Figure 2). The format of the original resource (e.g. video, image, presentation or document) was not made known to the participant in order to align the nature of the evaluation and to avoid biased judgments of the tag relevance based on non computer-understandable information. In addition to that, the participants were presented with ten tags to be evaluated. From the ten tags presented, five tags were originally added by the expert/author of the content, while the remaining five were the top ranked automatically generated ones. The tags were presented in a random order and their origin was not disclosed to the participants.

Each participant was then instructed to read the title and the description of the learning object and finally choose at least three tags from the set of ten suggested tags. Once the submission of the form is completed the participant was presented with a new object to be evaluated. We kindly asked for each participant to repeat the process for at least ten objects, however, we did not limited the maximum of their contribution to the study.

In order to compare the automatically generated tags against the ones provided by experts, we designed this experiment as a recommendation task, and used the recall measure, which is widely used to assess the recommendation quality [9] of recommender systems.

In this case recall is defined in Equation 5, as follows:

– Recall for a given author u and a learning object i is defined as:

$$recall(u,i) = \frac{|Tags(u,i) \cap TopN_{tags}(i)|}{|Tags(u,i)|}\ , \tag{5}$$

where $Tags(u,i)$ is the set of tags assigned by the user u to the learning object i and $TopN_{tags}(i)$ is the set of size N corresponding to the tags recommended to the user, either based on *α-TaggingLDA* or on the author's keywords. In this experiment we set $N = 5$.

As in Evaluation I, we averaged the values over all the participants. Using a fixed number of recommendations, precision is just the same as recall up to a multiplicative constant and thereby there is no need to evaluate precision.

4.4 Evaluation III: Free Choice User Study

The goal of this study was to collect evidence to evaluate if the automatic annotations are better candidate terms for assisting users in the tagging process than the keywords assigned by the LO's author.

Similarly to Evaluation II, in this user study, each participant was presented with the title and an abstract of a learning object. Once again, due to same reasons as presented before, the format of the original resource (e.g. video, image, presentation or document) was not disclosed to the participants.

Each participant was then instructed to read the title and the description of the learning object and finally input five tags she thinks to be relevant for describing the object, as depicted in Figure 3. Once the submission of the form was completed, the participant was presented with a new object to be evaluated. Each participant was asked to repeat the process for at least ten objects.

As in Evaluation II, in other to evaluate this experiment we cast it as a recommendation task and evaluate the recall measure. In this case $Tags(u,i)$ corresponds to the set of tags that would be recommended to the participant u for the given LO i. Note that, even though, the set of tags is not presented to the participant, it helps us to measure which terms are better for assisting users in the tagging process.

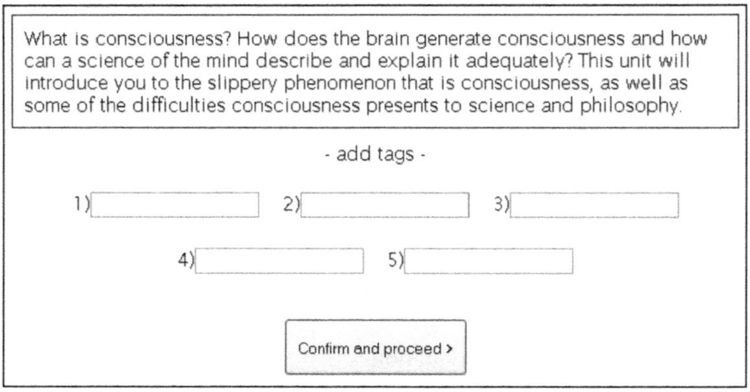

Fig. 3. Evaluation III: Free Choice User Study Interface. Participants were instructed input five tags they think to be relevant for describing the LO.

5 Results

Evaluation I aims to answer the research question *'Q1: At what extent do the LO's annotations assigned by the authors agree with the ones assigned automatically?'*. As an outcome for $|TopN_{tags}| = 10$, we obtained the following results, recall=0.26 precision=0.13 and f1=0.18. Table 2 shows the f1 measure for different sizes of $TopN_{tags}$.

Table 2. f1 measure for different sizes of $TopN_{tags}$

$TopN_{tags}$	1	2	3	4	5	6	7	8	9	10
f1	0.04	0.07	0.09	0.10	0.10	0.11	0.12	0.13	0.16	0.18

From the user study in Evaluation II we collected the feedback of 115 participants (43 female and 72 male), 100 of them explicitly stated to be students. Their average age was 24, ranging from 20 to 53 years old. In total the participants evaluated 1,134 objects covering 478 unique ones.

Also, in total 4,035 tags were chosen to represent the documents, in average each participant picked 3.56 tags per document. As explained in the setup of this evaluation, the tags exposed to the participants were originated from two different sources, the expert who created the learning material and a second set from the automatic tagging method.

It's important to note that the tags from each group were always presented to the participants in an equal distribution to preserve the fairness of the study. Additionally, when a participant chose a tag that was in both *α-TaggingLDA* set and the Experts' set we computed this choice as two tag assignments as outlined in Section 3. Although the participants chose 4,035 tags, in our experiments, we used a total of 4,939 tag assignments.

Out of the 4,939 tag assignments, 67.5% of them were originated by *α-TaggingLDA* and 32.5% by the experts.

The most straightforward analysis of these results shows a clear preference of the participants for the tags that were automatically added. Thus, it is also reasonable to conclude that these tags are more relevant to the participants, which answers our second question: '*Q2: From the user perspective, how relevant are the automatic generated tags comparing to the ones provided by experts?*'. The main reason is that the underneath approach generates tags that represents better the learners tagging behavior. Through the outcomes of this evaluation we interpret that the automatic generated tags are, in general, more descriptive and more useful for the learners than experts' tags. Additionally, it is reasonable to assume that these learners, when searching for one of these documents would (with a higher probability) use a tag that was automatically generated rather than the experts' tags. The same assumption is valid for the case of browsing resources in a hierarchical classification or in a facet browsing interface.

To validate the significance of the results achieved, for each participant, we took the averages of the distribution of *α-TaggingLDA* and Expert's tag sets. With two groups of 115 samples we performed a two-tailed t-test that confirmed a statistically significant difference of *α-TaggingLDA* mean (68.2%) and Expert's keywords mean (31.8%).

For Evaluation III, the participants evaluated 832 objects covering 454 unique one. In this phase, where the participants were instructructed to freely choose terms that best classify the objects, 4,745 tags were generated (1,868 unique tags).

Using these data we now have three different sets of tags: *α-TaggingLDA* tags, experts' tags and learners' tags. By validating the learners' generated tags against the other sets we found an overlap of 38.4% with automatic generated *α-TaggingLDA* tags and 20.7% with the experts' tags. Additionally, in only 8.9% of the cases, a tag occurs in all three sets. At this point we are just considering the whole sets of tags and not the precision of them regarding each resource. These results complement Evaluation II by firmly stating that on average the automatic generated tags are closer to the ones used by learners.

By considering only the results from those 100 participants that stated to be students, the numbers do not change significantly. The overlap with automatic generated tags increases slightly to 39.04% while the overlap with the experts' tags remains on the same levels (20.3%). These results help us to answer the third question – Q3, as the automatic annotations turned out to be the best candidate terms for assisting users in the tagging process.

Table 3. Tag Assignment (TAS) results for the user evaluations. The *Sets* rows show the number of TAS that were chosen by the participants that overlapped with TAS given by the experts, or with *α-TaggingLDA* method and the respective recall measure.

		Experiment II: Guided Choice		**Experiment III: Free Choice**	
	Participant's TAS	4939	-	4745	-
Sets	*α-TaggingLDA*	3336	67.5%	1824	38.4%
	Experts	1603	32.5%	983	20.7%

6 Discussion

The values of recall amd precision of Evaluation I (Section 5) suggest that the information captured by the automatic tag annotations partially agree with the expert keywords assigned to the LO. The values are not exceptionally high, which suggest that the automatic tag annotations tend to capture different information than the expert keyword assignments. The user studies conducted in Evaluation II and III exposed how additional information captured by the automatic annotations are perceived by the learners, and explore the usability improvements of the collaborative learning system. The results from the first user study setup (Evaluation II) clearly demonstrate the preference of the participants for tags produced by the automatic tagging method. This means that, the produced tags reflects better the participants' preferences in comparison to the experts' keyword assignments. The most probable reasons are, first, the aforementioned problem that a tag assignment is not always clear to users other than its creator. Second, learners usually have a viewpoint that differs from the experts, thus they are more prone to avoid terms that are too specific or that they would probably not remind later. Finally, the terms given by the automatic tagging, extracted from search results' snippets, represent better the wisdom of the crowd since these results are originally extracted from multiple resources. It is also important to remark that the search results themselves are consequence of ranking algorithms that exploit collective knowledge and preferences.

In principle, the results of the second user study (Evaluation III) support the same benefits. The goal of this phase was to prevent any possible biases in the first evaluation. We hypothesize that, when asking a participant to tag a learning object, we are implicitly observing which tags the participants would use in a collaborative social learning environment, and indirectly potential terms to query or browse for a learning object.

Bearing in mind the overall results obtained in the experiments, the most important consideration to highlight is the potential benefits produced by the information delivered by the automatic tagging method evaluated.

7 Conclusion

We have empirically demonstrated through a series of evaluations that the proposed *α-TaggingLDA* method produces quality metadata enhancement for the learning objects. First, by experimentally comparing against existing authors' tag annotations. Second, by running a user study comparing the participants' preference for automatically produced tags against the authors' tags. Finally, a last user study that demonstrated that *α-TaggingLDA* tags are the best candidate terms for assisting users in the tagging process.

The additional metadata that was automatically generated by our method can improve personal recommendations of learning objects and most notably it overcomes the 'cold-start' for objects that are not tagged, consequently isolated from the rest of the folksonomy.

Our approach faces some limitations, since we depend on the external resources provided by a search engine. One implication of this shortcoming is that the collection of the documents retrieved may not contain enough meaningful text for good topics to be generated. Another implication is that in some cases, there may be valuable documents for enriching the learning resources, but the documents may not be available if they are buried in the "Hidden-Web", i.e. documents that are not indexed by search engines. One potential solution to, at least, mitigate this limitation is to use multiple and heterogeneous sources for building the topic model. Heterogeneity would include the use of multiple search engines, and open information sources such as wikipedia. Future experiments are needed to examine heterogeneous sources, and we consider this in future work.

Additionally we plan to evaluate how this method could be further refined to assist authors in tasks of keyword assignment by recommending them relevant terms for the LO. Additionally we are interested in explore how automatically added tags can be incorporated in user's profiles and to what extent it can improve recommendations and discovery of new items. Although our work focuses on approaching the cold start problem, we are also interested in running an evaluation with learning objects that have already been enriched by an active community. This would provide us valuable insights to compare the automatic generated tags based on the general wisdom of the crowd and the focused learning community.

Acknowledgement. The authors would like to thank Katrina Maxwell at INSEAD for her support on the user evaluation conducted. This research has been co-funded by the European Commission within the eContentplus targeted project OpenScout, grant ECP 2008 EDU 428016 (cf. http://www.openscout.net).

References

1. Abel, F., Marenzi, I., Nejdl, W., Zerr, S.: Sharing distributed resources in learnWeb2.0. In: Cress, U., Dimitrova, V., Specht, M. (eds.) EC-TEL 2009. LNCS, vol. 5794, pp. 154–159. Springer, Heidelberg (2009)
2. Atkins, D.E., Seely, B.J., Allen, H.: A review of the open educational resources (oer) movement: Achievements, challenges and New Opportunities. Review Literature And Arts Of The Americas, 84 (2007)
3. Bischoff, K., Firan, C.S., Nejdl, W., Paiu, R.: Can all tags be used for search? In: CIKM 2008: Proceeding of the 17th ACM conference on Information and knowledge management, pp. 193–202. ACM Press, New York (2008)
4. Blei, D.M., Ng, A.Y., Jordan, M.I.: Latent dirichlet allocation. Journal of Machine Learning Research 3, 993–1022 (2003)
5. Diaz-Aviles, E., Georgescu, M., Stewart, A., Nejdl, W.: Lda for on-the-fly auto tagging. In: Proceedings of the fourth ACM conference on Recommender systems, RecSys2010, pp. 309–312. ACM Press, New York (2010)
6. Drachsler, H., Pecceu, D., Arts, T., Hutten, E., Rutledge, L., van Rosmalen, P., Hummel, H., Koper, R.: ReMashed – recommendations for mash-up personal learning environments. In: Cress, U., Dimitrova, V., Specht, M. (eds.) EC-TEL 2009. LNCS, vol. 5794, pp. 788–793. Springer, Heidelberg (2009)
7. Drachsler, H., Rutledge, L., van Rosmalen, P., Hummel, H.G.K., Pecceu, D., Arts, T., Hutten, E., Koper, R.: Remashed - an usability study of a recommender system for mash-ups for learning. iJET 5(S1), 7–11 (2010)

8. Griffiths, T.L., Steyvers, M.: Finding scientific topics. Proc. Natl. Acad. Sci. U S A 101(suppl. 1), 5228–5235 (2004)
9. Herlocker, J.L., Konstan, J.A., Terveen, L.G., Riedl, J.T.: Evaluating collaborative filtering recommender systems. ACM Trans. Inf. Syst. 22, 5–53 (2004)
10. Hotho, A., Jäschke, R., Schmitz, C., Stumme, G.: Information retrieval in folksonomies: Search and ranking. In: Sure, Y., Domingue, J. (eds.) ESWC 2006. LNCS, vol. 4011, pp. 411–426. Springer, Heidelberg (2006)
11. Krestel, R., Fankhauser, P.: Language models and topic models for personalizing tag recommendation. In: 2010 IEEE WIC ACM International Conference on Web Intelligence and Intelligent Agent Technology, pp. 82–89 (2010)
12. Krestel, R., Fankhauser, P., Nejdl, W.: Latent dirichlet allocation for tag recommendation. In: Proceedings of the third ACM conference on Recommender systems, RecSys 2009, pp. 61–68. ACM Press, New York (2009)
13. Lohmann, S., Thalmann, S., Harrer, A., Maier, R.: Learner-generated annotation of learning resources - lessons from experiments on tagging. In: International Conference on Knowledge Management (I-KNOW 2008), Graz, Austria (September 2008)
14. McCallum, A.K.: Mallet: A machine learning for language toolkit (2002), `http://mallet.cs.umass.edu`
15. Niemann, K., Schwertel, U., Kalz, M., Mikroyannidis, A., Fisichella, M., Friedrich, M., Dicerto, M., Ha, K.-H., Holtkamp, P., Kawase, R., Parodi, E., Pawlowski, J., Pirkkalainen, H., Pitsilis, V., Vidalis, A., Wolpers, M., Zimmermann, V.: Skill-based scouting of open management content. In: Wolpers, M., Kirschner, P.A., Scheffel, M., Lindstaedt, S., Dimitrova, V. (eds.) EC-TEL 2010. LNCS, vol. 6383, pp. 632–637. Springer, Heidelberg (2010)
16. Rendle, S., Balby Marinho, L., Nanopoulos, A., Lars, S.-T.: Learning optimal ranking with tensor factorization for tag recommendation. In: Proceedings of the 15th ACM SIGKDD international conference on Knowledge discovery and data mining, KDD 2009, pp. 727–736. ACM Press, New York (2009)
17. Rendle, S., Lars, S.-T.: Pairwise interaction tensor factorization for personalized tag recommendation. In: Proceedings of the third ACM international conference on Web search and data mining, WSDM 2010, pp. 81–90. ACM Press, New York (2010)
18. Sen, S., Vig, J., Riedl, J.: Tagommenders: Connecting users to items through tags. In: International World Wide Web Conference, Madrid, Spain, April 20. ACM Press, New York (2009)
19. Stewart, A., Diaz-Aviles, E., Nejdl, W., Marinho, L.B., Nanopoulos, A., Schmidt-Thieme, L.: Cross-tagging for personalized open social networking. In: Cattuto, C., Ruffo, G., Menczer, F. (eds.) Proceedings of the 20th ACM Conference on Hypertext and Hypermedia (Hypertext 2009), Torino, Italy, pp. 271–278. ACM Press, New York (2009)
20. Symeonidis, P., Nanopoulos, A., Manolopoulos, Y.: Tag recommendations based on tensor dimensionality reduction. In: Proceedings of the 2008 ACM conference on Recommender systems, RecSys 2008, pp. 43–50. ACM Press, New York (2008)

Educational Change through Technology: A Challenge for Obligatory Schooling in Europe*

Anusca Ferrari, Romina Cachia, and Yves Punie

Institute for Prospective Technological Studies (IPTS),
European Commission, Joint Research Centre,
Edificio Expo, C/Inca Garcilaso 3,
41092 Seville, Spain
{Anusca.Ferrari,Romina.Cachia,Yves.Punie}@ec.europa.eu

Abstract. This paper explores the gap between the innovative potential of technologies and their current use in classrooms in Europe. It argues that although significant progress has been made in implementing technologies in schools, ICT is still mainly used by teachers and students outside the school environment. The empirical basis of these findings come from a 2009 online survey with 7,659 primary and secondary school teachers, and from 80 interviews with educational stakeholders. Almost all the teachers surveyed are positive about the potential of ICT for learning and are using the internet as a resource for information and preparing lessons. The use of ICT in the classroom is less widespread. Few of the teachers surveyed recognized the benefits of interactive and collaborative technologies for learning. In addition, digital games and mobile phones are widely underestimated. These findings contrast with the potential of technologies to transform educational practices.

Keywords: Compulsory Education, Future of Learning, Educational change, Educational technology, schools and technology.

1 Introduction

Young people today are growing up surrounded by and immersed in technology. Video-games, mobile phones, social networking sites (SNS) and other digital media and ICT applications are shaping the daily lives of pupils and students. Young people are appropriating these rapidly changing technologies for leisure, entertainment and social interaction. They are shaping and negotiating their identities through the use of technologies [1, 2] and are often referred to as: *NetGen*, *Google Generation* [3], or *Digital Natives* [4].

The arrival of ICT in homes has mobilised material resources, skills, cultural values, social competences and capabilities as argued in the domestication theory. Technologies bring a new understanding of communication, information access, retrieval and meaning-making [5]. The technology-mediated world of young people

* The views expressed in this article are the sole responsibility of the authors and do not necessarily reflect the views of the European Commission.

C. Delgado Kloos et al. (Eds.): EC-TEL 2011, LNCS 6964, pp. 97–110, 2011.

creates certain expectations regarding their learning needs and experience, as learners need to acquire digital competence in schools to be prepared for life challenges. Nevertheless, these needs are currently not met due to differences in technology availability and use between the school and home environments [6]. While 86% of 15-year-olds who took part in the PISA 2009 survey declared they frequently used computers at home, only 55% claimed to use computers in school [7]. The relatively low use of technology in class has an influence on the way young learners build up a perception of the inadequacy of the current educational framework and format [8].

This paper explores the gap between the research on the potential of technologies and their current use in compulsory schooling. It does so by reporting data gathered by a 2009 study on the role of Creativity and Innovation in primary and secondary education in Europe (EU27). A major focus of this study (ICEAC)[1] was to investigate the use and understanding of technologies by teachers. The study combined experts' workshops, a literature review [9], the analysis of school curricula [10], an online survey with teachers [11], interviews with educational stakeholders [12], an analysis of good practices [13] and a final report [14].

This paper builds on the ICEAC study and limits its scope to the use of technologies in compulsory schooling. It elaborates on the data collected from the teachers' online survey and interviews with educational stakeholders. In this respect, the main research questions were: How are technologies being used by teachers? What are the opinions of teachers regarding the usefulness of technologies? How far is the potential of technologies for learning exploited in today's education?

The paper is structured as follows. After the introduction, Section 2 reviews foresight reports and studies on technology-enhanced environments to understand how research sees future learning settings. Section 3 gives an overview of the methodology that has been used for the study. Section 4 discusses the use of ICT in schools and the opinions of teachers on the importance of technologies for learning. Section 5 suggests some changes which are needed to align school practices with learners' and researchers' expectations. Finally, Section 6 offers conclusions.

2 Emerging Contours of the Ubiquitous Learning Paradigm

There is ample evidence that full immersion in a technology-rich environment enables new learning opportunities. The immediacy of information retrieval, the over-abundance of instantly available knowledge and resources, and the possibility to create, share and shape content are already changing the way young people think, learn and understand education [15]. Foresight studies coincide in seeing technology at the core of the future education [16, 17]. However, while research emphasises the benefits of technologies for learning, their uptake in schools needs to be supported by a shift in underlying pedagogies and teaching approaches.

---

[1] The ICEAC project was launched by the Institute for Prospective Technological Studies (JRC-IPTS) on the behalf of DG Education and Culture of the European Commission, see the project webpage: http://is.jrc.ec.europa.eu/pages/EAP/iceac.html

2.1 A Call for Disruptive Educational Change

Although various studies have demonstrated the potential of technology to sustain innovative education [18, 19], the implementation of technology in schools has not yet had the transformative impact that was expected [20, 21]. For Collins and Halversont [22], schools and technologies are incompatible because they are based on opposed learning theories and practices: they see school learning as traditional and rooted in the passive acquisition of knowledge and in testing of facts and notions; whereas learning through technologies implies interaction, experimentation, collaboration and questioning.

While schools are not seen as the best setting for learning through technologies, technologies are extensively explored and exploited for learning outside compulsory education [22]. In particular, collaborative ICT tools and services, such as serious games and simulations, interactive environments and Social Networking Sites (SNS) and communities are used by people of all ages to support both intentional and non-intentional learning [22-24].

The interactive, multi-sensorial and networked characteristics of technologies facilitate personalised, collaborative and enquiry-based learning. The convergence between new conceptions of learning and new technologies [25] underlines the affordances of ICT and media for the learning sphere. Current pedagogies see learning happening through interaction, networking, experimentation and enquiry, therefore technologies and media are perceived as the right tools to learn in an engaging and effective way [26]. In order to realise the full potential of ICT, and to enable new ways of learning, researchers advocate a paradigm shift in pedagogies [27], a disruptive change [28] and a technological revolution [22].

2.2 The Near Future of Learning

Technology is omnipresent in prospective studies on learning and education. Media and ICT are mentioned either directly by describing how technologies could be used for learning and educational purposes [7] or indirectly by stating that ICT and technical skills will be of high importance on the future job market [29, 30]. The general assumption is that learning in the future will be much more driven by internet-based and mobile social networking platforms, cloud-computing, simulations, augmented reality, serious gaming, and web-based research tools [17].

According to a New Zealand review of future-oriented programmes and projects across the world, a focus on ICT is prevalent in visions of education for the 21st century [31]. The 2011 Horizon report [17] ranked as a key trend the wish to be able to work, study, and learn wherever and whenever. Mobile phones, wireless internet connection, and cloud computing fuel the habit and expectation to be online 24/7. The report foresees that in the short term (within 12 months) electronic books and mobiles will be mainstreamed in educational institutions as a learning resource. Augmented reality and game-based learning will become mainstream in a second adoption horizon of 2 to 3 years.

The Beyond Current Horizons Programme [16] sees the coming two decades as being shaped by several trends towards, for example, a denser, deeper and diverse information landscape, a personal 'cloud' enabling people to constantly connect to their network, and a widespread presence of machines. In these trends, networking is seen as central for the future of the individual and of education. A similar emphasis on the network comes from the *European Internet Foundation* [32], which predicts that by 2025 the dominant educational paradigm will be constructed around "participative, digitally-enabled collaboration". Network learning is envisaged by Davidson and Glodberg [33], who see in the internet, in particular web 2.0 applications, and in mobile media the future trajectory of learning. Online social networking and technological progress are the two most influential factors in the changing ways of learning, according to the Learnovation Consortium [34]. The overall picture portrayed by these reports shows the pervasive presence of technologies for learning, the emerging role of educational games, the use of online networks and communities, and constant access to enable mobile and ubiquitous learning.

3 Method

One of the research strands of the ICEAC study was to investigate the current use and perception of ICT by teachers. Data was collected through an online survey which gathered 7,659 responses. Qualitative interviews with 80 educational stakeholders were used to contextualize results. The sections on technologies in the survey questionnaire and in the interviews' topic guide were designed in parallel.

3.1 The Online Survey

The survey was distributed via the eTwinning platform[2] and advertised through various European and national channels, in collaboration with European Schoolnet. It was made available online in the 23 official languages of the EU27 over a period of a month, from 15 September - 15 October 2009.[3] It included closed-ended questions and the majority of its items were five-point Likert-type scales. Out of 94 items, 40 were related to technologies, in particular focusing on the use of technologies in class and on opinions of the importance of certain technologies for education. The survey gathered 7,659 valid responses from compulsory schooling teachers from 27 European Union Member States (EU27). The majority of survey participants (70%) were aged between 36 and 55 and had teaching experience of more than 10 years (71%). Female respondents accounted for 77% of the total sample, reflecting the gender imbalance of the teaching population in Europe.

3.2 The Interviews

Interviews with 80 educational stakeholders were conducted by Futurelab [12] on behalf of IPTS, between November 2009 and April 2010. For each EU27 country, an average of three experts were identified among teacher trainers, school inspectors,

---

[2] www.etwinning.net

[3] The reader is referred to the survey report [11] for more information on the design and dissemination.

policy-makers and academics in the field of education, and a balance of profiles was provided for each country. The topic guide was developed by IPTS and refined by Futurelab. Particular attention was paid to the quality and use of ICT in schools, specifically focussing on the kind of access, availability of technologies and pedagogical practices behind the use of digital tools and media.

3.3 Limitations

This was an exploratory study, which aimed to take a snapshot of main uses of technologies across Europe. The following methodological limitations must therefore be acknowledged:

- Although evidence was collected from all Member States (EU27), the data cannot be considered as being representative for European teachers and educational stakeholders.
- The number and type of stakeholders consulted (both teachers and educational experts) are not representative, as neither of these groups was sampled.
- There was a strong southern-European bias in the survey, with 46% of respondents teaching in Italy, Greece and Spain.
- The fact that the survey was conducted online could have attracted respondents who are more prone to using ICT.
- The survey was answered on a voluntary basis.
- The level of certainty of the information collected from the interviews varies from personal opinions to direct and documented experience.

4 Present Learning Paradigm

Irrespective of the ubiquitous presence of technologies in western societies and their widespread usage especially amongst young people, current educational practices are not necessarily technology-enhanced. The rapid appropriation of technology by young people contrasts with the slow adoption rate in many schools, which, as Prensky [35] argues, leads students to become 'enraged' rather than 'engaged'. In this section, we will explore what is actually taking place in the classroom in terms of technological uptake and use, according to the teachers and educational stakeholders who took part in our study.

4.1 Teachers' Professional ICT Use

A high proportion of teachers in our survey acknowledge the importance of technology in their jobs, which is reflected in the fact that 85% of them agreed or strongly agreed that technology had improved their teaching.[4] When asked about how they used technologies for teaching, respondents claimed to use ICT mostly to update

[4] It should be noted that as this was an online survey, respondents are likely to be more proficient with technologies than the average teacher population.

their own knowledge for use in their lessons (90%), to prepare handouts and material (89%) and to search for teaching material (87%),[5] as shown in Graph 1.[6]

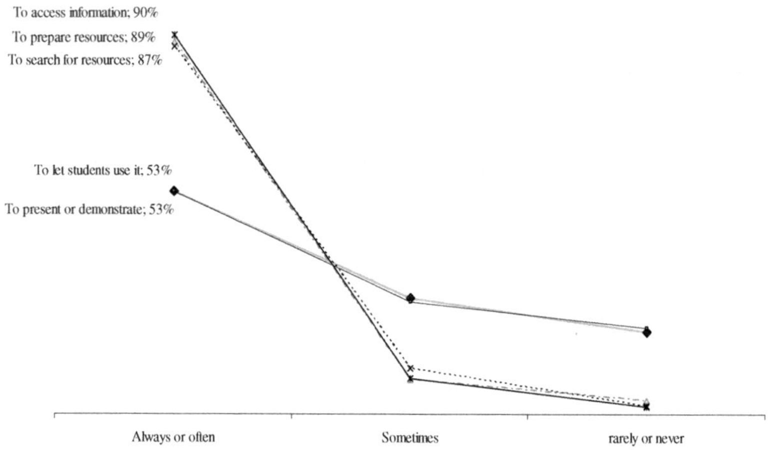

Graph 1. Professional use of technology by teachers. Source: JRC-IPTS 2009

The Internet has become an important professional tool for teachers, who rely on it as a major resource. Remarkably, only a very low percentage of respondents to our survey rarely or never use technology to prepare their lessons: i.e. between 2 and 3% of the total sample for the three items. This means that the internet has become a professional resource used by almost all the teachers surveyed, although mainly outside the classroom setting. A second observation relates to the use of technology in the classroom, exemplified in the item "I use a computer in class to present or demonstrate". Here, we see that 53% of teachers claim to do this always or often. This means that current professional teacher use of technologies seems to be more widespread outside than inside the classroom. As mentioned above, this is also the case for the students themselves. A third observation is that the use of technologies in the classroom to present or demonstrate appears to fit within a more traditional, frontal, teacher-led pedagogy. However, the item "I let my students use a wide range of technologies to learn (videos, mobiles, cameras, educational software, etc.)" also accounts for 53% of the teachers surveyed, indicating a different, and more student-led approach where students are encouraged to use technologies for learning. These observations are not contradictory, as the latter item does not allow us to differentiate between inside and outside the classroom. On the other hand, it is probable that those

[5] Percentages refer to the percentage of respondents who answered always or often to the question 'How do you use Information Communication Technology (ICT) for teaching and learning?'

[6] In the graph, the item label has been shortened.

teachers who use technologies for teaching also encourage their students to use them in the classroom or at home, for learning purposes.

The interviewees assert that technologies in class are often employed to reproduce old pedagogies and report cases where Interactive Whiteboards and projectors are used to deliver information in a frontal manner. According to the experts, lack of training in how to use technologies innovatively turns them into “expensive chalkboards and textbooks”. The interviewees also claimed that traditional teaching methods remain the norm in many European countries.

We also surveyed teachers' use of resources, both technological and non-technological, in class. Textbooks are the most common and widespread resource in compulsory schooling across the EU27. A high majority of survey respondents (85%) claimed they use textbooks always or often during lessons, as depicted in Graph 2, of whom 51% stated they always use textbooks. This finding was supported by the experts who thought that teachers relied on textbooks 'more than is necessary'. The next most used resources are those produced by the teachers themselves (81%). Internet is also becoming an increasingly important source of professional resources for teaching: 72% of respondents said they always or often use resources they download from the internet. As partially shown in Graph 2, there is a certain degree of acquiescence bias in responses, as teachers may have tended to inflate the frequency of use of some resources. This positive tendency is reversed in the case of digital games: only 17% of the sample claimed they often or always use digital games in class, and more than half the respondents (58%) rarely or never use them. Almost a third of teachers (32%) never use digital games in class.

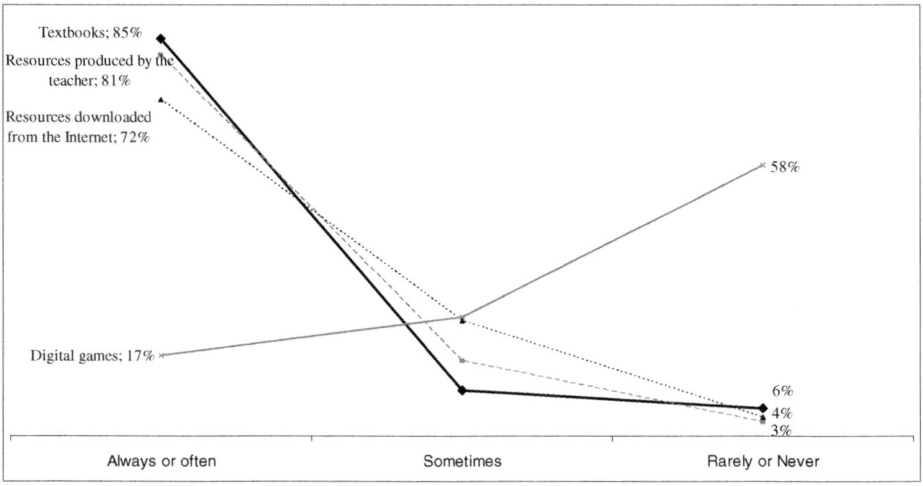

Graph 2. Resources used by teachers Source: JRC-IPTS 2009 'What resources are you using in your lesson?' Selection from a list of 15 resources (answers were originally on a 5 point Likert-type scale, from Always to Never)

4.2 Opinions on Importance of Technologies for Learning

Survey participants were asked to provide their opinions on the importance of specific technologies for learning. They ranked computers (98%),[7] educational software (93%), online collaborative tools (such as Wikipedia)[8] and videos (both 89%) as the top technologies. It can be noted, with some exceptions, that the more interactive, collaborative and user-led technologies were least regarded as important for learning, as shown in Table 1. Social media, such as SNS, podcasts, bookmarking and tagging sites, digital games and RSS feeds for learning were seen as less relevant than conventional technology by the teachers in this survey. The high proportion of teachers who neither agreed nor disagreed on the relevance of this group of technologies also suggests that many teachers were either not familiar with this specific tool or not able to see the link between the tool and the learning opportunities it could bring.

Table 1. Importance of technologies for learning

Technology	*Strongly Agree*	*Agree*	*Neither Agree nor Disagree*	*Disagree*	*Strongly Disagree*
Computers	66,9%	30,8%	1,9%	0,3%	0,1%
Educational software	51,9%	41,5%	5,7%	0,6%	0,3%
Videos	44,1%	44,8%	9,1%	1,4%	0,6%
Online collaborative tools[9]	41,4%	47,7%	9,2%	1,3%	0,3%
Virtual learning environments	43,3%	41,3%	13,5%	1,3%	0,6%
Interactive whiteboards	43,8%	39,9%	13,8%	1,7%	0,9%
Online free material	37,2%	49,1%	11,8%	1,5%	0,4%
Online courses	33,0%	47,1%	17,2%	2,1%	0,5%
Music/photo/video/slide sharing sites	19,5%	42,1%	30,7%	6,1%	1,7%
Blogs	19,3%	36,0%	33,7%	8,7%	2,4%
Social networking sites	13,9%	33,8%	38,4%	11,2%	2,7%
Podcasts	12,1%	27,8%	54,2%	4,0%	2,0%
Bookmarking and tagging sites	10,6%	30,1%	49,2%	7,9%	2,2%
Digital games	12,3%	34,8%	36,5%	12,0%	4,3%
RSS feeds	9,3%	25,9%	59,3%	3,6%	1,9%
Mobile phones	2,9%	11,0%	32,2%	37,7%	16,1%

Digital games and SNS, two technologies much preferred and used by students, are highly underestimated by teachers in terms of their relevance for learning. In contrast, research shows that social computing tools give a sense of ownership and responsibility with regards to learning [23] and that students motivation and skills greatly increase with the use of digital games in class [36, 37]. In a survey carried out by Pew Internet, 52% of teenagers claimed to have thought about moral issues during social online gaming [38]. Digital games, as shown in Graph 1, are seldom used

[7] Percentages refer to the percentage of teachers who answered "strongly agree" or "agree" to the question "Do you consider these technologies to be important for learning?"

[8] The use of online collaborative tools such as Wikipedia for learning has probably been understood by respondents as a passive one.

[9] For the online collaborative tool item Wikipedia was given as an example.

(17%) and their importance for learning is not fully recognised. However, the fact that over one third of the respondents (36.5%) neither agreed nor disagreed about the importance of digital games for learning suggests that awareness raising and training could help to shape the opinions of this group of undecided teachers.

Another remarkable observation is that mobile phones are perceived by our survey respondents as the antinomy of learning. More than a third (37.7%) disagreed that mobile phones were important for learning. 16.1% strongly disagreed that mobile phones were relevant for learning. This is by far the highest percentage of strong disagreement regarding technological usefulness for learning. As one interviewee stated: "It is shocking and noteworthy that portable interactive and digital devices are absent in terms of learning in classrooms. Mobile phones, for instance, which could be used so creatively in so many ways, are banned in many schools and generally switched off. There are obvious reasons why this might be necessary some of the time, but the total rejection of this technology is not in line with a spirit of innovation." These findings confirm Prensky's statement that educators see mobiles as a 'huge distraction' from education [39]. Although research acknowledges that mobiles can be a distraction, it also shows that they are endowed with great potential for learning and teaching [40]. Their wide range of attributes add new dimensions to educational processes [41]. The freedom of mobility also provides opportunities for learning outside the classroom [42] or bringing the real world to the classroom [43]. Moreover, mobile phones and hand-held devices are so central to young people's lives that they are perceived as brain extenders: "when you lose your mobile, you lose part of your brain" says a Japanese student, quoted by Prensky [39]. According to a study on mobile phone use by 13-19 year olds, 66% of the teens surveyed claim to want mobile phones which present opportunities to be educated anywhere in the world [44]. Engaging in teaching and learning with mobiles could be a way for teachers to start speaking the same language as their students. The contrast between the centrality of mobiles in young people's lives and teachers' reluctance to embrace mobiles as a learning opportunity and tool deserves further exploration and research.

5 Changing the School System

As can be noted from the above paragraphs, the possibilities offered by technologies for learning are currently not fully endorsed by teachers. Schools have not yet become innovative organisations, which are flexible and empowering, welcome ideas, tolerate risk, celebrate success, foster synergy and encourage fun [45]. In this section, we argue that major education policies have focused on access to technology, as a way of addressing the potential of technology to sustain innovative education. However, data from our study shows that unless such technology provision is understood in a wider sense and combined with support for teachers, the potential of technology for education will not be fully exploited.

5.1 Access to Technologies

In the future, learners will be able to access and interact with information anywhere, any time. This vision is not far from reality today, with the widespread use of technologies, such as smart phones and the increasing number of wifi zones in public areas. In this respect, European efforts to provide access to technologies in schools has been a significant first step towards the achievement of the information society: as the 2006 Benchmarking study reports [46], computer availability in European schools almost reached 100% saturation point. However, the European average of 11 computers per 100 pupils [46] shows that there is a long way to go to reach 1:1 computing. Netbook initiatives are on the rise in European schools, but these are still at the small-scale pilot level [47]. Technology is available in schools but, as an interviewee pointed out, it is present "in the most frustrating ways".

While access to ICT is an important focus for policies, ensuring that the ICT provided is of good quality and continuously maintained is equally important. Despite wide access across Europe, our survey reveals that only a quarter of the teachers agreed that the quality of the ICT in their schools was excellent, while 57% disagreed. The experts interviewed also highlighted pitfalls in the technological provisions of schools, lamenting that many schools still have to deal with slow internet connections or lack of computers that can handle fast connections. As the Benchmarking 2006 report shows, while internet is accessible from almost all European schools, significant differences can be noted in the type of connection [46]. Moreover, computers are not always distributed evenly in classes: experts claim that, in some countries, computers are concentrated in computer rooms and, when present in class, there is a lack of space or design to allow the efficient use of computers and internet by students. In many countries, computer rooms are either used only for certain subjects (e.g. informatics and ICT) or not used at all most of the time.

In addition, access to technologies should be understood in a wider sense. While computer availability and internet connections are fundamental tools, the possibilities offered by mobile and wireless technologies should not be disregarded. Access to a variety of media and ICT, and in particular to mobile or hand-held devices, enables a shift towards ubiquitous learning. Additionally, one-to-one computing should be mainstreamed and not limited to specific initiatives as it is now. Unless the ratio of learners to up-to-date and good quality technologies improves, educational institutions will not be able to adopt hands-on and learner-centred pedagogies.

5.2 Training and Support

Bottino [48] argues that schools have 'crammed' new technologies into their existing structures, as opposed to allowing the new technology to foster a new model. The provision of ICT infrastructure and training does not necessarily result in its effective pedagogical use. Interviewees confirm this claim, saying that in many countries government schemes have invested great amounts of funding in new technological hardware, with little funding for software, upgrading and maintenance and teacher training. As a result, in some cases resources have become outdated or have been left unused because teachers do not know how to embed them in their practices. Although significant efforts are being made in many countries in Europe, sustained teacher training strategy in the area of digital tools and technologies for teaching is still lacking: only 42% of our survey respondents claim to have received training on how to use ICT in class. The survey results show that there is a positive relation between

ICT training and teachers' opinions on the relevance of new technologies for learning. Teachers require different levels of training varying from enhancing personal digital competences, to pedagogic use of ICT.

But the support that teachers need goes beyond training. Three quarters of our respondents state they need more institutional support, and 36% are strongly convinced of this. Teachers are often isolated practitioners and can feel disconnected from the rest of the educational world [49]. The interviewees call for a shift in the educational culture and mindset of educational actors in order to achieve more innovative teaching practices. They claim that head-teachers should become, even more than now, 'pedagogic leaders', who are able, willing and ready to inspire educational change. However, school culture is not decided only at local level, but is also promoted through policy messages at regional and national level. Therefore, educational authorities should develop a holistic strategy for revising school education [14], promoting a spirit of innovation at all levels. A vision of learning through technologies cannot be limited to the provision of access and should be fostered through changes in the curriculum, assessment, pedagogies, school culture, and teacher training.

6 Conclusions

This paper explored the current gap between the innovative potential of technologies and the use of technologies by teachers in schools in Europe. It argued that significant progress has been made in implementing ICT in schools in Europe and that almost all the teachers surveyed are making professional use of ICT. However, while the teachers are positive about the potential of ICT for learning, they are mainly using the internet as an online resource for the preparation of their lessons and to download material to be used in their lessons. The use of ICT in the classroom is less widespread. In addition, conventional technologies seem to be preferred. Few teachers surveyed recognized the benefits of more recent interactive, collaborative and user-led technologies for learning. Digital games and especially mobile phones are underestimated as important tools for learning within the classroom environment.

These findings contrast with the high adoption of technologies by young people, who are embracing social media applications, digital games and smart phones in their everyday lives. The spread and use of recent digital technologies and social media bring with them the development of new skills and competences which are important for employability, competitiveness and participation in the labour market. Unless schools embrace and foster these new digital competences, not all learners will be able to be functional in our highly digitalised societies. The teachers who took part in our survey are positive about the contribution of technology to the quality of their teaching, but do not, however, fully endorse technology-enhanced teaching practice. Some barriers may be seen in the limited training they have received, in the lack of technical and institutional support and in the quality of current ICT provision in their schools.

Foresight studies on education continuously predict significant changes in the education sector. However, data from this study shows that there is still a gap between where we foresee changes and what is actually happening within the classroom. In order to close this gap, we argue that disruptive change is needed to realise the innovative potential of technologies to transform educational practices. School reforms cannot address one segment of education in the hopes of influencing the whole educational establishment. While access has been an important policy priority, teachers need technical and pedagogical support to experiment with ICT. School leadership and an open educational culture that embraces technological and pedagogical innovation are equally important. At the same time, reforms in the curriculum and in assessment practices should promote the use of technologies and the development of learners' digital competences. The challenge for reaping the benefits of technologies in obligatory schooling is to make sure these changes go hand in hand.

Acknowledgments. The authors would like to thank the following people who contributed to the study in different ways: Lieve Van Den Brande and Maruja Gutierrez-Diaz (DG Education and Culture); Patricia Wastiau and Caroline Kearney (EUN) for their collaboration during the questionnaire design and data collection of the survey; Isidro Maya Jariego and María José Jurado (University of Seville) for their collaboration during the analysis of the survey data; Shakuntala Banaji (IOE), Sue Cranmer, and Carlo Perrotta (Futurelab) for carrying out the interviews. The authors are also grateful to their colleagues at IPTS, in particular Kirsti Ala-Mutka, Christine Redecker and Patricia Farrer for providing feedback to this paper. We would also like to thank all the teachers and experts who took part in the survey and interviews.

References

1. Punie, Y., et al.: The Impact of Social Computing on the EU Information Society and Economy. In: European Commission-Joint Research Centre-Institute for Prospective Technological Studies. JRC-IPTS, Seville (2009)
2. Cachia, R.: Social Computing: Study on the Use and Impact of Online Social Networking. EC JRC IPTS, Sevilla (2008)
3. Herold, D.K.: Digital na(t)ives - Discourses of exclusion in an inclusive society. In: The good, the bad and the challenge. Cost, Copehangen, Denmark (2009)
4. Prensky, M.: Digital Natives, Digital Immigrants. On the Horizon, 9(5) (2001)
5. Silverstone, R.: Domesticating domestication. Reflections on the life of a concept. In: Berker, T., et al. (eds.) Domestication of media and technology, pp. 229–248. Open University Press, Maidenhead (2006)
6. Pedró, F.: The new Millennium Learners: Challenging our Views on ICT and Learning. OECD-CERI (2006)
7. Tuomi, I.: Learning in the age of networked intelligence. European Journal of Education 42(2), 235–254 (2007)
8. Selinger, M., et al.: The Future of School. Cisco Internet Business Solutions Group(IBSG) (2008)
9. Ferrari, A., Cachia, R., Punie, Y.: Innovation and Creativity in Education and Training in the EU Member States: Fostering Creative Learning and Supporting Innovative Teaching. Literature review on Innovation and Creativity in E&T in the EU Member States (ICEAC). JRC-IPTS, Seville (2009)

10. Heilmann, G., Korte, W.B.: The Role of Creativity and Innovation in School Curricula in the EU27: A content analysis of curricula documents. JRC-IPTS, Seville (2010)
11. Cachia, R., Ferrari, A.: Creativity in Schools: A Survey of Teachers in Europe. JRC-IPTS, Seville (2010)
12. Banaji, S., Cranmer, S., Perrotta, C.: Expert Perspectives on Creativity and Innovation in European Schools: Enabling factors and Barriers to Creativity and Innovation in Compulsory Education in Europe Based on Interviews with Educational Stakeholders. JRC-IPTS, Seville (2010)
13. Banaji, S., Cranmer, S., Perrotta, C.: Creative and Innovative Good Practices in Compulsory Education in Europe: Collection and Descriptive Analysis of Ten Good Practices of Creativity and Innovation in Compulsory Education in the EU27. JRC-IPTS, Seville (2010)
14. Cachia, R., et al.: Creative Learning and Innovative Teaching: Final Report on the Study on Creativity and Innovation in Education in EU Member States. JRC-IPTS, Seville (2010)
15. Punie, Y.: Learning spaces: An ICT-enabled model of future learning in the Knowledge-based Society. European Journal of Education 42(2), 185–199 (2007)
16. Facer, K.: Beyond Current Horizons. Futurelab (2009)
17. Johnson, L., et al.: The 2011 Horizon Report. The New Media Consortium, Austin, Texas (2005)
18. Blandow, D., Dyrenfurth, M.J.: Technology education in school and industry: emerging didactics for human resource development. Springer-Verlag in cooperation with NATO Scientific Affairs Division, Berlin (1994)
19. Ruiz i Tarrago, F.: Integration of information technology into secondary education: Main issues and perspectives. In: IFIP Working Group 3.1 (1993)
20. Ala-Mutka, K., Punie, Y., Redecker, C.: ICT for Learning, Innovation and Creativity. JRC-IPTS, Seville (2008)
21. Punie, Y., Zinnbauer, D., Cabrera, M.: A review of the Impact of ICT on Learning. JRC-IPTS, Seville (2008)
22. Collins, A., Halverson, R.: The second educational revolution: rethinking education in the age of technology. Journal of computer assisted learning 26(1), 18–27 (2010)
23. Ala-Mutka, K., Punie, Y., Ferrari, A.: Review of Learning in Online Networks and Communities. Learning in the Synergy of Multiple Disciplines, 350–364 (2009)
24. Ala-Mutka, K.: Learning in Informal Online Networks and Communities. JRC-IPTS, Seville (2010)
25. Sharples, M., Taylor, J., Vavoula, G.: Towards a theory of mobile learning. In: Proceedings of mLearn 2005, pp. 1–9 (2005)
26. Oblinger, D.G., Oblinger, J.L.: Educating the net generation (2005)
27. Simplicio, J.S.C.: Teaching classroom educators how to be more effective and creative teachers. Education 120(4), 675–680 (2000)
28. Christensen, C., Johnson, C.W., Horn, M.B.: Disrupting Class: How Disruptive Innovation Will Change the Way the World Learns. Mc Graw Hill, New York (2008)
29. CEDEPOF, Future skill supply in Europe. Office for Official Publications of the European Communities, Luxembourg (2009)
30. Redecker, C., et al.: The Future of Learning: New Ways to Learn New Skills for Future Jobs. Results from an online expert consultation. JRC-IPTS, Seville (2010)

31. Codd, J.A., et al.: Review of Future-focused Research on Teaching and Learning. Ministry of Education, Research Division (2002)
32. Linton, P., Schuchhard, P.: The Digital World in 2025. Indicators for European Action. European Internet Foundation (2009)
33. Davidson, C.N., Goldberg, D.T.: The future of learning institutions in a digital age. The MIT Press, Cambridge (2009)
34. Learnovation, Foresight Report (2008)
35. Prensky, M.: Engage Me or Enrage Me: What Today's Learners Demand. Educause Review 40, 60–65 (2005)
36. Sandford, R., et al.: Teaching with games: Using commercial off-the-shelf computer games in formal education. Futurelab (2006)
37. Wastiau, P., Kearney, C., Vanderberghe, W.: How are digital games used in schools? European Schoolnet, Brussels (2009)
38. Lenhart, A., et al.: Teens, video games and civics. Pew Internet and American Life Project (2008)
39. Prensky, M.: What can you learn from a cell phone? - Almost anything! Innovate (2004)
40. McFarlane, A., Triggs, P., Yee, W.: Researching mobile learning-Interim report to Becta Period, April-December 2007 (2008)
41. Kukulska-Hulme, A., Traxler, J.: Mobile learning: a handbook for educators and trainers. Routledge, London (2005)
42. Naismith, L., et al.: Literature review in mobile technologies and learning. Futurelab report No. 11, University of Birmingham (2004)
43. Ekanayake, S., Wishart, J.M.: Identifying the potential of mobile phone cameras in science teaching and learning: A case study undertaken in Sri Lanka. In: Mobile learning 2010. International Association for Development of the Information Society, Porto, Portugal (2010)
44. Harris Interactive. A generation unplugged - Research Report (2008)
45. Craft, A.: Creativity in schools: tensions and dilemmas. Routledge, London (2005)
46. empirica, Benchmarking Access and Use of ICT in European Schools 2006: Results from Head Teacher and A Classroom Teacher Surveys in 27 European Countries. European Commission, Brussels (2006)
47. Balanskat, A., Garoia, V.: Netbooks on the rise. European Schoolnet (2010)
48. Bottino, R.M.: ICT, national policies, and impact on schools and teachers' development. In: ICT and the Teacher of the Future, Melbourne, Australia (2003)
49. Baker, P.: Creating Learning Communities: the Unifinished Agenda.In: Pescosolido, B.A., Aminzade, R. (eds.) The social worlds of higher education: Handbook for teaching in a new century. Pine Forge Pr. (1999)

Activity-Based Learner-Models for Learner Monitoring and Recommendations in Moodle

Beatriz Florian[1,3], Christian Glahn[2], Hendrik Drachsler[2], Marcus Specht[2], and Ramón Fabregat Gesa[1]

[1] Institute of Informatics and Applications (IIiA), University of Girona, Girona, Spain
bflorian@eia.udg.edu, ramon.fabregat@udg.edu
[2] CELSTEC, Open University of the Netherlands, Heerlen, The Netherlands
{christian.glahn,hendrik.drachsler,marcus.specht}@ou.nl
[3] EISC, Universidad del Valle, Cali, Colombia
beatriz.florian@correounivalle.edu.co

Abstract. In technology-enhanced learning, activity-based learner models can provide evidence for competence assessment. Such models are the foundation for learning and teaching support, such as: adaptation, assessment, and competence analytics, recommendations, and so on. This paper analyses how to construct activity-based learner models based on existing data in the Moodle learning management system. Based on the activity theory model and the actuator-indicator model, aggregators of learner activities for different activity types were implemented in Moodle. This requires the consideration of the social roles in a course, in order to enable adaptive views for learners and instructors on the stored activity information. The implementation showed that Moodle stores information about course activities that requires filtering before it can get used for higher level processing. The social planes in Moodle reveal a higher complexity than it has been previously described by theories of classroom orchestration, such as actors who are no longer present in a course.

Keywords: Activity-based learner models, Moodle, Learners tracking, Learning analytics, Competence assessment, Indicators, TEL recommender systems.

1 Introduction

In Virtual Learning Environments (VLE), rich user models based on activity traces are required for different types of personalized learning support such as: analytics of competences' development, activity-based smart indicators and recommendations. The general process could be described as the record of data interactions and outcomes of activities, the semantic interpretation of collected data and their analysis to produce appropriate support responses.

The Learning Management System (LMS) Moodle records a broad range of individual learners' traces in real time. For adaptive systems these interaction footprints had been used to produce new learning paths [1]. However, this data is not easily accessible to Moodle users and the analytics based on this data are hidden to

C. Delgado Kloos et al. (Eds.): EC-TEL 2011, LNCS 6964, pp. 111–124, 2011.

students and poorly provided to teachers. This hinders practitioners to apply real time educational data in their practice. Although the increasing amount of learning analytics (LA) related papers published nowadays [2], research contributions about how automatically collected activity traces could be effectively used for supporting learning process are rare.

The research presented in this technical-design paper is the foundation for future work on applying the concepts of learning analytics for competence assessment and recommendations. This contribution focuses on the concept of social planes and analyses social perspectives for accessing Moodle's tracking data. The paper analyses the reuse of Moodle's tracking data for learner and group modelling. Moodle activity log is used to build rich learner models based on learner activities within a social context. Therefore, this paper addresses social facets of Moodle's log data and their implications for learning support. We have implemented an architecture that based on the Activity Theory model [4] and the actuator-indicator model [3] to have a flexible and extendable interface to Moodle's tracking data for different roles in learning processes. The resulting framework transforms the collected data into learning analytic information for the *social planes* "self", "peers" and "class". Furthermore, the implementation considered different *social perspectives* on the data. At this point these perspectives are coupled to the course roles "student" and "teacher". The concept of social perspectives on tracking data is useful to integrate aspects of privacy and data-protection while modelling learning analytics functions.

The social contexts of the initial implementation were grounded on social planes that have been identified by prior research on instructional design and on collaborative learning. First tests revealed that the given conceptualisations of social planes in education did not fully describe the tracking data of the Moodle logs.

This contribution has the following structure. Section 2 outlines the state-of-the-art about technology-supported competence-assessment. Section 3 presents the research objective. Section 4 analyses an implementation of activity-based learner models and learning analytics in Moodle. The implications of the prototypical implementation for technology-supported assessment are discussed in section 5. Section 6 presents the findings. Perspectives towards recommendations support are analysed in section 7. Finally, this paper concludes with an outlook on future research in section 8.

2 Background

Most VLEs already provide functions that can be used for supporting activity-centred learning, but the related information is commonly unavailable in a structured form. Semantically structured learner models are required in order to provide technological support for more activity-centred assessment types. An *activity-based learner model* creates a semantic structure of dynamically generated learner properties that reflect observed actions of a learner. Activity-based learner models are a prerequisite for activity-centred assessment and process support for competence development.

Contemporary competence models such as PALO [5] and EQF [6] describe proficiency levels of competences according to types of activities that learners are capable to perform. Previous research proposed [7] and implemented [8] the Adaptive Evaluation Engine Architecture (AEEA) for competence assessment. This architecture emphasizes the process factors for assessing competence developments over content-centred factors of conventional outcome-based assessment approaches.

Cheetham and Chivers [9] define a *competence* as knowledge- or theory-guided practice. This implies that a competence can be recognized only if it is demonstrated, reflected and used for guiding practice. In contrast to a *skill* that focuses on instrumental actions such as handling a specific tool, a competence requires more profound conceptual understanding of the underpinnings of the related practices as well as experiences in applying this understanding. Furthermore, a *competence* differs from a *competency* in so far that the former refers to the ability of linking knowledge with practices whereas the latter refers to knowledge about practice [9].

The *assessment of competence development* relies on evidence that learners are able to perform actions that are related to a competence. This perspective emphasizes the relevance of the process for its results. Previous research suggested *outcome-centered* testing as formative assessment of competence developments [10]. However, these approaches appear to be limited, because of the active nature of competence development.

The present study is grounded on two models: the Engeström's Activity Theory and the Actuator-Indicator model as pillars to implement an activity-based learner model in Moodle.

2.1 Activity Theory

Engeström's Activity Theory has its origins in modelling and analysing business processes [4]. The core underpinning of the activity theory is that activity cannot be limited to "means of getting to results" but needs to be analysed at the level of actions. The Activity Theory provides a system model to describe actions and their contextual constraints. This model has six components: A subject, an object, instruments, rules, social planes (community), and co-operative processes (division of labour). The interplay of these components leads to an outcome of an activity. The activity system can be separated into an action part and a context part. The relations in the action part describe the observable interplay of the elements in an activity. The subject, the object, and the instruments are part of the action part. The relations in the context part describe supporting and constraining factors for an activity. This part contains the rules, the social planes, and the co-operative processes.

The Activity Theory model describes the structural relations between the components of a single activity. Each element of this model may relate to individual activities that can be described with the model recursively. Additionally, the activity's outcome can trigger new activities. This allows the systematic description of complex processes. This model has been used to analyse the effectiveness and the efficiency of business processes for identifying potential improvements of work settings.

The Activity Theory has received some attention by TEL research, most notably in the context of Computer Supported Collaborative Learning (CSCL) [11]. The concepts of Activity Theory are attractive for educational-technology research because they share key aspects that have been identified by instructional design research [12]. The provided relationship model connects these aspects systematically. In educational settings the elements "teacher" and "learner" replace the "subject" and the "object" of the original model.

Dillenbourg [11], [13], [14] argues that social planes require consideration for orchestrating technology-enhanced learning. These planes are bound to the social connectedness of learners on the activity level and can include the individual, collaborative, collective (class wide) activities. Dillenbourg [13] identifies 5 generic social planes that structure or influence learning activities: the individual plane, the group plane, the class plane, the community plane and the world plane. The individual plane refers to solo activities. The group plane refers to activities in small groups that allow direct collaboration among all participants. The class plane includes activities that involve all participants from the same course. The community plane involves actors from other classes or courses on the same topic. Finally, the world plane refers to actions that involve unidentified actors, such as visitors of a public web-journal.

Glahn, Specht and Koper [15] have identified that activity information from other social planes influences the awareness and the self-regulation of learners. They identified that contrasting individual learning activities with the same information about activities on a different social plane enables learners to contextualize their own activities and stimulate the social awareness with regard to the activities undertaken on the other plane. This indicates that information of different social planes can support self-assessment activities in TEL.

The second aspect of the Activity Theory is that rules define and constrain an activity. This aspect focuses on the *contextualising factors* of an activity. In TEL rules on learning activities are commonly perceived as part of instructional design problems. This is mainly due to the fact that rules are an integral part of every instructional design [12], [16]. However, Verpoorten et al. [17] highlighted that rules in VLEs constraining learning activities can be located at several hierarchical control levels, namely, the system level, the organizational level, the teacher level, and the learner level. The hierarchy of these levels means that the rules at each level constrain the possible activities of the following levels. These levels also involve stakeholders such as system developers, technical administrators or organizational managers, who are typically ignored by TEL research.

While in Engeström's original model instruments are considered as passive mediators in an activity, the different types of rules directly affect these instruments in the activity system. In interactive information systems, actors often do not apply these rules directly. More commonly external rules constrain the possible use of an instrument, such as a VLE. These external rules can be inherent to an instructional design, hardwired into the logic of an information system, or configured as part of an organisational policy. These rules are included through the instruments that are used in an activity. In the same line of reasoning, these technology-related constraints can have a direct impact on collaboration and co-operation in learning processes. This technology-induced change suggests an extension of the original Activity Theory model that also considers the relations between procedural rules, instruments, and collaborative processes (Fig. 1).

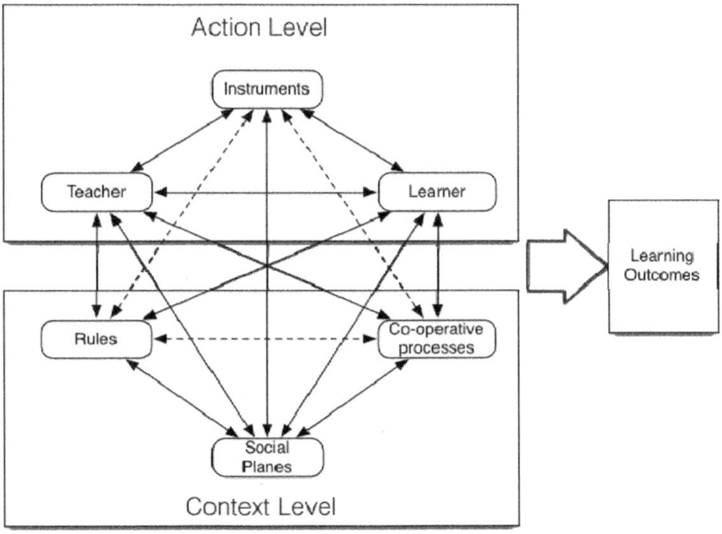

Fig. 1. Extended Activity Theory Model
(dashed lines refer to extended relations)

Outcome-based assessment focuses [10] on the results of an activity and tries to deduce the success of an activity by comparing expected and delivered learning outcomes. The activity itself remains a black box for such approaches. *Activity-based assessment* changes this perspective towards assessing the activities that lead to the outcomes. This includes the assessment of the appropriate applications of external rules, the interactions on and across social planes, and (if present) collaboration and co-operation among learners. All aspects of this kind of assessment contribute to the evidence that learners achieved the targeted competence levels.

From the perspective of the extended Activity Theory model the provisioning and exposure of analytical rules for accessing data in information systems remains a challenge for the effective application of learning analytics for supporting learners and teachers.

2.2 Actuator-Indicator Model

While the Activity Theory offers a well-structured model for analysing and conceptualizing learning and its assessment, it does not provide guidelines for implementing services for supporting learning or assessment. Zimmermann, Specht and Lorenz [3] have proposed a generic system architecture for adaptive and contextual systems. Further research [18], [19], [15] has extended this architecture with concepts of motivational research and applied it to different application areas of TEL [15], [20], [21].

The model proposed consists of four functional layers of an architecture. The core functional layers are sensor data management (sensor layer), context abstraction (semantic layer), the control of actuator output (control layer), and the indication of the output (indicator layer). Fig. 2 shows the information flow of the actuator-indicator model. The *sensor layer* is responsible for logging information about traces of learners' interactions and other contextual information.

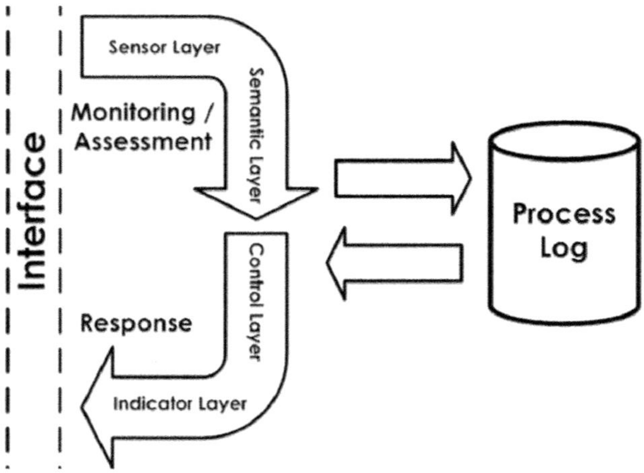

Fig. 2. Actuator-indicator model Zimmerman, Specht & Lorenz, 2005

The *semantic layer* collects the data from the log system and aggregates this data into higher-level information. An *aggregator* is a function that transforms sensor data from the log system. An *aggregator of activity* refers to how logs of a particular activity are semantically transformed. The aggregators respond differently depending on the context (social plane), in which they are called. The *control layer* is in charge of interpreting the response of aggregators through different *strategies*. A *strategy* determinates when and how to collect aggregator responses and how to present them to the user. In brief, the active strategy selects the representations and provides the aggregated information to them. To complement this definition, an *activity-based learner model* integrates the output of several aggregators. Finally, the *indicator layer* is in charge of transforming the returned data of the control layer into representations that are interpretable by humans.

For integrating learning analytics capabilities to complex legacy systems it is a challenge to identify existing functions and components along the information processing flow of this architecture.

3 Research Question

The main research question of this study addresses the need for structuring complex data resulting from activities in a learning environment. Moodle, like other VLEs, has only limited built-in support for learning analytics. The core components and extensions related to assessment focus on outcomes rather than the activities.

Therefore, the question is: how to introspect learning activities for competence assessment and recommendations?

Integrating the concepts of the actuator-indicator model with the Activity Theory approaches this question. This integration is an attempt of structuring learning analytics techniques for designing solutions for activity-based assessment and recommendations that can be used by teachers and instructional designers in TEL.

The core of this question is primarily related to the semantic layer of the actuator-indicator model. An aggregator in this layer can be defined in terms of the Activity Theory as a *rule* that enables *perspectives* on activities that are performed on one or many *social planes (s.)*. As such every aggregator can be verified regarding its meaning for a perspective on a social plane. Figure 3 highlights the previous concepts within the extended Activity Theory model.

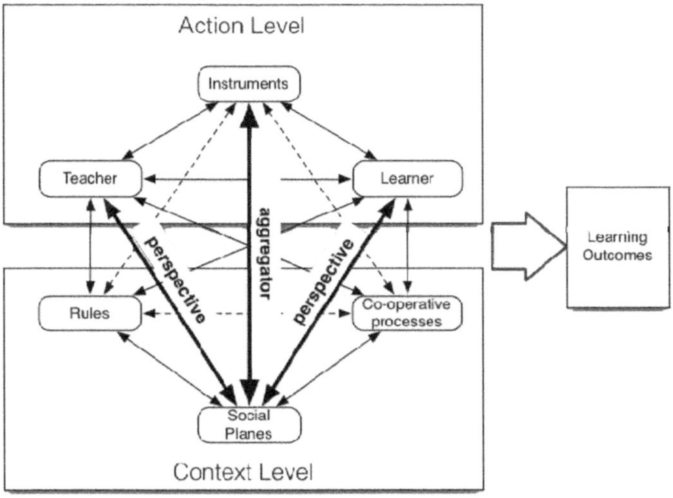

Fig. 3. Research scope in relation to the Extended Activity Theory Model

4 Architecture for Learning Analytics and Recommendations

The translation of Engeström's model [4] to TEL, described in section 2.1 along with the layered structure of [3], described in section 2.2, are permeated in this section to propose an architecture to support learning analytics and recommendations. Figure 4 shows the layout of the overall architecture. The architecture uses the context information present in Moodle and adds some other components. The architecture allows the construction of dynamic learner models based on perspectives over social planes in activities. The models need to be capable of reacting to actions during the learning process.

The proposed architecture builds on the layers proposed by [3]: Sensor layer, semantic layer, control layer and indicator layer. The activity-based learner model is related to the first two layers and the learning analytics solutions to the last two layers. In this section the components of the architecture are explained in relation to these two parts.

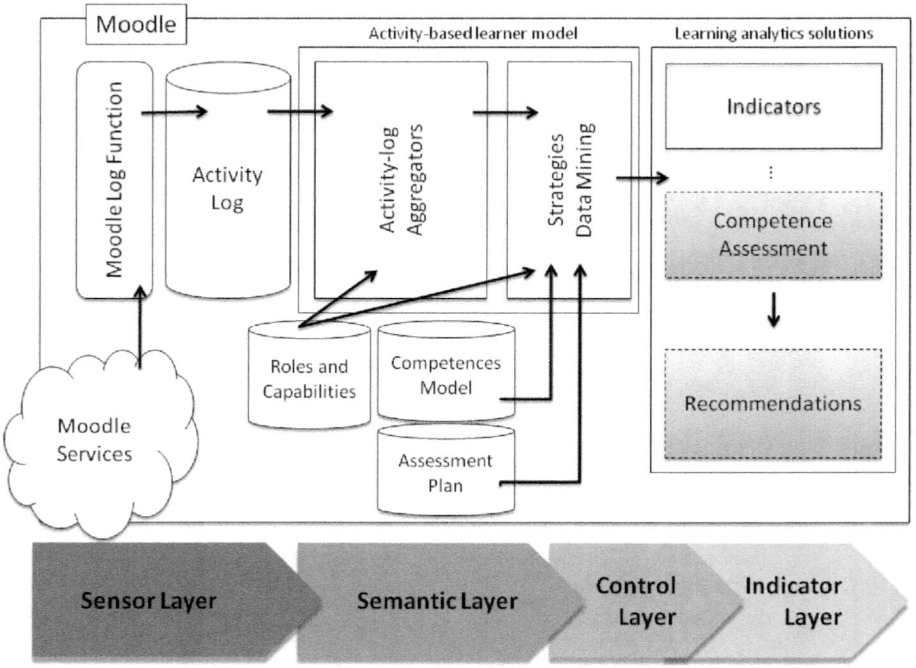

Fig. 4. Architecture to support activity-based learners models in Moodle

4.1 Activity-Based Learner Models for Moodle

The Sensor Layer. The purpose of this layer is to collect and to store traces of actions. Learners perform activities in Moodle. Moodle implements a detailed activity logging in its services. Consequently, it is not necessary to implement a separate sensor layer for tracking learner actions in Moodle, because the system already stores sufficient context information about the learners' interactions. Logs in Moodle are created by the Moodle Log Function and stored in mdl_log Moodle database, which stores all interactions and allows structured querying and filtering of this data. This data can be used for identifying complex activities by integrating the access time, the active user, and the performed action.

By default only system administrators and teachers have access to activity reports and basic statistics in Moodle. Some research about the use of Moodle Log Function were made previously by [17] and [20]. Verpoorten et al. in [17] delineated and documented a perspective on personalization based on the mirroring of personal tracked data to the user; [20] is a conceptual paper which analyses the underlying concepts for a system-architecture for device adaption for mobile learning, integrating Moodle into ubiquitous computing. In addition, tools for teachers and administrators feedback in Moodle are the report logs and the report statistics. The Moodle log reporting and statistics are drawn from the mdl_log database.

The Semantic Layer. This layer processes the data collected by the sensor layer into semantically meaningful information. At the level of the semantic layer several aggregators can be active to process the traces of learning activity. The following aspects constitute an aggregator.

- An Aggregation rule represents an SQL query that processes Moodle's user tracking database. Each aggregation rule returns the result data to the JSON format that can be easily interpreted by web-frontends. Each aggregation rule can get accessed through a distinct name that represents the analytic function of the rule.
- The context is used for filtering a social plane of the learners. The social planes implemented so far are: self, peer and class. The context "self" includes only the data of the learner, who requests the data from the Moodle system. The context "peers" includes the data of all other learners who are enrolled in the same courses as the learner excluding the data of the current user. The context "class" includes all learners who are enrolled and active in the course. The context is passed as a parameter to the aggregation rule.
- Role-based perspective on the data is automatically applied based on the current role of the requesting user. Students have only access to the contexts 'self' and – annonymised – to those of the 'peer' context. Teachers have access to all details of the aggregated information. When teachers make a request using the context 'self' or 'peer' an extra parameter is required for identifying the related student for whom this context will be applied.

The current system implements the semantic layer as a REST service through which the different aggregators have unique names and can be directly accessed through an URL. The implementation of the Strategy Design Pattern [22] is planned for future releases in order to access the aggregators through a facade. In summary, each aggregation rule can be limited to a perspective over a different social plane of the learner and to a specific course.

Other semantic information is stored in the Moodle database. For instance, in order to support competence development and competence assessment, the semantic layer requires a competence model and an assessment plan [7] [8]. Tables to express the competence model based on the European Qualification Framework [6] were created for the semantic layer. Similarly, tables to express the assessment plan were integrated. The competence model defines the ontology of competences, their levels of qualification and activities related to each level of competence. The assessment plan defines how actions in a course relate to the competence model and how they contribute to the evidence on the competence development of a learner. Other examples of semantic information stored in the Moodle database are the structure of roles and capabilities to classify the type of users and theirs permissions in the system. The capabilities in Moodle can be applied to many levels such as: activity, course, system and so on.

4.2 Learning Analytics Solutions

The *Control Layer*. This layer defines the arrangement of the aggregators and the visualizations that are used for mirroring. The control layer is implemented as a plug-in that provides several widgets that can be independently integrated into the user interface of a course. Each widget contains a set of aggregators and visualizations, which can be configured by the instructor of a course. Through a context parameter an instructor can define the scope of the data that is returned by the selected aggregator. In the case of recommendations an aggregator implements the data mining algorithms. In this layer the competence model and the assessment plan are data inputs to process the recommendation strategies and the indicators of competence analytics. The competence analytics will be implemented in further research. Section 5 analyses this aspect in depth. A recommender system is planned for further research. Using this architecture, these recommender systems will be base on learning analytics. Section 6 analyses this aspect.

The Indicator Layer. This layer provides different presentation modes for the data of the control layer. The indicator chooses the presentation mode based on the configuration of the indicator layer and receives the data from the control layer. So far the indicator layer shows smart indicators whose parameters are the context and the tracked activity. The indicator layer is embedded into the user interface of Moodle through a JavaScript.

5 Prototype Application

Prototypes of aggregators using the proposed architecture for Moodle were implemented. Each aggregator allows introspecting a particular activity type. The aggregators and related control rules are based on learning analytic functions that are accessible to the users through indicators [18]. An indicator offers different forms of "visualising" the data provided by one aggregator.

In the teacher's interface it is possible to configure and arrange indicators. This interface allows to select and to configure the aggregators that will be used with an indicator. These configurations are stored at the level of the control layer. A "yardstick" indicator also allows using two social planes for the same aggregator that enables students to compare their actions to those performed at the other social plane. The student interface reads the teacher configured indicators and displays the learning analytics information accordingly.

6 Findings

The initial version of aggregators implemented the data aggregation solitarily at the level of the log database. The core assumption at this level was that students in a course have an "enrolment" marker in their activities for the course, whereas teaching staff has no such marker. During tests of the aggregators on a fully deployed course the aggregators returned 22 students for the "class" plane, while Moodle's course administration reported only 15 students for the course. The first check revealed that no teaching staff was among the additional participants, which implies that the first assumption for filtering the activities in the log database was correct. A more focused analysis of the participants reported in the two interfaces together with the related course manager showed that the additionally reported students were former participants that were no longer enrolled in that course. However, for these participants no "unenrollment" marker could be found in the activity logs.

Based on these insights a second version of the aggregators have been implemented. In this version the social planes were based on the role assignment of a course. As expected this version returned data that was consistent with the data presented in the course administration.

On the first sight this test protocol appears like an ordinary bug related to a wrongly applied database query. From a learning analytics' viewpoint the results indicate that relying entirely on activity tracking in Moodle is not sufficient for providing accurate data that is related to the social planes that were identified by prior research. Instead of only active users the activity logs also reported information for former participants of the same course. These participants can be "drop-outs" that have failed course requirements at some intermediate point or "alumni" who have already completed and left the course. Both groups cannot be discriminated from the information provided by Moodle.

The analysis of the returned data indicates that the initial assumption regarding the "enrollment" marker was indeed correct because no participants with other (former) roles were returned. One important insight from this study is that prior research concerning social planes considered courses as stable social structures. This view is supported by the course administration of Moodle. However, the unexpected results of the initial tests highlight that the social planes that are found in Moodle systems are of greater complexity of social planes than it has been previously described by theories of classroom orchestration. Furthermore, this historical data can offer new perspectives towards learning analytics and recommendations. This opportunity is discussed as part of the following section.

7 Perspective towards Activity-Based Learning Analytics

This section outlines some possible extensions of the activity-based learner model for Learning Analytics solutions. We cluster this section along two questions (a) What kind of formative feedback could be delivered from activity-based learner model for running courses? And (b) How historical data of former students can help to produce new useful learning analytic solutions? For both questions we refer once again to the proposed architecture presented in Fig. 3.

Regarding question one, teachers and students can benefit from a combination of different indicators to help gauge whether or not a certain competence was successfully attained. Different kinds of comparative indicators about assessment results, expected knowledge levels versus achieved knowledge levels, and the develop of competences can be combined in order to draw a more complete picture of the learning process and assess the quality of the achieved competences. For example, a teacher could combine the outcome of a group project – e.g., a joined report on a particular topic – and also take into account the communication activities of the students from the group plane (e.g., forum discussions, shared files, amount of comments and annotations) that are reported and visualized in the group indicator. With this additional information the teachers not only receive the final result of the students, they can also value the group collaboration and the contribution of every single student to the joined report. This example only requires taking into account the current data created by the students of a running course.

The second question follows the idea to extend the data of a current course with historical data of students from former courses. Historical data in combination with learning analytics can be supportive to gain new insights into the running courses that would otherwise remain concealed. For instance, a drop-out detection system could recommend the teacher students that need special attention because they are in danger to withdraw from the course. It could be based on a classifier technique like decision trees, Bayesian classifiers or support vector machines [23] that are trained to learn drop-out patterns of students from former Moodle courses. The trained classifier can then be applied to running courses to identify students that show similar drop-out patterns. It could mark those students that show drop-out patterns in a list and give the tutor the opportunity to contact the students personally and ask them if they need any additional support for their studies. Alternatively, such a recommender system could offer motivating information and encouraging activities to help a student to break the drop-out pattern. Long term, such a system could potentially decrease the amount of drop-outs, improve the customer service of the University by needs driven support, and increase the amount of graduated students.

8 Conclusions and Future Research

This paper delineated and documented a perspective in activity-based learner models as semantically models for activity-centred assessment and recommendations. This approach advises the use of tracked data with social filters. The mechanics to aggregate information and its rules to deliver the response to the user interface are the core of activity-based user models. The paper contributed with a prototype that implements indicators as examples of learning analytic applications. The prototype-testing process indicated that the activity tracking of Moodle includes data about more complex social structures in the course of the VLE. In its last part, the paper contributed to the discussion of possible benefits of the approach in assessment, competence development and recommender systems. This discussion is accompanied with the analysis of the role of historical data for learning analytics and recommendations. Further elaboration of prototypes for other applications of learning analytics and recommender systems are in progress. This research contributes to the support of informative interventions during interlaced activities with dynamic learner models that are capable of reacting to actions during the learning process.

Acknowledgements. Authors would like to thank the LASPAU Program for the scholarship No. 20080847. Also, thanks to the Universidad del Valle, Colombia for co-funding the research of Beatriz Florian. The work presented is partly sponsored by the A2U@ project (TIN2008-06862-C04-01/TSI) funded by the Spanish Government.

The research presented in this paper has been partly sponsored by the GRAPPLE project (www.grapple-project.org) funded by the European Commission's 7th Framework Programme.

References

[1] Primus, N.J.C.: A generic framework for evaluating Adaptive Educational Hypermedia authoring systems - Evaluating MOT, AHA! and WHURLE to recommend on the development of AEH authoring systems. University of Twente, Twente (2005)

[2] Verbert, K., Duval, E., Drachsler, H., Manouselis, N., Wolpers, M., Vuorikari, R., Beham, G.: Dataset-driven Research for Improving Recommender Systems for Learning. In: Learning Analytics & Knowledge, Banff, Alberta, Canada, February 27-March 1 (2011)

[3] Zimmermann, A., Specht, M., Lorenz, A.: Personalization and Context Management. User Modeling and User-Adapted Interaction 15(3-4), 275–302 (2005), doi:10.1007/s11257-005-1092-2

[4] Engeström, Y.: Expansive Visibilization of Work: An Activity-Theoretical Perspective. In: Computer Supported Cooperative Work, vol. 8, pp. 63–93. Kluwer Academic Publishers, The Netherlands (1999)

[5] Najjar, J., Derntl, M., Klobučar, T., Simon, B., Totschnig, M., Grant, S., Pawlowski, J.: A Data Model for Describing and Exchanging Personal Achieved Learning Outcomes (PALO). International Journal of IT Standards and Standardization Research (IJITSR) 8(2) (2010)

[6] Commission of the European communities: Towards a European qualifications framework for lifelong learning, `http://ec.europa.eu/education/policies/2010/doc/consultation_eqf_en.pdf`

[7] Florian, B., Baldiris, S., Fabregat, R.: Adaptive Evaluation Based on Competencies. In: Proceedings of the Third Workshop Towards User Modeling and Adaptive Systems for All (TUMAS-A 2009): Modeling and Evaluation of Accessible Intelligent Learning Systems, held in conjunction with the 14th International Conference on Artificial Intelligence in Education (AIED 2009), Brighton, United Kingdom (July 2009)

[8] Florian, B., De La Hoz, A., Baldiris, S., Fabregat, R.: A set of software tools to build an author assessment package on Moodle. In: Proceedings of International Conference on Advanced Learning Technologies (ICALT) 2010, Sousse, Tunisia (July 2010)

[9] Cheetham, G., Chivers, G.: Professions, Competence and informal Learning. Edward Elgar Publishing, Northampton (2005)

[10] Crespo, R.M., Najjar, J., Derntl, M., Leony, D., Neumann, S., Oberhuemer, P., et al.: Aligning Assessment with Learning Outcomes in Outcome-based Education. In: EDUCON 2010: IEEE Annual Global Engineering Education Conference, Madrid, Spain, April 14-16 (2010)

[11] Dillenbourg, P., Tchounikinew, P.: Flexibility in macro-scripts for computer supported collaborative learning. Journal of Computer Assisted Learning 23, 1–13 (2007)

[12] Reigeluth, C.M.: Instructional Design: What is it and Why is it? In: Reigeluth, C.M. (ed.) Instructional-Design Theories and Models. An Overview of their Current Status. Erlbaum, Hillsdale (1983)

[13] Dillenbourg, P.: Framework for integrated learning. Report 2004 Kaleidoscope concepts and methods for exploring the future of learning with digital technologies. The European Commission, DG INFSO, under contract N°. IST 507838 (2004)
[14] Dillenbourg, P.: Integrating technologies into educational ecosystems. Distance Education 29(2), 127–140 (2008)
[15] Glahn, C., Specht, M., Koper, R.: Visualisation of interaction footprints for engagement in online communities (Special issue). Journal of Educational Technology & Society, 12(3), 44–57 (2009); In: Kalz, M., Koper, R., Hornung-Prähauser, V. (eds.)
[16] Van Merriënboer, J.J.G., Clark, R.E., de Croock, M.B.M.: Blueprints for Complex Learning: The 4C/ID-Model. Educational Technology, Research and Development 50(4), 39–64 (2002)
[17] Verpoorten, D., Glahn, C., Kravcik, M., Ternier, S., Specht, M.: Personalisation of Learning in Virtual Learning Environments. In: Cress, U., Dimitrova, V., Specht, M. (eds.) EC-TEL 2009. LNCS, vol. 5794, pp. 52–66. Springer, Heidelberg (2009)
[18] Glahn, C., Specht, M., Koper, R.: Smart Indicators on Learning Interactions. In: Duval, E., Klamma, R., Wolpers, M. (eds.) EC-TEL 2007. LNCS, vol. 4753, pp. 56–70. Springer, Heidelberg (2007)
[19] Glahn, C., Specht, M., Koper, R.: Reflection Support Using Multi-encoded Tag-clouds. In: Wild, F., Kalz, M., Palmér, M., Müller, D. (eds.) Proceedings of 2nd Workshop Mash-Up Personal Learning Envrionments (MUPPLE 2009). Workshop in conjunction with 4th European Conference on Technology Enhanced Learning (EC-TEL 2009): Synergy of Disciplines, Nice, France, September 29, 2009, pp. 89–97 (2009)
[20] Glahn, C., Specht, M.: Embedding Moodle into Ubiquitous Computing Environments. In: Montebello, M., et al. (eds.) 9th World Conference on Mobile and Contextual Learning, Valletta, Malta, October, 19-22 (2010)
[21] De Jong, T., Specht, M., Koper, R.: Contextualised Media for Learning. Journal of Educational Technology & Society 11(2), 41–53 (2008)
[22] Gamma, E., Helm, R., Johnson, R., Vlissides, J.M.: Design Patterns: Elements of Reusable Object-Oriented Software, 1st edn. Addison-Wesley Professional, Reading (1994)
[23] Haemaelaeinen, W., Vinni, M.: Classifiers for Educational Data Mining. In: Romero, C., Ventura, S., Pechenizky, M., Baker, R.S.J.d. (eds.) Handbook of Educational Data Mining. CRC Press, Boca Raton (2010)

Towards Responsive Open Learning Environments: The ROLE Interoperability Framework

Sten Govaerts[1], Katrien Verbert[1], Daniel Dahrendorf[2], Carsten Ullrich[3], Manuel Schmidt[4], Michael Werkle[4], Arunangsu Chatterjee[5], Alexander Nussbaumer[7], Dominik Renzel[6], Maren Scheffel[8], Martin Friedrich[8], Jose Luis Santos[1], Erik Duval[1], and Effie L-C. Law[5]

[1] Dept. of Computer Science, Katholieke Universiteit Leuven
Celestijnenlaan 200A, B-3001 Leuven, Belgium
{firstname.lastname}@cs.kuleuven.be

[2] imc information multimedia communication AG
Altenkesseler Str. 17/D3, 66115 Saarbruecken, Germany
daniel.dahrendorf@im-c.de

[3] Dept. of Computer Science and Engineering, Shanghai Jiao Tong University
1954 Huashan Road, 200050 Shanghai
ullrich_c@sjtu.edu.cn

[4] Festo Lernzentrum Saar GmbH
Obere Kaiserstraße 301, 66386 St. Ingbert, Germany
{mans,werk}@de.festo.com

[5] Dept. of Computer Science, University of Leicester
LE1 7RH Leicester, UK
{a.chatterjee,elaw}@mcs.le.ac.uk

[6] Lehrstuhl Informatik 5, RWTH Aachen University
Ahornstraße 55, 52056 Aachen, Germany
renzel@dbis.rwth-aachen.de

[7] Knowledge Management Institute, Graz University of Technology
Brückenkopfgasse 1, A-8010 Graz, Austria
alexander.nussbaumer@tugraz.at

[8] Fraunhofer Institute for Applied Information Technology FIT
Schloss Birlinghoven, 53754 Sankt Augustin, Germany
{firstname.lastname}@fit.fraunhofer.de

Abstract. In recent years, research on mash-up technologies for learning environments has gained interest. The overall goal is to enrich or replace traditional learning management systems (LMS) with mash-ups of widgets and services that can be easily combined and configured to fit the learner needs. This paper presents the implemented prototype of the ROLE interoperability framework and a business and an educational case study. The framework provides a common technical infrastructure to assemble widgets and services in Personal Learning Environments (PLEs). Evaluation results indicate that the perceived usefulness and usability is high for one case study in which a mature LMS was enriched with ROLE technology. In the second case study, an early mash-up prototype was deployed. The usefulness and usability of this early prototype were rated low, but the case study provides interesting insights for further research and development.

C. Delgado Kloos et al. (Eds.): EC-TEL 2011, LNCS 6964, pp. 125–138, 2011.

Keywords: personal learning environments, interoperability, mash-ups, widgets, standards and specifications.

1 Introduction

The development and proliferation of Web 2.0 technologies (e.g. wikis, blogs and social networks) has impacted the way users retrieve and use information and how they interact with each other [19,34,1]. An important feature of Web 2.0 services is that they experience an exponential growth of both users and content, leading to potentially viral social networking, collaboration, communication and knowledge sharing opportunities.

The abundance of web-based tools and content creates many opportunities for Technology Enhanced Learning (TEL) that aims to bring together new technological developments and learning models to support learning processes. The EU project ROLE[1] aims to exploit web-based tools and technologies to empower learners to construct their own personal learning environments (PLEs). The overall goal is to create flexible, web-based, open technologies for the federation and mash-up of learning services on a personal level. The project also targets critical transition stages of lifelong learning, e.g. due to shifts in learner interests or when leaving the university and entering a company. The vision of ROLE is to empower the learner to build her own responsive learning environment. Responsiveness is defined as the ability to react to the learner needs - i.e. through recommendation, adaptation or visual analytics services that support the learner to be aware of and reflect upon her own learning process [10].

This paper presents an implemented prototype of the ROLE interoperability framework, which is a technical infrastructure to assemble widgets within responsive open learning environments. The framework provides bundles of widgets with communication channels, authentication and authorisation mechanisms, services for activity tracking and analysis. This ensures that the bundles have access to the necessary information to react to learner needs. The research contribution of this paper is threefold:

- First, we present technical building blocks of the ROLE interoperability framework.
- Second, we present evaluation results of the usefulness and usability of the approach within two educational and business case studies.
- Third, we present a future vision on the integration of pedagogical models to leverage the framework.

The paper is organised as follows: we first discuss related work in Section 2. Section 3 presents an overview of the ROLE framework. The case studies are presented in Section 4. Finally, conclusions and future work are presented in Section 5.

2 Related Work

LMSs primarily focus on distributing learning content, organising the learning process, and serving as interface between learner and teacher. LMSs are prevalently used by

[1] The ROLE project web site, `http://www.role-project.eu`

many European institutions [24]. Popular examples of LMSs are Moodle, CLIX, Blackboard, WebCT, Sakai, ILIAS and .LRN. Dalsgaard [4] notes that in LMSs generally different tools, such as discussion forums, file sharing, whiteboards and e-portfolios, are integrated in a single system bundling all tools necessary to manage and run courses.

LMSs place a strong emphasis on how to centralise and standardise the learning experience [13]. Learning activities in an LMS-based course are organised within a centrally managed system, which is driven by the needs of the institution. In contrast to an LMS, a PLE takes a more natural and learner-centric approach and is characterised by the freeform use of a set of services and tools that are controlled and selected by individual learners.

In recent years, research on mash-ups has been elaborated, for example widget mash-ups have been deployed at Graz University of Technology [6]. The widgets are combined with JavaFX[2] technology to improve flexibility, e.g. to install widgets on the desktop. In addition, researchers have focused on augmenting traditional LMSs with widgets to provide live-updating and flexible applications. Wilson et al. [35] implemented widget support for Moodle. Their big challenge is logging student activities with the widgets, as there is no communication between the widgets and the LMS.

Our research builds on this existing work, but incorporates additional core technologies such as inter-widget communication (IWC), automated user activity tracking and authentication and authorisation services to protect data. This is the basis to enable real-time communication between widgets and users, and to automate user activity tracking from tools and services. The analysis of such data and IWC provides the basis to develop responsive systems that can react to learner needs in a coordinated way.

3 The ROLE Interoperability Framework

The prototype contains a common technical infrastructure to support the assembly of widgets in responsive open learning environments. The infrastructure is presented in Fig. 1 with the interconnectivity between the components. These building blocks provide open interfaces for widget functionality and the necessary information and technology to enable responsiveness.

The core of the infrastructure is the widget container that enables the assembly of various widgets. Learners and teachers use the Widget Store to select learning widgets. User activities with widgets and resources are tracked with the CAM widget.

Widgets can communicate locally in the PLE or remotely to widgets in other PLEs via XMPP [27] to foster collaboration. They can also access the CAM service to provide personalised recommendations [11] and visual analytics of CAM data [12] as a basis for self-reflection and awareness. The central identity provider allows single sign-on for the whole infrastructure. The remainder of this section describes the different components.

3.1 The Widget Container

The widget container is an environment for widget rendering as well as management of, and communication between widgets. It also provides a user-friendly way to organise

[2] JavaFX, `http://www.sun.com/software/javafx/`

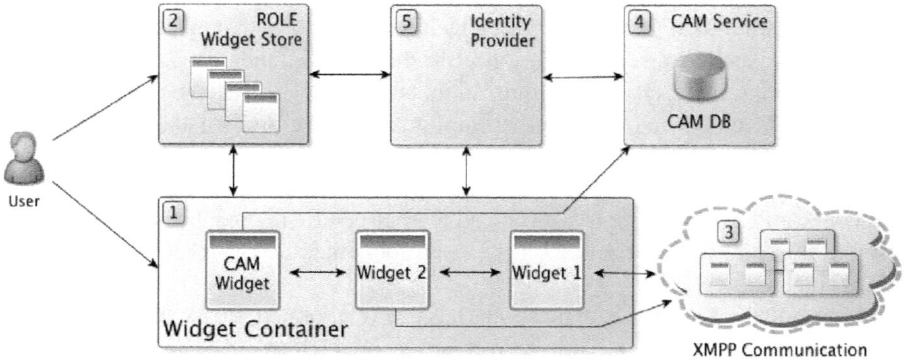

Fig. 1. The ROLE infrastructure

widgets visually, set preferences, navigate to the widget store for choosing additional widgets, etc. Such a widget container can be added in existing PLEs and LMSs [35].

There are two major widget specifications with very similar functionalities [26]: the OpenSocial [21] and the W3C widget 1.0 specification [3]. OpenSocial also provides a social API [21] to access social data from multiple web applications. Because of the higher availability of OpenSocial widgets, we currently employ OpenSocial compliant widgets, which run on major widget containers such as iGoogle[3] and Shindig[4] (for more info see [26]). We plan to also incorporate W3C widgets by using Wookie[5].

3.2 The Widget Store

The Widget Store provides a learning tool catalogue. Existing web widgets stores (e.g. iGoogle Directory[6]) do not focus on supporting learning processes. The ROLE Widget Store allows learners to search for fitting learning tools and rate them. Found widgets can be included in existing learning environments. Extensions of the store are planned on several levels. The assembly of tools and content for learning activities will be supported by intelligent recommender and social community features. Such a set of assembled widgets is defined as a bundle. Next to widgets, the store will also contain a variety of learning tools and will have features for social requirement engineering that will be the base for a community of practice for responsive open learning environments.

3.3 Inter-Widget Communication

IWC enables event-based communication between widgets following the Publish - Subscribe communication pattern [2,7]. We employ both local inter-widget communication

[3] iGoogle, `http://www.igoogle.com`

[4] Apache Shindig, `http://shindig.Apache.org/`

[5] Apache Wookie, `http://getwookie.org`

[6] iGoogle Directory, `http://www.google.com/ig/directory`

(LIWC) within a PLE and remote inter-widget communication (RIWC) among different users and computers.

LIWC is realised in the OpenApplication Event API [18,17] using the HTML5 Web Messaging standard [16] available in most major browsers, including backwards compatibility for the Google Gadget PubSub mechanism. Instead of 'hard-wiring' widgets with each other [33], all widgets within a PLE are notified of all events and then decide for themselves to react accordingly. The event payload format is designed for partial semantic interoperability, i.e. developers use a combination of established vocabularies in a simplified format with namespaced-properties (e.g. Dublin Core [5]).

RIWC enables communication among widgets in different browsers and on different machines in order to foster real-time remote communication and collaboration functionality. RIWC is realised with the Extensible Messaging and Presence Protocol (XMPP) [27,28], an open standard for real-time communication. The power of XMPP lies in its built-in federation capabilities and extensibility through XMPP Extension Protocols (XEPs), such as for Publish/Subscribe [20] and Multi-User Chat [29] as applied in responsive and collaborative learning scenarios [8]. Since no Javascript XMPP library with PubSub support was available, we extended the dojo XMPP library by a set of common PubSub operations. Users can discover nodes, retrieve subscriptions, create, configure, and delete nodes, subscribe and unsubscribe nodes, and publish/receive IWC events in an XML-based payload format across a federated network of XMPP servers. However, current libraries using XMPP over BOSH [23] are not applicable in public containers such as iGoogle or Google Wave due to cross-domain issues. Furthermore, they are rather unstable and unreliable [8]. Our experiments showed that the upcoming Web Socket API [15] for XMPP [22] outperforms BOSH with considerable performance and stability improvements and availability in all containers.

IWC enables more responsive, collaborative environments with real-time notifications and richer user experience, although attention to usability is required [18].

3.4 CAM Tracking Service

User activities are tracked [30] using the Contextualised Attention Metadata (CAM) format [36]. CAM describes the interactions of the users with their learning environment, i.e. which resources are used within which applications and in which contexts. These data can be used for analysis and computing of personal, social and contextual information about users and applications [30]. CAM can be exploited to provide personalised recommendations and thus serves as a basis for enabling responsiveness in ROLE. A second important goal of tracking such data is to enable the evaluation of ROLE services based on user activities that have been captured in real-world settings.

The CAM monitoring is implemented as a database-driven client-server architecture. A JSON REST service provides persistence and data access of the CAM data. Each monitored user action is immediately committed to the CAM database, so that the current status of the user can be retrieved by self-reflection and recommendation applications. Because CAM is privacy sensitive, the service is protected with authentication and authorisation mechanisms (see Section 3.5).

The CAM client widget tracks all user actions within the PLE and sends them to the CAM service. The widget uses the OpenApplication library [18] to listen to all the events broadcasted by all widgets in the PLE. Performance test showed that the service and database worked reliable in real-world scenarios [9]. Implementing the tracker as a widget and not integrating it in the widget container achieves container independence.

3.5 The Authentication and Authorisation Service

CAM contains sensitive and personal data protected by law. The data access has to be trusted and allowed by the users. The data communication occurs at two different levels: service-to-service and widget-to-widget communication.

Service-to-service communication can occur when for example a recommendation service requires CAM relevance feedback on resources. Data can be transferred across institutions and countries with different laws. Thus, we decided to leave the decision of service-to-service authentication and authorisation (A&A) up to the service developers.

Widget-to-service communication occurs when for example a self-reflection widget wants to query the CAM service. Here some usability issues apply: the user needs an account for every service in his PLE and needs to login to all services. We surveyed many solutions, including Facebook Connect, Shibboleth, OAuth, OpenID and Mozilla Account Manager. We selected OAuth[7] because the technological maturity, support within OpenSocial widget containers, available libraries and usability (i.e. users are warned when an application wants to access protected data).

To solve the usability issues with multiple accounts and logins, we propose to use a central identity provider for Single Sign On, using OpenID[8] in combination with OAuth. This work is currently in the design stage and will be implemented soon.

4 Case Studies

This section briefly introduces the widgets developed for the case studies. Then, two case studies in a company (Festo) and a university (SJTU) are presented. In addition, evaluation results on the usability and usefulness of the prototypes are described.

4.1 ROLE Widgets for the Case Studies

Many widgets have been developed that are compliant with ROLE technology, including language learning [26], self-reflection [12] and chat widgets [8].

To address the requirements of test beds that were used to evaluate first ROLE prototypes, two additional widgets were developed. We briefly present the widgets in this section. Case studies that assess the usefulness and usability of the deployment of these widgets in PLEs are presented in Section 4.2 and Section 4.3.

---

[7] The OAuth 2.0. Protocol Draft, `http://tools.ietf.org/html/draft-ietf-oauth-v2-15`

[8] OpenID specification, `http://openid.net/developers/specs/`

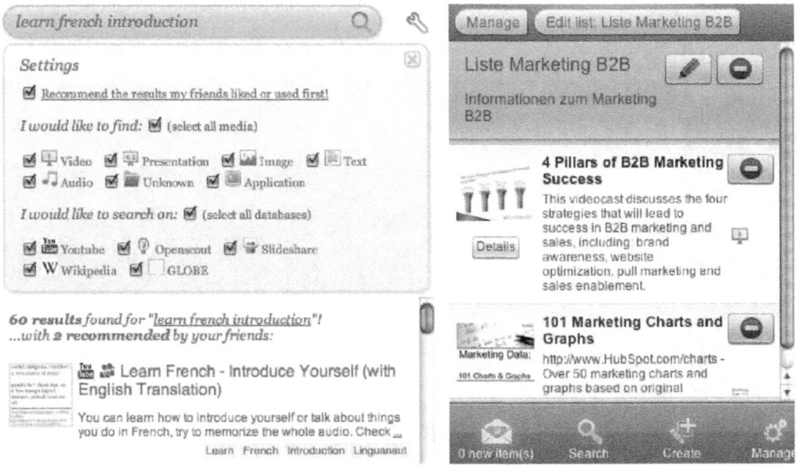

(a) The federated search widget (b) The media list widget

Fig. 2. Screenshots of the media list and federated search widgets

The Federated Search and Social Recommendation Widget. We developed a federated media search widget (see Fig. 2a) that provides access to a wide variety of learning content sources. The widget currently searches on YouTube.com, SlideShare.net, Wikipedia, OpenScout[9] and GLOBE[10] repositories.

Social recommendations are integrated that suggest resources based on preferences of peers. These recommendations are generated based on data that is captured by the CAM tracking service (cf. Section 3.4). The recommendation algorithm uses both explicit (ratings) and implicit (i.e. previews and selection of the results) relevance data [11]. In both case studies, the use of the widget has been evaluated in a real-world setting.

The Media List Widget. When finding interesting results, users might want to save and group them for later referral. The media list widget (Fig. 2b) stores and groups results for later use and allows sharing of the results with peers. The widget works in combination with the federated search widget. Local inter-widget communication (cf. Section 3.3) is used to enable communication between the widgets. When the user saves a result in the federated search widget, an OpenApplication event is sent. This event is intercepted by the list widget and stored. The federated search can also search in the media list database to find relevant results saved by peers. The media list widget was used in the Festo study to assist the sales department with re-occurring product searches.

[9] The OpenScout repository, `http://www.openscout.net/`

[10] The GLOBE alliance, `http://www.globe-info.org`

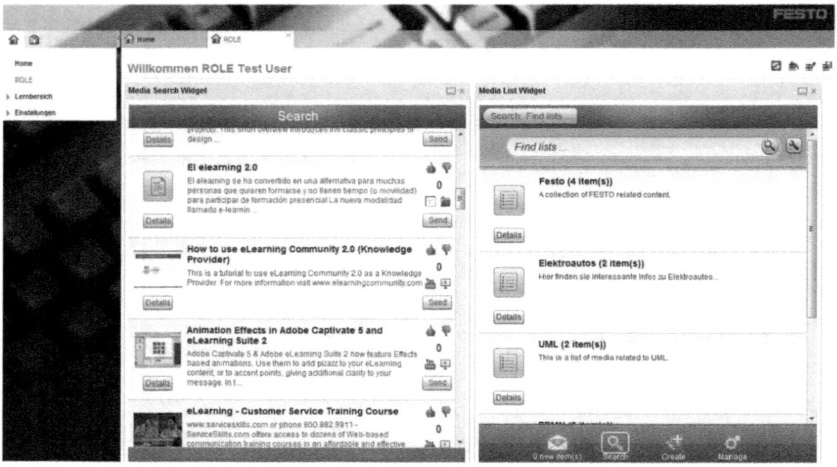

Fig. 3. Festo PLE example

4.2 The Festo Case Study

The Festo[11] test bed provides a responsive learning environment for further education activities in a company. The learners in this test bed are very heterogeneous. They differ in age, education, job-roles, learning requirements and learning preferences. This test bed provides a platform for federating and mashing-up different ROLE learning services within their LMS, the Festo Virtual Academy (FVA). The FVA focuses more on institutional learning than on self-regulated learning and is more or less restricted to internal resources. To support self-regulated learning it is necessary to open up LMSs and introduce features where learners can adapt the system according to their needs.

Festo uses CLIX[12] for running the FVA. We extended CLIX to support widget container functionality (cf. Section 3.1) and inter-widget communication (cf. Section 3.3). In this case study, we provided learners with a single access point to different learning resources (internal and external) and enable collaborative organisation and sharing of these resources. To achieve this functionality, the federated search and media list widgets were evaluated in the FVA (see Fig. 3). The graphic design of the federated search widget was adapted to the corporate style.

In a first stage, we evaluated the acceptance of a widget-enriched system by learners who are used to a classic LMS. The testing took place on the live FVA. The test group consisted of 25 employees and took place from March 9th 2011 until March 31th 2011. The usage of the system was demonstrated to the test users in a screencast. After using the system, the participants filled out a questionnaire that was targeted to assess the perceived usefulness and usability. The questionnaire was divided into three question blocks: demographics, assessment of the widgets and feedback for improvement.

The preliminary evaluation showed that the majority of the users liked the look-and-feel as well as the usability (see Fig. 4a). For both criteria only a few people wanted

[11] Festo, `http://www.festo.de`
[12] http://www.im-c.de

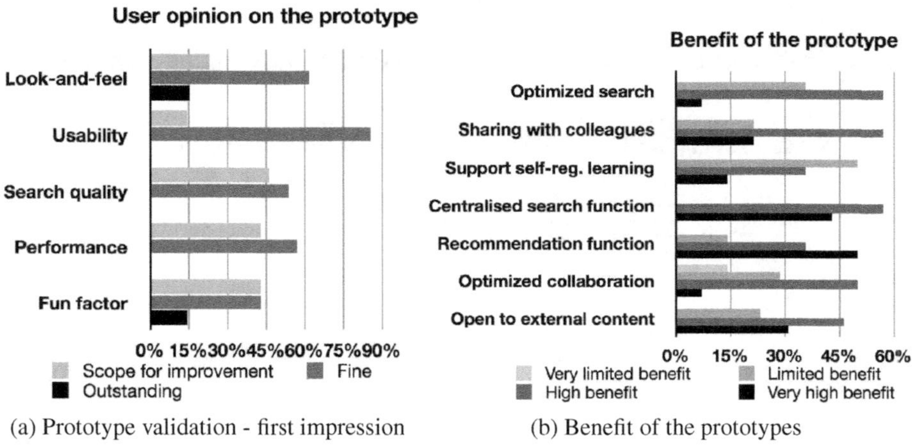

Fig. 4. Evaluation results

improvements (23% for look-and-feel, 14% for the usability). People preferred to have higher quality search results (46%) and better performance (43%). This is in line with the earlier evaluation of the federated search widget in [11]. To solve this issue, we will add more repositories (e.g. Festo product databases) and adapt the federated search so that it shows results immediately after it has received initial results.

Most users see a high or very high benefit of the offered prototypes (see Fig. 4b). People benefited most from centralised search (90% high or very high benefit) and recommendations (86% high or very high benefit). The optimised information search was also quite beneficial, but could probably be improved by adding more repositories. Sharing media and openness of the LMS is important for the users. Some people see collaboration as slightly less important. The benefits of self-regulated learning were not clear for many users (50%). Section 5 elaborates on our plans to address this issue.

Overall, the use of the two widgets in the widget environment was positive. 88% of the users would recommend the tools to their colleagues. The perceived usefulness and effectiveness was evaluated by a question whether the offered services and the ROLE approach would help the users to work more effectively in their job, compared to today. 31% fully agreed with this point, 50% agreed and only 6% denied that they would not be more effective. This is also a positive result for this early prototype. In the future, we want to evaluate the usefulness for their job in more detail with more employees.

4.3 The SJTU Case Study

The Shanghai Jiao Tong University (SJTU) School of Continuing Education (SOCE) offers higher education to adults (24.000 in 2010) with a full-time job. The students form a very heterogeneous group caused by age, command of English and digital literacy. In ROLE, SOCE serves as a test bed to understand PLE usage in a context that is characterised by limited study time and low digital literacy. SOCE implements blended learning where the lectures can be attended in person, remotely by watching a broadcast

Fig. 5. SJTU PLE example

using SOCE's LMS, which has limited support for active and self-regulated learning. PLE technology has the potential to overcome these limitations by enabling the integration of tools and recommendation services. To investigate this, SJTU set up an experimental PLE, configured by teachers and researchers, in Liferay[13]. Most of the functionality is offered by webpages or existing OpenSocial widgets (cf. Section 3.1) that are tracked by the CAM service (cf. Section 3.4).

In contrast to the evaluation at Festo with a mature learning environment and experienced users, the evaluation at SJTU was done with an early prototype of a learning environment that was entirely new to the students.

The evaluation focused on the use and usefulness of the PLE and the social recommendations of the federated search widget. It took place in the three on-going English language courses that started in February 2011 with 100-300 students each.

The setup was very similar for each class and consisted of the following pages:

- "Practice" (part of it is shown in Figure 5): a language resource browser; a translator; a vocabulary trainer; an Etherpad widget with exercises (Etherpad is an easy-to-use collaborative text editor); and the CAM widget.
- "Self-Assessment": a widget for testing the English level; and the CAM widget.
- "Recommendations": the federated search widget.

The students had to complete a task list, focused on exploration of the PLE and embedded tools, with an estimated duration of 15 minutes. After that, students were prompted

[13] Open source portal system Liferay, `http://www.liferay.com/`

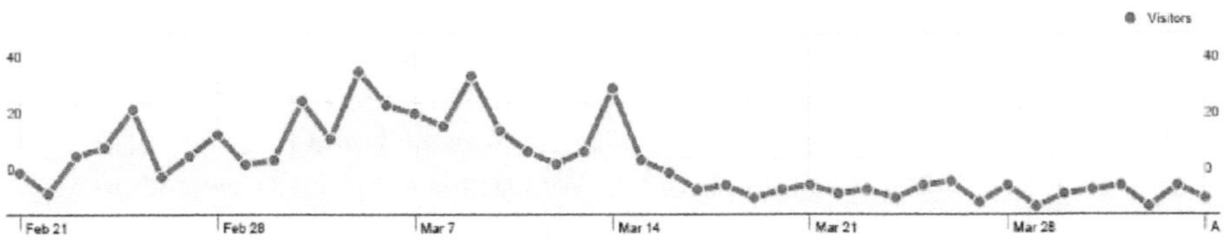

Fig. 6. The absolute unique visitors of SJTU test bed over the evaluation period

to complete an online survey in Chinese. The evaluation ran during 3 weeks. Data about the PLE usage was collected through the survey, student interaction recordings with the PLE and web analytics data. The survey was designed to collect user feedback about the federated search widget and the overall usefulness and usability of the PLE. The questionnaire included 13 items pertaining to recommendations adapted from the ResQue (Recommender systems' Quality of user experience) framework [25] utilising the constructs accuracy, relevance, interface adequacy and intention of future use. The usability section of the questionnaire consisted of 14 items encompassing usefulness, navigational consistency, overall ease-of-use, satisfaction and enjoyment [32,31]. A total of 13 completed responses were obtained.

Among the respondents, 69% used the federated search widget. Overall, 56% of the federated search widget users found the recommendations relevant to the course. Among the federated search widget users, 56% also reported that the recommendations were well presented and the widget UI was easy to understand. However, only 38% of these students wanted to continue using the federated search widget. The PLE as a whole was useful for the course for 23% of the respondents and 39% felt that it was easy to use. Only 15% of the users found the PLE to be navigationally consistent, which could have significantly influenced the usefulness and ease-of-use measures. In spite of the difficulties, 46% of the respondents were satisfied with the PLE. Additional comments provided via open-ended questions point towards performance issues. One such comment, "I like it for translation, but it works so slowly, I almost can't bear it", reflects the significant loading times that affect the user experience reducing future intention of use. This may relate to the slower Internet connections in China as opined by the teachers and because most widgets were remotely hosted across the globe.

Mouse tracking is a technique for monitoring and visualising mouse movements on a web interface to gain insights and discover usability issues. The Userfly[14] mouse tracking service was integrated in the PLE. The recordings provide some useful insights. For example, it confirms the findings from the survey pertaining to the lack of navigational consistency and specifically performance issues like loading time.

The web analytics for the first six weeks of the course (February 21 until April 4) show a total of 444 visits (10 visits a day), with a spike of 77 visits on the day the PLE was jointly used in the computer room.

The absolute unique visitors are shown in Fig. 6. After the first three weeks (the time in which the researcher attended class and motivated the students to use the system) the numbers declined. On average, users accessed 5.59 pages per visit. This number is

[14] Userfly, `http://userfly.com`

slightly higher as expected, as each lecture consisted of 2 or 3 pages. A detailed look at the data reveals that 18% visited a single page (the login page), 14% two pages (the login page and the first course page), 17% three and 50% more than three pages. Two third visited the site more than once. On average they spent 8 minutes on the PLE, which is lower than expected to finish the task list. We assume some visits were short and of an exploratory nature. The most frequently visited pages are the "Practice" pages.

Almost 1 out of 5 users did not get further than the login page. The analysis of the recordings of the user interactions highlighted that students used login credentials from another system. This demonstrates the importance of single-sign-on solutions.

In the future, we plan to enable the mouse tracking for a limited duration during lab-based sessions, followed by Retrospective Think Aloud [14] sessions with a few learners to extract some more detailed explanations. User interaction analysis of a larger group could also be achieved with the CAM data, which we currently have not done yet. Rather than merely triangulating data obtained from a survey, the proposed approach will help yield additional qualitative data for future design and development decisions.

5 Conclusion and Outlook

This paper presented the prototype implementation of a infrastructure for responsive open learning environments with support for widget assembly, inter-widget communication, authentication and authorisation services, and services for activity tracking. The infrastructure allows widget bundles to react to learner needs in a coordinated way.

Future work will integrate the infrastructure with the psycho-pedagogical integration model (PPIM) [10] for self-regulated learning. The learning process in the model is described by a four-phase cycle based on work by Zimmermann [37], where the learner: (1) (re-)defines his profile information, (2) finds and selects learning tools, (3) uses learning tools, (4) reflects and reacts on strategies, achievements and usefulness.

In the presented case studies, phases three and four are supported. Learners can use their widget bundles and receive feedback – i.e. by visual analytics of CAM data [12]. For now, phase two is done by the teacher who preselected widgets for the student. Phase one is discarded because the user profile service was not available.

The next step is to provide full support of the PPIM model within the ROLE infrastructure. Learners will be able to set up a user profile including the current and competencies to be achieved, preferences, learning history and progress. Suggestions will be automatically generated by analysing CAM data and can be revised by the learner.

In phase two, the Widget Store can recommend widget bundles for the defined learning goals based on the current competency profile. This requires a semantic machine-readable description of the available tools, content and bundles. Social recommendations will be also important in this phase. Our ongoing work in this area is described in [11]. After integrating the PPIM model, the infrastructure will be evaluated with the presented case studies and two new academic test beds.

Acknowledgements. This research is funded by the European Commission's Seventh Framework Programme (FP7/2007-2013) under grant agreement no 231396 (ROLE). Katrien Verbert is a Postdoctoral Fellow of the Research Foundation - Flanders (FWO).

References

1. Ashley, H., Corbett, J., Jones, D., Garside, B., Rambaldi, G.: Change at hand: Web 2.0 for development. Participatory Learning and Action 59, 8–20(13) (2009)
2. Birman, K.P., Joseph, T.A.: Exploiting virtual synchrony in distributed systems. In: SOSP 1987: Proc. of the 11th ACM Symp. on Operating Systems Principles, pp. 123–138 (1987)
3. Cáceres, M.: Widget packaging and configuration, W3C working draft, March 22 (2011), http://www.w3.org/TR/widgets/
4. Dalsgaard, C.: Social software: E-learning beyond learning management systems. European Journal of Open, Distance and E-learning (2) (September 2006)
5. DCMI Usage Board: DCMI Metadata Terms. DCMI Recommendation, Dublin Core Metadata Initiative (December 2006), http://dublincore.org/documents/2006/12/18/dcmi-terms/
6. Ebner, M., Taraghi, B.: Personal learning environment for higher education - a first prototype. In: World Conference on Educational Multimedia, Hypermedia and Telecommunications 2010, pp. 1158–1166. AACE, Chesapeake (2010)
7. Eugster, P.T., Felber, P.A., Guerraroui, R., Kermarrec, A.M.: The Many Faces of Publish/Subscribe. ACM Computing Surveys 35(2), 114–131 (2003)
8. Friedrich, M., Wolpers, M., Shen, R., Ullrich, C., Klamma, R., Renzel, D., Richert, A., von der Heiden, B.: Early results of experiments with responsive open learning environments. Journal of Universal Computer Science 17(3), 451–471 (2011)
9. Friedrich, M., Wolpers, M.: CAM schema and library, https://sites.google.com/site/camschema/home
10. Fruhmann, K., Nussbaumer, A., Albert, D.: A psycho-pedagogical framework for self-regulated learning in a responsive open learning environment. In: Proc. of the International Conference eLearning Baltics Science (eLBa Science 2010), Rostock, Germany (July 2010)
11. Govaerts, S., El Helou, S., Denis, G., Duval, E.: A federated search and social recommendation widget. In: Proc. of the 2nd Int. Workshop on Social Recommender Systems (SRS 2011) in conjunction with ACM CSCW 2011 (March 2011)
12. Govaerts, S., Verbert, K., Klerkx, J., Duval, E.: Visualizing activities for self-reflection and awareness. In: Luo, X., Spaniol, M., Wang, L., Li, Q., Nejdl, W., Zhang, W. (eds.) ICWL 2010. LNCS, vol. 6483, pp. 91–100. Springer, Heidelberg (2010)
13. Guo, Y., Rui, J., Zhou, H.: Pervasive and personal learning environment using service-oriented architecture: A framework design. In: International Conference on Networking and Distributed Computing, pp. 153–155 (2010)
14. van den Haak, M.J., de Jong, M.D., Schellens, P.J.: Evaluation of an informational web site: Three variants of the think-aloud method compared. Tech. Comm. 54(1), 58–71 (2007)
15. Hickson, I.: The Web Sockets API. Tech. rep., W3C, W3C Working Draft (December 2009)
16. Hickson, I.: HTML5 web messaging. Working draft, W3C (March 2011), http://www.w3.org/TR/2011/WD-webmessaging-20110317/
17. Isaksson, E., Palmér, M.: OpenApplication specification, https://docs.google.com/View?id=dfb5qpz7_4cp5nzbcn
18. Isaksson, E., Palmér, M.: Usability and inter-widget communication in PLEs. In: Proceedings of the 3rd Workshop on Mashup Personal Learning Environments (September 2010)
19. Maness, J.M.: Library 2.0 theory: Web 2.0 and its implications for libraries. Webology 3(2) (June 2006)
20. Millard, P., Saint-Andre, P., Meijer, R.: XEP-0060: Publish-Subscribe. Tech. rep., XMPP Standards Foundation(July 2010), draft Standard, http://xmpp.org/extensions/xep-0060.html

21. Mitchell-Wong, J., Kowalczyk, R., Roshelova, A., Joy, B., Tsai, H.: Opensocial: From social networks to social ecosystem. In: Digital EcoSystems and Technologies Conference, DEST 2007. Inaugural IEEE-IES, pp. 361–366 (2007)
22. Moffit, J., Cestari, E.: An XMPP Sub-protocol for WebSocket. Tech. rep., Internet Engineering Taskforce (December 2010), draft Standard, http://tools.ietf.org/html/draft-moffitt-xmpp-over-websocket-00
23. Paterson, I., Smith, D., Saint-Andre, P., Moffitt, J.: XEP-0124: Bidirectional-streams Over Synchronous HTTP (BOSH). Tech. rep., XMPP Standards Foundation (July 2010), http://xmpp.org/extensions/xep-0124.html
24. Paulsen, M.F.: Experiences with learning management systems in 113 European institutions. Educational Technology & Society 6(4), 134–148 (2003)
25. Pu, P., Chen, L.: A user-centric evaluation framework of recommender systems. In: ACM Conf. on Recommender Systems (RecSys 2010), Workshop on User-Centric Evaluation of Recommender Systems and Their Interfaces (UCERSTI 2010), pp. 14–21 (2010)
26. Renzel, D., Höbelt, C., Dahrendorf, D., Friedrich, M., Mödritscher, F., Verbert, K., Govaerts, S., Palmér, M., Bogdanov, E.: Collaborative development of a PLE for language learning. International Journal of Emerging Technologies in Learning (iJET) 5(S1), 31–40 (2010)
27. Saint-Andre, P.: RFC 3920 – Extensible Messaging and Presence Protocol (XMPP): Core. Tech. rep., Jabber Software Foundation (October 2004), http://www.ietf.org/rfc/rfc3920.txt
28. Saint-Andre, P.: RFC 3921 – Extensible Messaging and Presence Protocol (XMPP): Instant Messaging and Presence. Tech. rep., Jabber Software Foundation (October 2004), http://www.ietf.org/rfc/rfc3921.txt
29. Saint-Andre, P.: XEP-0045: Multi-User Chat. Tech. rep., XMPP Standards Foundation (July 2008), draft Standard, http://xmpp.org/extensions/xep-0045.html
30. Schmitz, H.-C., Scheffel, M., Friedrich, M., Jahn, M., Niemann, K., Wolpers, M.: CAMera for PLE. In: Cress, U., Dimitrova, V., Specht, M. (eds.) EC-TEL 2009. LNCS, vol. 5794, pp. 507–520. Springer, Heidelberg (2009)
31. Scott, B., Neil, T.: Designing Web Interfaces: Principles and Patterns for Rich Interactions. O'Reilly, Sebastopol (2009)
32. Shneiderman, B., Plaisant, C.: Designing the User Interface: Strategies for Effective Human-Computer Interaction. Addison-Wesley, Reading (2004)
33. Sire, S., Paquier, M., Vagner, A., Bogaerts, J.: A messaging API for inter-widgets communication. In: Proceedings of the 18th International Conference on World Wide Web. WWW 2009, pp. 1115–1116. ACM, New York (2009)
34. Ullrich, C., Borau, K., Luo, H., Tan, X., Shen, L., Shen, R.: Why Web 2.0 is Good for Learning and for Research: Principles and Prototypes. In: Proceedings of the 17th International World Wide Web Conference, pp. 705–714. ACM Press, New York (2008)
35. Wilson, S., Sharples, P., Popat, K., Griffiths, D.: Moodle Wave: Reinventing the VLE using widget technologies. In: Proceedings of 2nd Workshop Mash-Up Personal Learning Envrionments (MUPPLE 2009). Workshop in conj. with 4th European Conference on Technology Enhanced Learning (EC-TEL 2009): Synergy of Disciplines, pp. 47–58. Springer, Heidelberg (2009)
36. Wolpers, M., Martin, G., Najjar, J., Duval, E.: Attention metadata in knowledge and learning management. In: Proceedings of I-Know 2006: 6th International Conference on Knowledge Management, September 2006, pp. 1–8 (2006)
37. Zimmerman, B.J.: Becoming a self-regulated learner: An overview. Theory Into Practice 41(2), 65–70 (2002)

Deriving Knowledge Profiles from Twitter

Claudia Hauff* and Geert-Jan Houben

WIS, Delft University of Technology, Delft, the Netherlands
{c.hauff,g.j.p.m.houben}@tudelft.nl

Abstract. E-learning systems often include a personalization component, which adapts the learning content to the learner's particular needs. One obstacle to personalization is the question of how to obtain a learner profile for a learner who just starts using an E-learning system without overwhelming her with questions or unsuitable learning material. One possible solution to this problem lies in the social Web. If a learner is active on the social Web, a considerable amount of information about her is already available. Depending on the social Web service(s) the learner uses, her tweets, photos, bookmarks, etc. are publicly accessible. We investigate if it is feasible to exploit the social Web, more specifically the social Web service Twitter, to infer a learner's knowledge profile in order to overcome the "cold-start" problem in E-learning systems.

1 Introduction

Platforms that facilitate E-learning have become increasingly prevalent in recent years. Due to the ubiquitous nature of the Internet, learning online at your own pace and at your own time has never been easier. E-learning systems may include a personalization and adaptation component, which adapts the learning content to the learner's needs and capabilities. Adapting an E-learning system based on a learner's profile can increase learner satisfaction and decrease learner frustration. For example, learning units the learner is already knowledgeable about can be automatically removed by the system, while content the learner is unfamiliar with can be covered in greater depth. One obstacle to personalization is the question of how to obtain a learner profile for a learner who is new to an E-learning system. Although it is possible for the system to derive the learner's knowledge profile over time or by posing a series of test questions, the learner may be unwilling to spend a lot of effort on this procedure. By the time an adequate knowledge profile of the learner has been aggregated, the learner might have already given up on the system.

One potential solution to this problem lies in the social Web. The rise of the social Web has made people not merely consumers of the Web, but active contributors of content. Widely adopted social Web services, such as Twitter[1], Flickr[2]

* This research has received funding from the European Union Seventh Framework Programme (FP7/2007-2013), grant agreement no ICT 257831 (ImREAL project).

[1] `http://www.twitter.com/`

[2] `http://www.flickr.com/`

C. Delgado Kloos et al. (Eds.): EC-TEL 2011, LNCS 6964, pp. 139–152, 2011.

and Delicious[3], are frequented by millions of active users who add, comment or vote on content. If a learner is active on the social Web, a considerable amount of information about her is available on the Web. Depending on the social Web service(s) the learner uses, her blog entries, tweets, bookmarks, etc. are publicly accessible. For the future, we foresee the following scenario: a learner who uses an E-learning system for the first time, is asked by the system to list her handles of the social Web services she is active on. Then, based on the learner's "online persona", aggregated from the social Web, the system can automatically infer a basic profile of the learner's knowledge in the desired domain. It needs to be stressed, that we envision this approach to be mostly applicable in "cold-start" situations, that is, when the system has no other information available about the learner.

The assumption behind this vision is of course, that it is possible to extract a basic learner's knowledge profile from the social Web. In this paper we investigate this very assumption. More specifically, we investigate if it is feasible to infer a learner's knowledge profile from her activities on the micro-blogging platform Twitter. Our motivation for exploiting Twitter is based on the fact that it is a highly popular platform, used by millions of people[4]. Moreover, most Twitter users make their microblog posts (tweets) publicly accessible, and thus there will be few privacy concerns. The E-learning system only needs to query the learner for her handle on Twitter, no further information (that is, no login information) is required. As many people use Twitter throughout the day, we postulate that among all the posts a user publishes, at least some of them will be pertinent to the user's work and study. Examples of tweets that we envision to be useful are:

- *The Microwave Toolbox for Scilab v0.3 Available for Scilab and Scicoslab.*
- *In algebra 4*
- *Confident I did perfect on my algebra 2 test.*

These tweets on the one hand allow us to infer what the learner is currently learning (*In algebra 4*), but they also allow us to build to some extent a knowledge profile of the learner; the tweet *Confident I did perfect on my algebra 2 test.* implies a high level of knowledge in this particular study area according to the learner's self-assessment.

Of course, not all tweets provide us with useful information. On the contrary, the majority of tweets may be focused on day to day activities, news, sports, holidays, etc. These observations lead to two research questions:

1. Are there enough utilizable tweets to build a knowledge profile?
2. And if this is the case: How can we filter out these non-informative tweets that add noise to the profile?

Ideally, at the end of the filtering process, we would only be left with tweets that are relevant to the learner's knowledge profile.

[3] `http://www.delicious.com/`

[4] In September 2010 more than 145 million accounts were registered with the service: `http://blog.twitter.com/2010/09/evolving-ecosystem.html` (URL last accessed in June 2011)

In order to empirically investigate these questions, we have collected tweets from people that use Twitter as well as one of three social bookmarking services, namely CiteULike[5], Bibsonomy[6] and LibraryThing[7] (examples of each service are shown in Figure 1). CiteULike and Bibsonomy are academic bookmarking services which let users bookmark scientific papers. LibraryThing is a general book management service, from which we extract the bookmarked scholarly books. For each user in our data sets, we derive the knowledge profile from one of these bookmarking services (the ground truth profile) and investigate how well it can be approximated by the user's tweets.

It should be emphasized that we solely focus on the question to what extent a knowledge profile can be constructed from Twitter data alone. We do not use the derived profiles in an application.

The rest of the paper is organized as follows. In Sec. 2 related work is presented. Then, in Sec. 3 we outline our methodology. The experiments and results are presented in Sec. 4, followed by conclusions in Sec. 5.

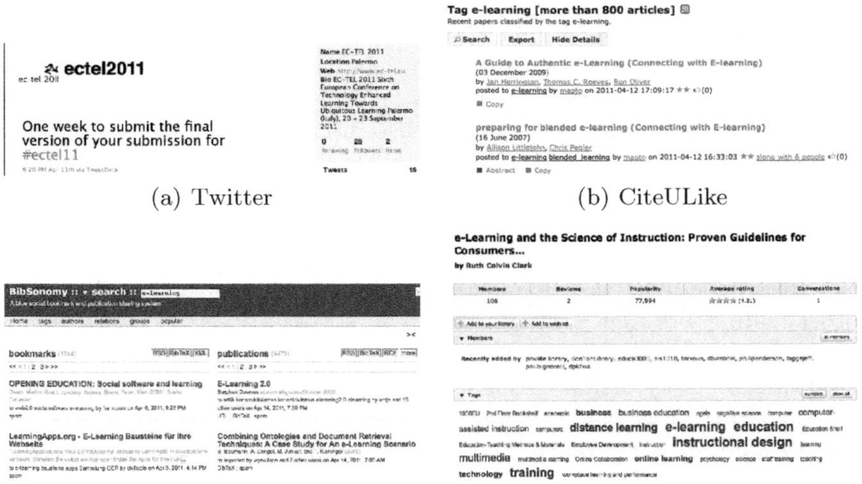

(a) Twitter (b) CiteULike

(c) Bibsonomy (d) LibraryThing

Fig. 1. Examples of the social Web services used in this study

2 Related Work

In this section, we report on related work that discusses different aspects of Twitter: (i) the motivation for people to use Twitter and what they use it for, (ii) how scholars utilize Twitter, and, (iii) how Twitter is used in learning. While

[5] `http://www.citeulike.org/`

[6] `http://www.bibsonomy.org/`

[7] `http://www.librarything.com/`

a considerable number of works investigate Twitter and news (e.g., news recommendation [15], real-time event detection [19], information spread [8]), fewer works focus on Twitter as a learning aid or a source of information about a user's knowledge.

Two important questions that have been investigated by a number of researchers are why do people use Twitter and what do they tweet about. The authors in [5] developed four broad categories of tweets: daily chatter (most common use of Twitter), conversations, shared information/URLs and reported news. Naaman et al. [14] identified nine different categories: information sharing, self promotion, opinions, statements and random thoughts, questions to followers, presence maintenance, anecdotes about me and me now. Moreover, they also found that the vast majority of users (80%) focus on themselves ("Meformers"), while only a minority of users are driven largely by sharing information ("Informers"). Westman et al. [20] performed a genre analysis on tweets and identified five common genres: personal updates, direct dialogue (addressed to certain users), real-time sharing (news), business broadcasting and information seeking (questions for mainly personal information). Zhao et al. [21] interviewed people about their motivations for using Twitter; several major reasons surfaced: keeping in touch with friends and colleagues, pointing others to interesting items, collecting useful information for one's work and spare time and asking for help and opinions. These studies show that a lot of tweets are concerned with the user herself; we hypothesize that among these user centered tweets, there is also useful ones for the derivation of the learner's knowledge profile.

In [16] the Twitter posting behaviour of academics was investigated. The authors conducted a study with twenty-eight faculty, postdocs and doctoral students to determine the extent of scholars tweeting citations. About 6% of the tweets with hyperlinks were found to be citations to peer-reviewed resources. Another finding was that a large percentage (40%) of tweeted citations appear within a week of the cited resource's publication date. Tweets related to a particular scientific activity, namely conference tweets, were investigated in [9]. Here, tweets related to three conferences (identified by the conferences' official hashtags) were analysed according to how scientific information is spread on Twitter. It was found that the users mainly tag for the benefit of their own network, they do not target the wider audience. The tweets are focused on announcing future events, links to slides, publications and other related information. Although not useful to the general public, these kind of tweets are elements we consider useful in the generation of a knowledge profile from Twitter posts.

A number of Twitter studies also attempt to predict user characteristics from tweets. While we are aiming to extract knowledge profiles, Michelson et al. [12] derive topic profiles from Twitter users. In their approach, the named entities (e.g., *Barack Obama*, *David Beckham*) are extracted from tweets, they are disambiguated and linked to their corresponding Wikipedia page and then a topic profile is build. We do not follow this approach as we compare the Twitter-based profile to a ground truth profile which is based on free text (abstracts of scientific papers). In [4,13,17] elementary user characteristics are inferred from Twitter, including gender, age, political orientation, regional origin and ethnicity.

In the context of learning, first studies have begun to appear that investigate the usability of Twitter as a learning help. For instance, Borau et al. [2] utilized Twitter as part of a course on English as a foreign language for Chinese university students. The students were instructed to tweet regularly as part of actively using English. The majority of students indicated after the experiment that using Twitter made them less shy when communicating in English. McWilliams et al. [11] used Twitter as a collaborative writing tool. Twitter was also found to enhance the instructor's credibility among students, when Twitter was not only used as an academic tool, but also as a social tool by the instructor [6].

3 Method

Twitter users tweet short messages with up to 140 characters about anything they choose. They can be followed by other users and themselves follow users in order to receive their tweets. Tweets can be directed (*@user*) and tweets can contain hashtags (*#ectel11*).

In order to investigate how well we can derive a knowledge profile from a user's Twitter data, for each user in the data set, we perform the following procedure:

1. Calculate a knowledge profile from a scholarly bookmarking service; this is the *ground truth profile*.
   (a) Index the user's bookmarks.
   (b) Derive a term vector $\boldsymbol{b}$ as knowledge profile.
2. Calculate a knowledge profile from Twitter data.
   (a) Select a number of the user's tweets for indexing.
   (b) Index the selected tweets.
   (c) Derive a term vector $\boldsymbol{t}$ as Twitter-based knowledge profile.
3. Calculate the cosine similarity between $\boldsymbol{b}$ and $\boldsymbol{t}$: $sim = \frac{\sum_{i=1}^{n} b_i \times t_i}{\sqrt{\sum_{i=1}^{n} b_i^2} \times \sqrt{\sum_{i=1}^{n} t_i^2}}$ where $sim \in [0, 1]$ and n is the number of terms in the term vector.

In the best case, that is, when the two vectors have the same direction, $sim = 1$, while at worst, $sim = 0$. The higher the similarity between the Twitter based knowledge profile and the ground truth, the better the Twitter-based knowledge profile approximates the ground truth profile.

Step 2(a), the selection of tweets for indexing, is the crucial step in our experiments - not all tweets are equally useful for the derivation of a learner's knowledge profile. Here, we investigate if it is possible to rely on simple rules to select those tweets that are useful. In the following paragraphs, we present a number of filtering options and the underlying hypotheses we have.

Language Identification: We cannot expect all tweets to be in English, often, tweets in English are mixed with tweets in other languages, in particular if the user is not a native English speaker. At the same time, we expect the ground truth profiles to consist largely of terms in the English language, as English is the main language in science. Thus, excluding non-English tweets is hypothesized to reduce the noise in our data sets. A simple, yet effective, approach to language identification has been proposed in [3]. Given sets of training texts in different languages, N-grams [10] are derived for each training language and an unknown text is then assigned to the language its N-gram distribution it matches best.

Weekdays versus Weekend: We hypothesize that tweets made during week days are more likely to be work and study related than tweets made during the weekend. Thus, we filter out tweets that are posted on Saturday or Sunday.

Style: We can also consider a multitude of tweet features for training a Naive Bayes classifier [18] that determines for each tweet if it should be classified as an informative or noisy tweet with respect to the knowledge profile. A total of 19 features are derived, including whether or not the tweet is a retweet or a directed tweet, the number of words and characters in the tweet, the number of exclamation marks and question marks, the number of hashtags, the number of smileys in the tweet as well as the number of letters repeated four or more times (e.g., "oooooh" or "sooooo"). First, for a number of tweets, it is determined whether they are suitable for the knowledge profile or not (the training data). Then, the features are extracted from these tweets, a model is learnt based on the training data and this model is applied to predict whether a tweet in the test data should be included in the Twitter-based profile.

External Documents: Finally, instead of removing tweets, we can also include extra information. Tweets are short and although they can be informative by themselves, often a tweet contains a link and a very short expression of interest or an explanation. Thus, we also consider the external documents that are linked to as an additional potential source for the knowledge profile.

4 Experiments

4.1 Data Set Overview

As we investigate how accurately we can derive a knowledge profile from users' Twitter messages, we require a ground truth. The goal of our work is to extract user profiles that are utilizable in E-learning applications, thus we focus on the users' knowledge profiles in scholarly subjects. We decided to collect these ground truth knowledge profiles automatically, as it allows us to conduct our experiments on a larger scale. To this end, we rely on existing social Web services that are aimed at or include the organization and sharing of scholarly works. In particular, we relied upon:

- **CiteULike**: a service for organizing and sharing scientific publications.
- **Bibsonomy**: an alternative to CiteULike, that allows the bookmarking of scholarly papers as well as Web pages, and,
- **LibraryThing**: a service to catalogue books (a virtual bookshelf).

We collect the data of Twitter users, for whom we are able to link their Twitter account to one of the three bookmarking/cataloguing services listed above. We found the users in our data sets by crawling Google Profiles[8] and claimID[9], where users can list and manage their online identities on various social Web services.

Table 1 gives an overview of the collected data. The users found for each bookmarking services are treated as a separate data set. We crawled up to the latest 3150 tweets per user (Twitter limits the maximum amount of accessible tweets). In total, we found 73 users of Twitter and CiteULike, 47 users of Twitter and Bibsonomy and 122 users of Twitter and LibraryThing. The LibraryThing service is not focused on scholarly works, any book can be added. Since we are interested in the scholarly books, we only consider a subset of all bookmarks, namely those that can be found at Amazon[10] under the following categories: *Textbook*, *Science*, *Religious Studies*, *Social Sciences*, *Computer Science* and *Engineering*. The number of users in our data sets are very small compared to the total numbers of users of Twitter and the bookmarking services. Listing one's various social Web accounts on Google Profiles and claimID is voluntary and many users may either not know these services or may not feel the need nor want to publicly list their handles. The advantage of relying on Google Profiles and claimID is that the data is provided by the users themselves, thus we do not need to infer a linkage between accounts of different social Web services.

While in CiteULike and LibraryThing users bookmark publications and books only, in Bibsonomy users can bookmark publications as well as web pages.

From the CiteULike and LibraryThing services, we indexed the titles, the abstracts and descriptions (if available) as well as the tags assigned to the bookmark by the users in our data sets. In the case of Bibsonomy, we also indexed the bookmarked web pages.

Before indexing the Twitter posts, in each tweet, if applicable, the user names (*@user*) and hyperlinks (`http://bit.ly/kxreiG`) were removed. Hashtags (#educationOnline) were split according to a simple capital letter rule. Thus, a post such as "*@Tom E-learning courses start in April #educationOnline #course*" will be transformed into "*E-learning courses start in April education online course*".

All bookmarks and tweets were indexed with the Lemur Toolkit[11] with Krovetz stemming applied [7]; stopwords were removed. A user's knowledge profile is simply a vector of terms with weights according to the frequency each term occurs in the bookmark index or Twitter index. In order to avoid overestimating the similarity between the ground truth profile and the Twitter-based profile, terms that occur in more than 1% of a newspaper corpus (from the years 1995-1997) were removed; examples of removed terms are *just*, *love* and *weather*. We refrained from calculating vector elements according to TF.IDF [1], as such weights are not useful here: if a tenth of the CiteULike articles in our index for example would include the term *genetics*, it would receive a low weight, although it may actually represent the user's knowledge profile very well.

[8] `http://profiles.google.com/`
[9] `http://claimid.com/`
[10] `http://www.amazon.com/`
[11] `http://www.lemurproject.org/`

Table 1. Overview of the derived data sets (σ is the standard deviation). CiteULike and LibraryThing do not allow the bookmarking of web pages. The rows marked "Ext. URLs/Tweet" indicate the average, median, etc., number of hyperlinks in a tweet.

	CiteULike +Twitter	Bibsonomy +Twitter	LibraryThing +Twitter
#Users	73	47	122
Average #Publications	490.1 ($\sigma = 689.4$)	291.3 ($\sigma = 374.3$)	63.5 ($\sigma = 82.3$)
Median #Publications	32	244.0	162.5
Minimum #Publications	2.0	0.0	1.0
Maximum #Publications	4397.0	1651.0	460.0
Average #Webpages		542.5 ($\sigma = 1006.1$)	
Median #Webpages		198.0	
Minimum #Webpages		0.0	
Maximum #Webpages		4038.0	
Average #Tags/Bookmark	3.7 ($\sigma = 2.4$)	3.2 ($\sigma = 1.6$)	1.6 ($\sigma = 1.9$)
Median #Tags/Bookmark	3.4	2.9	0.9
Minimum #Tags/Bookmark	0.0	0.5	0.0
Maximum #Tags/Bookmark	16.5	7.3	7.4
Average #Twitter Posts	1607.0 ($\sigma = 1235.1$)	808.7 ($\sigma = 1011.2$)	1909.4 ($\sigma = 1182.8$)
Median #Twitter Posts	3150.0	3095.0	3150.0
Minimum #Twitter Posts	23.0	1.0	1.0
Maximum #Twitter Posts	3150.0	3150.0	3150.0
Average #Ext. URLs/Tweet	0.5 ($\sigma = 0.3$)	0.5 ($\sigma = 0.3$)	0.4 ($\sigma = 0.3$)
Median #Ext. URLs/Tweet	0.4	0.5	0.3
Minimum #Ext. URLs/Tweet	0.0	0.0	0.0
Maximum #Ext. URLs/Tweet	1.8	1.0	1.5

Examples of the top weighted terms of typical user profiles derived from the four social Web services are shown in Table 2. Note, that they are all from different users; shown are the stemmed terms. The CiteULike profile clearly focuses on bioinformatics and genetics, whereas the Bibsonomy profile contains terms typical for the Semantic Web. This is a general trend we found across these two services in our data sets: CiteULike is frequented by users coming from the biomedical domain, while Bibsonomy is used by users whose profile indicates work in Computing Science and related areas. The LibraryThing profile shown is mixed: on the one hand, it contains terms that indicate knowledge in areas of computer science (*visualize*, *internet*, *web*), but on the other hand it also contains terms such as*obama* and*barack*, indicating an interest in the political domain. Since the LibraryThing service lets users add books they have read (or might read), instead of scientific papers, it is possible to a lesser degree to make a distinction between a user's general interest and his scholarly knowledge.

Furthermore, the categories we rely on to filter academic books play a role here as well, as books may belong to different categories. The Twitter profile in the last column shows that this user has tweeted the most about current news.

Table 2. The top weighted terms of example user profiles drawn from each data set

CiteULike	Bibsonomy	LibraryThing	Twitter
connotea	ontology	web	rt
csb	stlab	navigation	wikileak
interaction	web	findable	assange
bacillu	semantic	obama	cablegate
gene	workshop	morville	libya
cell	owl	barack	leak
stochastic	dns	interface	guardian
yeast	descriptionsandsituations	visualize	google
genetic	dolce	internet	twitter
biology	ontologydesign	designer	wiki

4.2 Results

We first present the results of the baseline (how well can the knowledge profile be approximated by utilizing all tweets of a user?) and the results of the upper bound (if we have an oracle, that tells us which tweets are the right ones, what is the best possible knowledge profile we can achieve?). Then, the results of the tweet selection and expansion experiments are reported. We will show, that Twitter is a useful source for deriving knowledge profiles, though predicting which tweets aid in profiling is a difficult task.

Baseline & Upper Bound. The baseline is derived by calculating the similarity between the Twitter-based profiles and the ground truth (bookmarking service based) profiles. We report the mean, median, minimum and maximum cosine similarity across the users of each data set. The results are reported in Table 3. CiteULike and Bibsonomy reach an average similarity of 0.18 and 0.2, respectively, while for the LibraryThing data set the similarity is considerably lower (0.07). This shows, as expected, that profiles based on all tweets, on average, are not suitable for deriving a learner's knowledge profile. However, when we consider the maximum cosine similarity, that is, the similarity reached by the user whose tweets match the ground truth profile the most, in the CiteULike data set, the similarity is greater than 0.9, whereas in the other two data sets the similarity reaches ≈ 0.5. Thus, indeed there are some users, whose tweets are very much related to their professional life, while for most users, those tweets are either not existing or hidden among the non-informative tweets.

In order to investigate if users simply do not tweet about their study or work life or if it is indeed a case of overbearing non-informative tweets, we experimentally derived the highest cosine similarity possible with our model and our data. We conducted the following experiment: at step $k = 0$, we start with a set T which contains all tweets $t_1, ..., t_m$ of a user and an empty set S. We then iteratively add tweets to S: in each step k, we add the tweet t_x to S such that $S \cup t_x$ together form the Twitter-based profile having the highest cosine similarity with the ground truth profile. Specifically, in the first step ($k = 1$), the tweet t_i (among all tweets in T) is selected which itself has the highest cosine similarity with the ground truth profile. Tweet t_i is added to the empty set S and removed from set T. In step $k = 2$, the next selected tweet from T is the one that together with tweet t_i from set S forms the profile that is most similar to the ground truth profile. And so on for $k = 3, 4, ..., m$. At step m, the set T is empty and set S contains all tweets. At each step k, we record the cosine similarity the tweets in S reach with respect to the ground truth profile. The greedy upper-bound is then the maximum similarity we record across all k[12]. The upper-bound results for each data set, are also reported in Table 3. On average, if the "right" tweets are selected to represent the user's knowledge profile, the average cosine similarity reaches between 0.4 (LibraryThing) and 0.6 (CiteULike), which are substantial improvements over the baselines. The minimum similarity is still low; there are Twitter users that offer very few or no suitable information in their tweets. On the other hand, the maximum similarity with the ground truth reaches 0.8 or higher across all data sets. These results show, that if we would be able to select the right tweets, we could derive useful knowledge profiles from Twitter.

Table 3. Baseline results and greedy upper bound. Reported is the mean (standard deviation), median, maximum and minimum cosine similarity between the Twitter-based profile vectors and the bookmarking service-based profile vectors across the users of each data set.

Data Set		**Mean**	**Median**	**Min.**	**Max.**
CiteULike	baseline	0.176 ($\sigma = 0.171$)	0.143	0.002	0.933
	upper-bound	0.608 ($\sigma = 0.241$)	0.654	0.008	0.994
Bibsonomy	baseline	0.203 ($\sigma = 0.130$)	0.192	0.003	0.550
	upper-bound	0.545 ($\sigma = 0.253$)	0.555	0.006	0.919
LibraryThing	baseline	0.075 ($\sigma = 0.070$)	0.059	0.000	0.473
	upper-bound	0.412 ($\sigma = 0.185$)	0.411	0.022	0.838

In Figure 2(a) we plot the development of the cosine similarity of set S for the tweets of three random users. Although the absolute cosine similarity differs, the process is similar across all of them: initially, adding tweets to set S increases the cosine similarity, but early on a peak is reached (the reported upper-bound) and adding more tweets reduces the cosine similarity again. Across the data sets, the average, minimum and maximum number of tweets at the peak are reported in Table 4. Thus, although for the majority of users we have 3150 tweets in our data set, the highest similarity with the ground truth is reached after $30 - 40$ tweets.

To provide a better impression of the difference between the baseline and upper-bound for each individual user, consider the plots in Figure 2(b), 2(c) and 2(d). Here, in each plot, we sorted the users in the data set according to the cosine similarity of their upper-bound from high to low and plotted the corresponding

[12] The upper bound is "greedy" due to the iterative process. It is an approximation of the true upper-bound, which would require calculating all possible combinations of tweets in S, which is computationally not feasible.

Table 4. Mean, minimum and maximum number of tweets at the greedy upper-bound

	Mean	Minimum	Maximum
CiteULike	42.1	1	229
Bibsonomy	31.3	1	113
LibraryThing	35.8	1	133

baseline. It is evident, that across all data sets the gained improvements are large. It is also apparent, that even if the baseline similarity is low, if the right tweets are selected, a very good knowledge profile can be generated.

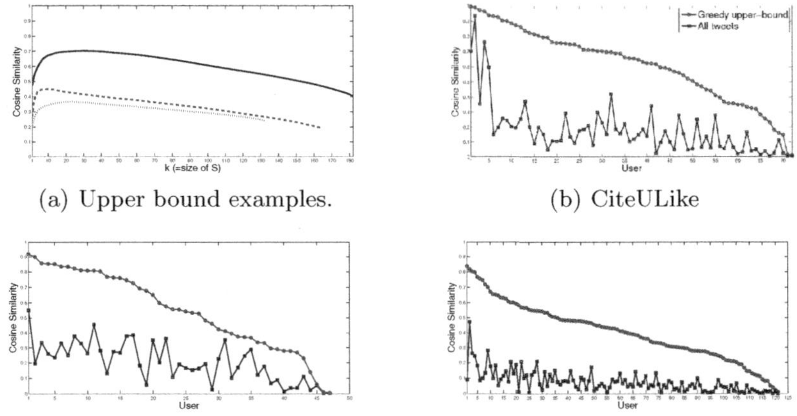

(a) Upper bound examples. (b) CiteULike

(c) Bibsonomy (d) LibraryThing

Fig. 2. Examples of the upper bound over a range of k are shown in Fig. 2(a). The cosine similarity for the baseline and upper bound across all users of each data set are plotted in Fig. 2(b)-2(d).

Tweet Filtering & Expansion. In the next step, we implemented the heuristics for tweet filtering and expansion introduced in Sec. 3. First, we filter out all tweets and bookmarks that are not in English. While the number of Twitter posts decreased by 14% when filtering out non-English posts, the three bookmarking services had fewer entries identified as non-English (CiteULike 1%, Bibsonomy 8%, LibraryThing 0.5%). In a second experiment, we removed all tweets posted at the weekend from the Twitter profiles. In a third experiment, we expanded the Twitter index by including documents that are linked from tweets. Finally, we built a Naive Bayes classifier and performed 5-fold cross validation: each data set was split into five equal parts and four parts were used for training the classifier and one was used for testing. This procedure was repeated 5 times (each time, a different part was held out for testing) and the results on the test data were averaged. The labels for the training data were derived automatically: tweets that have a high cosine similarity with the ground truth profile were labelled as informative (to be selected) while tweets with a similarity of ≈ 0 with the ground truth profile were labelled non-informative (not to be selected). The classifier was applied on the test data and each tweet was classified as informative or not and only the informative tweets were included in the Twitter-based knowledge profile.

The results of all experiments are shown in Table 5. Underlined entries indicate an improvement over the respective baseline (Table 3). Selecting only tweets posted during weekdays does not yield improvements over the baseline for any data set when considering the mean, our most important measure. Excluding non-English tweets has a positive effect on the CiteULike and LibraryThing data sets, though only marginally. Expanding the Twitter based profile by including documents whose links were tweeted has a drastic effect on the Bibsonomy data set: while the baseline mean cosine similarity is 0.20, the mean cosine similarity of the expanded profile is 0.35, a 75% increase. Notable is also the increase in the maximum of the CiteULike data set for this experiment: while in the baseline, the maximum similarity is 0.55, in the expanded profile, it reaches 0.85, a 55% increase. We strongly suspect that this result is due to the user group that we found to mostly make up our Bibsonomy user set: users bookmarking papers in areas of computer science. Due to the nature of the field, there is a lot of relevant information on the Web and thus hyperlinks posted on Twitter by such users may often refer to aspects of computer science.

Table 5. Overview of the cosine similarity when performing tweet selection and including linked external documents

Filtering	Data Set	Mean	Median	Minimum	Maximum
English Only	CiteULike	0.184 ($\sigma = 0.178$)	0.145	0.002	0.942
	Bibsonomy	0.202 ($\sigma = 0.129$)	0.188	0.004	0.536
	LibraryThing	0.077 ($\sigma = 0.070$)	0.060	0.000	0.482
Weekdays Only	CiteULike	0.173 ($\sigma = 0.167$)	0.143	0.000	0.921
	Bibsonomy	0.200 ($\sigma = 0.134$)	0.190	0.004	0.549
	LibraryThing	0.074($\sigma = 0.070$)	0.058	0.000	0.477
Including External Documents	CiteULike	0.169 ($\sigma = 0.118$)	0.167	0.005	0.474
	Bibsonomy	0.350 ($\sigma = 0.254$)	0.332	0.011	0.846
	LibraryThing	0.079 ($\sigma = 0.068$)	0.064	0.000	0.339
Naive Bayes	CiteULike	0.192 ($\sigma = 0.168$)	0.151	0.003	0.936
	Bibsonomy	0.195 ($\sigma = 0.128$)	0.170	0.000	0.557
	LibraryThing	0.081 ($\sigma = 0.070$)	0.064	0.000	0.489

The results of the Naive Bayes classifier are inconsistent, we found the largest improvement, 9%, over the baseline in the CiteULike data set (mean cosine similarity); however, in the Bibsonomy data set, the baseline outperformed the classifier based tweet selection. This implies, that while predicting which tweets are concerned with aspects of a user's work or study is to some degree influenced by the style of a tweet, it should not be the only source of information. The tweet content and the Twitter network structure of the user are likely to play a significant role as well.

Across all experiments, the LibraryThing data set performs less well than the CiteULike and Bibsonomy data sets. It will require further investigation to determine the differences and similarities between them. One potential explanation may be, that the description of the bookmarked books are often short and may not always contain the key concepts. The bookmarked scientific publications in CiteULike and Bibsonomy on the other hand do mostly also contain the abstract of the work, which by its nature contains the keywords and topic words that we expect in the profile vectors.

5 Summary and Future Work

In this work, we set up a framework that acts as a testbed for further exploration of learner profile gathering on the social Web. We investigated how well a profile built from a user's tweets can approximate a user's knowledge profile. We offered a methodology of how to collect relevant data sets automatically, by considering users that explicitly link their Twitter account and a scholarly bookmarking service account together. We found that indeed a large number of users tweet not only about news, sports, etc. but also about aspects of their professional life. By determining the greedy upper-bound between the ground truth profiles and Twitter-based profiles, we could show that if the right tweets were selected, a good approximation of the ground truth profile is possible for the majority of users. We further found that for different user groups, different aspects of Twitter posts are useful, in particular the Bibsonomy data set, which includes many users with knowledge related to computer science, profited considerably from the inclusion of documents that were linked to in Twitter posts. Finally, we observed that predicting which tweets to include in the Twitter based profile is a difficult task.

These results leave a lot of potential for future work. A limitation of our work, that we have not yet adressed is the question of whether the users in our data set resemble the average Twitter users. Based on the genres, types of tweets and their distribution [14,20], we will conduct a qualitative analysis of a set of random tweets in our data sets. Additionally, we will further investigate the tweet selection problem by increasing the number of features, but also by taking Twitter's network structure into account. Finally, we also plan to increase the data set size further, and to investigate different user groups within each data set such as professionals versus students, males versus females and user groups of different geographic origin.

References

1. Baeza-Yates, R.A., Ribeiro-Neto, B.: Modern Information Retrieval. Addison-Wesley Longman Publishing Co., Inc., Amsterdam (1999)
2. Borau, K., Ullrich, C., Feng, J., Shen, R.: Microblogging for language learning: Using twitter to train communicative and cultural competence. In: Spaniol, M., Li, Q., Klamma, R., Lau, R.W.H. (eds.) ICWL 2009. LNCS, vol. 5686, pp. 78–87. Springer, Heidelberg (2009)

3. Cavnar, W., Trenkle, J.: N-gram-based text categorization. In: SDAIR 1994, pp. 161–175 (1994)
4. Hecht, B., Hong, L., Suh, B., Chi, E.H.: Tweets from justin bieber's heart: the dynamics of the location field in user profiles. In: CHI 2011, pp. 237–246 (2011)
5. Java, A., Song, X., Finin, T., Tseng, B.: Why we twitter: understanding microblogging usage and communities. In: Proceedings of the 9th WebKDD and 1st SNA-KDD 2007 workshop on Web mining and social network analysis, pp. 56–65. ACM Press, New York (2007)
6. Johnson, K.: The effect of Twitter posts on students perceptions of instructor credibility. Learning, Media and Technology 36(1), 21–38 (2011)
7. Krovetz, R.: Viewing morphology as an inference process. In: SIGIR 1993, pp. 191–202 (1993)
8. Lerman, K., Ghosh, R.: Information contagion: An empirical study of the spread of news on Digg and Twitter social networks. In: ICWSM 2010, pp. 90–97 (2010)
9. Letierce, J., Passant, A., Breslin, J., Decker, S.: Understanding how Twitter is used to widely spread Scientific Messages. In: WebSci 2010: Extending the Frontiers of Society On-Line (2010)
10. Manning, C.D., Schütze, H.: Foundations of statistical natural language processing. MIT Press, Cambridge (1999)
11. McWilliams, J., Hickey, D., Hines, M., Conner, J., Bishop, S.: Using Collaborative Writing Tools for Literary Analysis: Twitter, Fan Fiction and The Crucible in the Secondary English Classroom. Journal of Media Literacy Education 2(3), 238–245 (2011)
12. Michelson, M., Macskassy, S.A.: Discovering users' topics of interest on twitter: a first look. In: AND 2010, pp. 73–80 (2010)
13. Mislove, A., Lehmann, S., Ahn, Y.-Y., Onnela, J.-P., Rosenquist, J.N.: Understanding the Demographics of Twitter Users. In: ICWSM 2011 (2011)
14. Naaman, M., Boase, J., Lai, C.-H.: Is it really about me?: message content in social awareness streams. In: CSCW 2010, pp. 189–192 (2010)
15. Phelan, O., McCarthy, K., Smyth, B.: Using twitter to recommend real-time topical news. In: RecSys 2009, pp. 385–388 (2009)
16. Priem, J., Costello, K.L.: How and why scholars cite on twitter. In: ASIS&T 2010, pp. 75:1–75:4 (2010)
17. Rao, D., Yarowsky, D., Shreevats, A., Gupta, M.: Classifying latent user attributes in twitter. In: SMUC 2010, pp. 37–44 (2010)
18. Rish, I.: An empirical study of the naive Bayes classifier. In: IJCAI 2001 Workshop on Empirical Methods in Artificial Intelligence, pp. 41–46 (2001)
19. Sakaki, T., Okazaki, M., Matsuo, Y.: Earthquake shakes Twitter users: real-time event detection by social sensors. In: WWW 2010, pp. 851–860 (2010)
20. Westman, S., Freund, L.: Information interaction in 140 characters or less: genres on twitter. In: IIiX 2010, pp. 323–328 (2010)
21. Zhao, D., Rosson, M.B.: How and why people twitter: the role that micro-blogging plays in informal communication at work. In: GROUP 2009, pp. 243–252 (2009)

Orchestration Signals in the Classroom: Managing the Jigsaw Collaborative Learning Flow

Davinia Hernández-Leo, Raul Nieves, Ernesto Arroyo, Andrea Rosales, Javier Melero, Pau Moreno, and Josep Blat

GTI, ICT Department, Universitat Pompeu Fabra, Roc Boronat 138, 08018 Barcelona, Spain
{davinia.hernandez,raul.nieves,ernesto.arroyo, andrea.rosales,javier.melero,pau.moreno,josep.blat}@upf.edu

Abstract. The orchestration of collaborative learning processes in face-to-face physical settings, such as classrooms, requires teachers to coordinate students indicating them who belong to each group, which collaboration areas are assigned to each group, and how they should distribute the resources or roles within the group. In this paper we present an Orchestration Signal system, composed of wearable Personal Signal devices and an Orchestration Signal manager. Teachers can configure color signals in the manager so that they are transmitted to the wearable devices to indicate different orchestration aspects. In particular, the paper describes how the system has been used to carry out a Jigsaw collaborative learning flow in a classroom where students received signals indicating which documents they should read, in which group they were and in which area of the classroom they were expected to collaborate. The evaluation results show that the proposed system facilitates a dynamic, visual and flexible orchestration.

Keywords: computer-supported collaborative learning, orchestration, ubiquitous wearable devices, flexible learning flow.

1 Introduction

Physical spaces, such as classrooms or the playground, have a relevant role in collaborative learning since they can bring students together and shape their interactions [1, 2]. The characteristics of a particular space can encourage experimentation, exploration, collaboration, and discussion. The introduction of technologies in physical educational spaces has brought new possibilities that are transforming the learning experiences [3]. Computational artifacts such as media representation systems, remote interaction systems, room-scale peripherals and devices such as handhelds have moved from being conceived as means to support distance communication and learning to be elements embedded in augmented physical spaces that can enrich face-to-face learning experiences [4, 5]. Teachers can design new learning strategies according to their perceived affordance regarding the properties of these technologies [6].

C. Delgado Kloos et al. (Eds.): EC-TEL 2011, LNCS 6964, pp. 153–165, 2011.

Technology-enhanced educational spaces go beyond the desktop computing by using interactive artifacts and computing facilities derived from three fields: tangible user interfaces, ubiquitous computing and augmented reality [7]. Tangible user interfaces involve explicit contact with the computing artifacts such as tabletops, smartboards, multitouch screens and tangible building blocks [8, 9, 10, 11]. Ubiquitous computing deals with situating and embedding devices within a space so that computational power is available everywhere and the interaction with the devices is mediated through this space. This is now possible due to improvements in computing power, hardware size, wireless communications, power management, and software architectures. Ubiquitous computing offers new possibilities for helping people organize and work collaboratively, mediating social interactions in technology-rich spaces. Ubiquitous computing devices used to support learning settings include light-weight and roomware awareness tool devices [12], mobile phones, QR codes, radio-frequency identification tags and GPS [13, 14, 15]. The devices can incorporate sensors, actuators or both, and can also be network linked. These tangible and ubiquitous devices are augmenting the reality, in the sense that they overlay and add digital information to real objects or integrate computer power into them [5, 16].

In this paper, we introduce a system that adds digital orchestration information to ubiquitous devices that can be worn by students. This orchestration information refers to coordination aspects of collaborative learning processes [17], such as group formation indicators, signals to indicate the distribution of resources during the activity, etc. While the orchestration problem has been to a large extent solved in the context of PC-oriented learning environments (see for example the collaborative learning flows created with Collage and run in IMS LD compliant systems [18]), no solutions have been proposed to provide coordination information to students in wearable devices so that the use of a PC is not required and, therefore, more agile dynamics in different spaces are enabled. We have considered low-cost wearable devices in contrast to mobile phone-based approach because, on the one hand, phones tend to be more expensive, sometimes it is difficult for a teacher to ensure that every student will own one which is compliant with the system requirements, and students can lose the concentration on the activity if they play with other mobile applications. On the contrary, wearable devices could be designed so that they are more visual and generic and can be used by students at any educational level in the classroom, the playground, etc.

The system, named Orchestration Signal system, has been used and evaluated in a real learning situation where 27 students are expected to follow a Jigsaw collaborative learning flow [19] for the collaborative analysis of three cases. According to the Jigsaw pattern, in a first phase students read individually one of the cases, in the second phase they meet in expert groups with other students that have read the same cases, and finally, in the third phase, the students join Jigsaw groups composed of students that have read different cases so as to solve a common problem that required the knowledge studied in the three cases. The research questions explored in the evaluation are: Does the orchestration signals enable/facilitate the coordination of the Jigsaw learning flow in the classroom? Are the orchestration signals flexible enough to deal with unexpected situations? Are the characteristics of the prototype usable for the purposes of the Jigsaw learning flow orchestration? What aspects need/can be improved?

The remainder of the paper is structured as follows. Section 2 describes the design, functioning and modules of the Orchestration Signal system. Then, section 3 explains the activity and educational context where the system has been used for its evaluation. The evaluation results are presented in section 4. Finally, section 5 is devoted to conclude the paper and indicate the future lines of research derived from this research work.

2 The Orchestration Signal System Prototype

The *Orchestration Signal system* prototype includes multiple *Personal Signal devices* (PS-device), which have visualization module and a communication module, and the *Orchestration Signal manager* (OS-manager), a graphical user interface to monitor and control the experience. As illustrated in Fig. 2, the PS-device visualization module displays several color combinations associated to signals that teachers would like to send students for indicating orchestration aspects of the collaborative learning flow, such as resources distribution or group formation. It consists of 4 leds (red, green, blue and yellow), which can be turned on and off individually or in pairs trough a communication module. This module includes a transceiver RF12B that allows the PS-device to be remotely controlled by a central computer from up to 100 meters away. A central computer (e.g., the PC in the classroom or the teachers' laptop) runs the OS-manager where teachers can configure the orchestration signals to be transmitted to the PS-devices.

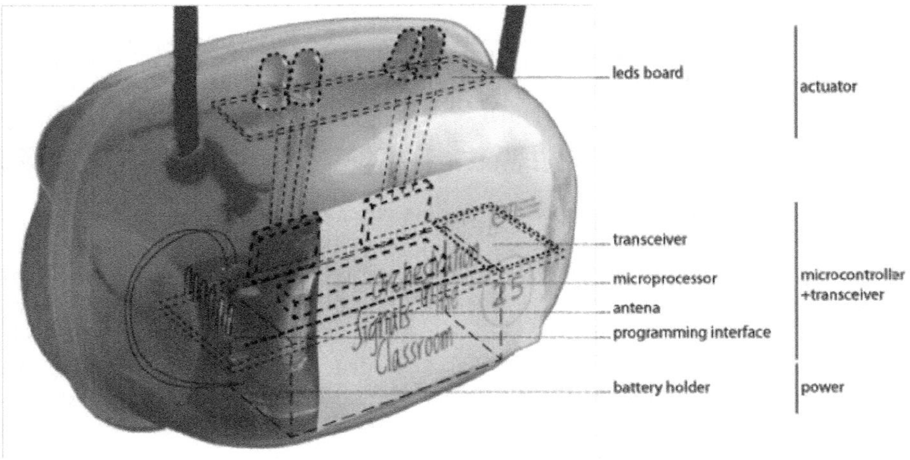

Fig. 1. Personal Signal device (PS-device)

The hardware used in the development of the PS-devices is based on JeeNodes, a low-cost Arduino clone board [20]. The board is powered by 3 AA batteries and includes an ATmega328 microcontroller which supports embedding programmed logic.

The system includes a master node that relays commands between the computer with the OS-manager and each PS-device. The communication is unidirectional in order to avoid message sequencing and bottlenecks. Moreover, the data sent is coded into only 1 byte in order to optimize communication speed data transfer rate.

The central computer hosts the OS-manager with a uni-directional serial link with the master node. The OS-manager interface visualizes a canvas box associated to each PS-device (see Fig. 2). In each canvas teachers can configure two possible types of signals (a color or a combination of two colors) to be sent to each device. Besides the OS-manager has three buttons for controlling batch message transfer. Two of the buttons send either the first or second color combination that each PS-device is setup to. The third button turns all units off.

Fig. 2. Orchestration-Signal manager (OS-manager)

A unique number that matches the internal configuration and external labeling of each device identifies each PS-device in the OS-manager. The graphical box representing each device in the OS-manager also holds an input text-field to be filled with the name of the student for quicker identification. Bellow each box, a small button allows for individual signal transfer, in order to enable testing, individual correction, and group-membership readjustment. Each box can be freely dragged and dropped within the canvas, so that teachers can order them for a comfortable use (for example, arranging them to emulate the physical classroom arrangement).

The physical casing of the PS-devices was selected to be physically and visually unobtrusive so as to minimally disrupt the user's activity. The devices can be moved and rotated freely around the participants' neck to allow for a better viewing angle, or to share and match their visual indicator to that of their partners. The visual signal indicator is also located on a surface oriented to optimally display the illuminated led lights when seen from above. Finally, the casing is low-cost for a prototype and can be easily replaced.

3 Activity Based on the Jigsaw Collaborative Learning Flow

The activity proposed for testing the Orchestration Signal system prototype is framed in the context of a master seminar on Education & Media Communication. A total of 27 students, with 12 different nationalities – 6 men and 21 women, were enrolled in the seminar. Most of them (20) had a media communication or journalism background, 3 are pedagogues, and the remainder had a diverse background. All of them are interested in the educational field, however their use of educational technologies is limited (for example, only 4 have used the Moodle platform, and 1 has used the Blackboard management system). Table 1 summarizes the design of the activity following the structure of the Jigsaw collaborative learning flow pattern.

Table 1. Design of the activity according to the Jigsaw collaborative learning flow

Phases of the Jigsaw Collaborative Learning Flow Pattern [19]	Specific activity in a scenario for the collaborative reading of three cases on the use of ICT in Education	Distribution in the classroom and signal required
Initial phase: Jigsaw Groups are formed in order to collaboratively solve a global problem or task. This problem is divided into sub-problems. Each student in a Jigsaw Group studies a sub-problem.	Since there are three cases (A, B, C), the Jigsaw groups need to be formed by a minimum of three members (each of them having read a different case). Since 27 students are enrolled in the course, it is expected that 9 students will read each case and, therefore, 9 Jigsaw groups will be formed. In this phase each student reads the assigned case (A, B or C) and answers a number of proposed questions about the case.	In this initial phase, since the activity is individual, the members of each Jigsaw group do not need to be physically close in the classroom, however they should pick one case (out of three) so that in each member of a Jigsaw group reads a different case. *Orchestration signal required*: indicating the case to pick
Expert phase: Students having worked on the same sub-problem meet, forming Expert Groups, in order to exchange ideas about their sub-problem.	In order to have Expert Groups of a reasonable size, a total of 6 Expert groups will be formed (there will be two Expert Groups on the same case, each of them with 4 or 5 students having read the same case). The members of each Expert group will meet in order to reflect on the case and discuss their answers to the questions.	Expert groups will meet in a specific work area of the classroom so that they are close to each other. These areas should be as much separated as possible from each other. *Orchestration signal required*: indicating expert groups and group working areas
Jigsaw phase: Students of each Jigsaw Group meet again and each member contributes with their expertise in order to solve the global problem.	The three members of each Jigsaw group will meet and compare the cases from the perspective of the proposed questions (which are common to the cases). The group must complete an on-line form with an agreed description of the differences identified in the cases for each question.	Jigsaw groups will meet in a specific work area of the classroom so that they share a PC and are close to each other. These work areas should be as much separated as possible from other Jigsaw groups. *Orchestration signal required*: indicating Jigsaw groups and group working areas

The activity consisted in the collaborative reading of three cases explaining different real scenarios that apply ICT to enhance learning. The cases included a narrative describing the scenarios and a set of questions that students had to answer. Table 2 also specifies the expected number of groups and members as well as the requirements regarding the distribution of resources and spaces in the classroom and, in consequence, the signals needed in order to indicate students the orchestration aspects of the activity.

The signals associated to individuals (students) were distributed using the Orchestration Signal system. However, the signals needed to identify the cases and the group working areas or spaces were built using color cardboards, so that they matched the LED colors, such as those in Fig. 3.

Fig. 3. Color cardboards to signal the collaboration working areas

4 Evaluation

To evaluate the research questions posed in the introduction regarding the facilitation of the orchestration and usability of the approach, we followed a mixed evaluation method [21] combining quantitative and qualitative data and the use of several data gathering techniques. These data are triangulated [22] in order to provide trustworthy results. The data gathering techniques were: observations collected by 2 researchers – they noted down information regarding timing, incidents, use of devices, etc.; post-questionnaires with closed and open questions for students; and post-questionnaires with open questions for teachers. Two teachers of the seminar completed this questionnaire. They did not belong to the research team proposing the Orchestrating Signal system.

Table 2 describes the actual enactment of the experience in comparison to the expected situation summarized in Table 1. The table lists incidents of different nature that occurred in the different phases of the activity, the resulting composition of expert and Jigsaw groups and the time that was needed for accomplishing the orchestration aspects and performing the tasks. Fig. 4 shows some images illustrating the use of the PS-devices.

Table 2. Actual enactment of the experience (as annotated by the observers)

Phases	Incidents	Actual enactment vs. what was planned	Actual timing
Initial phase	- 3 students did not attend the class (total number of students in the class: 24) - The red led of PS-device n° 24 did not work	- The **distribution of cases** in the S-manager had to be **changed so that each case was read by a balanced number of 8 students** (not 9 as expected). The **configuration of groups** were also **changed** so that expert groups were composed of 4 members, and every Jigsaw group (8, not 9 as expected) included 3 members, each of them expert in a different case - 23 students picked and read the correct case, pre-assigned by the teacher in the S-manager, according to the signals received in their PS-device - **1 student picked and read an incorrect case** (case B), not the one pre-assigned by the teacher in the S-manager (case A), since the student only saw the green signal in the PS-device (not the green-red signal as expected) - Case A was read by only 7 students, case B was read by 9 students, case C was read by 8 students	Presenting the whole activity: 10 minutes Orchestration: 1 minute (teacher sends signal and students receive signals), 2 minutes (students pick their cases marked with colors – corresponding to the signals), total of 3 minutes Task (reading the case): 12 minutes
Expert phase	- Student with PS-device n° 15 had to leave the class during the expert phase	- Because of the problem with PS-device 24, **1 of the expert groups** was composed of **3 members** and **another expert group** was composed of **5 members** - The student with PS-device 15, leaving the class, was a member of the group with 3 members, therefore this **group** finished with activity **with only 2 members** - At the end of the expert phase, only 6 students were expert in case A	Orchestration (receiving the signal and joining the expert group in the area of the classroom indicated with a similar signal): 2 minutes Task (discussing the case): 15 minutes
Jigsaw phase	- No incidents	- The **Jigsaw groups composition** was **changed** again in the PS-manager so that all the groups had at least one member expert in every case. A total of **6 Jigsaw groups** were defined, **5 of each composed of 4 members and 1 formed of 3 members**. While each had a member expert in case A, 5 of the groups have 2 experts either in case B or C	Orchestration (receiving a new signal and joining the expert group in the area of the classroom indicated with a similar signal): 1 minute Task (explaining cases, completing on-line form): 20 minutes

The analysis of the data shows that the experience was successful in the orchestration of the Jigsaw collaborative learning flow (see Table 3). The system enabled a distribution of signals to the personal devices worn by the students, so that students knew automatically, and without "social indications" by the teachers, which case they should read and to which group they belong. As a result, the teachers' orchestration workload decreased as compared to their previous experiences (S1-2, T1-2). Besides configuring the OS-manager, the orchestration tasks carried out by the teacher were limited to explaining the meaning of the signals, distributing the cardboard signals to identify collaboration areas in the classroom and noting when the phases finished (the colors of the cardboards matched with the group signals received in the PS-devices; S3, O12).

Fig. 4. Pictures taken during the experience

Most of the students found the approach useful (84% rated it as quite or very useful). Those students not finding it useful indicated that the system was not indispensable to carry out the activity (S4-5-6). A critical element that led to the successfulness of the orchestration was the flexibility supported by the approach. Despite the unexpected incidents (O4-5) that occurred during the experience, teachers were able to re-configure the design of the orchestration transparently to the students (T3-4). Though it is true that the activity could have been carried out without the use of the system, the provided transparent flexibility and the decrease in the orchestration workload represent important added values of the approach. The relevance of these values is higher if we consider a higher number of students involved in the whole activity. Moreover, students and teachers highlighted the agile, dynamic and engaging collaboration achieved using the system when compared to their previous experiences (S7-15, T5). The timing reported in Table 2 also shows the agile enactment of the activity.

The personal signals were seen and understood fairly well (see Table 4). Though it was not critical for the experience, it is interesting to note that they could see the signals of their classmates (though not as clear as theirs). Depending on the position of the students in the classroom, the cardboards indicating working spaces in the classroom were seen better or worse. Teachers' comments and observations pointed out that the students get familiar to the signals and devices very quickly and the process followed to distribute the signals is easy and agile (T6-7, O6-8).

Table 3. Findings: Facilitating flexible orchestration in the classroom

Findings	**Supporting data** (S, comments of students; T, comments of teachers; O, observations by researchers)
The collaborative learning flow was followed as desired according to the Jigsaw intrinsic constraints (every expert group had more than 2 members, and every Jigsaw group included at least a member expert in each case). The Orchestration Signal system was not indispensable to achieve the orchestration, however, it decreased the teachers' workload and required attention to the orchestration of students when distributing the cases and forming the groups.	- The constraints of the Jigsaw collaborative learning flow were respected, and the flow of activities and distribution of groups in the classroom were achieved as desired (see Table 2) - **46%** of the students rated the PS-devices as quite useful, **38%** as very useful, **12%** as somehow useful, **4%** as not useful - Decrease the teachers' workload *"The system may enable to create different dynamics without the need that the teacher is close to you explaining the next step to follow..."* **(S1)** *"It avoids that the teacher decide the compositions of the groups... If a student is not happy in her group, she could not blame the teacher... "* **(S2)** *"I didn't need to indicate students in every moment what case each of them should read. Students were autonomous identifying their groups and task to accomplish. The group distribution was easier and agile since I didn't need to pay attention to where each student were going..."* **(T1)** *"I can pay more attention to the tasks themselves and not that much to the organization"* **(T2)** - Limited help of the teacher in the orchestration tasks (explaining meaning of the signals, tasks descriptions and moving the "signals" for the classroom collaboration areas) *"Teachers needed to explain how to interpret the signals in the devices"* **(S3)** *"The teacher changes the position of the color cardboards to indicate the new group working areas"* **(O1)** *"The teacher shout that the Expert phase finished ..."* **(O2)** - The system is not indispensable *"The device is not indispensable"* **(S4)** *"This can be also done with papers of different colors, though with children using the devices may be funny"* **(S5)** *"It's helpful but it also depends on how expensive the devices are..."* **(S6)** *"The teacher explains the activity at the beginning but the students ask her to remind them every task along the phases of the Jigsaw"* **(O3)**
Despite the unexpected incidents, teachers could easily rearrange the configuration of the orchestration transparently to the students. Teachers largely appreciated the flexibility supported by the system.	-The Jigsaw flow was followed meaningfully despite the incidents without spending extra time (see Table 2). - Teachers' comments and observations regarding flexibility aspects included: *"The process for sending signals was easy; there was even a student that left the class during the second phase, and it didn't occasion a problem..."* **(T3)** *"The system is very helpful, because it allows me to make changes during the activity in the signals to send..."* **(T4)** *"In one of the devices (n° 24) the red led is not working"* **(O4)** *"One of the students left in the minute 30 of the activity"* **(O5)**
When compared to previous experiences of students and teachers, the system showed to facilitate a more organized and dynamic collaboration and a more engaging experience.	- **77%** of the students experienced similar collaboration situations in the past - Collaboration more organized and dynamic: *"The devices speed up the dynamic"* **(S7)** *"The devices facilitated the organization of the activities"* **(S8)** *"Very helpful using the devices because they enables a complete organization"* **(S9)** *"It appeared to be a very well organized activity!"* **(S10)** *"I value that along the whole activity we keep the rhythm of the dynamic"* **(S11)** *"The system supported the group formation and the changes of groups were more rapid than previous years..."* **(T5)** - The system facilitates the movement and mixture of students: *"It's good that not always the groups are formed by the same people"* **(S12)** *"The movement in the classroom is motivating and favors collaboration and motivation"* **(S13)** - More engaging: *"The devices open our interest and raise expectations of what will be the next signal"* **(S14)** *"It's funny to see your color and then look for the place you need to go..."* **(S15)**

Table 4. Findings: Usability of the system and directions for improvement

Findings	Supporting data
Globally, the signals were seen and understood fairly well and quickly.	- Signals: **38%** of the students said that they could see their signals in the PS-devices quite well, **35%** very well, **12%** not very well, **0%** bad **62%** of the students said that they could see the signals of their classmates quite well, **15%** very well, **23%** not very well, **0%** bad **35%** of the students said that they could see the "cardboard furniture signals" quite well, **42%** very well, **19%** not very well, **4%** bad - Observations and comments regarding the global usability were: *"The process followed for sending signals was easy"* **(T6)** *"The students get familiar with the device very quickly because it is very easy to use"* **(T7)** *"Students identify very quickly their colors"* **(O6)** *"All of the students saw the signals almost at the same time"* **(O7)** *"In the third phase students appeared to be used to the devices and understood very quickly what to do..."* **(O8)**
More than a 70% of the students said that if they were to organize a similar activity, they would like to use the Orchestration Signal system. When asked about the positive aspects of the system they talked about dynamism, visual indicators, and engagement.	- **73%** of the participants said that if they were to organize a similar activity, they would like to use the signal system, **8%** said that they won't like to use it and **5%** indicates that it would depends on the situation. - More positive aspects indicated by the students (dynamism, visual, engaging) *"Enabling a more dynamic class"* **(S16)** *"Facilitating a rapid group formation"* **(S17)** *"Fluid organization"* **(S18)** *"Fosters students mobility in the classroom"* **(S19)** *"New, motivating, funny..."* **(S20)** *"Raise expectations, curiosity, engagement..."* **(S21)** *"The organization of the dynamic is highly visible... you do not need to read continuously the description of the "logistics"..."* **(S22)** *"All the students pay a lot of attention to the device, expecting the signals"* **(O9)**
Since the devices were prototypes, their design was not optimal in terms of size, weight and robustness. Some of the students did not were the PS-devices as expected and the students receiving only one color were confused thinking that they might need to see a second color. Future work proposed by the teachers include, the addition of intelligent functionalities to the PS-manager, being able to send signals also to furniture or locations in the classroom, and enabling students to send signals to the teacher from their PS-devices.	- Prototype *"It is important that all the devices work well"* **(S23)** *"They may be too delicate for children"* **(S24)** *"It's big..."* **(S25)** *"In the use of colors you should consider the people suffering from color-blindness"* **(S26)** *"Sometimes the colors were confusing"* **(S27)** *"It is a prototype... for normal use the hardware would need to be more robust"* **(T8)** *"Make more comfortable devices... some students just had them in their hands..."* **(T9)** *"The manager tool could be more "intelligent" performing automatic groupings or suggesting what changes to do in case of incidents"* **(T10)** *"It would be wonderful if I could also assign signals to the places in the classroom, instead of using paper indicators"* **(T11)** *"It would be interesting if students could also send signals to the teachers from their devices, for example to indicate that they need help..."* **(T12)** *"Some students didn't wear the devices and just put them on the table..."* **(O10)** *"After picking their cases, only 3 (out of 6) men and 6 (out of 18) women wore the devices"* **(O11)** *"Some students with only a color were expecting for a second color, considering that they might have a pair of colors"* **(O12)** - Classroom space *"The spaces were not too comfortable for discussing in groups"* **(S29)** *"Everything was fine, but the room was not very good... "* **(S30)**

A 73% of the students said that if they were to organize a similar activity, they would like to use the Orchestration Signal system. They highlighted the dynamism (S16-19), visual indicators (S22), and engagement effect (S20-21, O9) of the approach as its more positive aspects. Regarding the aspects for improvement, it became clear that the robustness, size and weight of the PS-devices are important characteristics that need to be improved towards a lighter, more compact device (S23-25, T8-9, O10-11). No communication problems in the transmission of the signals appeared during the experience. Finally, the use of mono- and bi-color signals seemed to be confusing, since some students receiving a mono-color signal were waiting during a brief moment for an eventual second color (S26-27, O12). Additional facilities proposed by the teachers to be supported by the system were the incorporation of intelligent functionalities to the PS-manager (T10), being able to send signals also to furniture or locations in the classroom (T11), and enabling students to send signals to the teacher from their PS-devices (T12).

5 Conclusions and Future Work

This paper has introduced a system that uses network linked ubiquitous computing devices to distribute signals to students indicating orchestration aspects in face-to-face settings. Examples of orchestration aspects that can be indicated with the signals include coordination indicators regarding group membership, collaboration areas assigned to each group, or distribution of resources and roles within groups. The prototype is named Orchestration Signal system and includes two components: an Orchestration Signal manager, which enables teachers to configure color signals to be transmitted, and a set of Personal Signal devices, which can be worn by students and display the transmitted orchestration signals.

The Orchestration Signal system has been used in a real classroom activity based on the Jigsaw collaborative learning flow pattern. Students received signals in their personal devices indicating which documents they should read, in which group they were and in which area of the classroom they were expected to collaborate. The findings, analyzed following a mixed evaluation method, show that the system enabled and facilitated the Jigsaw activity with a positive impact in decreasing the teachers' workload regarding the orchestration tasks. When compared to previous similar experiences, participants also indicated that the system supported a more dynamic, agile, organized and engaging activity. The system proved to flexibly support changes derived from unexpected incidents that emerged during the experience.

Future work include conducting new experiences in classrooms and other educational spaces, such as the playground, implementing diverse collaborative learning flow patterns, such as the Pyramid or the Simulation or Roleplay [19]. These new experiences will incorporate revised versions of the prototype towards lighter and more compact PS-devices and an easier-to-use OS-manager extended with additional facilities for random composition of groups, consideration of the intrinsic constraints of the flow patterns, etc. Moreover, we plan to incorporate a sound signal to indicate the change of phases and add a new component to the system that will be easily bound to classroom furniture so that it can also receive and visualize configurable signals. Furthermore, we are also working towards supporting active interaction between the personal devices and the manager and between several personal devices, in a mesh node interacting on a multipoint fashion, so as to facilitate collaborative learning activities where the management of the orchestration could be also controlled by the students.

Acknowledgments. This work has been partially funded by the Learn 3 project, (TIN2008-05163/TSI). The authors would also especially like to thank Mar Pérez-Sanagustín, Patricia Santos and other members of the GTI research group for their support and ideas.

References

1. Ciolfi, L.: Understanding spaces as places: extending interaction design paradigms. Cognition, Technology & Work 6(1), 37–40 (2004)
2. Gee, J.P.: Semiotic social spaces and affinity spaces: from The Age of Mythology to today's schools. In: Barton, D., Tusting, K. (eds.) Beyond Communities of Practice: Language, Power and Social Context, pp. 214–232. Cambridge University Press, Cambridge (2005)
3. Oblinger, D.: Leading the transition from classrooms to learning spaces. EDUCAUSE, Quaterly 28(1), 14–18 (2005)
4. Suthers, D.: Technology affordances for intersubjective learning: A thematic agenda for CSCL. In: 2005 conference on Computer support for collaborative learning, Taipei, China, pp. 662–671 (2005)
5. Dillenbourg, P., Huang, J., Cherubini, M.: Interactive Artifacts and Furniture Supporting Collaborative Work and Learning. Springer Science, New York (2009)
6. Kirschner, P.: Can we support CSCL? Educational, social and technological affordances for learning. In: Kirschner, P. (ed.) Three worlds of CSCL: Can we support CSCL?, pp. 7–47. Heerlen, The Netherlands (2002)
7. Huang, J., Cherubini, M., Nova, N., Dillenbourg, P.: Why would furniture be relevant for collaborative learning? In: Dillenbourg, et al. (eds.) Interactive Artifacts and Furniture Supporting Collaborative Work and Learning, pp. 1–13. Springer, NY (2009)
8. Arroyo, E., Ardaiz, O., Righi, V., Blat, J.: Distributed multi-touch virtual collaborative environments. In: 2010 International Symposium on Collaborative Technologies and Systems, Illinois, USA, pp. 635–636 (2010)
9. Bachour, K., Kaplan, F., Dillenbourg, P.: An interactive table for supporting participation balance in face-to-face collaborative learning. IEEE Transactions on Learning Technologies 3(3), 203–213 (2010)
10. Kemp, R., Kemp, E., Mohanarajah, T.: Supporting group learning using a digital whiteboard. Computer Human Interaction 3101, 594–598 (2004)
11. Jermann, P., Zufferey, G., Dillenbourg, P.: Tinkering or Sketching: Apprentices' Use of Tangibles and Drawings to Solve Design Problems. In: 3rd European Conference on Technology Enhanced Learning, Maastricht, The Netherlands, pp. 167–178 (2008)
12. Bachour, K., Alavi, H.S., Kaplan, F., Dillenbourg, P.: Low-resolution ambient awareness tools for educational support. In: CHI 2010 Workshop: Next Generation of HCI and Education, Atlanta, GA (2010)

13. Dix, A., Rodden, T., Davies, N., Trevor, J., Friday, A., Palfreyman, K.: Exploiting space and location as a design framework for interactive mobile systems. ACM Transactions on Computer-Human Interaction 7(3), 285–321 (2000)
14. Roschelle, J., Pea, R.: A walk on the WILD side: How wireless handhelds change computer-supported collaborative learning. International Journal of Cognition and Technology 1(2), 145–168 (2002)
15. Pérez-Sanagustín, M., Ramírez-González, G., Hernández-Leo, D., Muñoz-Organiero, M., Santos, P., Blat, J., Delgado-Kloos, C.: Discovering the campus together: a mobile and computer-based learning experience. Journal of Network and Computer Applications (in press), `http://dx.doi.org/10.1016/j.jnca.2011.02.011`
16. Santos, P., Pérez-Sanagustín, M., Hernández-Leo, D., Blat, J.: QuesTInSitu: from tests to routes for assessment in situ activities (submitted)
17. Dillenbourg, P., Järvelä, S., Fischer, F.: The evolution of research on computer-supported collaborative learning. In: Balacheff, N., Ludvigsen, S., Jong, T., Lazonder, A., Barnes, S. (eds.) Technology-enhanced learning, pp. 3–19. Springer, Netherlands (2009)
18. Hernández-Leo, D., Villasclaras-Fernández, E.D., Jorrín-Abellán, I.M., Asensio-Pérez, J.I., Dimitriadis, Y., Ruiz-Requies, I., Rubia-Avi, B.: Collage, a Collaborative Learning Design Editor Based on Patterns. Educational Technology & Society 9(1), 58–71 (2006)
19. Hernández, D., Asensio, J.I., Dimitriadis, Y., Villasclaras, E.D.: Pattern languages for generating CSCL scripts: from a conceptual model to the design of a real situation. In: Goodyear, P., Retalis, S. (eds.) E-learning, design patterns and pattern languages, pp. 49–64. Sense Publishers (2010)
20. JeeNode low-cost combination of radio and microcontroller board, `http://jeelabs.net/projects/hardware/wiki/JeeNode`
21. Creswell, J.W.: Research design: qualitative, quantitative, and mixed method approaches. John W. Creswell. Sage Publications, Thousand Oaks (2009)
22. Guba, E.G.: Criteria for assessing the trustworthiness of naturalistic inquiries. Educational Communication and Technology: A Journal of Theory, Research, and Development 29(2), 75–91 (1981)

Adult Self-regulated Learning through Linking Experience in Simulated and Real World: A Holistic Approach

Sonia Hetzner[1], Christina M. Steiner[2], Vania Dimitrova[3], Paul Brna[3], and Owen Conlan[4]

[1] University of Erlangen-Nuremberg, Germany
[2] Graz University of Technology, Austria
[3] University of Leeds, UK
[4] Trinity College Dublin, Ireland

Abstract. This research considers the application of simulated environments for adult training, and adopts the view that effective adaptive solutions for adults should be underpinned by appropriate adult learning theories. Such environments should offer learning experiences tailored to the way adults learn: self-directed, experienced-based, goal- and relevancy oriented. This puts andragogy and self-regulated learning at the heart of the pedagogical underpinnings of the intelligent augmentation of simulated environments for experiential learning. The paper presents a holistic approach for augmented simulated experiential learning. Based on andragogic principles, we draw generic requirements for augmented simulated environments for adult learning. An extended self-regulated learning model that links experiences in simulated and real world is then presented. A holistic framework for augmenting simulators - SRL-A-LRS - is presented and illustrated in the context of the ImREAL EU project. This points at a radically new approach for augmenting simulated systems for adult experiential learning.

Keywords: Simulated Environments for Learning, Self-regulated Learning, Andragogy, TEL Requirements.

1 Introduction

Experiential learning environments (e.g. simulations, serious games, immersive activities) create a practical, and often social, context in which novel skills can be learned, applied and mastered. These environments are increasingly popular as a means to turn experience into knowledge, and are being applied in a variety of domains and learning contexts. A class of simulated learning environments – **simulated situations for learning** – provide the means for learners to immerse themselves in simulated situations by performing activities that resemble actual job activities. In this paper such environments will be referred to as **simulated environments** or **simulators**. The popularity of these environments for experiential learning is growing; and there is a strong expectation that they will become one of

C. Delgado Kloos et al. (Eds.): EC-TEL 2011, LNCS 6964, pp. 166–180, 2011.

tomorrow's key learning technologies. Furthermore, we believe that successful simulated environments will need to be adaptive. The research described here considers the application of simulated environments for adult training, and adopts the view that effective adaptive solutions for adults should be underpinned by appropriate adult learning theories. Such environments should offer learning experiences tailored to the way adults learn: self-directed, experienced-based, goal- and relevancy oriented. This puts self-regulated learning (SRL) and andragogy at the heart of the pedagogical underpinnings of the intelligent augmentation of simulated environments for experiential learning. Offering experiences that are personalised to each learner's abilities involves balancing their need for support with their capabilities to self-regulate. Personalisation can span from fully automated adaptation to user-defined adaptability. As a user gains experience they will require less adaptive scaffolding.

A key factor for experiential learning in simulators is how well the learner connects learning to the real world. However, current intelligent simulators for learning suffer from a major deficiency because they incorporate a limited understanding of the learner usually based on the skills and knowledge only required within the simulated world. Consequently, such environments may be designed to satisfy the perceived needs of learners but they are often totally disconnected from the learners' real job experiences. For example, attempting to teach trainee doctors to communicate a life threatening diagnosis when they have no clinical experience at all highlights such a disconnect. This often hinders learners' engagement and motivation to undertake training since: (a) assessment in the learning environment is distinctly different from the consequences of performing the real job; (b) skills developed in the simulated learning environment are not effectively connected to the skills used in real job practice; (c) tacit and peer knowledge is buried in practical experience and is not capable of being utilised for skills development and effective assessment with the simulation; (d) because of the disconnect, it is not possible to relate feedback and guidance to real life experiences with the result that the feedback and guidance received can often be ignored by learners as irrelevant or even wrong.

Therefore, the key challenge is:

How to effectively align the learning experience in the simulated environment with the real world context and the day-to-day job practice where the skills are deployed.

This challenge is addressed in the ImREAL[1] (Immersive Reflective Experience-based Adaptive Learning) project. A novel approach is proposed that broadens the scope of the work to include activities that take place both within and around the simulation. We call this approach **augmented simulated experiential learning**. The approach also seeks to connect the learner's experiences both in the real and simulated worlds to others who have related experiences. This approach goes beyond normal work with simulations in training and requires technological innovations together with the development of theories tailored and adapted from those that currently exist.

The paper presents a psycho-pedagogical framework that underpins the design of innovative intelligent services to augment simulated environments for learning. Starting with generic andragogic principles, Section 2 focuses on self-regulated

[1] http://www.imreal-project.eu

learning (SRL), which is encompassed in andragogic principles, and presents an extended SRL model for augmented simulated experiential learning. This extended model underpins the ImREAL framework for adult self-regulated learning through linking experiences in the real and simulated world (SRL-A-LRS), which is presented in Section 4. A comparison with relevant TEL work is given in Section 5, while Section 6 concludes and points to future work.

2 Simulator Augmentation

ImREAL adopts a holistic approach to improving a simulator and facilitating its integration in a training environment. This requires improvement of the main components of a simulator.

2.1 Simulator Components

Simulators provide opportunities for a learner to engage in simulated situations that resemble real world job situations. The content in the simulator is created around a **simulated scenario**. This includes selected aspects of real world job situations driven by learning objectives and target competences. The simulator is based on an internal simulator structure, called hereafter the **simulation model**, which includes a *simulation graph* (a directed graph with the key steps the learner goes through and the choices given at each step) and *assessment decisions* associated with the learner's choices. As the learner progresses through the simulation graph, certain characteristics about the learner (e.g. knowledge, skills, competences) are observed, and represented in a **learner model**. This model can be used to select individualized paths through the simulation graph or to provide learner-tailored feedback. The simulators can vary based on the media used to represent simulated situations (e.g. graphics, audio, video, virtual reality) or the interaction means provided to the learner (e.g. menu choices, free text input, speech input). To abstract from the specific simulator implementation, we will focus on the key models common for a broad range of simulators - the simulation model and the learner model.

During their deployment and repurposing simulators usually undergo a series of improvements. Often, a simulator is re-used in different training contexts, which can require **tuning the simulation model**, e.g. by changing the simulation graph by adding more simulated situations (graph nodes) or suppressing some situations. Furthermore, intelligent functionality is added to **improve the effectiveness of the simulator**. This usually includes improving the *adaptation* and improving the *feedback* features.

The simulator is usually integrated in a training environment, which can include traditional courses, online training, or practical job activities. Notably, the learners using the simulators usually have **pre-simulator** and **post-simulator practical experience**, where they may be engaged in real world situations similar to the situations practiced within the simulator. Intelligent functionality can be added in the simulator to help learners link their experience in the simulator with 'real-world' job practice.

We argue that simulators should be augmented in a **holistic** and **cost-effective** way by developing appropriate augmentation services. Importantly, the augmentation is considered **outside the simulator** in the form of intelligent services that can be:

- *plugged into a simulator* to augment simulator components; or
- *used by simulator developers or tutors* to facilitate the simulator integration in a training environment.

ImREAL aims at developing such intelligent services for simulator augmentation, and will validate them via example simulators.

2.2 Simulators Used in ImREAL

Intelligent services for simulator augmentation developed in ImREAL will be illustrated in two simulators developed by consortium partners:

- the ASPIRE simulator developed by EmpowerTheUser (ETU)[2] Ltd; and
- a simulator developed by Imaginary[3] Srl.

ImREAL focuses on extending **simulated environments for dialogic interactions**. These simulators offer branching scenarios consisting of a series of steps (presented using text, graphics, videos) together with some interactive method for the learner to respond via menu selection. Such simulators provide affordable technological solutions suited to the trends in the growing adult learning market where demand is increasing, while budget is decreasing. ImREAL will aim at a generic approach for augmenting simulators developed by project partners, which will also be applicable to simulators developed outside the project.

This generic approach will be applied to illustrative use cases within a common training domain: interpersonal communication. We mainly focus on *dyadic conversations* (e.g. doctor-patient interview; job interview between an applicant and interviewer; interaction between two buddies from different cultures). Interpersonal communication is a highly complex, ill-defined domain. In such domains, training solutions are extremely challenging, and are mostly based on practical role-based experience (which makes the domain particularly suited to simulated environments for learning). We focus on selected key competences related to recognition, awareness, and use of verbal and non-verbal signals.

The design of intelligent services for simulator augmentation is underpinned by the ImREAL pedagogical framework, presented next.

---

[2] http://etu.ie/

[3] http://www.i-maginary.it/

3 Andragogic Principles and the Extended SRL Model

After developing the andragogic principles needed and outlining the relevance of self regulated learning, the need for an extended SRL model is introduced.

3.1 Andragogic Principles

Andragogy has been defined as the art and science of helping adults learn and the study of adult education theory, processes, and technology to that end [1, p. 19]. Andragogy, as counterpart to pedagogy (children's learning), has gained popularity through the work of Knowles (1970, 1984, 2005) [2,3,4] who introduced his theory of adult learning (i.e. andragogy). Knowles' theory is focused on the development of a set of assumptions and principles that reflect specific aspects of adult learning in the work context (primarily). Knowles defines adult learners as independent and self-directing, i.e. learners that have accumulated a great amount of experience, which is a rich resource for learning. Furthermore, Knowles describes adults as persons who value learning that integrates well with the demands of their everyday life and are more interested in immediate, problem-centred learning approaches.

Andragogy provides a set of **principles** to encourage adult learning. These are summarized, as follows:

- Learning situations are seen as directly relevant to the real world job context
- Learners need to know what they need to learn and why
- Training experiences have to be aligned with the learner's own goals
- Learning supports the learner's own sense of self, respecting individual differences
- Learning situations provide intrinsic motivation
- The learners are self-directing: they set their own agenda and learning path; assess their learning experience

The foundation of Knowles' principles is the perception of a learner that recognizes the relevance of learning, and takes the initiative as well as the responsibility for their own decisions in the learning process. The learner is able to diagnose his/her learning needs, formulate learning goals, identify resources for learning, select and implement learning strategies, and evaluate learning outcomes [5]. Andragogy underlines the importance of supporting the learner to become a competent and self-directed learner, which implies the development of learning environments that guide and support the learner in his/her learning process without taking away learner control. This brings forth two **key issues** for the design of learning environments for adult learners: (a) *learner context* linking learners' experiences and contexts of action (e.g. work, study) with the learning process; and (b) support in terms of *scaffolding* which assists learners in reaching levels beyond their current abilities [6].

Following the andragogic principles discussed above, two **generic requirements** for augmented simulated experiential learning can be drawn:

- Firstly, the need to *provide a simulated environment for learning that aligns the virtual, simulated world with the 'real-world' job experiences.* This supports the leaner to recognise the relevance of the learning activity and to transfer created knowledge to the real world as well as to integrate real world experiences in his/her learning process. Experiences are personal, from the learner him/herself as well as from others (peers, tutors, colleagues, experts)
- Secondly, the necessity to *develop and integrate concepts and services for supporting the learner to develop and enhance self-directness and self-regulation* in the learning process. This means supporting the learner to have control over his/her learning process, to be aware of his/her goals and strategies, and to be capable of monitoring, re-thinking and reflecting on his/her learning strategies.

3.2 Self Regulated Learning

As emphasised by andragogy, a key characteristic of adult learning is the need to learn to be self-directed and self-regulated. Self-regulated learning is a current focus of educational research as well as educational practice. The tradition of SRL considers learning as an active and constructive cognitive process, in which learners take responsibility and control over their own learning, e.g. [7,8]. SRL is carried out in a proactive manner with individuals regulating their own cognitive, meta-cognitive, and motivational processes within educational settings. SRL involves aspects of cognition, metacognition, motivation, affect and volition [9]. **Self-regulated adult learners use:**

- **Cognitive strategies,** such as elaboration, rehearsal, and organizational strategies [10].
- **Metacognitive strategies**, which are essential components of self-regulated learning, for enabling the formation of knowledge or beliefs about what factors or variables affect the course and outcome of cognitive enterprises [11].
- **Motivational strategies,** which have a strong positive impact on learning, on enhancing learners' attention to their learning, on progressing the task itself and on their satisfaction and affect [11, 12]. Aspects of motivation, such as "expectancy of success, interest, utility, and task value", influence learners' task engagement [11].

Furthermore, **learning includes feelings, emotions, and attitudes:** metacognition has an effect on cognition both through the cognitive regulatory loop and also through affect [9]. Volition strategies relate to emotions, which influence the learning process, such as activity-related emotions (interest and boredom) or outcome-related emotions (anger, pride, etc.). There is another category – "metacognitive feelings" – which are focused on cognition and result from monitoring (e.g. the feeling of knowing).

A range of different theories and models have been devised to distinguish and model individual phases of SRL, most of them representing SRL as a cyclic process (for an overview see e.g. [7]). One of the most influential models has been devised by Zimmerman [8] and postulates that the cyclical phases of self-regulation are comprised of:

- **Forethought,** which involves activities that *precede and prepare for learning actions*, like goal setting and strategic planning. It also involves processes of self-motivation based on self-efficacy beliefs, outcome expectations, intrinsic interest and values.
- **Performance,** which refers to the *actual process of learning* and involves strategies aimed at fostering the quality and quantity of learning performance through self-instruction, self-control, and self-observation.
- **Self-reflection,** which involves processes of *self-evaluation, causal attribution* (i.e. beliefs about cause of error and success), and *self-reaction.* This influences, in turn, the forethought phase of subsequent learning efforts.

3.3 Augmented Simulated Experiential Learning Needs an Extended SRL Model

The benefits of self-regulated learning need to be transferred to the context of experiential learning environments for work-related contexts. In simulated worlds, learners need to be put into situations that engage their self-regulated learning skills and provide high relevance for real-world experiences. To achieve this, adaptive scaffolding of SRL is very promising [13,14]. However, ImREAL goes beyond current adaptive scaffolding of SRL focused on learning and performance within primary and secondary educational systems. To minimise the gap between the virtual and real world requires a holistic approach explicitly considering, harmonizing, and combining simulated and real-world learning. As a result, the SRL model [8] needs to be extended to account for the integrated consideration of learning processes in the virtual world and in the real world (this makes ImREAL unique). Taking the two worlds and the experiences therein, learning can be regarded as consisting of two SRL cycles, one for the simulated world and one for the real world, see Figure 1. Activities of forethought, performance, and reflection occur in both the virtual environment and the real world. Each of these two cycles is understood in a dynamic manner, i.e. with the individual phases being recursive and triggering and influencing each other. While current simulated environments commonly suffer from the deficiency that these two cycles are rather detached and separated from each other, **ImREAL aims at connecting and integrating the two SRL cycles**. The two SRL cycles influence each other, i.e. the learning process in the context of the simulation may trigger a certain SRL phase in the real world. The most straightforward case is that reflection in the virtual world triggers forethought in the real world. However, the opposite link may also happen, e.g. during forethought in the real-world a learner may recognize skill gaps, which can trigger a SRL cycle in the virtual world first, before going over to performance in the real world.

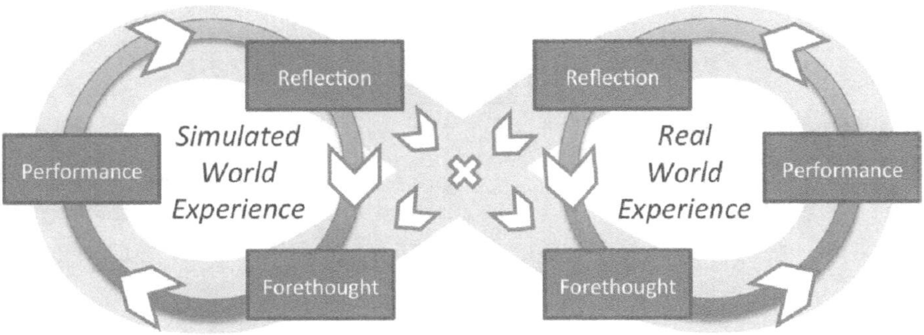

Fig. 1. Extended SRL model for learning in simulated and real worlds

For an even stronger integration between simulated and real world learning, **the forethought and the reflection phase of the real world SRL cycle can be shifted to the simulation** (see the next section for the approach taken in ImREAL). The real world performance phase naturally can only happen within the real world, but planning and reflecting on the real world activity can realistically be done also in the context of the virtual environment – at least to some extent (see Figure 2). This can be realized by prompting learners to reflect on their real world experiences, the knowledge and competence needed and acquired and to record this kind of information in the simulation. These real world records can be exploited for user modelling in the simulated environment in order to refine the picture of a learner's characteristics and prior knowledge, which in turn may serve the improvement of the simulated learning by being better able to select suitable learning situations and paths. The learner may use his/her records of prior experiences as a basis for future forethought phases i.e. for planning further learning activities.

The view from individual learning is further broadened to peer experiences (see Figure 2). The real world experiences that a learner has recorded in the context of the simulation may not only serve his/her own reflection and further planning, but the respective information can also be shared with peers and serve as a valuable source of information for other learners. Conversely, the learner may benefit from getting to know about the experiences of others, by contrasting them with his/her own experience or by using them as a source of information for planning upcoming learning tasks in the simulation. A learner may even look up peer experiences made in the real world and recorded in the simulated environment without actually doing the simulation itself, a form of vicarious learning. In this way, the link between virtual and real world learning can be considerably strengthened by integrating real world data and experiences into the simulation. Through an open learner model approach [15] practical and peer experiences can be made more tangible among learners and can be utilised for skill development and assessment in the simulation. This allows increased individualization and better tailoring of the virtual experience to the individual learner.

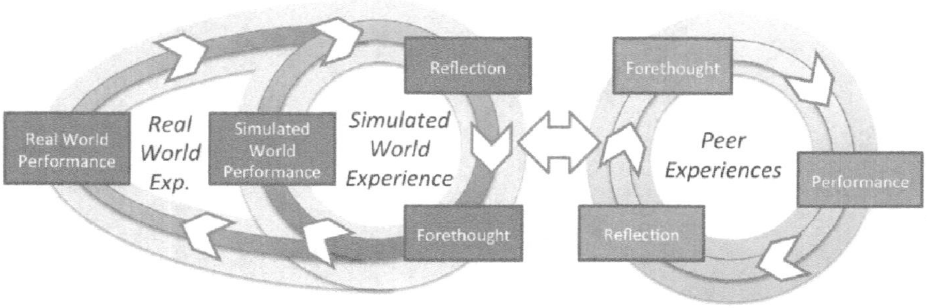

Fig. 2. An extended SRL model integrating simulated and real world learning and incorporating peer experiences

4 The ImREAL Approach

The ImREAL model integrates the extended andragogic principles for simulation design with the extended self-regulated learning model. The principles of andragogy are key requirements for the ImREAL services and for the way ImREAL services are integrated into the (sequential) self-regulated learning model.

4.1 ImREAL Framework: SRL-A-LRS

Following the andragogic principles and the extended SRL model presented in the previous section, we derive a framework for Adult Self-Regulated Learning through Linking Real and Simulated world experiences (SRL-A-LRS). The framework underpins the design of intelligent services for simulator augmentation by enabling the connection between the performances in the virtual and the real world in a self-regulated learning and experiencing activity.

The SRL-A-LRS framework includes three main aspects:

- The adapted and extended andragogic principles;
- The self-regulated learning model linking real and virtual performance phases; and
- The integration of real world experiences from the learner and from others (peers, colleagues, experts).

The framework requires/embeds services that support the realisation of the andragogical model for manifold learning experiences with simulation environments as key elements of the learning activity. Figure 3 illustrates the relation between the ImREAL model and the ImREAL services.

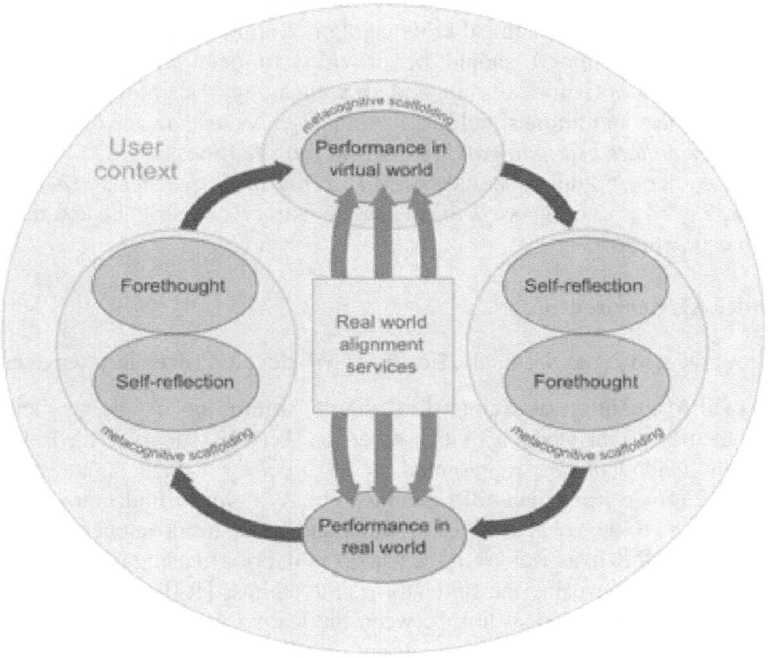

Fig. 3. The ImREAL SRL-A-LRS framework and the ImREAL services

4.2 Design Requirements

Based on the SRL-A-LRS model, we derive **design requirements** for intelligent services for simulator augmentation:

[R1] Appropriate services should be provided to support trainers and simulator developers to *identify key aspects of real-world job situations* that can be included in the simulator.

[R2] Appropriate services should be provided to facilitate trainers and simulator developers to *find authentic, relevant examples with real-world job experience* related to specific simulated situations.

[R3] Appropriate services should be provided to facilitate trainers and simulator developers to *find relevant user characteristics* in order to specify simulator usage scenarios and decide on feedback.

[R4] Learners should be helped to see *relationships between the experience in the simulator and real world job situations* they can engage in.

[R5] Feedback and prompts should be *related to relevant real world experiences of the learner or of peers* who have similar characteristics.

[R6] Intelligent services should be developed to *identify what a learner needs to learn* based on his/her experiences both in the simulated and real world.

[R7] The learner should be offered *help to see what he/she needs to learn* based on his/her experiences both in the simulated and real world.

[R8] Appropriate support should be provided to help the learner *set learning goals and agenda* (before, during, and after engaging in a simulator).

[R9] Intelligent support should be provided to *help the learner assess their learning experiences* (before, during, and after engaging in a simulator).

[R10] Learners and tutors should be facilitated to *become aware of the relationship between the simulated situation and real-world job situations*.

[R11] Learners and tutors should be *made aware of the learner's experience in the simulator*, e.g. by providing ways to review the learner's experience and the skills the learner has practiced.

4.3 ImREAL Services

To address the above requirements, ImREAL will develop **three key services**.

Real World Modelling and Semantic Content Annotation. Applying social science methods to model real world activities based on Activity Theory [16], ImREAL will develop an *activity model*, represented as an *ontology*, which captures real world experiences derived from many different sources. We aim to find innovative ways of providing access to these experiences to tutors and simulator developers (i.e. to address **[R1]**, **[R2]**, and **[R10]**), as well as to the learner both prior to using the simulator, within the simulator and after using the simulator (i.e. to address **[R4]** and **[R5]**). This service represents a vital andragogical link between the learner's learning and how they may wish to apply it in real world settings. From an SRL perspective, the reflection in the simulation should lead to improved performance in the real world.

Augmented User Model. Following recent trends in user modelling[4], we will address the challenge of *user model augmentation*. ImREAL services will make an adaptive learning system, such as a simulator, better by selecting relevant external data about real-world experiences, and providing ways to integrate this data into the user model such that the user model becomes richer (for the purposes of adaptation). Hence, the augmented user modelling services developed in ImREAL will help address requirements **[R6]**, **[R7]**, and **[R9]**. Furthermore, deriving social profiles of similar users (user stereotypes), we aim to provide key user characteristics to simulator developers, i.e. to address requirement **[R3]**.

Meta-cognitive Scaffolding. ImREAL will take the approaches for providing adaption in simulators to the next level by providing meta-cognitive scaffolding in an adaptive manner [17]. Specifically, we focus on how to provide salient, timely services to support the metacognitive processes of self-regulated learning within the framework of experiential training in a cognitively sensitive (non-invasive) manner related to, but not embedded within, e-Learning simulation execution environments. Through appropriate meta-cognitive scaffolding the learner will enhance key traits, such as planning, comprehension and evaluation, to allow them to discern, regulate and manage their learning. The meta-cognitive scaffolding services will utilise the real world model, semantic content annotation, and augmented user model, to augment the simulator addressing requirements **[R4]**, **[R5]**, **[R6]**, **[R7]**, **[R8]**, **[R9]**, **[R10]**, **[R11]**.

---

[4] E.g. see workshop on Augmented User Modelling: http://wis.ewi.tudelft.nl/aum2011/

5 Related Work

Immersive learning experiences such as successful simulator-based training and digital educational games (DEGs) are designed to be inherently motivational with narratives that engage the learner. A key challenge lies in maintaining this motivation whilst making the experience more meaningful for each individual learner. In DEGs, such as ELEKTRA[5] and 80Days[6], the challenge of realizing real-time, non-invasive adaptations were addressed through leveraging the Adaptive Learning In Games through Non-invasion (ALIGN) [18] system. This proved successful, but DEGs are typically employed within a controlled curriculum where the desired learning objectives are usually known a priori and adaptive hinting is employed to assist the learner in achieving these objectives. The majority of adaptive systems developed work within such a closed and prescriptive corpus and domain.

Applying the principles of andragogy implies that the adult learner will have a much higher degree of freedom, compared to that offered in typical DEGs, and that they will expect the learning experience to be closely correlated to their needs. Whilst the idea of SRL is increasingly taken up in the context of TEL the challenge of maintaining the flow of experience and supporting the learner in a non-intrusive manner remains [19]. Many learning systems are designed and built in order to support explicit (and somewhat intrusive) self-regulation by providing tools and supporting different self-regulatory processes [20, 21, 22, 23]. In recent years, the idea of utilising adaptive technologies for supporting SRL has emerged and approaches and computational frameworks have been devised in order to exploit intelligent tutoring and adaptation for measuring self-regulated learning [24] scaffolding metacognition [14, 17, 25]. However, the challenge of incorporating real-world experiences to augment and ground the metacognitive gain has not been tackled.

SRL and the use of adaptive mechanisms might seem a contradiction at first sight, in particular when thinking of adaption in terms of automated customization and strict guidance. However, through striking a balance between the learner's experience and the degree of adaptive support provided, a continuum of personalization may be achieved. By monitoring observable evidences of cognitive, metacognitive, and motivational processes while using a learning environment, self-regulated learning can be measured in an unobtrusive way [24, 26]. One potential way for realizing such non-invasive assessment and adaptive scaffolding is the so-called micro-adaptivity approach of Competence-based Knowledge Space Theory ([27]. This approach, which has been developed in the context of DEGs and to date has been primarily focusing on supporting domain skill measurement and acquisition, can be adopted for realizing non-invasive assessment of SRL skills and intelligent adaptive support as reaction. During learning, user actions in the learning system (e.g. the use of certain tools or actions within a tool) can be interpreted in terms of available and lacking skills based on competence assessment. Through continuous, unobtrusive monitoring of a learner's interactions with the system, an image of the SRL competence of a learner can be built up. This model can be used to trigger prompts about forethought,

[5] http://www.elektra-project.org

[6] http://www.eightydays.eu

performance and reflection specifically tailored to the learner and enhancing the acquisition of lacking but necessary skills [27].

6 Conclusions and Future Work

This paper presents a novel approach to augment simulated environments for adult experiential learning by linking the learner experience in the simulated world with experience (of the learner or peers) in the real world. A **holistic view of augmentation** is adopted, aiming at (a) *supporting the current practice of developing simulators for adult learners* by facilitating tutors and simulator developers to become aware of relevant real world experiences and identify relevant user profiles; (b) *extending the simulators* with intelligent features to provide feedback and promote reflection in a motivating manner; and (c) *supporting the training environment in which simulators are integrated* by extending the simulator's understanding of users and providing support for the learner's experience before, during, and after the simulator.

The paper makes the first step towards addressing the challenge of '*how to link experiences in simulators (and in virtual training environments in general) to experience in real-world job practice*', which is paramount for developing effective TEL solutions for adult training. The work presented here makes the following **novel contributions** to TEL research:

- Links andragogic principles and design requirements for simulated environments for adult learning;
- Presents an extended model of self-regulated learning that links experience in the real and simulated world; and
- Presents a framework (SRL-A-LRS) for developing intelligent services for augmenting simulator based on adult self-regulated learning through linking experience in real and simulated worlds;
- Outlines three key services (real world modelling and semantic content annotation, augmented learner modelling, and meta-cognitive scaffolding) to implement the SRL-A-LRS framework within the ImREAL project.

We are currently implementing the first versions of the three key services, which are independent from simulators but will be illustrated in two use cases with the simulators used in ImREAL. The ASPIRE simulator for medical training (training doctors to interview patients in distress, e.g. with depression) is being extended with adaptive meta-cognitive scaffolding augmenting the performance in the simulator and the self-reflection after the experience in the simulator [19]. The augmentation also includes extending the user model based on user characteristics from social web environments [28]. The Imaginary simulator is being developed for job interview training (training interviewers to interview applicants) and will then be extended to link verbal and non-verbal signals with multi-cultural awareness. For this, we use an activity model ontology developed following Activity Theory and using intuitive ontology engineering tools [29]. We have annotated user comments on job interview examples (following a youtube style) to provide authentic content about activity aspects [30].

Our next step is to test the services within realistic settings against the requirements of the SRL-A-LRS framework. We are also designing intelligent prompts in line with the extended SRL model, to be integrated in both the meta-cognitive scaffolding services and semantic content assembly services.

Acknowledgements. The research leading to these results has received funding from the European Union Seventh Framework Programme (FP7/2007-2013) under grant agreement no ICT 257831 (ImREAL project).

References

[1] Krajinc, A.: Andragogy. In: Titmus, C.J. (ed.) Lifelong Education for Adults: An International Handbook, pp. 19–21. Pergamon, Oxford (1989)
[2] Knowles, M.S.: The modern practice of adult education. Andragogy versus pedagogy. Association Press, New York (1970)
[3] Knowles, M.: Andragogy in Action. Applying Modern Principles of Adult Learning. John Wiley and Sons Ltd., Chichester (1984)
[4] Knowles, M.S., Holton, E.F., Swanson, R.A.: The Adult Learner: The Definitive Classic in Adult Education and Human Resource Development. Elsevier Science & Technology (2005)
[5] Knowles, M.S.: The Adult Learner: A Neglected Species. Gulf Publishing Company, Houston (1975)
[6] Dennen, V., Burner, K.: The Cognitive Apprenticeship Model in Educational Practice. In: Spector, J., et al. (eds.) Handbook of Research on Educational Communications and Technology, New York (2007)
[7] Puustinen, M., Pulkkinen, L.: Models of Self-regulated Learning: A Review. Scandinavian Journal of Eduational Research 45, 269–286 (2001)
[8] Zimmerman, B.J.: Becoming a self-regulated learner: An overview. Theory into Practice 41, 64–70 (2002)
[9] Efklides, A.: The role of metacognitive experiences in the learning process. Psicothema 21, 76–82 (2009)
[10] Pintrich, P.R., De Groot, E.V.: Motivational and Self-regulated learning components of classroom academic performance. Journal of Educational Psychology 82, 33–40 (1990)
[11] Pintrich, P.R.: The role of motivation in promoting and sustaining self-regulated learning. International Journal of Educational Research 31, 459–470 (1999)
[12] Zimmerman, B.J., Schunk, D.H.: Motivation. An essential dimension of self- regulated learning. In: Schunk, D.H., Zimmerman, B.J. (eds.) Motivation and self- regulated learning. Theory, research, and applications, pp. 1–30. Lawrence Erlbaum Associates, New York (2008)
[13] Bannert, M., Mengelkamp, C.: Scaffolding Hypermedia Learning through Metacognitive Prompts. In: Azevedo, R., Aleven., V. (eds.) International Handbook of Metacognition and Learning Technologies. Springer Science, Heidelberg
[14] El Saadawi, G., Azevedo, R., Castine, M., Payne, V., Medvedeva, O., Tseytlin, E., Legowski, E., Jukic, D., Crowley, R.: Factors affecting feeling-of-knowing in a medical intelligent tutoring system: The role of immediate feedback as a metacognitive scaffold. Advances in Health Sciences Education 15, 9–30 (2010)
[15] Kay, J.: Learner control, User Modeling and User-Adapted Interaction, vol. 11, pp. 111–127 (2001)

[16] Engeström, Y.: Enriching activity theory without shortcuts. Interacting with Computers 20(2), 256–259 (2008)

[17] Macarthur, V., Conlan, O.: Using Psychometric Approaches in the Modelling of Abstract Cognitive Skills for Personalization. In: Proc. Lifelong User Modelling Workshop (LLUM 2009), pp. 22–26 (2009)

[18] Peirce, N., Conlan, O., Wade, V.: Adaptive Educational Games: Providing Non-invasive Personalised Learning Experiences. In: Proceedings of the Second IEEE International Conference on Digital Games and Intelligent Toys Based Education, pp. 28–35 (2008)

[19] Moore, A., Conlan, O., Dagger, D., Wade, V.: Towards Non-invasive Adaptive Meta-Cognitive Support for Online Training. In: Proc. 11th IEEE International Conference on Advanced Learning Technologies (ICALT 2011), Athens, USA (July 2011) (accepted for Publication)

[20] Azevedo, R., Hadwin, A.F.: Scaffolding self-regulated learning and metacognition – Implications for the design of computer-based scaffolds. Instructional Science 33, 367–379 (2005)

[21] Bannert, M.: Promoting self-regulated learning through prompts. Zeitschrift für Pädagogische Psychologie 23, 139–145 (2009)

[22] Dabbagh, N., Kitsantas, A.: Supporting self-regulation in student-centered web-based learning environments. International Journal on E-Learning 3, 40–47 (2004)

[23] Narciss, S., Proske, A., Körndle, H.: Promoting self-regulated learning in web-based learning environments. Computers and Human Behavior 23, 1126–1144 (2007)

[24] Winne, P.H.: Improving measurements of self-regulated learning. Educational Psychologist 45, 267–276 (2010)

[25] Conati, C., VanLehn, K.: Toward computer-based support of meta-cognitive skills: a computational framework to coach self-explanation. International Journal of Artificial Intelligence in Education 11, 389–415 (2000)

[26] Aleven, V., Roll, I., McLaren, B.M., Koedinger, K.R.: Automated, unobtrusive, action-by-action assessment of self-regulation during learning with an intelligent tutoring system. Educational Psychologist 45, 224–233 (2010)

[27] Kickmeier-Rust, M.D., Albert, D.: Micro adaptivity: Protecting immersion in didactically adaptive digital educational games. Journal of Computer Assisted Learning 26, 95–105 (2010)

[28] Abel, F., Gao, Q., Houben, G.-J., Tao, K.: Contextualization, User Modeling and Personalization in the Social Web. In: Proceedings of Extended Semantic Web Conference (ESWC), Heraklion, Greece. Springer, Heidelberg (2011)

[29] Denaux, R., Dolbear, C., Hart, G., Dimitrova, V., Cohn, A.G.: Supporting Domain Experts to Construct Conceptual Ontologies: AHolistic Approach. Journal of Web Semantics (in press)

[30] Despotakis, D.: Multi-perspective Context Modelling to Augment Adaptation in Simulated Learning Environments. In: Konstan, J.A., Conejo, R., Marzo, J.L., Oliver, N. (eds.) UMAP 2011. LNCS, vol. 6787, pp. 405–408. Springer, Heidelberg (2011)

Can Erroneous Examples Help Middle-School Students Learn Decimals?

Seiji Isotani[1], Deanne Adams[2], Richard E. Mayer[2], Kelley Durkin[3], Bethany Rittle-Johnson[3], and Bruce M. McLaren[1]

[1] Carnegie Mellon University, U.S.A.
[2] University of California, Santa Barbara, U.S.A.
[3] Vanderbilt University, U.S.A.
sisotani@icmc.usp.br, deanne.adams@gmail.com, mayer@psych.ucsb.edu, {kelley.l.durkin, bethany.rittle-johnson}@vanderbilt.edu, bmclaren@cs.cmu.edu

Abstract. This paper reports on a study of learning with erroneous examples, mathematical problems presented to students in which one or more of the steps are incorrect. It is hypothesized that such examples can deepen student understanding of mathematics content, yet very few empirical studies have tested this in classrooms settings. In a classroom study, 255 6th, 7th, and 8th graders learned about decimals using a web-based system under one of three conditions – erroneous examples, worked examples, and partially-supported problem solving. Although students' performance improved significantly from pretest to posttest the learning effect for erroneous examples was not better than the other conditions, and unlike some earlier empirical work, the higher prior knowledge students did not benefit more from erroneous examples than from worked examples or problem solving. On the other hand, we were able to identify certain key decimal misconceptions that are held by a high percentage of students, confirming earlier mathematics education studies. Also, the incidence of misconceptions declined over the course of the lesson, especially for the worked example group. Overall, these results could indicate that erroneous examples are simply not as effective for learning as we (and other) researchers hypothesize. The results could also indicate that the manner in which erroneous examples were presented to the students in this study somehow missed the mark in promoting learning. It is also possible that erroneous examples, like some other e-learning techniques, do not work as well in classroom as they do in a laboratory setting. We discuss these possibilities and how we are redesigning the treatments to be more focused and appealing to learners for a subsequent study.

Keywords: erroneous examples, decimal, math education.

1 Introduction

A U.S. National Math Panel Report emphasizes the importance of students mastering decimals [1]. Although conceptual understanding of decimals is critical for most of

C. Delgado Kloos et al. (Eds.): EC-TEL 2011, LNCS 6964, pp. 181–195, 2011.

later mathematics, the Panel reports that in general, students receive very poor preparation in decimals. In fact, it is well documented that students often have difficulty understanding and mastering decimals [2, 3, 4]. Indeed, even adults are known to have trouble with decimals [5].

One way to possibly remedy this situation is to present students with an approach that falls outside of the classroom norm: erroneous examples. An erroneous example is a step-by-step description of how to solve a problem in which one or more of the steps are incorrect. Erroneous examples are seldom used in classrooms or empirically investigated as a means for teaching students mathematics. On the other hand, some researchers have argued that confronting students with mathematical errors can be valuable; particularly when students are sufficiently prepared to deal with errors. Further, some empirical research has demonstrated that erroneous examples can particularly benefit higher-prior knowledge learners in mathematics [6, 7].

To evaluate the effects of learning with erroneous examples in the domain of decimals we developed web-based instructional materials that help students learn by reflecting upon and self-explaining errors. We compared erroneous examples with more typical instructional materials, namely worked examples and problem solving, in a classroom study with 255 subjects from 6th, 7th and 8th grades. Our hypothesis was that including erroneous examples would be a better instructional method to help students learn both cognitive and metacognitive skills and the positive effects would be particularly pronounced for higher-prior knowledge students.

The empirical findings of this work provide new insights into the controversial use of erroneous examples. With regard to learning gains, they suggest that erroneous examples can be as effective as, but not necessarily better than, more traditional approaches. However, these findings may also suggest that somehow our designed materials missed the mark in promoting learning.

In the following sections we discuss the materials, the study design and the results. We also present ideas for re-designing our materials for a subsequent study, developing a more interactive, more focused, and less verbose version of our initial erroneous examples materials.

2 Erroneous Examples

An erroneous example (ErrEx) is a step-by-step problem solution in which one or more of the steps are incorrect. Some theory and research in mathematics education has explored the phenomenon of erroneous examples and provides anecdotal evidence that studying errors can help student learning (e.g., [8]). For example, Borasi argues that mathematics education could benefit from the discussion of errors by encouraging critical thinking about mathematical concepts, by providing new problem solving opportunities, and by motivating reflection and inquiry. In an OECD report released in 2001, the highly-publicized TIMSS studies showed that Japanese math students outperform their counterparts in the U.S., as well as the rest of the western world, with a key difference cited that Japanese educators present and discuss incorrect solutions and ask students to locate and correct errors [9]. The critical point is that these theoretical analyses suggest that directly confronting and reflecting upon errors may lead to the eradication of those errors, similar to what has been shown in much of the learning research on misconceptions [10].

On the other hand, another view is that allowing errors to occur, or exposing learners to errors, hurts learning, presumably because errors are then more likely to be retrieved during later problem solving. For instance, B. F. Skinner's experiments, and the studies of other behaviorists, showed that the lack of a "penalty" when undesirable behavior was exhibited led to repetition of this behavior [11]. Many teachers are ambivalent about – or even hostile to – discussing errors in the classroom [12], quite possibly in reaction to the powerful impact that behaviorism has had, not just in academic research but in everyday education as well.

Thus, the question of how – and if – erroneous examples are beneficial to learning is still very much open and controversial. Despite this, there have been few controlled, empirical studies that have studied the effects of ErrExs on learning. For example, in the aforementioned study conducted by Grosse & Renkl [6], with 118 German university students in the domain of probability, it was identified that providing correct and incorrect solutions of problems fostered far transfer performance if learners had high prior knowledge; for learners with low prior knowledge, only correct solutions were beneficial. In another study with 87 subjects from the 3rd and 4th grade, Siegler [13] identified that students who self-explained both correct and erroneous (incorrect) examples learned mathematical equality better than those who explained only correct examples or who explained their own answers. According to Siegler, the learning benefit of explaining comes from strengthening correct knowledge about the domain and weakening incorrect knowledge. In a study with 93 middle school students in the domain of fractions, Tsovaltzi et al. [7] identified significant metacognitive learning gains for weaker (i.e., lower prior knowledge) students who learned from erroneous examples, as well as cognitive and conceptual learning gains for higher-prior knowledge students when additional help was provided.

We are aware of only a few other studies that have investigated the benefits of learning with erroneous examples [14, 15, 16]. Therefore, there is ample opportunity and need to conduct further empirical work to better understand the impact of erroneous examples on learning.

3 Design of Erroneous Examples in the Domain of Decimals

To design erroneous examples that address students' difficulties in learning decimals, we focused on the most common and persistent errors committed and misconceptions held by students. For example, students often treat decimals as if they are whole numbers (e.g. they think 0.25 is greater than 0.7, since 25 is greater than 7). Some students believe that adding a zero on the end of a decimal increases its magnitude, as with whole numbers. These misconceptions interfere with conceptual understanding of decimals and manifest themselves in various ways [4].

Through an extensive mathematics education literature review, covering 42 published papers and extending as far back as 1928 (e.g., [4, 17, 18, 19, 20, 21]), we found that most of the published papers address either a single misconception or a small set of related misconceptions. We evaluated the prior results and organized them into a taxonomy of decimal misconceptions as shown as reported in [22] and shown in Figure 1.

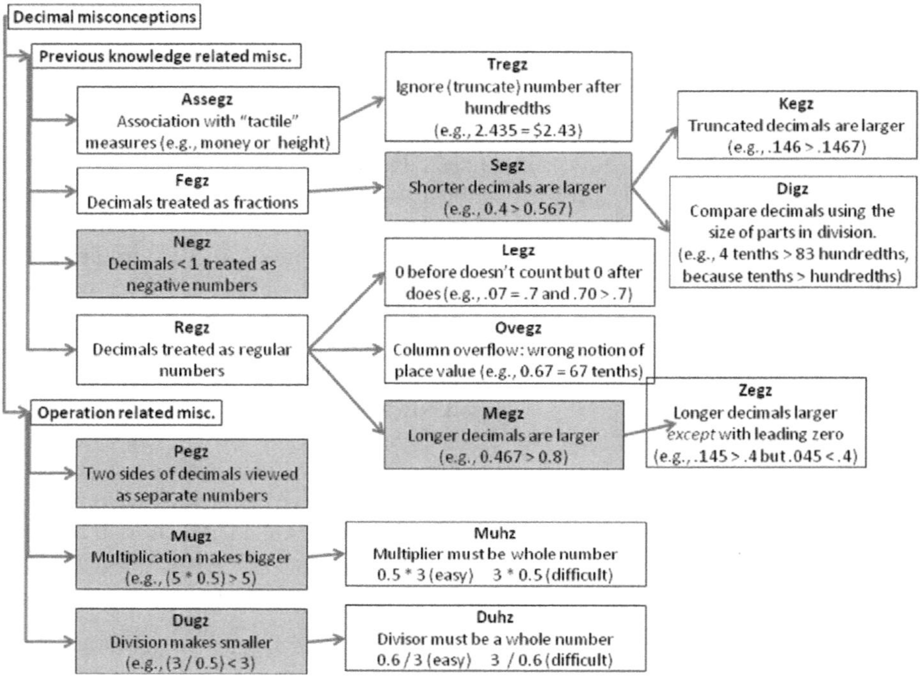

Fig. 1. A portion of the decimal misconceptions taxonomy. The gray boxes are the misconceptions included in the study described in this paper.

To categorize and create meaningful clusters of misconceptions, we relied on the quantitative data reported by past researchers. Currently the taxonomy has 45 nodes and classifies the misconceptions according to (a) the prior knowledge the student may have had before learning decimals (e.g., whole numbers, fractions) and (b) the operations associated with the misconception. For example, many students who learn fractions before decimals believe that shorter decimals are larger than longer ones. This misconception occurs because students think that 0.3 is like 1/3 and 0.367 is like 1/367 and, thereby, 0.3 > 0.367. Conversely, students who only have been exposed to whole numbers are more likely to believe that longer decimals are larger (as per the example in the first paragraph of this section). With respect to decimal operations, the taxonomy includes misconceptions like "multiplication makes bigger."

By organizing and relating misconceptions we intended to provide a unique resource, useful both to ourselves as developers of decimal materials, and for other researchers. We also identified key decimal misconceptions that have been reported in different papers as highly frequent among middle school students. Here is a brief description of the seven most common misconceptions:

- **Decimals are like whole numbers** (we call this *regz – Regular numbers misconception*). A fundamental misconception that underlies other misconceptions. Students believe that decimals should be treated as whole numbers and, therefore, they ignore the decimal point [3, 19].
  - Example: 0.07 is equal to 7.
- **Longer decimals are larger** (we call this *megz – Mega numbers misconception*). The most frequently observed misconception. In this case, students think they can compare decimals by counting the digits of each decimal [23].
  - Example: 0.25 is greater than 0.7.
- **Shorter decimals are larger** (we call this *segz – Shorter numbers misconception*). Next to Megz, this is the most frequently observed misconception. Students use knowledge about fractions to evaluate decimals. Thus, when comparing decimals if you increase the number after the decimal point then the decimal will become smaller [4].
  - Example: 0.4 is greater than 0.567 because 0.4 is a fraction like 1/4 and 0.567 is a fraction like 1/567
- **Decimals less than 1 confused as negative** (we call this *negz – Negative numbers misconception*). After megz and segz, this is the next most frequently observed misconception, but its occurrence is far less frequent. Students believe that decimals starting with a zero are smaller than zero (usually students have not learned the concept of negative numbers) [5].
  - Example: 0.25 is less than zero
- **Two sides of decimals viewed as separate numbers** (we call this *pegz – Misconception on either side of the "peg"*). Along with negz, this is the next most frequently observed misconception, the first that involves operations. Students think that the numbers on the left side of the decimal point are completely separate from the right side. Thus, although they can add whole numbers correctly, they won't get the correct answer if they add decimals and need to "carry" a digit to the next column on the left of the decimal point [20].
  - Example: 1.5+3.9 = 4.14
- **Multiplication makes bigger** (we call this *mugz – Multiplication misconception*). One of the most common misconceptions with decimal operations. It builds upon other misconceptions (e.g. regz) since, for instance, the belief that decimals are like integers multiplication should always yield larger results. Students believe that the product of a multiplication should always be larger than the leftmost factor of the operation [24].
  - Example: 5 multiplied by 0.5 is bigger than 5
- **Division makes smaller** (we call this *dugz – Division misconception*). Along with mugz, one of the most common misconceptions with decimal operations. Students believe that the quotient of a division should always be smaller than the leftmost factor of the operation (dividend) [18].
  - Example: 3 divided by 0.5 is smaller than 3

These persistent misconceptions evident in students' decimal knowledge must be addressed so that students can master decimals and move on to more advanced mathematics. The erroneous examples developed in this work focus on 6 of the 7 misconceptions above (megz, segz, negz, pegz, mugz, dugz); the seventh misconception, regz, is encountered throughout all of the others, often the underlying cause of the others.

To give students the opportunity to practice their cognitive and metacognitive skills (vs. practicing procedures) the erroneous examples were designed to (a) prompt students to compare incorrect solutions with correct solutions; (b) guide them in self explaining why a solution is incorrect, and, finally, (c) prompt them to self explain how to correctly solve a problem. Such an approach offers the opportunity for students to be aware of common mistakes and learn from them. We also believe it will encourage critical thinking and motivate reflection and inquiry. This three-step process with erroneous is also consistent with Ohlsson's theory on learning from performance errors [25]. Ohlsson claims that learning from errors requires one to first perceive inappropriate actions or facts while performing a task. Then, one must understand why performed actions/facts are incorrect. And finally, the correct knowledge and skills are employed to fix the problem.

Figure 2 shows an erroneous example focused on the *mugz* (multiplication makes bigger) misconception.

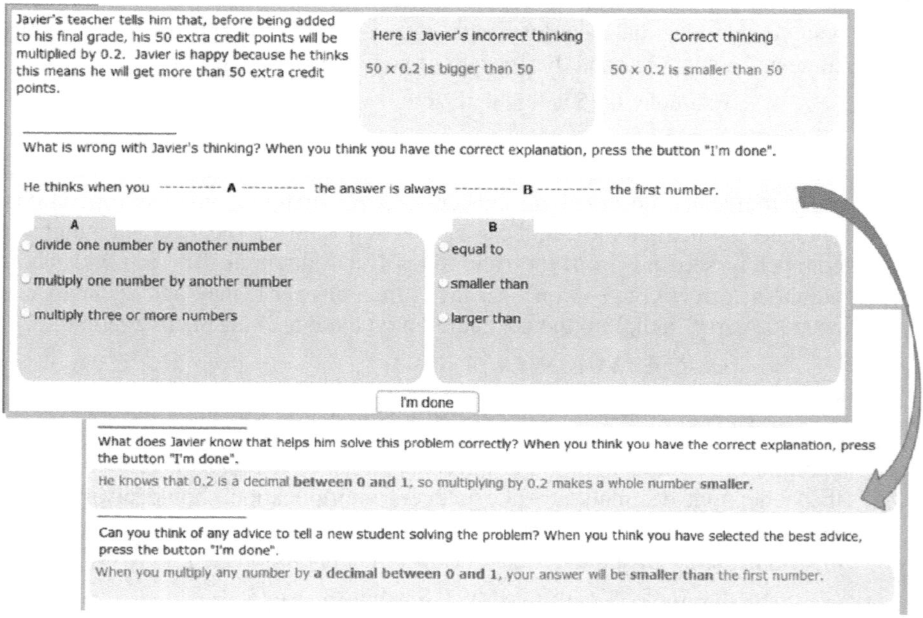

Fig. 2. Example of ErrEx for *mugz* (multiplication makes bigger)

Initially students only see the problem statement (top-left) and the incorrect solution (red-box in the top-center). By clicking on a button ("Show me the answer") the correct answer appears in a green box on the top-right of the screen. Students are prompted to compare the correct and incorrect solutions and, using a multiple-choice menu, they next try to self explain the misconception by building a sentence that says why the first solution (in the red-box) is incorrect. If they build the sentence correctly the sentence turns green and a second question appears; otherwise the sentence turns red and the student must change to the correct sentence to continue. The second question asks students to give advice to another student with difficulties. The intention is to motivate the student to practice the correct knowledge in order to fix the erroneous solution (i.e., in the red box). Only after answering both questions correctly can the student proceed to the next problem.

The erroneous examples were developed using an intelligent tutoring system (ITS) authoring tool, known as CTAT [26]. CTAT supports rapid development of web-based learning materials and produces exercises that track and log students' interactions with the software. Thus, it is possible not only to see whether a student's answer is correct or not, but also to check the student's overall behavior in tackling a problem (e.g., how many times they get a particular item wrong; whether they "game the system" by trying all possible answers [27]).

4 Experiment

Due to the previous findings about the effects of erroneous examples, i.e., that erroneous examples can lead to positive learning effects, particularly for higher prior-knowledge learners, we determined that our experiment be guided by the following **hypothesis**: Erroneous examples will help students improve both their cognitive and metacognitive skills. With respect to metacognitive skills, the students will especially benefit by learning to diagnose bugs/errors in problem solving, a type of metacognitive practice that is not readily available, but for which the erroneous examples provide just such an opportunity. In turn, students will refine their own problem solving skills through such practice. Furthermore, prompting students to reflect on potential errors will lead them to more carefully and deeply reflect upon and improve their conceptual understanding and problem solving in the domain of decimals. The positive effects on cognitive and metacognitive skills will be particularly pronounced for higher- prior knowledge students, as they are better prepared than weaker students to handle the more complex and challenging erroneous examples.

4.1 Method

Participants. Two hundred and fifty-five (255) middle school students Pittsburgh, Pennsylvania, U.S., participated; 86 students from 6th grade, 125 students from 7th grade and 44 students from 8th grade. The study materials were used as a complement for normal class work and the eight participating teachers received the students' scores for all tests in addition to two reports summarizing students' and classroom's performance on each item of the tests (pretest, posttest, delayed posttest).

Materials and Procedure. The complete study design is shown in Figure 3.

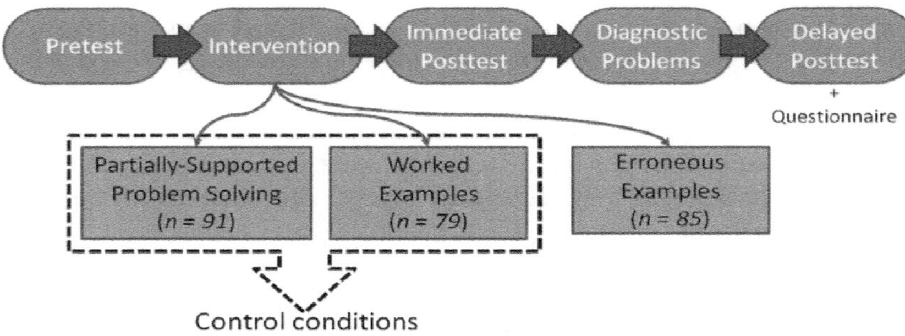

Fig 3. Study design. Overview of the sequence of activities

All students were randomly assigned to one of three conditions: Partially-Supported Problem Solving (PS condition); Worked Examples (WE condition) and Erroneous Examples (ErrEx condition).

(1) *Partially-Supported Problem Solving (PS)*: In this condition, the student is presented with problems to solve. Incorrect answers turn red and correct answers turn green. After the student gets a problem wrong at least one time, a button is displayed on the screen with the label "Show me the Answer." From this point on the student can select the button to see the correct solution to the problem. (The combination of green/red feedback and the "Show me the Answer" button is why this condition is called "partially-supported" problem solving, rather than pure problem solving, in which no feedback or help would be provided.)

(2) *Worked Examples (WE)*: In this condition, the student was presented with a word problem and a *correct example* to solve it. Similar to Figure 1 the student was prompted to answer two self-explanation statements. However in this case the first and second questions help students to build sentences that explicitly show which concepts were used to solve the problem correctly;

(3) *Erroneous Examples (ErrEx)*: In this condition the student was presented with materials like those in Figure 1. That is, they were presented with a word problem and *both an incorrect and correct example* to solve it. As previously discussed, the student then had to answer two self-explanation questions, one to explain why the incorrect example is wrong and one to provide advice to a fellow student on how to correctly solve the problem;

It is worth noting that the interfaces in each condition were similar and the problem statements were the same. Only the manner in which the problems were presented to the students (e.g. correct solution, incorrect solution, or problem to solve) and the interaction pattern with the interfaces were a bit different, as described above. The PS condition had 91 subjects; the WE condition 79 subjects; and the ErrEx condition 85 subjects. Each condition had 30 treatment problems, 5 problems targeted at each of the 6 misconception types of interest (Table 1 – The *, **, *** are indications of an identical item across conditions). The goal of the first problem of each set of 5 was to familiarize students with the misconception type. In the PS condition this first problem was a partially-supported problem to solve, while in the WE and ErrEx condition it was a worked example with the same problem statement. The rationale behind using a worked example as the first problem in the ErrEx condition was that an ErrEx problem might be too overwhelming and confusing as the first problem. Problems 2 and 5 in each sequence of 5 problems were embedded test problems to check whether a student shows evidence of the particular misconception and whether the intervention helped him/her to overcome it. Finally the problems 3 and 4 were the treatment problems unique to each condition. The problem statements were the same across conditions, however, the way students interact and solve the problems was different across the conditions. For example, the fourth problem for the *megz* misconception cluster was: "*Jorge has 3 cups. The first cup holds 0.47 L, the second holds 0.5 L and the third holds 0.613 L. Jorge's friend asks him to pick the smallest cup. Which cup should Jorge choose?*"; in the PS condition students had a multiple choice menu to select one of the three cups as smallest; in the WE condition students were shown the correct answers (0.47 L cup) and had to build two self-explanation sentences; and in the ErrEx condition students were shown an incorrect answer, related to the misconception type *megz* (0.5 L, because this cup is the shortest among the three), then they could click on a button to see the correct answer, and finally, they had to build two self-explanation sentences.

Table 1. Study design with an overview of each condition. *n* is the number of students per condition.

Condition 1 (*n* = *91*): Problem Solving	**Condition 2 (*n* = *79*): Worked Examples**	**Condition 3 (*n* = *85*): Erroneous Examples**
Megz: "Longer is larger" group of 5 problems		
1: Problem to solve	1: Worked example*	1: Worked example*
2: Embedded Test 1**	2: Embedded Test 1**	2: Embedded Test 1**
3: Problem to solve 1	3: Worked example 1	3: Erroneous example 1
4: Problem to solve 2	4: Worked example 2	4: Erroneous example 2
5: Embedded Test 2***	5: Embedded Test 2***	5: Embedded Test 2***
Segz: "Shorter is larger" group of 5 problems		
Negz: "Decimals seen as Negative Numbers" group of 5 problems		
Pegz: "Two sides of decimal viewed as separate numbers" group of 5 problems		
Mugz: "Multiplication always makes bigger" group of 5 problems		
Dugz: "Division always makes smaller" group of 5 problems		

Test and diagnostic problems were created to target the six misconceptions treated in the study. The pretest, posttest and delayed posttest were three isomorphic and counterbalanced tests (A, B, C) with 42 problems on each test. Thus, one third of the students used version A of the test as the pretest while other students used version B or C as pretest (similarly with the other tests). Each test took 25 to 50 minutes to complete and all of the materials took students between 160 and 220 minutes to complete. Students complete the pretest, intervention, immediate posttest and diagnostic problems in four consecutive periods (42 minutes each). The delayed posttest was applied in class a week after the immediate posttest. Students were also asked to fill a questionnaire that contained demographic questions (e.g. gender and age) and questions about their confidence on working with decimals and questions about their familiarity with computers. The diagnostic problems focused on checking metacognitive understanding which required students to verify whether a solution to a problem is correct or not and explain the concepts and/or processes used to answer the problem correctly/incorrectly.

All of the materials were developed based on problems used in earlier studies reported in the mathematics education literature. In particular, we selected problems that: (a) have been reported as difficult to solve in prior studies; (b) students' answers can lead to common misconceptions (one of the six discussed above); and (c) require more than procedural knowledge to answer correctly. All of the materials were made available through the MathTutor website (*https://mathtutor.web.cmu.edu/*). All participants received private user-IDs and passwords and worked alone at a computer at their own pace. Students were allowed to stop when desired to continue during the next period, except for tests in which all the problems had to be completed in a single period.

5 Results and Discussion

Table 2 shows the mean (and SD) for each of the three groups (and all of the groups combined) on the pretest, immediate posttest, and delayed posttest. To examine whether students' performance improved from undergoing the intervention, t-tests were conducted and revealed that there were significant differences between the scores on the pretest and the immediate posttest, $t(254)= -.7928$, $p < .001$, and between the pretest and the delayed posttest scores, $t(254)= -7.725$, $p < .001$, demonstrating the students' performance improved significantly overall in all conditions.

Table 2. Percentage correct on pretest, posttests, and diagnostic decimal problems

	Pretest		Immediate Posttest		Delayed Posttest		Diagnostic Problems	
Condition	M	SD	M	SD	M	SD	M	SD
PS	0.5	0.15	0.56	0.16	0.56	0.17	0.57	0.17
WE	0.54	0.17	0.6	0.18	0.6	0.19	0.61	0.2
ERREX	0.53	0.16	0.58	0.18	0.58	0.19	0.59	0.2

To test the general hypothesis that the intervention condition would affect learning outcomes, an ANCOVA was conducted using pretest score as a covariate. No significant differences were found between the three conditions for either the immediate posttest, $F(2,251)=.253$, $MSE=.003$, $p=.777$, or the delayed posttest, $F(2,251)=.178$, $MSE=.002$, $p=.837$. To examine whether condition affected performance on the diagnostic problems, an ANCOVA with pretest as a covariate revealed no significant differences between the three conditions, $F(2,251)= .119$, $MSE = .002$, $p = .888$. Looking at performance on the embedded test problems within the intervention, an ANOVA found no significant differences on overall embedded test performance between the three conditions, $F(2,252)=1.023$, $MSE=138.37$, $p= .36$.

In terms of how long it took the students to complete the intervention, there was a significant difference in the total amount of time if took for the three conditions, $F(2,252)=117.623$, $MSE=72802881.01$, $p<.001$. Tukey post-hoc comparisons of the three condition revealed that the problem solving condition took significantly less time compared to the worked example condition, $p < .001$, and the erroneous examples condition, $p < .001$. However the worked examples condition and erroneous examples did not differ from each other significantly in terms of the amount of time taken to complete the intervention, $p=.805$.

Table 3 shows the mean number of misconceptions (and SD) for each group (and all groups combined) on the pretest, immediate posttest, and delayed posttest.

Table 3. Misconception percentages per condition for pretest, posttest, and delayed posttest

	Problem Solving			Worked Exampled			Erroneous Examples			Overall		
	Pre	Post	Delay	Pre	Post	Delay	Pre	Post	Delay	Pre	Post	Delay
Regz	21%	16%	21%	21%	17%	16%	19%	18%	18%	20%	17%	18%
Megz	36%	28%	28%	34%	25%	21%	33%	25%	24%	34%	26%	25%
Segz	29%	25%	23%	30%	25%	21%	29%	23%	19%	29%	24%	21%
Negz	21%	15%	18%	22%	16%	14%	21%	14%	17%	21%	15%	17%
Pegz	40%	34%	39%	36%	31%	35%	41%	35%	36%	39%	33%	36%
Mugz	37%	32%	31%	35%	32%	30%	34%	31%	30%	35%	32%	31%
Dugz	31%	30%	32%	32%	30%	27%	29%	29%	27%	31%	30%	29%
Overall	**31%**	**26%**	**28%**	**30%**	**26%**	**24%**	**30%**	**26%**	**25%**	**31%**	**26%**	**26%**

Similar to previous research, students displayed many of the misconceptions targeted by this intervention with misconceptions representing 31% of the overall answers given on the pretest, posttest, and delayed postest. For each problem, answers were scored according to whether the answer was correct, incorrect, or incorrect due to a misconception (e.g. On a Negz problem did the student treat the decimal like a negative number?) Examining the difference between the number of misconceptions displayed, there was a significant improvement between the pretest and posttest, $t(254)=8.921$, $p<.001$, and between the pretest and delayed posttest, $t(254)=9.956$, $p<.001$. For the pretest, there were not significant differences between the three conditions for number of misconceptions made, $F(2,252)=1.45$, $MSE=29.17$, $p= .24$ Examining how the three learning conditions affected the overall number of misconceptions displayed there were no significant differences between the three groups for posttest, $F(2,252)=.606$, $MSE=12.71$, $p= .546$. On the delayed posttest, however, there was a significant difference for number of overall misconceptions made, $F(2,252)=3.72$, $MSE=89.72$, $p= .026$. Post hoc Tukey analysis revealed that the worked examples group ($M= 9.62$, $SD= 4.7$) made significantly fewer misconceptions than the problem solving group, ($M= 11.57$, $SD= 5.12$) $p = .028$, but not the erroneous examples group, ($M= 10.09$, $SD= 4.87$) $p= .881$. There was also no significant difference between the misconceptions on the delayed posttest between the erroneous example group and the problem solving group, $p= .116$.

To examine whether the prior knowledge affected the use of the intervention, participants were first placed in high and low prior knowledge groups using a median split based on pretest performance. As predicted, students classified as having high prior knowledge scored higher on both the immediate, $F(1,249)=3.5$, $MSE=220.82$, $p<.001$, and the delayed posttest, $F(1,249)=3.53$, $MSE=178.87$, $p<.001$. There was no significant interaction between high/low prior knowledge and condition for the immediate, $F(2,249)=.845$, $MSE=.013$, $p=.431$, or delayed posttest, $F(2,249)=.915$, $MSE=.018$, $p=.402$. ANOVAs examining the effect of high/low prior knowledge and total misconceptions made revealed those individuals with higher prior knowledge were less likely to show misconceptions on the pretest, $F(1,249)=268.83$, $MSE=2641.256$, $p<.001$, immediate posttest, $F(1,249)=139.54$, $MSE=1888.75$, $p<.001$, and delayed posttest, $F(1,249)=165.17$, $MSE=2405.8$, $p<.001$. No significant interaction was found for condition and prior knowledge level on overall number of misconceptions made on any of the tests.

Our main hypothesis that erroneous examples would lead to positive learning effects, particularly for higher prior-knowledge learners, was partially confirmed. Students in the erroneous examples group learned as much as students in the two control groups. Similar to previous findings in the literature our results indicate that erroneous example is a good instructional tool to help students learn and the view that errors can be dangerous and potentially hurt learning is not supported. However, conversely to what has been reported in other studies, high prior knowledge students did not learned more than low prior knowledge ones and no significant learning difference was found between the three conditions. A possible explanation is that erroneous examples are not as effective as hypothesized. In fact, students in problem solving condition spent considerably less time to learn the same content compared to students in worked examples and erroneous examples conditions; thus, problem solving was a more efficient learning treatment.

Another explanation for the results may lie in how the three-step learning process discussed in section 3 was implemented in our materials. While in the problem solving condition whenever students make a mistake the "Show me the answer" button appears and they could check the correct answer immediately, in the erroneous examples and worked examples condition to self-explain an incorrect or a correct solution students have to *read* all alternatives and select the best one to build the correct self-explanation sentence. Furthermore, no hint is given to students if they fail to build the correct sentence. Thus, we hypothesize that students were overwhelmed and that our erroneous examples and worked examples materials were too wordy and difficult for middle school students to read. It is also possible that these materials did not draw enough attention of students to motivate them to truly reflect on the correct/incorrect solutions because the tests were not used as a class grade. Therefore, there is a chance that some of the benefits of erroneous examples were lost. A third hypothesis is that perhaps the erroneous examples effect, like other effects (e.g., politeness effect), is much harder to achieve in the lively, buzzing confusion of real-life classroom (vs. lab) settings where most learning actually occurs.

To answer some of these questions and hypotheses we are preparing new materials that will simplify the self-explanation sentences and animate the examples to draw attention to important steps. We will also ask teachers to use materials as a class grade, so the students have more motivation to solve the problems. It also part of our plan to conduct think-aloud pilot studies with the new materials to have a better understanding about how the treatment might work in classrooms.

6 Conclusions

This paper discusses a classroom study with 255 subjects from 6th, 7th and 8th grades to better understand the impact of erroneous examples on decimal learning. To evaluate the effects of erroneous examples we developed web-based materials that use decimal misconceptions to present errors to students and help them learn by reflecting upon errors using a three-step process. We also compared erroneous examples with more typical instructional materials namely, worked examples and problem solving.

Our study produced mixed results. First, similar to previous research, students displayed many misconceptions representing 31% of the overall answers given on the pretest, posttest, and delayed posttest. The most frequent ones were *pegz* - two sides of decimals viewed as separate numbers - and *mugz* – multiplication makes bigger. Second, unlike earlier empirical work that showed the benefits of erroneous examples for higher-prior knowledge learners, our results indicate that higher prior knowledge students did not benefit more from erroneous examples than from worked examples or problem solving. Furthermore, although the incidence of students' misconceptions declined significantly in all conditions, it was more accentuated for the worked examples group. These results suggest that erroneous examples are not as effective for learning as we – and other – researchers have hypothesized. It also could indicate that erroneous examples, like some other e-learning techniques, do not work as well in the rough and tumble classroom as they do in a laboratory setting. Nevertheless, there are some changes in the manner in which erroneous examples were presented that could improve learning from errors. Thus, we are currently developing new erroneous examples materials that will potentially provide better learning outcomes and planning a new study to provide more precise answers about the effectiveness of learning from erroneous examples.

Acknowledgments. The U.S. Department of Education (IES), Award No: R305A090460, provided support for this research. We also thank the Pittsburgh Science of Learning Center, NSF Grant # 0354420, for technical support of our work.

References

1. National Math Panel Report (2008), `http://www2.ed.gov/about/bdscomm/list/mathpanel/report/final-report.pdf`
2. Glasgow, R., Ragan, G., Fields, W.M., Reys, R., Wasman, D.: The decimal dilemma. Teaching Children Mathematics 7, 89–93 (2000)
3. Rittle-Johnson, B., Siegler, R.S., Alibali, M.W.: Developing conceptual understanding and procedural skill in mathematics: An iterative process. Journal of Educational Psychology 93(2), 346–362 (2001)

4. Stacey, K., Steinle, V.: Refining the classification of students' interpretations of decimal notation. Hiroshima Journal of Mathematics Education 6, 49–69 (1998)
5. Putt, I.J.: Preservice teachers ordering of decimal numbers: When more is smaller and less is larger! Focus on Learning Problems in Mathematics 17(3), 1–15 (1995)
6. Grosse, C.S., Renkl, A.: Finding and fixing errors in worked examples: Can this foster learning outcomes? Learning and Instruction 17(6), 612–634 (2007)
7. Tsovaltzi, D., Melis, E., McLaren, B.M., Meyer, A.-K., Dietrich, M., Goguadze, G.: Learning from Erroneous Examples: When and How Do Students Benefit from Them? In: Wolpers, M., Kirschner, P.A., Scheffel, M., Lindstaedt, S., Dimitrova, V. (eds.) EC-TEL 2010. LNCS, vol. 6383, pp. 357–373. Springer, Heidelberg (2010)
8. Borasi, R.: Exploring mathematics through the analysis of errors. For the Learning of Mathematics: An Int. Journal of Mathematics Education 7(3), 2–8 (1987)
9. OECD. Knowledge and Skills for Life. First Results from PISA 2000, Paris (2001)
10. Bransford, J., Brown, A.L., Cocking, R.R.: How People Learn: Brain, Mind, Experience, and School. National Academy Press, Washington, D.C (1999)
11. Skinner, B.F.: The Technology of Teaching. Appleton-Century-Crofts, New York (1968)
12. Tsamir, P., Tirosh, D.: In-service mathematics teachers' views of errors in the classroom. In: International Symposium: Elementary Mathematics Teaching (2003)
13. Siegler, R.S.: Microgenetic studies of self-explanation. In: Granott, N., Parziale, J. (eds.) Microdevelopment, Transition Processes in Development and Learning, pp. 31–58. Cambridge University Press, Cambridge (2002)
14. Durkin, K., Rittle-Johnson, B.: The effectiveness of comparing correct and incorrect examples for learning about decimal magnitude. The Journal of Experimental Child Psychology (submitted)
15. Kopp, V., Stark, R., Fischer, M.R.: Fostering diagnostic knowledge through computer-supported, case-based worked examples: Effects of erroneous examples and feedback. Medical Education 42, 823–829 (2008)
16. Melis, E.: Design of erroneous examples for ActiveMath. In: Proceedings of the Int. Conference on Artificial Intelligence in Education, pp. 451–458. IOS Press, Amsterdam (2005)
17. Brueckner, L.J.: Analysis of Difficulties in Decimals. Elementary School Journal 29, 32–41 (1928)
18. Graeber, A., Tirosh, D.: Multiplication and division involving decimals: Preservice elementary teachers' performance and beliefs. Journal of Mathematics Behavior 7, 263–280 (1988)
19. Hiebert, J.: Analysis of arithmetic for mathematics teaching. In: Mathematical, Cognitive, and Instructional Analyses of Decimal Fractions, ch. 5, pp. 283–322. Lawrence Erlbaum, Mahwah (1992)
20. Irwin, K.C.: Using everyday knowledge of decimals to enhance understanding. Journal for Research in Mathematics Education 32(4), 399–420 (2001)
21. Resnick, L.B., Nesher, P., Leonard, F., Magone, M., Omanson, S., Peled, I.: Conceptual bases of arithmetic errors: The case of decimal fractions. Journal for Research in Mathematics Education 20(1), 8–27 (1989)
22. Isotani, S., McLaren, B.M., Altman, M.: Towards Intelligent Tutoring with Erroneous Examples: A Taxonomy of Decimal Misconceptions. In: Aleven, V., Kay, J., Mostow, J. (eds.) ITS 2010. LNCS, vol. 6095, pp. 346–348. Springer, Heidelberg (2010)

23. Bell, A., Swan, M., Taylor, G.: Choice of operations in verbal problems with decimal numbers. Educational Studies in Mathematics 12, 399–420 (1981)
24. Fischbein, E., Deri, M., Nello, M., Marino, M.: The role of implicit models in solving verbal problems in multiplication and division. Journal of Research in Mathematics Education 16, 3–17 (1985)
25. Ohlsson, S.: Learning from performance errors. Psychological Review 103(2), 241–262 (1996)
26. Aleven, V., McLaren, B.M., Sewall, J., Koedinger, K.R.: The Cognitive Tutor Authoring Tools (CTAT): Preliminary Evaluation of Efficiency Gains. In: Ikeda, M., Ashley, K.D., Chan, T.-W. (eds.) ITS 2006. LNCS, vol. 4053, pp. 61–70. Springer, Heidelberg (2006)
27. Baker, R.S., Corbett, A.T., Koedinger, K.R., Wagner, A.Z.: Off-Task Behavior in the Cognitive Tutor Classroom: When Students "Game The System". In: Proceedings of ACM CHI Computer-Human Interaction, pp. 383–390 (2004)

Simulating LEGO Mindstorms Robots to Facilitate Teaching Computer Programming to School Students

Torsten Kammer[1], Philipp Brauner[2], Thiemo Leonhardt[1], and Ulrik Schroeder[1]

[1] RWTH Aachen University, Lehr- und Forschungsgebiet Informatik 9
Ahornstr. 55, 52074 Aachen, Germany
{torsten.kammer,thiemo.leonhardt,ulrik.schroeder}@rwth-aachen.de
[2] RWTH Aachen University, Human Technology Centre
Theaterplatz 14, 52056 Aachen,
brauner@humtec.rwth-aachen.de

Abstract. Programmable robots like Lego Mindstorms have proven to be an effective mediator to teach computer programming to school children. Therefore several projects that aim at increasing the interest in computer programming and computer science in general use robots as a cornerstone in their course concepts. Handing out robotic kits to the school students who have participated in the courses is not feasible, thus the learning content cannot be repeated and enhanced at home. We developed a flexible multi-user simulation environment for LEGO Mindstorms NXT robots which is closely integrated into our pedagogical teaching scenarios. User tests show that this environment can be successfully used to increase the long-term outreach of our courses.

Keywords: Simulation Based Learning, CS0, Simulator, LEGO Mindstorms, NXT.

1 Introduction

Despite the economic crisis, demand for IT-professionals continues to be strong. Therefore, one goal to achieve is to optimize teaching at schools in quality and quantity in order to improve young peoples interest in science, technology engineering and mathematics (STEM). The PISA study revealed a high deficit in German school education especially in international comparison. The lack of mastery of basic skills due to lack of references to applications and networking of teaching has been criticized, as well as rather static teaching scenarios. Another aspect is the low number of female students in STEM subjects. Already in school, girls show less interest in topics of computer science in Germany[12].

To increase the participation in STEM in general and the participation of women in particular a lot of projects have been initiated. One popular concept is to use robots as a tool to increase interest in computer programming and computer science [6,9,5]. In a controlled experiment we found that programming

C. Delgado Kloos et al. (Eds.): EC-TEL 2011, LNCS 6964, pp. 196–209, 2011.

robots resulted in significantly better learning outcomes, increased motivation, and increased interest in STEM compared to programming visual objects on the computer screen [2].

Therefore we implemented a series of robotic workshops mainly for girls in the age of eleven to twelve. These workshops last two days and are carried out in local schools instead of the regular lessons. The students learn the basics of computer programming using Lego Mindstorms NXT robots. This hardware allows users to build a wide variety of robots and offers (among others) the *C*-like programming language NXC (Not Exactly C) to develop software for them. Our course concept has run successfully for two years, and approximately 1250 school students have participated in our robotic courses. Our evaluation shows that students like the courses and that their interest in STEM in general and computer science in particular has risen after participation in a workshop [8].

However, one drawback of our course concept is the limited sustainability. As each robotic set costs roughly 350 Euros, most participants do not own one privately. Thus participants usually cannot repeat the subject matter individually or deepen their programming knowledge on their own after the workshop has ended. This limits the benefit of the project, because the participants would probably learn more if they had more time to experiment with the robots. To overcome this problem we developed a simulator that is closely integrated into our course concept and can easily be used by individual students or groups of students after the workshop has ended.

The simulator provides a three dimensional view on a standard LEGO Mindstorms NXT robot and its environment. The simulated robot model can be modified by adding or removing sensors or by changing the positions of the sensors. The environment is easily reconfigurable with walls and arbitrarily shaded floor tiles so that a magnitude of learning scenarios can be created. The simulator can be downloaded for free from our InfoSphere website[1].

We focused on the learning experience with regard to computer science education. Thus we didnt provide a physically accurate simulation of the environment, the robot and its components, but rather an abstraction as learning tool. However, the simulated robot resembles the behavior of a real robot very closely. Additionally, the simulator is able to communicate with other instances of the simulation environment over a network. This allows users to implement and test their robot programs collaboratively in a shared virtual environment similar to collaborative scenarios of our go4IT!-workshops.

The simulator directly executes code compiled for the NXC platform. This allows users to develop software using the same tools as used in the workshops for the actual device and test the software without access to the hardware, both at school and at home. As the simulator executes native code, it can also be used with arbitrary third party tools like a programming environment or a programming language different from the ones we used in our courses.

[1] `http://schuelerlabor.informatik.rwth-aachen.de/simulator`

2 Course Concept

During the first day students build a simple LEGO Mindstorms robot which can move on wheels and can utilize various sensors. The students have been familiar with LEGO blocks since kindergarten, so there are no problems to build a working robot. After that the students actively explore the first simple commands to control the robots movement (forward, backward, different speeds, square, circle, stop for obstacles or at certain distances) and to write corresponding programs in NXC. At the end of the first day the robots control the touch sensor and play self-composed and programmed songs. These goals are usually achieved by all participants.

During the second day students learn to control additional sensors (audio, ultrasonic and light sensors). Thus, students come up with more advanced tasks they want to implement. Potential projects are discussed by the participants and are implemented mostly without help of the tutors. Individual solutions of a team are presented to all students and successful students are encouraged to help the others as experts in their problem-solving. At the end of the second day, all participants work on a common task for the whole group such as dance performances and choreographies for interacting robots. In this case, all robots of all teams are involved in the common performance.

The workshop ends with the presentation of this team performance in front of classmates, teachers, parents, or sometimes the invited press. The students gain recognition for their work through the feedback of the audience. At the end of the workshop all students have successfully solved tasks independently and developed programs to let a robot perform a certain task. The analysis and solution of complex problems enhance the self-efficacy of the students (*"The robot does what I want!"*).

Fig. 1. Workshop impression. The robot is programmed to follow the black line.

3 Related Work

The field of simulation is a wide one, ranging from simulators for individual physical attributes of single components all the way to fully integrated simulations of industrial machinery. Simulating what is essentially a toy, and with little physical accuracy, this simulator falls on the low end of this scale.

3.1 Robot Simulators

Simulators for Lego Mindstorms already exist, although they have different goals and limitations as the one developed here. RobertaSim [15] is a simulator for the RCX microcontroller (the deprecated LEGO Mindstorms RCX sets) for Microsoft Windows platforms. It was developed for courses specifically aimed at teaching programming to girls. The goal is to have a robot move through a virtual environment executing the same code that works on real robots. This makes it similar to the go4IT!-simulator developed in our project. However, it cannot work as a replacement, as it simulates a different hardware generation. Furthermore, RobertaSim focuses on physically accurate simulation, which makes it more complex to use and thus less suitable for our computer science teaching scenarios. In our teaching scenarios the focus lies on the understanding of algorithms (e.g. path finding algorithms) while the increased accuracy leads of a simulation usually leads to different kinds of problems that must be solved (e.g. a robots sensor can be caught by a wall tile). Additionally it lacks network support, thus it cant be used in collaborative learning scenarios.

"LMS" (Lego Mindstorms simulator) simulates Mindstorms NXT controllers [13]. It requires the programs to be written for a custom firmware. This simulation aims at complex scenarios and offers a fairly extensive user interface. While sharing many of our goals on the technical side, this simulator is far more complicated than ours and was not designed for teaching school children. LMS also requires a custom firmware to be installed on the robots which limits its compatibility with the development tools we use in our courses.

At the higher end, simulation of robots can become extremely precise and complex. An example for this is SimRobot [14]. Here, complex robots can be defined including sensors, various joints and even simulated cameras. Users can add more elements by implementing their own modules in *C++*. There is also a wide variety of scenarios that can be modeled through this extensibility. This simulator is clearly more advanced than the one described here, but this complexity also makes it unsuitable for the basic teaching task that the simulator is meant to fulfill in our scenario.

3.2 Other Microworld Approaches

The Java Hamster programming model or the Greenfoot microworld [1,7] follow a similar approach in teaching the basics of programming to children. Here, users program a cybernetic beastie such as a virtual turtle on the computer screen using a subset of the Java language. The feature set of the hamster is limited to turning, moving in a straight line, detection of walls and similar high-level concepts. It differs from our simulator firstly in that the code cannot be used on any real-life robot. It also offers a more abstract and high-level approach. In our simulator the motors have to be turned at different speeds to execute a turn while the hamster simulator has a direct turn instruction. This is, in fact, an adoption of Seymour Paperts microworld approach [10] for teaching computer science to school children.

In our course concept we intentionally use a programming model that requires the users to program individual sensors and actuators in a *C*-like syntax without hiding the actual programming complexity by providing abstractions through pre-defined commands like "`turnLeft()`" or puzzle programming metaphors like in Scratch [11]. We found that the participants of our courses enjoy presenting the more difficult looking *C*-like code to their classmates and their parents and that this leads to an increased self-concept in dealing with and interest in technology.

A number of approaches have been made to simulate robots in general and Lego Mindstorms in particular, but none fulfill our requirements for simple teaching scenarios as described in the following.

4 Requirement Analysis

Our main goal was the development of a simulator for LEGO Mindstorms NXT robots that can easily be used by all participants of our robotic courses. That led to a number of requirements from a technical and a usability perspective.

4.1 Hardware Requirements

First, we strived for a platform independent solution that operates on Microsoft Windows, Mac OS X and Linux systems. An important goal of our Simulator project is to have an appealing 3D output. To guarantee cross-platform compatibility we decided to implement all of this using OpenGL.

The participants of our courses come from different schools and backgrounds. Therefore, we cant assume modern computing machinery to be available to all school students. We expected that many of the target systems will have relatively little processing power, so we decided to use as little OpenGL features as possible while still maintaining a clean code base. To ensure this, only features included in the 1.5 version of the OpenGL specification were used. This offers features such as vertex buffer objects for management of geometry data, but does not yet include any form of pixel- and vertex shaders that may cause problems on older hardware, for example with integrated GPUs. Furthermore, only features included in OpenGL ES 1.1, a subset of OpenGL 1.5 specifically aimed at embedded systems, were used. In the future this might allow us to explore new teaching scenarios that incorporate simulation based learning and mobile devices.

4.2 Usability Requirements

The simulator should use the same development environment as used for real robots in the courses. Consequently, it should read the same bytecode that is produced by the development environment. It should also be possible to easily launch the simulator directly from the development environment that is used in our courses.

The simulator should be easy to deploy. Consequently the simulator needs to be executable without the need for an installation routine. No settings may be stored in the Windows Registry or other local configuration files. The application should not need extra privileges that would require an administrator or root account.

After the end of the course the simulator and the other development tools should be handed out to the participants. They must be able to easily run the development and simulation environment when at home.

The number of user interface elements should be kept minimal and they should require little or no explanation. To make the user interface that simple, it contains only five buttons by default. The only more complex part is a sensor selection view that can be opened to configure the robot (see Figure 3). There is no modality: While the configuration view is opened, all other functions of the user interface remain fully usable and the program is not paused. Apart from the window title required by the operating system, the user interface does not contain any text at all. Thus the program can be used regardless of languages the user may speak and without needing any localization for the content.

4.3 Simulation Requirements

As we focus on teaching computer science concepts, it is not necessary to simulate many different robot models or even allow full Lego construction. However, as different sensor placements obviously have a profound impact on the algorithms used to solve a given problem (e.g. path finding algorithms), it should be possible to change the sensor placements easily. Likewise, the virtual 3D environment should be easy to edit, so that the algorithms can be tested in different environments (like different maze setups). As the focus should be on simulating programs, not physics, physical simulations can and should have low accuracy.

The simulator is able to emulate a reasonably large subset of all programs for the NXT system. Nevertheless some rarely used commands of the NXT bytecode need to be implemented. Of course all language features taught in our robotic courses must work within the simulator.

4.4 Additional Features

Our robotic courses are group learning experiences in which two students share a laptop and a robot and work together and different teams and their robots also interact with each other in common tasks. This collaborative group learning experience should also be possible for students using the robot simulator, thus we implemented a network mode that allows multiple robots to be simulated in a shared environment. As easy deployment was a key concern, the simulator uses network auto discovery to discover other instances of the robot simulator running on the local network. If one is found, the simulator automatically connects to this instance and the environment is shared. If no other simulator is found on the local network, the client automatically becomes a server and starts to announce itself to future clients. The network layer is targeted at local environments where auto discovery usually works. There is currently no user interface to manually configure network settings to allow connections to arbitrary IP-addresses (e.g. connections over the Internet), even though that would be technically possible.

5 Simulator

The go4IT!-simulator is an interactive program with a GUI and 3D display of simulated robots. It is written entirely in *C++* to support as many platforms as possible. It uses OpenGL for drawing, OpenAL for sound out- and input and SDL for the interaction with the operating system. At the core of the system, an interpreter executes the compiled Lego NXT bytecode.

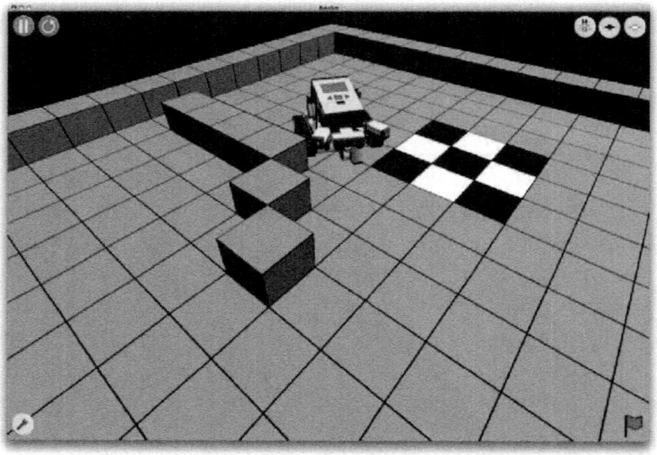

Fig. 2. Screenshot of the simulator with a single robot, a wall and shaded floor tiles

5.1 Environment

The environment of the simulated robot is designed to be primarily simple to edit, allowing to create a wide range of scenarios in little time. In our simulator the environment is a grid of 20×20 cells by default. Grid-based approaches have worked very well for this in various forms, for example by using height maps to display the ground in large outdoor scenes, e.g. in [4]. Each cell can have one of two states: Wall, which the robot cannot pass through, and floor, which the robot can drive on. Touch and ultrasound sensors will detect contact or distance to the walls. To allow the light sensor to work with meaningful inputs, each cell has an arbitrary shade of grey (rendered in the user interface as black, gray or white).

All the attributes of a cell can be changed at any time by clicking on it. Through the user interface, one can change between height mode (a click toggles whether the cell is wall or floor), lighten mode and darken mode (Figure 2, upper right corner). This makes it easy to create mazes for the robot to navigate through or to create colored lines that the robot should follow. Both tasks are typical for our robotic courses.

The editor does not allow the cells at the edges of the map to be changed to floor cells, although their colors can be altered. This prevents the robot from steering into the void.

5.2 Robot

There is only a single robot model available, using two of the three motor outputs for propulsion (a model with two wheels is used). The robot is simulated by assuming essentially infinite acceleration and ignoring masses, which is a seemingly imprecise approach. However, the mass of a real robot is low enough compared to the total power output of the robot that this proves to be an adequate approximation. This is also aided by the fact that the speed of the motors on the real robot is electronically regulated to reach a given speed as soon as possible and then keep it approximately constant. As a result, the difference to a more realistic simulation is small.

The robot can have up to four sensors registering touch events, light values, noise levels or distances to an obstacle. Any combination of these sensors is possible, and the sensors can be freely placed on a circle around the robot. A sensor may point forward or downwards. Its configuration can be changed at any time by opening the sensor configuration panel (see Figure 3). Opening this panel does not pause the execution of the program. Anything that can be done while the configuration panel is not shown can also be done when it is shown (e.g. altering the environment by adding a wall).

The execution of a robot program can be paused, which also stops the robot. It is also possible to reload the program of the robot, which also restarts the execution from the beginning. A robot can be picked up and placed somewhere else via drag and drop. This feature is useful in case the robot is stuck inside a maze and the students want to check if the algorithm is working correctly for a different position on the map. In case of a networked scenario, all this only applies to the robot controlled by this particular computer. To signify which computer owns which robot, each robot has a uniquely colored flag. A flag of the same color is displayed in the user interface of the computer controlling that robot (see Figure 2, lower right corner).

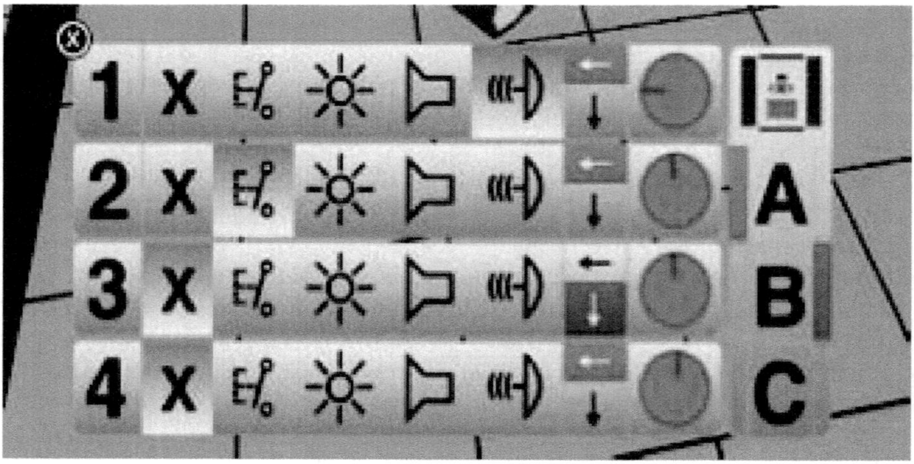

Fig. 3. Sensor configuration panel. Sensor 1 is configured as ultra-sound sensor pointing sideways. Sensor 2 is configured as a touch sensor pointing ahead. Sensors 3 and 4 are disabled.

5.3 Sensor Values

Generating accurate sensor values is highly important, as this is the only means of input available to programs running inside of the simulator. The simulator offers the four types of sensors that are part of the LEGO Mindstorms NXT sets: Touch sensors, light sensors, sound sensors, and ultra-sound sensors.

Typical programs running on a LEGO Mindstorms NXT robot use polling to read sensor values. For example, a program that lets the robot follow a dark line is usually implemented as follows: At first it will turn on the motors. Then it will constantly check the light sensor until a certain threshold is exceeded (i.e. `until(SensorLight(IN_3) > 40);`). This would trigger several thousand recalculations of the virtual sensor values per frame of the simulation run-loop. Hence the simulator calculates the value of each sensor once on every iteration of the run-loop, regardless of whether and how often it is actually requested, and caches it until the next iteration of the run loop. As environment and positions only change once per run through the run-loop at most, this delivers correct results without calculating sensor data numerous times.

Each sensor is mounted to the robot at a specific angle, has a certain offset, and may point forward or downwards. This combination creates a specific sensor position, based on which the results are read. All sensors work the same no matter if they point forward or downwards, only with different directions. However, as with real LEGO Mindstorms NXT robots it hardly makes sense to set the touch, ultrasound, or sound sensor to point downwards.

Ultrasound Sensor. To find the value for the ultrasound sensor, a ray is cast from the sensors position. It is checked against the environment and other robots, and shortest distance to any object hit by this ray is used as the sensors value.

If no hit is found within the maximum distance, a value of 255 is used instead, matching the real sensor.

Touch Sensor. For the touch sensor we used collision detection methods. Based on the sensor position, a bounding box for the active part of the sensor is generated and tested against the environment and other robots. The robot doesnt get moved if such a collision occurs, but the sensor value is set correctly.

Light Sensor. The light sensor also casts a ray through the environment and reports the shade of the cell hit, or 0, for black, if no cell is hit. For calculating the light sensor value, any robot in the way is ignored, as it is generally not easy to find a color value for them. In our go4IT!-tasks, light sensors are thus only use pointing downward to read color information on the floor.

Sound Sensor. The sound sensor does not directly interact with the environment. Instead, the simulator calculates a value, based on the position of the sound sensor, the positions and volumes of all robots playing sounds and the noise reported in from an attached audio source like the computer microphone. This allows the users of the simulator to build applications that let the simulated robot react on real-world events like clapping in the hands in front of the computer. The sound sensor also reacts to sounds emitted by other robots connected in network mode (see below).

5.4 Network Mode

It is possible to run the simulator alone, with a single robot, or together with others over the network, so that every robot gets shown on every computer. Each computer in the network simulation executes its own code and transmits the state changes to the server. Conceptually, there is a difference between the single server, which calculates the physical simulation for all robots, and clients, that just execute code, transmit its output and then display robots at the positions given by the server, but from the users view, both work the same.

5.5 Collision Detection

As robots cant pass through walls or other robots, the simulator uses collision detection. For this an oriented bounding box of the robots is calculated. These bounding boxes are also used to provide input values for the touch sensor and the ultra-sound sensor. In that case a ray is cast from the sensors. In the case of the ultra-sound sensor the reported distance is the distance from the starting point of the ray to the nearest point where a bounding box intersects with the casted ray. As the robot model is not a perfect box, this can result in situations in which a collision is reported even though the ray might pass the robot model but not the bounding box. In practice this is not a problem as the area of the robot is comparably large.

If a robots bounding box has collided with a wall in the environment the robot is moved away from the wall. This is done without changing the direction of the robot thus, if the robot hits a wall at an angle, it will slide along that wall. If two robots collide, both robots are moved so that they no longer collide. This is done without changing the driving direction.

All this happens in 2D only, as the environment design does not allow any robot to leave the ground floor. In the special case of a robot being picked up to move it somewhere else, all collision detection is disabled.

Using this method, no momentum, energy, elastic or plastic properties are regarded. If a robot hits a wall, it simply stops moving any further, and if it hits the wall at an angle, it will glide along the wall.

It is possible to implement a more accurate collision model for the simulator, however, a more accurate representation of physics would require users to not only develop programs, but also keep in mind the physical properties of the robot. This higher grade of complexity, however, would increase the extraneous load of the learners and would hinder their ability to focus on the development and implementation of algorithms [3].

5.6 Integration into the Development Environment and Process

The simulator described above works well as a stand-alone simulation engine for Lego Mindstorms NXT bytecode. However we wanted to closely integrate the simulator into the development process the school students learned and used during the courses. There the students usually apply an iterative development cycle in which they start off with defining a goal (e.g. let the robot follow a dark line). Then they implement the corresponding program and test it. If the program does not work as expected – which is usually the case –, they go back to the implementation phase. In our courses this iterative development process has rather quick cycles, as the deployment of an adjusted program on the robot can be done within seconds over a Bluetooth connection.

To allow equally fast development cycles when the simulator is used, we closely integrated the simulator in the BricxCC[2] integrated development environment we are using in the courses. This modification allows the students to automatically deploy a program on the simulator once it is compiled.

6 Evaluation

There were two main parts of the testing. First of all, the simulator was tested on a number of different computers and different operating systems to ensure that it is platform independent and easy to deploy. The simulator was tested with a number of robot programs that the school children developed during our courses. The programs ranged from rather simple programs that let the robot drive simple geometric shapes without reacting to sensory input to non-trivial maze solvers like Pledges Algorithm.

[2] `http://brixcc.sourceforge.net/`

Secondly, a user test with school students was carried out to see how well they understand the user interface and whether they integrate the simulator into their development process.

Up to now, the simulator has not been used in our actual go4IT!-workshops and just put on the web site for download. Thus, there have not been any evaluations of its utilization as a follow-up learning tool.

6.1 Technical Evaluation

To evaluate if the simulator works platform independent, it was tested with different computers and using different NXC programs. It was tested successfully on multiple Mac OS X versions (10.5 and 10.6) using either a PowerPC CPU or an Intel CPU. It was also tested with Windows XP and Windows 7 running on an Intel CPU. It worked well with GPUs by ATI and NVIDIA, as well as with an emulated GPU used in a virtual machine.

A number of programs from the robot workshops were gathered. Their operations in the simulator were compared to a real Lego Mindstorms NXT robot. Indeed the programs running on simulated robots worked similar to programs running on real robots.

Finally, testing was also done with network support between all testing target systems, as well as locally between two instances running on the same machine.

On all platforms, the simulator was able to execute all test programs accurately and produce results corresponding closely to the ones observed with the real robot. Network sessions also worked between any of the systems, with each of the systems working either as server or client. Networking was also tested with multiple clients and worked as expected there.

No precise performance measurements were gathered, but the simulator worked without any noticeable slowdowns on all target systems, including in network mode. The graphics results were generally identical and comparable to the observations in the real world.

6.2 User Testing

Finally we invited a group of school students of grade ten to twelve to use the simulator. In addition to the normal BricxCC IDE and a LEGO Mindstorms NXT robot, the simulator was offered to the students. They were asked about any problems they noticed.

The students confirmed that such a simulator is a very useful aid because it allowed faster write-test cycles. They also agreed that the simulator is useful to repeat the subject matter from the workshops at home.

The students also discovered a number of smaller issues and a few usability flaws. In particular they criticized:

- It is not possible to remap the motor ports, so code written for a real robot using different ports for the motors does not work directly.

- The robot can only be picked up, not turned, which would be desirable in some situations.
- The sensor configuration should persist between starts of the program.
- Some more advanced operations of the NQC programming language were not implemented.

These issues have been fixed shortly after the user test.

7 Summary and Future Work

The developed simulator is able to execute a wide variety of NXC programs and allows creating environments for simulated Lego robots to test these programs. Following the requirements, more complex programs cannot be simulated accurately, although it should be possible to extend the amount of supported operations in the future. Networking works as expected, and there is no functional difference between any of the supported platforms. With that, the simulator fulfills all requirements as a teaching tool.

The simulator is developed to be a teaching tool, and as such the simulation need not to be exceedingly precise. Also it is not necessary that the full NXC bytecode and all advanced features are supported. For future work, the tools scope could be expanded to make it a fully featured simulator for all Lego Mindstorms robots and programs. This could allow both teaching more advanced programming techniques and using it to simulate other projects faster. To do so, first of all, current limitations that were acceptable within the scope of this project would need to be removed. For example, it should be possible to support a larger portion of the byte code. In addition, the current simple physical simulation could be replaced by a more robust one, possibly using libraries such as Bullet, adding realism, but making programming harder as users would have to account for physical reactions. In such a situation, it would also be desirable to have more robots to choose from.

The environment systems could be extended in various ways. Other objects for the robot to interact with would allow for new, more complex scenarios that are currently impossible. The same would apply to an environment with more actions possible, such as ascending to a higher point by means of ramps. This would necessarily make it more complex to modify the environment and possibly require a completely different approach to editing.

For such a tool, more efficient networking and Internet support might also be desirable. This would require making the network protocol more efficient or replacing it completely, and adding measures against high transmission delays and improved synchronization techniques.

Since all the libraries used in the simulator are available on other operating systems, it should be possible to port the simulator to systems such as GNU/Linux rather easily if desired.

References

1. Boles, D.: Programmieren spielend gelernt mit dem Java-Hamster-Modell. Teubner (2008)
2. Brauner, P., Leonhardt, T., Ziefle, M., Schroeder, U.: The effect of tangible artifacts, gender and subjective technical competence on teaching programming to seventh graders. In: Hromkovič, J., Královič, R., Vahrenhold, J. (eds.) ISSEP 2010. LNCS, vol. 5941, pp. 61–71. Springer, Heidelberg (2010)
3. Chandler, P., Sweller, J.: Cognitive Load Theory and the Format of Instruction. Cognition and Instruction 8(4), 293–332 (1991)
4. De Boer, W.H.: Fast Terrain Rendering Using Geometrical Mipmapping (2000), http://www.flipcode.com/tutorials/tut_geomipmaps.shtml
5. Eggert, D.W.: Using the Lego mindstorms NXT robot kit in an introduction to C programming class. J. Comput. Small Coll. 24(6), 8–10 (2009)
6. Hartmann, S., Schecker, H.: Bietet Robotik Mädchen einen Zugang zu Informatik, Technik und Naturwissenschaft? – Evaluationsergebnisse zu dem Projekt Roberta. Zeitschrift für Didaktik der Naturwissenschaften 11, 7–19 (2005)
7. Kölling, M., Henriksen, P.: Game programming in introductory courses with direct state manipulation. In: Proceedings of the 10th annual SIGCSE conference on Innovation and technology in computer science education. ITiCSE 2005, pp. 59–63. ACM Press, New York (2005)
8. Leonhardt, T., Brauner, P., Siebert, J., Schroeder, U.: Übertragbarkeit singulärer MINT-Interesse-initiierender außerschulischer Maßnahmen. In: INFOS 2011, 14. GI-Fachtagung Informatik und Schule (September 2011) (accepted)
9. Leonhardt, T., Schroeder, U.: go4IT!: Initiierung und nachhaltige Förderung von Interesse an MINT-Fächern bei Mädchen. In: Informatische Bildung in Theorie und Praxis, Beiträge zur INFOS 2009, 13. GI-Fachtagung - Informatik und Schule, Berlin (2009)
10. Papert, S.: Mindstorms: Children, Computers, and powerful Ideas. Basic Books, Inc., New York (1980)
11. Resnick, M., Maloney, J., Kafai, Y., Rusk, N., Eastmond, E., Brennan, K., Millner, A., Rosenbaum, E., Silver, J., Silverman, B., et al.: Scratch: programming for all. Communications of the ACM 52(11), 60–67 (2009)
12. Schinzel, B.: Informatik und Geschlechtergerechtigkeit in Deutschland Annäherungen. Reihe Gender Studies (Gender and Science, Perspektiven in den Natur- und Ingenieurwissenschaften), 127–146 (2007)
13. Schmalzbauer, J., Scheel, O.: Lego Mindstorms Simulator (2011)
14. Siems, U., Herwig, C., Röfer, T.: SimRobot, ein System zur Simulation sensorbestückter Agenten in einer dreidimensionalen Umwelt. No. 1/94 in ZKW Bericht, Zentrum für Kognitionswissenschaften. Universität Bremen (1994)
15. Theidig, G., Börding, J., Petersen, U.: Roberta - Der Simulator RobertaSim. Fraunhofer IRB Verlag, Stuttgart (2006)

Adaptive Domain-Specific Support to Enhance Collaborative Learning: Results from Two Studies

Anastasios Karakostas and Stavros Demetriadis

Department of Informatics, Aristotle University of Thessaloniki
POBOX 114, 54124, Thessaloniki, Greece
{akarakos,sdemetri}@csd.auth.gr

Abstract. Although "adaptivity" and "intelligence" have built a long tradition in technology systems for individual learning, it is only relatively recently that these ideas have entered the domain of computer-supported collaborative learning (CSCL). In order to investigate how and when adaptive interventions in CSCL settings are effective, we implemented two experimental research studies, using a freely available open source tool (LAMS), trying to identify the effectiveness of a particular adaptive domain-specific support strategy. In both studies, treatment student groups were supported by a domain-specific adaptive intervention, where control groups were supported by "fixed" support methods. The results from both studies revealed that treatment students outperformed those in the control group in domain knowledge acquisition. Overall, this paper presents the positive results of these two studies providing evidence that by implementing techniques of adaptive domain-specific support during a collaborative activity, instructors can substantially improve learning outcomes.

Keywords: Computer-Supported Collaborative Learning, Adaptive Collaboration Support, Collaboration Scripts.

1 Introduction

Collaborative learning has been proved rather important for students both for cognitive and social reasons [1]. However, collaborating students usually fail to engage in productive learning interactions when left without teachers' consistent support and scaffolding (e.g. [2], [3]). Currently, there have been reported efforts from various research groups towards the adaptive operation of CSCL (computer-supported collaborative learning) systems (e.g. [4]). These efforts advance the tradition of *Intelligent Tutoring Systems* and *Adaptive Hypermedia Environments* (which were basically meant for individual learning) toward CSCL and expand the perspective of the field while setting innovative research agendas. In general, intelligent and adaptive collaboration support techniques aim to model the major aspects of the collaborative activity (such as peer interactions, collaboration activity structure, problem-solving tasks, student/group profile, domain knowledge, etc.) and activate learner/group support interventions when needed and in the form it is needed [5]. There have been reported some encouraging first results regarding the learning

C. Delgado Kloos et al. (Eds.): EC-TEL 2011, LNCS 6964, pp. 210–219, 2011.

impact of these interventions (e.g. [6]). However, there are also implementations that do not prove that such type of interventions lead to enhanced domain-related learning outcomes (e.g. [7]).

Based on these facts, this paper presents two experimental studies. The first one (study 1) investigates the capacity of an adaptive support strategy to further improve learning outcomes in the context of a scripted collaborative activity. The second study (study 2) investigates if the same adaptive support strategy would lead to better learning outcomes compared to a fixed and informationally equivalent support mechanism, also in the context of a scripted collaborative activity. Both of these studies were conducted in LAMS environment, which is a freely available open source tool for designing and deploying collaborative activities. In the following, we present briefly (a) the theoretical background of our research, (b) the main findings of the two studies, and (c) a discussion analyzing the learning impact of the adaptive collaboration support method.

2 Theoretical Background

Collaborative learning has been proved significant for students for social, cognitive and meta-cognitive reasons [1], [8]. In CSCL, the group interactions among individuals are mediated by computer environments [9]. However, when students are engaged in computer-supported collaborative learning, they need significant support and guidance since they are rarely engaged in productive interactions such as asking each other questions or reflecting upon their knowledge [2], [3]. There have been also reported studies that identified patterns of ineffective collaboration in free (non-supported) collaboration conditions (e.g. [3]).

Current collaborative learning research focuses on the effects of giving *fixed* support to groups of students, while more recent research efforts investigate the impact of *adaptive* collaboration support mechanisms.

2.1 Fixed Collaboration Support

In order to trigger peer interactions, researchers have suggested various forms of "fixed" collaboration support. By "fixed" we refer to predefined structure and guidance that work the same for all participants in a collaborative learning activity, without employing any modeling (of learner, situation, etc.) that would enable the dynamic adaptation of the form, the content or other features of the support.

One important and complex form of fixed collaboration support is the collaboration script [10], [11]. Collaboration scripts are didactic scenarios that specify the way in which learners interact with one another [12]. In other words, a script is a set of guidelines prescribing how the group members should interact, how they should collaborate and how they should solve a problem [13]. Scripts structure the collaborative process by defining sequences of activities, by creating roles within groups and by constraining the mode of interaction among peers or between groups [14]. In this way the script increases the probability of productive student-student and student-teacher learning interactions, promoting cognitive and social processes which otherwise might not occur. Lately, the considerable interest that the scripting approach has gained in the CSCL community has motivated efforts for the formalization of collaboration scripts and the development of computer-based environments for supporting scripted collaborative learning (CSCL scripting) [15], [16].

Implementing CSCL scripts has been reported to result in improved learning outcomes (e.g. [11], [17]). However, CSCL scripting has been criticized for its loss of flexibility (difficulty of modifying a script in run time according to the needs of the instructional situation) [14], and also the danger of "over-scripting" collaborative activity (the pitfall of overemphasizing script imposed interactions and constraining natural collaboration) [13].

2.2 Adaptive Support

Current CSCL efforts have focused on supporting groupwork through the use of adaptive and/or intelligent systems. Adaptive and intelligent interventions tailor the collaborative learning process to the needs of the individual students or groups. The target of the adaptive/intelligent interventions varies and can be classified into 2 main categories: 1) peer interaction support (i.e. help peers to "learn to collaborate" by encompassing interaction rules in the medium), 2) domain knowledge support (i.e. help peers to deepen their domain understanding by monitoring and regulating the interactions) [5].

Peer interaction support refers to the actions taken by the system in order to help learners improve their interaction and develop domain-general knowledge and skill (for example, argumentation, peer tutoring, and peer reviewing). The objective of adaptive peer interaction support is to track the progress of the collaboration process and make supportive interventions, based on patterns of well-known productive interaction sequences and indicators. Several studies from this area indicate the difficulty to model the student dialogue or student interactions in general and therefore to compare the current level of collaboration to a desired one in order to support the collaboration (e.g. [18], [19]).

For the purpose of this study, we focus on the second category of support. *Domain-specific support* refers to the actions taken by the system in order to help learners understand the domain better. This kind of support concerns the aspects of users and groups (and their activities) that have to be modeled and can be inferred or observed in system/user interaction in order to support group learning (e.g. [7], [4]).

In general, three major issues emerge from the area of adaptive and intelligent collaboration support area [20]: (1) systems are in an early stage of development and evaluation and relevant studies most often do not report clear learning benefits (although there are some first encouraging results, (e.g. [6]) (2) the systems are strongly related to the target domain and thus they can not in general be deployed easily in another domain, (3) there is a lack of coherence in assessing the learning impact, since no common benchmarks have been agreed upon, making almost impossible to compare the efficiency of using different methods for supporting the same target of intervention.

2.3 Research Motivation

Current research indicates that although collaboration scripts have positive effects on various aspects of collaborative learning [12], they do not necessarily have beneficial impact on domain learning (e.g. [21]). Additionally, it is a fact that few intelligent and adaptive collaborative support systems have been implemented and even less evaluated [4]. A great deal of them is research prototypes that demonstrate possible system architectures and they are not widely available outside the research laboratory. Therefore, it is difficult not only to evaluate and determine the effect of these systems on students' collaboration and learning, but also to deploy them in every day classroom conditions. Finally, there is no large-scale evidence available that proves the effectiveness of the adaptive collaboration support techniques regarding domain-specific learning outcomes.

The above drawbacks provide the research motivation for the current paper, which poses the following key questions: is it possible to significantly improve students' domain learning by integrating adaptive domain-specific support in a scripted collaborative activity? Moreover, is the adaptive intervention possible when employing easy to use and freely available contemporary technology systems?

To this end, we conducted two experimental studies. The first one focused on the impact of a domain-specific adaptive support mechanism compared to the support provided by a simple collaboration script [20]. In the second one we examined if the same domain-specific adaptive support mechanism is more learning beneficial compared to a fixed and informationally equivalent support mechanism [22]. In both studies we used the "Learning Activity Management System" (LAMS) tool [23], an open source freely available system to support online collaborative learning activities. LAMS provides a simple yet highly intuitive graphical user interface that allows teachers to drag and drop activity tools into the workspace and use connecting arrows to organize the activities into a sequential workflow [24].

In the following, we present briefly the method and the main findings for each study.

3 Study 1: Adaptive Domain-Specific Support vs. Collaboration Script

3.1 Method

We conducted a 2-hour experimental lab study comparing the two conditions: a) students who were supported only by the collaboration script (*control condition*), and b) students who were supported both by the collaboration script and the adaptive prompting method (*treatment condition*). We expected the students in the treatment condition to perform significantly better in a post-test regarding their domain learning.

The instructional domain of the activity was "Multimedia Learning". This particular subdomain was part of the course "Learning theories and educational software" that the participants followed during the semester.

The study employed 36 (16 females) undergraduate computer science students in their 3rd (out of 4) year of studies. All students were domain novices and they had never before engaged in online collaborative learning activity. Based on a prior domain knowledge questionnaire, 18 mildly heterogeneous dyads were formed and they were distributed in the two conditions (treatment and control) stratified by student's domain knowledge.

A collaboration script orchestrated the whole activity and provided fixed form of support to all the learners by guiding their collaboration. The script included three phases. The assigned task for the groups in each phase was to provide answers to an open-ended domain question using the LAMS chat tool (text-based chat). These were essentially "learning questions", that is, provided the opportunity for structured peer interaction. However before answering each learning question, dyads were asked to discuss and agree on theory keywords that are relevant to the subject under investigation. Thus, in each of the 3 phases the dyads had to answer first one keyword question (KQ) and then the relevant learning question (LQ).

The script provided guidance on the roles (author and reviewer) that the students had to follow during the activity. In each phase one of the students was assigned the author role (responsible for introducing an initial answer/comment for the LQ) and the other one the role of reviewer (to review and propose improvements for the suggested answer). LAMS environment informed each student individually, through a notice board, about the specific role (e.g. guidelines, goal of the role, possible actions) that he/she had to "play" during the next collaborative task (LQs). Students were then encouraged to further discuss freely their common answer, improve it if necessary and submit it. Peers exchanged their roles when they moved to the next phase.

Students in the treatment group were supported additionally by an adaptive prompting mechanism. The system was monitoring the keywords that the students provided in each keyword question and compared them with keywords that we had pre-declared as being the most important and relevant for the subject under discussion. In case there was no match the system responded with a proper prompt that included short information (up to 30-40 words) about the keyword the students had missed in their discussion. Dyads in the treatment mode were prompted after each keyword question in each collaboration phase. In case more than one missing keywords were detected the system presented the prompts sequentially (not all of them at once).

The study procedure included two collaborative activity sessions lasting 2 h each. The treatment group worked in the first session, and the control group in the second. The students in the two conditions were not informed that they would be treated differently. Each dyad partner was assigned to a different laboratory room where he/she was able to work online and log in to the LAMS environment. As they logged in the system, a notice board announced the working dyads. Both groups (treatment and control) followed the same script. Additionally, the treatment group was supported by the adaptive domain-specific prompting mechanism.

3.2 Results

After the collaborative activity ended, the students were asked to individually answer a post-test during a 30 minutes interval. The post-test comprised two sections focusing on: (a) students' improvement of basic domain knowledge, and (b) students' ability to individually answer domain open-ended questions (2 questions). The first section of the questionnaire included the same questions as the pre-test questionnaire. The second section included 2 open-ended questions, on parts of the learning material that students collaboratively worked.

For the statistical analyses, a level of significance at 0.05 was chosen. We conducted a Mann–Whitney test to compare the means of the students' performance in the post-test (treatment and control). We used Mann–Whitney because of the uneven distribution of the sample into the two conditions. Tests of normality (Shapiro–Wilk) showed that the conditions were not normally distributed, while Levene test of homogeneity of variance showed that the assumption of homogeneity has been met. Mann-Whitney test results showed that the students who were prompted outperformed the students in the control condition. Significant differences were identified in both dependent measures (section 1 and section 2 of the post-test instrument) between the treatment and control condition (section 1: $U= 96.00$, $p= .018$, $r= -.38$ and section 2: $U= 95.50$, $p= .017$, $r= -.35$) (Table 1).

Table 1. Mann-Whitney post-test results

Post test analysis	Control Group (n=18)	Treatment Group (n=18)	Mann-Whitney
Section 1 (closed-type questions domain basic knowledge)	*Mdn*= 6.00	*Mdn*= 8.00	$U= 96.00$, $p= .011$, $r= -.38$
Section 2 (open-ended questions domain understanding)	*Mdn*= 2.50	*Mdn*= 5.00	$U= 95.50$, $p= .017$, $r= -.35$

4 Study 2: Adaptive Domain-Specific Support vs. Fixed Domain-Specific Support

4.1 Method

We conducted a 2-hour experimental lab study comparing the two conditions: a) students who were supported by a fixed method (*control condition*), and b) students who were supported also by an adaptive prompting method (*treatment condition*). Furthermore, a peer-tutoring collaboration script also supported both conditions. We expected that the students in the treatment condition would learn more than the students in the control condition because of the adaptive supportive mechanism.

The instructional domain of the activity was "Multimedia Learning" which was the same with the first study. 40 undergraduate informatics students (19 females) participated in the study. Students were in their 3rd (out of 4) year of studies and they all had enrolled in a Multimedia Learning course. Following the same procedure with study 1, 20 mildly heterogeneous dyads were formed based in learners' answers to a pre-test questionnaire. The groups were distributed in the two conditions (treatment and control) stratified by student domain knowledge. Furthermore, the groups in both conditions followed the same collaboration script. The script was the same with the script in study 1. The only difference was that this script implementation completed in two phases (and not in three as in study 1).

Apart from the collaboration script, students in the treatment group were supported by the adaptive domain-specific support mechanism (the same mechanism with study 1). The goal of the adaptive domain prompting mechanism is first to identify missing domain conceptual knowledge (as documented by analyzing the peer dialogue) and second to provide the needed information and help peers develop a more accurate, shared, individual, mental representation. Consequently this would help group members to collaborate more efficiently (for example, during the next discussion). Therefore, the triggering of the prompting technique depended on peer interaction and also the objective of the intervention was to help partners in their next task.

Students in the control condition also followed the same collaboration script with the students in treatment condition. Moreover, they were supported by a fixed method: after each KQ the system informed students that they could follow a link to see all the keywords with their extended definitions that have been pre-defined by the instructor as the answer to this KQ. This list of keywords is considered as a fixed support mechanism, informationally equivalent to the adaptive support mechanism, since it included all the prompts possibly presented to students in the treatment condition. As students clicked on the link, a new window opened with all the seven prompts. Students could study this list and then continue with the next task.

Students in study 2 followed the same procedure with the procedure of study 1. Consequently, the two main differences of the two studies were a) the study 2 script included two instead of three phases and b) the students in control condition of study 2 were also supported by the fixed method.

4.2 Results

In order to measure the individual domain learning we conducted a post-test comprising two parts: (a) the first part included the same closed-type questions as the pre-test questionnaire (section 1). (b) The second part of the questionnaire included three open-ended questions, which referred to sections of the learning material that students collaboratively worked on. This part focused on assessing the students' understanding of the domain (section 2).

Table 2. Post test ANCOVA analysis

Post test analysis	**Control n=20**	**Treatment n=20**	**ANCOVA**
Section 1 (closed-type questions domain basic knowledge)	$M = 5.80$ $SD = 2.42$	$M = 7.30$ $SD = 1.49$	$F(1, 5.49)$, $\mathbf{p < .05}$
Section 2 (open-ended questions domain understanding)	$M = 3.19$ $SD = 2.61$	$M = 5.00$ $SD = 2.43$	$F(1, 4.92)$, $\mathbf{p < .05}$

We proceeded to apply parametric statistics to our data as, in contrast with the study 1, the normality and the homogeneity of variance criteria were satisfied. Analysis of covariate (ANCOVA) was applied to students' post-test data with the pre-test as covariate. The results revealed that the treatment group outperformed the control group in both measures of domain basic knowledge (close type questions) and domain understanding (open ended questions). ANCOVA indicated that the difference was statistically significant (Table 2).

5 Discussion and Conclusions

This paper presented the results from two studies that investigated the extent to which it is possible to effectively implement a domain-specific collaboration support strategy of dynamic format. More specifically, the first study focused on and eventually compared an adaptive intervention by identifying missing domain keywords in the peer discussion log file and presenting reminding prompts to partners accordingly, to the guidance and the support provided by a collaboration script. The second study compared the same adaptive intervention to a fixed support by giving the students the option to see all the domain keywords of a question afterwards. The adaptive method indeed resulted in improved individual learning outcomes in both studies, as indicated by the statistical analysis of the post-test.

More specifically, concerning the individual learning, in both studies the first section of the post-tests showed that students in the treatment mode outperformed the students in the control mode in recalling conceptual knowledge (learning at "basic" level of Bloom's taxonomy). Additionally, results from the second section of the post-test indicated that the treatment students used much more efficiently the conceptual domain knowledge to a new problem situation than the students in the control mode ("understanding" and "application" levels of Bloom's taxonomy).

We believe that the improved outcomes of the treatment group can be explained by considering that the students are exposed to remedial domain-specific information right after they discuss the relevant domain issues. The adaptive mode of presentation enabled students to easily integrate missing information in the domain model they constructed by activating three key cognitive processes: "selection" (focus on relevant information), "organization" (organize new information in a coherent model) and "integration" (link it to their previous domain knowledge) [25].

By contrast students in the control condition did not achieve comparable performance level as students in the treatment group, in both studies. This means that the additional prompt-based support offered to treatment students was necessary for achieving the higher performance and had greater impact than the fixed support mechanisms (the collaboration script in the first study or the fixed support mechanism in the second).

We argue that the result of the two studies highlights an important perspective in computer-supported collaborative learning. Even when peer interactions are triggered by collaboration scripts and/or supported by a fixed domain-specific support, the individual domain models of the partners might not be as rich as necessary to result to an elaborated common domain model for learners. In this case, adaptive supportive mechanism (dynamic form of support) that help students "repair" their incomplete domain models can result to significantly improved learning outcomes. This perspective of collaborative learning is in line with the paradigm of sharing individual mental representations [26]. The two collaborators aim to establish shared knowledge by exchanging ideas based on their internal mental representations. However, when these representations fail to meet certain criteria then the computer-based partner intervenes to the dialogue externalizing its own representations and contributing to the shared knowledge. The two studies showed clearly that the remedial information provided by the adaptive system could improve students' domain models and lead to better performance in problem solving.

Overall, this paper provides encouraging evidences that dynamic forms of support (as opposed to fixed forms) can be implemented during a collaborative activity and result in improved collaborative learning outcomes. We believe that by implementing adaptive collaboration support mechanisms instructors can substantially improve learning outcomes in various learning settings. Finally, we consider that our work provides incentive to further exploring the impact of more complex and thoughtful adaptive support mechanisms in the context of collaborative learning.

References

1. Slavin, R.E.: Research on cooperative learning and achievement: What we know, what we need to know. Contemporary Educational Psychology 21, 43–69 (1996)
2. Hewitt, J.: Toward an understanding of how threads die in asynchronous computer conferences. The Journal of the Learning Sciences 7(4), 567–589 (2005)
3. Liu, C.C., Tsai, C.C.: An analysis of peer interaction patterns as discoursed by on-line small group problem-solving activity. Computers & Education 50(3), 627–639 (2008)
4. Walker, E., Rummel, N., Koedinger, K.R.: CTRL: A research-oriented architecture for providing adaptive collaborative learning support. User Modeling and User-Adapted Interaction 19(5), 387–431 (2009)
5. Magnisalis, I., Demetriadis, S., Karakostas, A.: Adaptive and Intelligent Systems for Collaborative Learning Support: A Review of the Field. IEEE Transactions on Learning Technologies 4(1), 5–20 (2011)
6. Kumar, R., Rosé, C.P., Wang, Y.C., Joshi, M., Robinson, A.: Tutorial dialogue as adaptive collaborative learning support. In: Luckin, R., Koedinger, K.R., Greer, J. (eds.) Proceedings of Artificial Intelligence in Education, pp. 383–390. IOS Press, Amsterdam (2007)
7. Baghaei, N., Mitrovic, T., Irwin, W.: Supporting collaborative learning and problem solving in a constraint-based CSCL environment for UML class diagrams. International Journal of Computer-Supported Collaborative Learning 2(2-3), 159–190 (2007)
8. Dillenbourg, P., Baker, M., Blaye, A., O'Malley, C.: The evolution of research on collaborative learning. In: Spada, H., Reimann, P. (eds.) Learning in Humans and Machine: Towards an Interdisciplinary Learning Science, pp. 189–211. Elsevier, Oxford (1995)
9. Stahl, G., Koschmann, T., Suthers, D.: Computer-supported collaborative learning: An historical perspective. In: Sawyer, R.K. (ed.) Cambridge handbook of the learning sciences, pp. 409–426. Cambridge University Press, Cambridge (2006)
10. Fischer, F., Kollar, I., Haake, J., Mandl, H.: Introduction. In: Fischer, F., Kollar, I., Mandl, H., Haake, J. (eds.) Scripting Computer-Supported Collaborative Learning, pp. 1–10. Springer, New York (2007)

11. Rummel, N., Spada, H.: Can People Learn Computer-Mediated Collaboration by Following A Script? In: Fischer, F., Kollar, I., Mandl, H., Haake, J. (eds.) Scripting Computer-Supported Collaborative Learning, pp. 39–55. Springer, New York (2007)
12. Kollar, I., Fischer, F., Hesse, F.W.: Computer-supported collaboration scripts-a conceptual analysis. Educational Psychology Review 18, 159–185 (2006)
13. Dillenbourg, P.: Over-scripting CSCL: The risks of blending collaborative learning with instructional design. In: Kirschner, P.A. (ed.) Three Worlds of CSCL. Can we support CSCL?, pp. 61–91. Open Universiteit Nederland, Heerlen (2002)
14. Dillenbourg, P., Tchounikine, P.: Flexibility in macro-scripts for computer-supported collaborative learning. Journal of Computer Assisted Learning 23(1), 1–13 (2007)
15. Kobbe, L., Weinberger, A., Dillenbourg, P., Harrer, A., Hämäläinen, R., Häkkinen, P., Fischer, F.: Specifying computer-supported collaboration scripts. International Journal of Computer-Supported Collaborative Learning 2(2), 211–224 (2007)
16. Hernández-Leo, D., Villasclaras-Fernández, E.D., Asensio-Pérez, J.I., Dimitriadis, Y., Jorrín-Abellán, I.M., Ruiz-Requies, I., Rubia-Avi, B.: COLLAGE: A collaborative Learning Design editor based on patterns. Educational Technology & Society 9(1), 58–71 (2006)
17. Weinberger, A., Fischer, F., Mandl, H.: Fostering computer supported collaborative learning with cooperation scripts and scaffolds. In: Stahl, G. (ed.) Computer support for collaborative learning: Foundations for a CSCL community. Proceedings of the conference on computer support for collaborative learning, Boulder, CO, pp. 573–574 (2002)
18. Constantino-Gonzalez, M.A., Suthers, D., Escamilla de los Santos, J.: Coaching web-based collaborative learning based on problem solution differences and participation. Artificial Intelligence in Education 13(2-4), 263–299 (2003)
19. Tedesco, P.: MArCo: Building an artificial conflict mediator to support group planning interactions. International Journal of Artificial Intelligence in Education 13(1), 117–155 (2003)
20. Karakostas, A., Demetriadis, S.: Enhancing collaborative learning through dynamic forms of support: the impact of an adaptive domain-specific support strategy. Journal of Computer Assisted Learning 27, 243–258 (2011)
21. Weinberger, A., Stegmann, K., Fischer, F.: Computer-supported collaborative learning in higher education: Scripts for argumentative knowledge construction in distributed groups. In: Koschmann, T., Suthers, D., Chan, T.W. (eds.) Computer Supported Collaborative Learning 2005: The Next 10 Years! pp. 717–726. Lawrence Erlbaum Associates, Mahwah (2005)
22. Karakostas, A., Demetriadis, S.: Adaptive vs. Fixed Domain Support in the Context of Scripted Collaborative Learning: Targeting the Grounding Process (in preparation)
23. LAMS International: LAMS, Learning activity management system, `http://www.lamsinternational.com/`
24. Bower, M., Wittmann, M.: Pre-service teachers' perceptions of LAMS and Moodle as learning design technologies. In: Proceedings of the 4th International LAMS Conference. LAMS Foundation, pp. 28–39 (2008), `http://lamsfoundation.org/lams2009sydney/papers.htm`
25. Mayer, R.E.: The promise of multimedia learning: using the same instructional design methods across different media. Learning and Instruction 13(2), 125–139 (2003)
26. Stahl, G., Hesse, F.: Paradigms of shared knowledge. International Journal of Computer-Supported Collaborative Learning 4(4), 365–369 (2009)

On the Way to a Science Intelligence: Visualizing TEL Tweets for Trend Detection

Peter Kraker[1], Claudia Wagner[2], Fleur Jeanquartier[1], and Stefanie Lindstaedt[1]

[1] Know-Center, Inffeldgasse 21a, 8010 Graz, Austria
{pkraker,fjean,slind}@know-center.at
[2] Joanneum Research, Steyrergasse 17, 8010 Graz, Austria
claudia.wagner@joanneum.at

Abstract. This paper presents an adaptable system for detecting trends based on the micro-blogging service Twitter, and sets out to explore to what extent such a tool can support researchers. Twitter has high uptake in the scientific community, but there is a need for a means of extracting the most important topics from a Twitter stream. There are too many tweets to read them all, and there is no organized way of keeping up with the backlog. Following the cues of visual analytics, we use visualizations to show both the temporal evolution of topics, and the relations between different topics. The Twitter Trend Detection was evaluated in the domain of Technology Enhanced Learning (TEL). The evaluation results indicate that our prototype supports trend detection but reveals the need for refined preprocessing, and further zooming and filtering facilities.

Keywords: science 2.0, trend detection, social media, qualitative analysis.

1 Introduction

Twitter has high uptake in the scientific community. According to a recently published survey, personal email, Twitter, Skype, and project mailing lists are the most popular applications used for disseminating information by Semantic Web researchers [14]. The main motivations for publishing and sharing content on Twitter named by survey participants were: (1) to share knowledge about their field of expertise, (2) to communicate research results, and (3) to expand their network. The fact that the two main reasons for researchers to use microblogging services are communicating their research results and sharing information about their field of expertise makes Twitter a rich source of information, which can be exploited to detect research trends. It seems to be a reasonable assumption that the results of this study can be transformed to other technology-rich research fields such as Technology Enhanced Learning.

This paper presents an adaptable system for detecting trends based on the micro-blogging service Twitter, and sets out to explore to what extent such a

C. Delgado Kloos et al. (Eds.): EC-TEL 2011, LNCS 6964, pp. 220–232, 2011.

tool can support researchers. In the context of this work we define a "trend" as a term belonging to a topic which gains considerable interest during a certain period of time. Similar to the TF/IDF measure from Information Retrieval, the interestingness of an item can be defined as the number of occurrences of that item in time interval i out of a larger interval j [6].

Our research revealed that there is a need to have a means of extracting the most important topics from a Twitter stream. According to our evaluation (see Section 3), there are too many tweets to read them all, and there is no organized way of keeping up with the backlog. Finding something interesting is more of a coincidence than the result of a structured search, even with tools that allow for various lists of users and hashtags. What makes it even worse is the large amount of noise generated by superfluous postings. Twitter's trending topics do not help with that as they are not related to research.

Following the cues of visual analytics, we use visualizations to show either the temporal evolution of topics, or the relations between different topics. We developed a focused Twitter Crawler which uses the Twitter Streaming API [20]. The crawler can be adapted to any domain, by either (a) specifying a taxonomy of keywords, (b) specifying a list of users, or (c) a combination of both. On the client-side, our system provides two explorative visualization components: a streamgraph for analyzing trends over time and a co-occurrence network for analyzing semantic networks of terms which may for example reveal which topics are popular at the moment or which topics are strongly correlated.

The Twitter Trend Detection was evaluated in the domain of Technology Enhanced Learning (TEL). The system was adapted to the domain using a taxonomy of 30 hashtags, and a list of 450 expert users from TEL. First, we conducted semi-structured interviews involving the use of the system with researchers from the domains of TEL and knowledge management. The outcomes from these interviews were used to further improve the system. It was then evaluated a second time at the 2nd STELLARnet Alpine Rendez-Vous, where the visualizations were used as a means of support. The evaluation results indicate that our prototype supports trend detection but reveals the need for refined preprocessing, and further zooming and filtering facilities.

1.1 Related Work

Past research, such as ThemeRiver [9], was exploring solutions for automatically detecting emerging trends from collections of documents. With the rise of user generated content, researchers started exploring to what extent social media can be used to detect and monitor emerging trends. Fukuhara [7], for example, presents a system which generates a daily trend graph of weblog articles containing any given keyword. Glance et al. [8] introduce a tool called BlogPulse which allows monitoring trends in weblogs. They show a correlation between "blog and real world temporal data" such as temperature and news articles.

Hotho [11] presents an approach for discovering topic-specific trends within folksonomies, by adapting PageRank algorithm to the triadic hypeWeighted Graph structure of a folksonomy.

Cheong et al. [5] describe an approach of analyzing trend patterns on Twitter. They explore the properties and features of a trending topic and the properties of the users (or 'trend setters') that contribute 'tweets' to a trending topic, which makes them a part of the trend. For analyzing trending topics versus non-trending topics, they used only the last 1500 tweets which one obtains from the Twitter search API when conducting a keyword search. Mathioudakis [15] introduces TwitterMonitor, a system that discovers "bursty" keywords[1] in Twitter streams and uses them as 'entry points' for trend detection. Finally, Benhardus [2] also presents a trend detection system based on Twitter. He uses term frequency-inverse document frequency analysis and relative normalized term frequency analysis to identify the trending topics.

Our work differs from previous work by focusing on supporting researchers during their daily work with domain-specific information. Most existing work deals with general trends on Twitter which are largely irrelevant to researchers. To the best of our knowledge, this is the first system that can be adapted to a certain domain. The prototype's UI is a web-based one that relies on web standards. Therefore the visualization can be displayed in a standard web browser, but also be easily integrated with any system that allows for widgets adhering to the W3C standard. This design is far more adjustable compared to existing solutions.

2 System

2.1 Twitter Stream Analysis

We developed a focused Twitter Crawler which uses the Twitter Streaming API [20]. The crawler can be adapted to any domain, by either (a) specifying a taxonomy of keywords, (b) specifying a list of users, or (c) a combination of both. The Twitter Crawler then logs only tweets which were either authored by a certain user or which contain at least one keyword of the taxonomy. Next, we lexically preprocess the logged tweets in order to extract "informative" tokens (mainly nouns and hashtags) from it, using a part-of-speech tagger (POS Tagger), namely TreeTagger [16]. Finally we store the tweets, their metadata, and their associated informative tokens in a Solr index [18] which is used by our Visualization Dataservices. Fig. 1 illustrates this architecture.

The Visualization Dataservices are REST-ful Webservices which enable client-side, lightweight visualizations to fetch and reload the data on demand. Since we provide two different types of visualizations, streamgraphs and weighted graphs, we also implemented two different types of Visualization Dataservices.

---

[1] Keywords which are frequently used within a short period of time.

Both Visualization Dataservices take the following parameters as input: (1) a query consisting of one or several search terms (e.g., "conferences") (2) a time period of interest (e.g., 10.4.2010 - 15.5.2010), (3) the maximum number of co-occurring terms, and (4) the type of co-occurring terms (either nouns, hashtags, or users). Besides, the Streamgraph Dataservice optionally takes the number of time intervals as additional parameter. Both Dataservices translate the search query into a Solr query and preprocess the Solr result in different ways: the Streamgraph Dataservice focuses on analyzing the temporal evolution of topics over time; the Weighted Graph Dataservice focuses on relations between different topics.

The Streamgraph Dataservice splits the time-period of interest into the specified number of intervals. For each interval, the Streamgraph Dataservice returns the most frequent topics which co-occur with the query term within said interval. The Weighted Graph Dataservice returns the most important topics which co-occur with the query terms and the number of times they co-occurred.

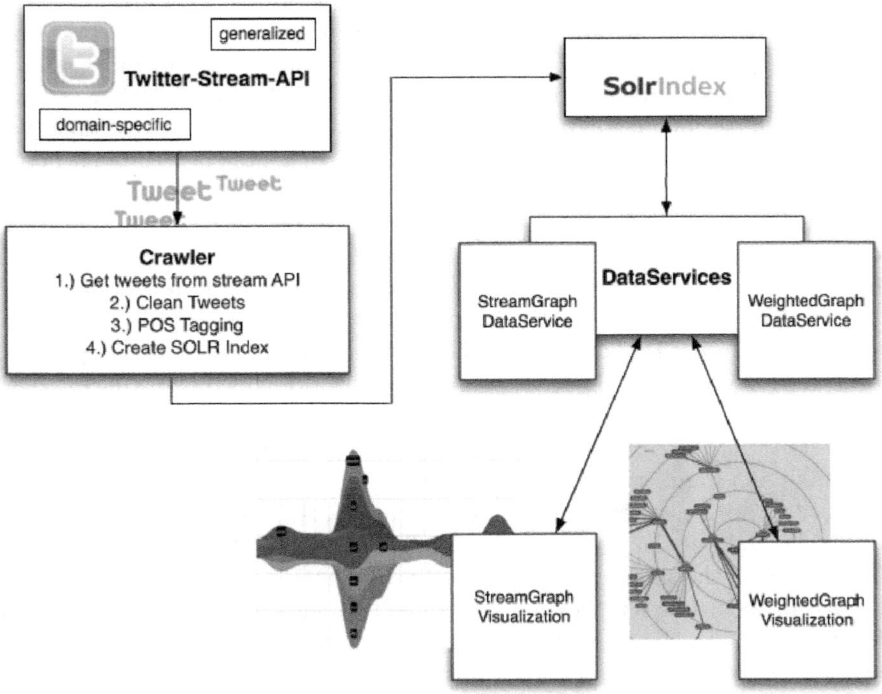

Fig. 1. Architecture

2.2 Visualizations

Following the cues of visual analytics [12], we use visualizations to show both the temporal evolution of topics, and the relations between different topics. First, we studied existing visualization techniques to identify those that are suitable for our purpose.

Heer et al. [10] describe a stacked graph as a classic method for visualizing change in a set of items, where the sum of the values is as important as the individual items. While such charts have proven popular in recent years, they do have some limitations such as the fact, that stacking may make it difficult to accurately interpret trends that lie atop other curves. The authors in [9] describe a more enhanced visualization technique that is suitable for visually describing thematic variations over time within a large collection of documents. This technique makes use of a river metaphor for revealing patterns and trends. One of the major advantages of this technique is little dependence on the number of documents; furthermore, the stacked areas are suitable for observation and comparisons. Inspired from the visualization technique mentioned before, Byron and Wattenberg further describe the Streamgraph design, a unified approach to stacked graph geometry and algorithms [4]. We therefore identified this special stacked graph design technique called "streamgraph" as one suitable approach to visualize Twitter-data over time. Lamping et al. [13] describe a focus+context visualization technique that is intended for displaying hierarchies, which is the basis for our second visualization approach: visualizing the relations between topics.

We further extended our visualization study to comparing existing web script libraries for the two visualization techniques described above that could be used in our prototype. The Streamgraph Visualization builds on top of the Grafico javascript library [22] while the Weighted Graph Visualization is based on the Javascript InfoVis Toolkit [1]. An HTML-based frontend offers the possibility to type in one or several search terms, furthermore specifying a facet such as "hashtags that occurred with the term above". Furthermore, the UI provides a date selection to narrow the search to a specific date range, and a field to specify the maximum number of co-occurring terms.

Fig. 2 shows a screenshot of the Streamgraph Visualization, displaying the co-occurring hashtags for the query "conferences" from 20/2/2011 to 14/04/2011. On the x-axis, the time intervals are outlined, whereas on the y-axis, the relative number of occurrences is shown. Each colored stream represents one co-occurring hashtag. The visualization shows that the hashtag for the South-by-Southwest conference (#sxsw) is trending around the actual event on March 15.[2] The #pelc11 hashtag was trending around April 7, with the Plymotuh E-Learning conference taking place from April 6-8. Another conference that is trending is the PLE Conference in Southhampton which has not taken place yet (#PLE_SOU). The other co-occurring hashtags are not tied to a certain conference (such as #mlearning and #edchat), but they denote hashtags in the TEL area which

[2] The conference took place from March 13-20.

contain a large amount of tweets about conferences. These hashtags could therefore be used to find out about other conferences in the area. An example for the Weighted Graph Visualization is shown down below in section 3.2.

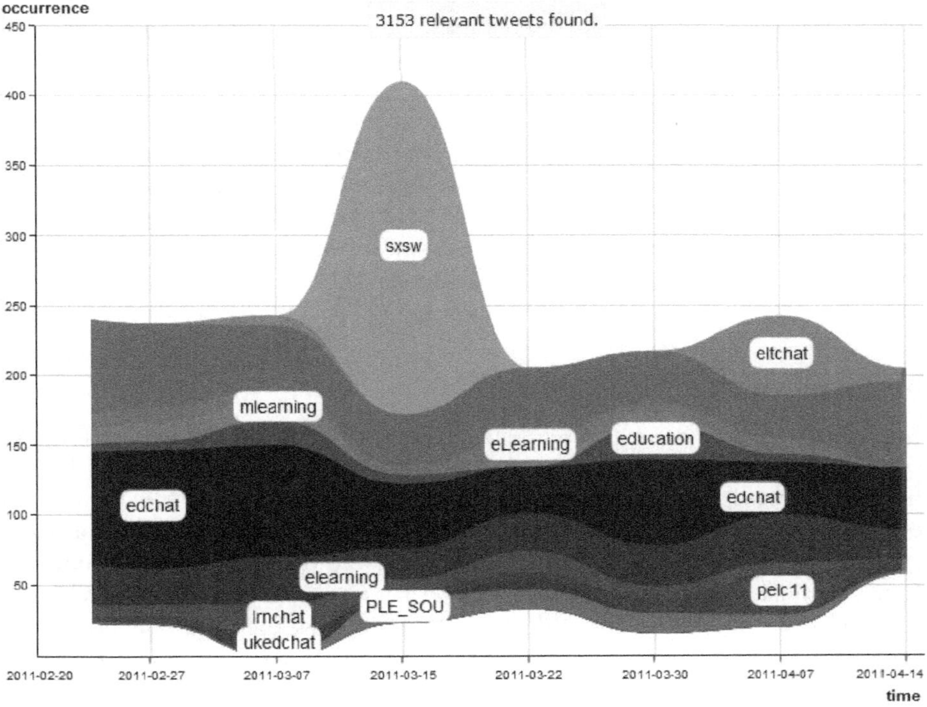

Fig. 2. Streamgraph Visualization

3 Evaluation

The Twitter Trend Detection was evaluated in the domain of Technology Enhanced Learning (TEL). The system was adapted to the domain using a domain-specific taxonomy of 30 hashtags. That removes the noise that is generated when taking all tweets into account, or when using general keywords such as "learning" and "training". In an early prototype using general keywords, we found that there is a strong correlation between "training" and "dog"; this is interesting as a fact, but irrelevant to the domain of TEL. The taxonomy was created (1) by analyzing hashtags that occur with the more general keywords, and (2) by searching Twapperkeeper [19] for relevant archives. Besides the taxonomy, we also created a list of 450 user accounts from various lists related to Technology Enhanced Learning which belong to well-known domain experts, researchers, and students. The system is part of the STELLAR Science 2.0 infrastructure [21]. The visualizations are also integrated as early-stage widgets on the TELeurope platform [17].

The evaluation of the Twitter Trend Detection is based on the creative thinking method PMI [3]. PMI stands for Plus, Minus, and Interesting. Instead of asking participants to take an objective stance in the evaluation, their attention is guided to think separately about (a) the positive aspects of the system (Plus), (b) the negative aspects of the system (Minus), and (c) the neutral but noteworthy aspects of the system (Interesting). This terminology was not only used in the instruments, but also later in the analysis, guiding the qualitative coding scheme.

The Twitter Trend Detection was evaluated in two settings: first, we held semi-structured interviews involving the use of the system with five researchers from the domains of Technology Enhanced Learning and knowledge management. The new system was then further evaluated at the 2nd STELLAR Alpine Rendezvous where the visualization was used as a means of support. We utilized them in a reflection session in one of the workshops; this was accompanied by three in-depth interviews with conference participants.

3.1 Evaluation 1

In the first evaluation, we held semi-structured interviews involving the use of the system with five researchers from the domains of Technology Enhanced Learning and knowledge management based in Austria. Among the participants were one professor, three senior researchers, and one PhD student. Two had a background in computer science, two in psychology, and one in business administration. Participants were interviewed about their use of Twitter in research, and specifically in relation to trend detection. Afterwards, they were introduced to the visualizations. Following a short tutorial, we asked participants to search for trends in their area of interest. The interviews were recorded on tape, and later transcribed. We qualitatively analyzed the transcripts using a reducing and interpreting approach. Codes were divided into three sections, following the PMI terminology: Plus, Minus, and Interesting. Each of these sections was subdivided into the codes "General", "Weighted Graph", and "Streamgraph". To capture the general remarks on Twitter, we added "Usage of Twitter", "Advantages of Twitter", and "Disadvantages of Twitter" to the scheme. Initially, the transcripts were coded with these general codes. In a second iteration, we refined these codes to paraphrase the content of marked statements. In the last step, we merged similar paraphrases to remove redundancy in the scheme.

The goal of the first evaluation was to collect feedback on the usability and the general applicability of the system to trend detection. In this phase, the system contained three different visualizations: two competing streamgraph visualizations, and one Hypertree Visualization. Table 1 shows the top two recommendations that resulted from this first evaluation. The evaluation showed that users were struggling with the interface. It proved to contain too many parameters that were labeled with technical terms. Furthermore, users had a hard time interpreting the co-occurring terms, and they were unclear about the underlying data. As a result, the user interface was completely overhauled, and we started to display the first 100 analyzed tweets alongside the visualizations.

Table 1. Recommendations from first evaluation

General:
Redesign the user interface to make it more accessible to the user
Include more metadata for the co-occuring terms
Streamgraph:
Keep Grafico streamgraph as it is more clearly laid out and has a better usability
Highlight meaning of the axis to the user
Hypertree:
Replace visualization with a version that is more clearly laid out
Position search term more prominently

The initial Hypertree Visualization was replaced by the Weighted Graph Visualization, and from the two streamgraphs, only the Grafico Streamgraph was developed further. Axis descriptions were added to the Grafico Streamgraph, and in the Weighted Graph Visualization the search term was highlighted.

3.2 Evaluation 2

The new system was further evaluated at the 2nd STELLAR Alpine Rendez-vous, where the visualizations were used as a means of support. Fig. 3 shows the Weighted Graph of hashtags for the main conference hashtag "arv11" from 27/03/2011 to 14/04/2011. This covers the conference which took place from 28/03 to 01/04 as well as the discussion afterwards. The size of the circles indicates the number of occurrences, while the line thickness indicates, how often two hashtags co-occurred. The visualization is centered on "arv11". The hashtags on the first level are directly related to arv11, namely "ngtel", "arvmupemure", "multivocalanalysis", "datatel11", and "arv3t". They all represent different workshops that were held during the conference. "jtelws11" represents the JTEL Winter School which was co-located with the Alpine Rendez-vous. For each of the workshops, as well as the winter school, a number of co-occurring hashtags are identified on the second level that tell a bit more about the individual workshops.

In the dataTEL workshop, we presented the corresponding visualizations in a reflection session on the workshop. During the course of the presentation and the ensuing discussion, participants were asked to list positive (Plus), negative (Minus), and neutral but noteworthy (Interesting) aspects on post-its. This evaluation was complemented with three in-depth interviews with participants from the Alpine Rendez-vous involving the visualizations. The interviewees were all tweeting in the course of the conference, and were specifically asked about the usefulness of the system. Among them were one professor, one senior researcher, and one PhD student. Two had a background in education, and one in computer science. Participants came from Europe, the United States, and Canada. The results were recorded, transcribed, and analyzed in the same manner as described in the first part of the evaluation above.

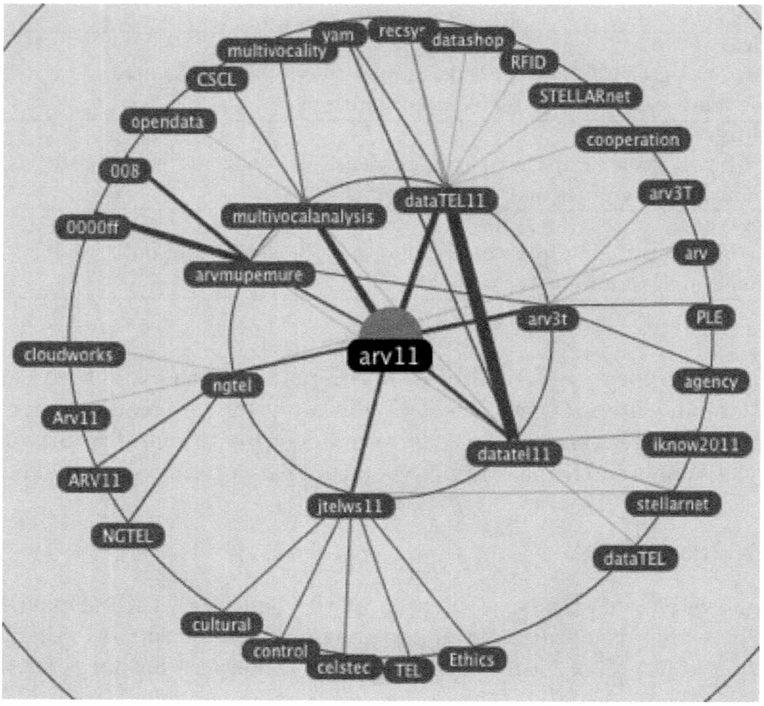

Fig. 3. Weighted Graph Visualization

The evaluation showed that Twitter is regarded as an important means of communication among the participating TEL researchers. Interviewees found it interesting (a) to follow and to contribute to the backchannel discussion in their own workshop, thus enriching their experience, and (b) to follow what is going on in other workshop. They also used Twitter to document parts of the workshop, and to keep their teams at home up-to-date, sometimes even using designated hashtags for that.[3] Among the uses outside of conferences were (1) to use it as a source of information, (2) to ask for feedback on one's own work, and (3) to directly communicate with other researchers.

What had already surfaced in the first evaluation, was also repeatedly noted in the second evaluation: there is a need to have a means of extracting the most important topics in a Twitter stream. According to the participants, there are too many tweets to read them all, and there is no organized way of keeping up with the backlog. As one of the interviewees put it so aptly, "If I get up to get coffee, I could have already missed something important." For the interviewees, finding something interesting is more of a coincidence than the result of a structured search, even with tools that allow for various lists of users and hashtags. What

[3] One of these hashtags "#yam", can be seen in Fig. 3 as a co-occurring hashtag of "#dataTEL11".

makes it even worse in the eyes of the participants is the large amount of noise generated by superfluous postings ("I am having breakfast now"). Twitter's trending topics do not help with that as they are not related to research.

3.3 Discussion

Participants liked the looks of the visualizations, and the idea behind them. In both evaluations, they noted that the interface is visually appealing. They also noted on several occasions that the system might be a useful way to deal with the backlog in their Twitter streams. The two visualizations are complementing each other very well; participants were interested in the connections between topics as well as the temporal evolution. Participants noted that both of them condense a lot of information in one view. They enjoyed the fact that the visualizations operate on live data with the ability to go back in time. Another feature that was well received was the consistency between the two visualizations, as the visualizations always operate on the same set of tweets for a given search term and a given time range.

As for the Weighted Graph, people were easily able to understand the basic visual metaphor, albeit it's sometimes crowded nature, and the fact that the weighted Graph is not always centered on the initial search term. Most interviewees could instantly interpret the size of the nodes and the thickness of the edges correctly. It was only the different levels that were hard to grasp in some cases. Participants noted that edges between topics are useful to determine the connection between the two, and the kind of clustering that is provided by the Weighted Graph in that way. They recognized topics from the discussions in their workshops, as well as users which participated in the backchannel discussion. An additional use that they saw was to get a quick overview of a field that they were not familiar with.

In the case of the Streamgraph, participants that were not accustomed to the visual metaphor needed a bit more time to understand the concept. Especially the alignment and the color-coding were often mis-interpreted. After a short introduction though, they liked that topics are shown in such a way that one can clearly see the bursty terms. They were able to reconstruct the time wise evolution of certain discussions from their workshops. One participant said it would be interesting to have such a visualization running on several screens at a conference to keep everyone updated about the current sessions. Another possible use that was mentioned is the detection of pivotal moments in online discussions.

Despite all the positive feedback, the evaluation also pointed out several shortcomings of the visualizations. First and foremost, users would like to be able to zoom into the results. They would like to be able to click on a co-occurring term to see its metadata, and they want to be able to not only see a full list of tweets but rather a filtered one. This revealed the need to implement zooming and filtering and the need to provide more details on demand. This should be complemented by a short help page on either visualization to make the metaphor crystal clear. In addition, users demanded more meaningful terms. They found the co-occurring terms to be too broad and generic. In addition, we need to get better at filtering out hashtags and mentions, and ignoring the case in the output. This revealed the need for refined preprocessing.

On a meta-level, participants sometimes criticized Twitter as a data base, as they trust only certain experts with trends. To address those concerns, we need to find a way to include only certain user accounts in the search. Finally, users would like to be able to integrate the visualizations with their existing Twitter applications.

4 Outlook

The evaluation results indicate that our prototype supports trend detection, but we still need to address several issues. First and foremost, we will provide more insight into the data in connection with further filtering mechanisms to allow users to view only a portion of that data. We will also improve the data quality by changing our crawler so that it does not count users and hashtags as nouns, and by applying lowercase to all output terms. To weed out the generic terms, we will look into applying the TF-IDF measure, and/or blacklists of common and broad terms.

Moreover, we will look into ways of integrating the visualizations with existing platforms emerged. We already took a first step into that direction by creating a W3C compliant widget, which can be included into any system that allows for such widgets; a first version has already been deployed to the social network TELeurope. Furthermore, it would be interesting to have a kind of self-evolving taxonomy of hashtags and users which semi-automatically adds new pieces of information to the taxonomy. On the front-end side, we are constantly looking into new meaningful visualizations, which can be adapted for system.

With the ever increasing amount of tweets, scalability becomes an issue; in a little over three months, we have collected over 500.000 tweets. Nevertheless, Solr has proven to be able to go way beyond that number of documents. The system is currently tailored towards Technology Enhanced Learning, but it could easily be adapted to other fields of research. The only precondition is to produce a list of hashtags and/or users from the field. If such a taxonomy does not exist, the system itself can help by detecting co-occurrent hashtags starting even from a single high-frequency hashtag.

Acknowledgements. We would like to thank all participants in the evaluation. This work was carried out as part of the STELLAR Network of Excellence, which is funded by the European Commission (grant agreement no. 231913). The Know-Center is funded within the Austrian COMET program - Competence Centers for Excellent Technologies - under the auspices of the Austrian Federal Ministry of Transport, Innovation and Technology, the Austrian Federal Ministry of Economy, Family and Youth, and the State of Styria. COMET is managed by the Austrian Research Promotion Agency FFG. Claudia Wagner is a recipient of a DOC-fForte fellowship of the Austrian Academy of Science.

References

1. Belmonte, N.G.: JavaScript InfoVis Toolkit, http://thejit.org/
2. Benhardus, J.: Streaming Trend Detection in Twitter. Tech. rep., University of Colorado at Colorado Springs (2010)
3. Bono, E.D.: De Bono's thinking course. Pearson Education, London (2006)
4. Byron, L., Wattenberg, M.: Stacked graphs–geometry & aesthetics. IEEE Transactions on Visualization and Computer Graphics 14(6), 1245–1252 (2008)
5. Cheong, M., Lee, V.: Integrating web-based intelligence retrieval and decision-making from the twitter trends knowledge base. In: Proceeding of the 2nd ACM workshop on Social web search and mining - SWSM 2009, pp. 1–8. ACM Press, New York (2009)
6. Dubinko, M., Kumar, R., Magnani, J., Novak, J., Raghavan, P., Tomkins, A.: Visualizing tags over time. ACM Transactions on the Web 1(2) (August 2007)
7. Fukuhara, T.: Analyzing concerns of people using Weblog articles and real world temporal data. In: Proceedings of the 14th International Conference on World Wide Web - WWW 2005 (2005)
8. Glance, N., Hurst, M., Tomokiyo, T.: Blogpulse: Automated trend discovery for weblogs. In: WWW 2004 Workshop on the Weblogging Ecosystem: Aggregation, Analysis and Dynamics, vol. 2004, Citeseer (2004)
9. Havre, S., Hetzler, E., Whitney, P., Nowell, L.: ThemeRiver: visualizing thematic changes in large document collections. IEEE Transactions on Visualization and Computer Graphics 8(1), 9–20 (2002)
10. Heer, J., Bostock, M., Ogievetsky, V.: A tour through the visualization zoo. Communications of the ACM 53(6), 59 (2010)
11. Hotho, A., Jäschke, R., Schmitz, C., Stumme, G.: Trend detection in folksonomies. In: Avrithis, Y., Kompatsiaris, Y., Staab, S., O'Connor, N.E. (eds.) SAMT 2006. LNCS, vol. 4306, pp. 56–70. Springer, Heidelberg (2006)
12. Keim, D., Kohlhammer, J., Santucci, G., Mansmann, F., Wanner, F., Schäfer, M.: Visual Analytics Challenges. In: eChallenges 2009, Istanbul, Turkey (2009)
13. Lamping, J., Rao, R., Pirolli, P.: A focus+context technique based on hyperbolic geometry for visualizing large hierarchies. In: Proceedings of the SIGCHI Conference on Human Factors in Computing Systems - CHI 1995, May 1995, pp. 401–408. ACM Press, New York (1995)
14. Letierce, J., Passant, A., Breslin, J., Decker, S.: Understanding how Twitter is used to spread scientific messages. In: Web Science Conference, Raleigh, NC (2010)
15. Mathioudakis, M.: TwitterMonitor: Trend Detection over the Twitter Stream. In: Proceedings of the 2010 International Conference on Management of Data, pp. 1155–1157 (2010)
16. Schmid, H.: Probabilistic part-of-speech tagging using decision trees. In: Proceedings of the International Conference on New Methods in Language Processing, Manchester, UK (1994)
17. STELLAR Network of Excellence. TELeurope.eu, http://www.teleurope.eu/
18. The Apache Software Foundation. Apache Solr, http://lucene.apache.org/solr/
19. Twapperkeeper.com. Twapper Keeper, http://twapperkeeper.com/

20. Twitter Inc., Twitter Streaming API, http://dev.twitter.com/pages/streaming_api
21. Ullmann, T.D., Wild, F., Scott, P., Duval, E., Parra, G., Reinhardt, W., Heinze, N., Kraker, P., Fessl, A., Lindstaedt, S., Nagel, T., Gillet, D.: A Science 2.0 Infrastructure for Technology-Enhanced Learning. In: Wolpers, M., Kirschner, P., Scheffel, M., Lindstaedt, S., Dimitrova, V. (eds.) Sustaining TEL: From Innovation to Practice. Proceedings of the 5th Conference on Technology Enhanced Learning, pp. 590–595. Springer, Heidelberg (2010)
22. Valkhof, K.: Grafico javascript charting library, http://grafico.kilianvalkhof.com/

Adding Weights to Constraints in Intelligent Tutoring Systems: Does It Improve the Error Diagnosis?

Nguyen-Thinh Le and Niels Pinkwart

Clausthal University of Technology,
Germany
{nguyen-thinh.le,niels.pinkwart}@tu-clausthal.de

Abstract. The constraint-based modeling (CBM) approach for developing intelligent tutoring systems has shown useful in several domains. However, when applying this approach to an exploratory environment where students are allowed to explore a large solution space for problems to be solved, this approach encounters its limitation: It is not well suited to determine the solution variant the student intended. As a consequence, system's corrective feedback might be not in accordance with the student's intention. To address this problem, this paper proposes to adopt a soft computing approach for solving constraint satisfaction problems. The goal of this paper is two-fold. First, we will show that classical CBM is not well-suited for building a tutoring system for tasks which have a large solution space. Second, we introduce a weighted constraint-based model for intelligent tutoring systems. An evaluation study shows that a coaching system for logic programming based on the weighted constraint-based model is able to determine the student's intention correctly in 90.3% of 221 student solutions, while a corresponding tutoring system using classical CBM can only hypothesize the student's intention correctly in 35.5% of the same corpus.

Keywords: intelligent tutoring systems, constraint satisfaction problems, weighted constraint-based model, cognitive diagnosis, evaluation.

1 Introduction

An intelligent tutoring system (ITS) is used to support students individually to solve problems. Research on ITSs has made considerable progress and some systems have been integrated into regular classrooms (e.g., [6]). Constraint-based modelling (CBM) [15] is one of the promising approaches for building ITSs. This technique assumes that a cognitive skill can be acquired in form of declarative knowledge. This approach has been applied successfully to the domains of natural language [12], SQL [13], and for database design [14]. The strength of this approach is that the space of correct solutions is modelled by using constraints to represent a number of domain principles and properties of correct solutions instead of enumerating every correct solution. Furthermore, it is not necessary to

C. Delgado Kloos et al. (Eds.): EC-TEL 2011, LNCS 6964, pp. 233–247, 2011.

anticipate errors possibly made by students. The goals of this paper are twofold. First, we argue that the pure CBM approach is limited in providing appropriate feedback to solutions for problems whose solution space is relatively large, e.g., in programming. Second, we propose and evaluate a weighted constraint-based model which adopts a soft computing approach for solving constraint satisfaction problems: each constraint is enriched with a weight value indicating the importance of the constraint. The weighted constraint-based model is superior to the pure CBM approach with respect to the capability of error diagnosis.

2 A Limitation of a CBM Tutoring System

We illustrate a limitation of the CBM approach in the example domain of programming. According to [9], a programming problem might have several solution strategies and each strategy can be implemented in different solution variants. The sample task *Investment*: *"Calculate the return after investing an amount of money at a constant yearly interest rate"* can be solved by applying different solution strategies, including the following two:

1. *Analytic strategy*: The profit of investing a sum of money with a yearly interest rate is calculated based on mathematical geometric series.
2. *Tail recursive strategy*: A variable can be used to accumulate the sum of investing money and its interest after each year.

In logic programming, the analytic strategy and tail recursive strategy can be implemented as follows:

Analytic strategy:

```
invest(Money,Rate,Period,Return):-Return is Money*(Rate+1)^Period.
```

Tail recursive strategy:

```
inv(M,_,P,Ret):- P=0,M=Ret.
inv(M,R,P,Ret):-P>0, NM is M*R+M, NP is P-1, inv(NM,R,NP,Ret).
```

Each solution strategy can be implemented in many solution variants. For instance, the tail recursive strategy can be implemented in many ways by varying the order of the two clauses or the order of the second and third subgoal in the second clause, by choosing one of two unification techniques (implicit and explicit), or by using the commutative and distributive laws in mathematics to transform arithmetic expressions. As a result, there exist several thousands of correct implementations for the problem *Investment* in total. How can constraints be used to model such a large solution space?

Ohlsson [15] proposed using only constraints to model specific properties of correct solutions. If a solution violates a constraint, the solution does not satisfy a semantic requirement of correct solutions. If the analytic solution strategy above should be modelled using constraints, we have to identify all possible pro-perties required to implement this strategy. For instance, the following constraint represents a property which requires the correct implementation of the exponent term of the analytic formula. Note, for simplicity, this constraint assumes that the order of the argument positions in the clause head is fix.

IF a calculation subgoal S exists **AND** one multiplication term T_m exists on the right hand-side of S **AND** T_m consists of two product factors **AND** one factor is a variable unified with the 1st position of the clause head **AND** the 2nd factor is an exponential term T_e

THEN the exponent of T_e is unified with the 3rd position of the clause head **AND** the exponent basis is a sum of the value 1 and a variable unified with the 2nd position of the clause head

This constraint is specified with five propositions in the relevance part (IF-clause). Such a constraint, whose relevance part contains many conditions, tends to fail in erroneous situations, because the relevance part is not robust against minor deviations from the specified situation. A complex constraint with a conjunction of conditions in the relevance part becomes irrelevant for a student solution if a single one of the conjuncts fails. For example, the following student solution, which implements the analytic strategy, would not match the relevance part of the constraint above, because in the clause body of the student solution the multiplication term consists of one product factor (namely `(Rate+1)^Period`), while the relevance part of the constraint requires two.

A sample erroneous student solution:

```
invest(Money,Rate,Period,Return):-Return is (Rate+1)^Period.
```

Thus, the constraint above can be satisfied even though this undesired result might have been caused by another error elsewhere in the student solution (e.g., one product factor is missing). The potential that complex constraints might become useless is obvious when specifying constraints for the domain of programming. Because the relevance part of a constraint should describe a problem state, it makes no sense to make the relevance part simpler. (If the satisfaction part of a constraint is highly specific, then the constraint can be broken into several simpler constraints according to the rule: $A \rightarrow B \wedge C \equiv (A \rightarrow B) \wedge (A \rightarrow C)$). The approach of using constraints as the only means to model correct solutions produces not only complex but also task-specific constraints. Every time new tasks need to be integrated into a CBM system, it is necessary to specify new task-specific constraints. This is not an easy undertaking task for authors who are not familiar with the constraint representation.

Instead of specifying task-specific requirements in constraints, Ohlsson and Mitrovic [16] suggested using an *ideal solution* to capture the characteristics of correct solutions and defining constraints to compare the necessary components of the ideal solution with a student solution. The approach of using an ideal solution to encapsulate the semantic requirements of solutions has the advantage that manually created complex constraints can be avoided to a certain extent. Furthermore, it is not necessary to specify new constraints for a new task because task-specific requirements are contained in the ideal solution, assuming that the existing set of constraints can cover the learning domain sufficiently. However, choosing an ideal solution among many alternatives for a programming problem is not an easy task. Assuming that an ideal solution for the problem *Investment* is available, similar to the way of defining constraints using an ideal solution, we might define constraints to cover different solution variants in logic programming. For instance, to solve the sample task *Investment*, the tail recursive strategy requires implementing a guard which checks whether the investment period is positive before it can be decremented recursively. The guard can be implemented in a variety of ways: e.g., `P>0`, `P>=1`, `0<P`, `1=<P`. If we choose the implementation of the tail recursive strategy as an ideal solution, the following constraint checks the correctness of an arithmetic expression and its variants, where $\lhd, \lhd_s \in \{=<, >=, >, <\}$ and r is a function which finds a reverse operator for an arithmetic comparator according to the following rules: $r(>) \rightarrow <$, $r(<) \rightarrow >$, $r(=<) \rightarrow >=$, $r(>=) \rightarrow =<$.

IF In the ideal solution, there exists an arithmetic test $X \lhd Y$ **AND** $SX \lhd_s SY$ is a corresponding subgoal in the student solution
THEN $\lhd_s$ is identical to $\lhd$, and (SX, SY) correspond to (X, Y) **OR** $\lhd_s = r(\lhd)$, and (SX, SY) correspond to (Y, X)

This constraint is useful for capturing different solution variants of a guard which is required by the tail recursive strategy. However, this constraint is meaningless if student solutions implementing the analytic strategy are diagnosed, because this constraint is intended to check the guard (specified in the IF-clause) and the analytic strategy does not require one. The problem would even be more serious if a constraint is specified to require the existence of a guard like the following.

IF In the ideal solution, there exists an arithmetic test $X \lhd Y$
THEN a corresponding $SX \lhd_s SY$ exists in the student solution
HINT A guard is missing.

If this constraint is used to evaluate a student solution which implements the analytic strategy, then it would be violated and the error diagnosis results in a feedback "A guard is missing" which is obviously misleading. This is because the solution strategy the student intended to implement is not the same as the one the constraints are based on (tail recursive strategy). This problem has also been identified in the domain of SQL in [11, p. 43] and [7, p. 321], as is discussed in [18, p. 85]. This problem raises the need to hypothesize the solution strategy underlying a solution during the process of diagnosing errors. Once the solution strategy of the student has been identified, it makes sense to evaluate constraints in the context of that specific solution strategy only.

To determine the most plausible hypothesis about the student's solution variant in the context of language learning, Menzel [12] proposed to count the number of constraint violations for each hypothesis. The hypothesis which causes least constraint violations is assumed to be plausible. Yet, the author indicated that this kind of measure is too gross, because constraints which represent grammar rules might not have the same level of importance. For example, if the sentence "These fish stinks" is diagnosed, two hypotheses can be generated: 1) a constraint (C1) requiring the agreement of determiner-noun is violated; 2) a constraint (C2) representing the agreement of noun-verb cannot be satisfied. If only the number of violated constraints is used to evaluate each hypothesis, this might result in inaccurate diagnoses, because researchers in language learning might consider C1 as more or less severe than C2.

As a result, we conclude that the classical CBM approach is limited in providing appropriate feedback to solutions of problems whose solution space is relatively large, e.g., in programming. In the next section, we introduce a weighted constraint-based model which adopts the idea of a *probabilistic* approach for solving constraint satisfaction problems to improve the capability of error diagnosis of CBM tutoring systems.

3 A Weighted Constraint-Based Model for ITS

In order to coach the student, an ITS needs to identify shortcomings in the student solution and to provide appropriate feedback according to the student's intention. Hence, the student solution needs to be analyzed correctly. In the approach proposed in this paper, a weighted constraint-based model (WCBM) serves this purpose. The model includes four modeling components: a *semantic table*, a set of *constraints*, constraint weights, and a set of *transformation rules*. The semantic table represents solution strategies for a given problem. Each solution strategy is described by a set of semantic components. Constraints are used to establish a mapping between the student solution and requirements in the semantic table and to check general well-formedness conditions. Domain-specific transformation rules can be exploited to extend the coverage of different solution variants. A constraint weight represents the importance of each constraint. The error diagnosis process is controlled using these four components.

3.1 Semantic Table

Instead of using a single ideal solution to capture task-specific requirements, the WCBM approach presented in this paper is built on a so-called *semantic table* which serves two proposes: 1) modeling several solution strategies, and 2) representing model solutions in a relational form. The first characteristic serves to hypothesize the most plausible strategy underlying a student solution. The second one has the advantage that solution variants (e.g., created by changing the order of solution components) can easily be covered. Table 1 illustrates a partial semantic table for the problem *Investment*, covering two alternative solution strategies (more could easily be added), where the first row represents the generalized description of the analytic solution strategy and the remaining ones are the generalized description of the tail recursive solution strategy.

3.2 Constraints

In this framework, constraints are employed to model the solution space for each problem and to identify errors based on this model. We distinguish between two types of constraints: *general* and *semantic* constraints.

General Domain Principles: A domain is characterized by certain principles, which every solution variant of any problem must adhere to and which are independent of task-specific requirements. For instance, in logic programming, it

Table 1. An partial semantic table for the example problem *Investment*

Strategy	CI	Head	SI	Subgoal	Description
Analytic	1	p(S,R,P,Ret)	1	Ret is S*(R+1)^P	Using a formula
Tail recursive	1	p(M,_,P,Ret)	1	P=0	Recursion stops
Tail recursive	1	p(M,_,P,Ret)	2	Ret=M	Recursion stops
Tail recursive	2	p(M,R,P,Ret)	1	P>0	Check period
Tail recursive	2	p(M,R,P,Ret)	2	NM is M*R+M	Calculate new money
Tail recursive	2	p(M,R,P,Ret)	3	NP is P-1	Update period
Tail recursive	2	p(M,R,P,Ret)	4	p(NM,R,NP,Ret)	Recur with new period

CI: clause index; SI: subgoal index

is required that all variables on the right hand-side of an arithmetic calculation subgoal (e.g., `X is A+B`) must be instantiated. Otherwise, the evaluation of this arithmetic expression will fail. Such domain principles can be modeled by means of *general constraints*. These constraints can be instantiated by constraint schemas of type (1), where the problem situation X and the condition Y can be composed of many elementary propositions using conjunction or disjunction operators. A problem situation X describes a static state of solution components required to solve an arbitrary task in the domain. If a student solution violates one of the general constraints, the student did not consider the corresponding principle of the domain.

Type **(1)**: **IF** problem situation X is relevant **THEN** condition Y must be satisfied

Semantic Correctness: Using task-specific information specified in the semantic table, constraints can be specified to check the semantic correctness of a student solution (STS). We refer to this kind of constraints as *semantic constraints* which have the following schemas. Constraint schemas (2.1) and (2.2) check for missing or superfluous components in the student solution, respectively.

Type **(2.1)**: **IF** in the semantic table, a component X exists
THEN in the STS, a component corresponding to X exists

Type **(2.2)**: **IF** in the STS, a component Y exists
THEN in the semantic table, a component corresponding to Y exists

If the goal is to check a required value of a component, the following schema can be applied:

Type **(2.3)**: **IF** in the semantic table, a component Z exists and has value A
THEN in the STS, a component corresponding to Z has value A

To model different solution variants for a concept (e.g., an arithmetic comparison), constraint schema of type (3) can be applied to cover two possible variants using the OR-operator.

Type **(3)**: **IF** in the semantic table, a component X exists
THEN in the STS, a component corresponding to X **OR** a variant of X exists

3.3 Transformation Rules

To extend the coverage of the solution space for a problem, various transformation rules which are domain-specific can be exploited, for instance: program transformation in the domain of programming, model transformation in the domain of computational modeling, or mathematical transformation rules.

3.4 Constraint Weights

As discussed in Section 2, classical CBM tutoring systems might provide misleading diagnostic results and are not well suited to decide on the most plausible hypothesis about the student's solution variant, because constraints are solely based on a binary logic (constraint is either violated or not). We need a means to enhance the error diagnosis capability of classical CBM tutoring systems. To serve this purpose, we adopt approaches to softening constraints in a constraint satisfaction problem (CSP), because constraint-based error diagnosis is a CSP whose goal is to identify inconsistencies between an erroneous solution and a constraint system.

The most popular approaches to softening constraints include fuzzy CSP [1], partial CSP [4], cost-minimizing CSP[1] [17], and probabilistic CSP frameworks [2]. The probabilistic CSP approach is most appropriate for constraint-based error diagnosis, because it does not evaluate a constraint system partially (like the partial CSP framework), nor is it necessary to specify constraints with possible instantiations of constraint variables in advance (like fuzzy and cost-minimizing CSP frameworks). This approach has been applied successfully to enhance the quality of error diagnosis, e.g., in the context of a natural language parser [3]. In the approach proposed in this paper, we adopt the probabilistic approach to enhance the diagnosis capability of classical CBM tutoring systems by attaching a *weight*, indicating the measure of importance, to each constraint.

Searching the most plausible hypothesis about the student's solution variant, we need to evaluate the plausibility of all possible hypotheses. For this purpose, adopting the probabilistic approach, we apply a multiplicative model: constraint weights are taken from the interval $[0; 1]$. The value 1 represents the weight for least important constraints and 0 indicates the weight for constraints which model the most important requirements. Constraints of the latter type can be considered *hard constraints*. The importance of a constraint is determined based on the role of the components being investigated. Constraints checking a component which contributes more information to the overall correctness of the solution should receive a weight close to the value 0. The determination of the importance level for constraints resembles the assessment of written examinations by a human tutor: if a solution contains more important components, then it receives a better mark. Constraint weight values need to be adjusted manually to yield acceptable diagnostic results.

---

[1] This term is also referred to as *weighted* CSP in the literature.

3.5 Error Diagnosis

Given a student solution, the error diagnosis process generates hypotheses about it by matching the student solution against each of the solution strategies specified in the semantic table (left part of Fig. 1). The matching process initializes *global mappings* representing hypotheses about the strategy underlying the student solution. Then, the hypothesis generation process continues to generate hypotheses about the student's solution variant by matching components of the student solution against the ones representing the selected solution strategy. That is, highest-level components of the student solution are matched against components of the same level specified for each solution strategy in the semantic table, similarly for middle-level and lowest-level components. The matching process results in *local mappings* which represent hypotheses about the student's solution variant. They are used to complete the global mappings. Depending on the structural hierarchy of a solution, local mappings are generated according to the levels of that hierarchy.

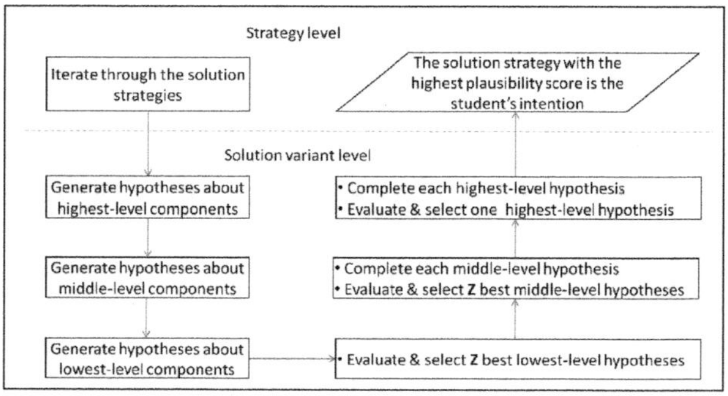

Fig. 1. The process of error diagnosis

After hypotheses on the lowest component level have been generated, they are evaluated with respect to the plausibility, and the best ones are selected using the Beam criterion Z (the right part of Fig. 1). The plausibility of each hypothesis is computed by multiplying the weights of constraints violated by that hypothesis according this formula, where W_i is the weight of a violated constraint: $Plausibility_{Prod}(H) = \prod_{i=1}^{N} W_i$.

The plausibility of each generated hypothesis can be used as a Beam criterion to restrict the space of hypotheses to the most plausible ones: Only a fraction of most plausible hypotheses whose plausibility score is higher than Z $(0 \leq Z \leq 1)$ are selected for each level. The best selected hypotheses on the lower level are used to multiply the space of hypotheses on the next higher level. This procedure of generating hypotheses, selecting the best ones using the Beam criterion, and multiplying the space of hypotheses of the next level continues up to the highest one of the structural hierarchy. At this level, the process of generating strategy hypotheses is completed. The error diagnosis process is finished by evaluating the strategy hypotheses with respect to their plausibility and choosing the best hypothesis for the selected solution strategy. Diagnostic information is derived from constraint violations resulting from the plausibility computation of the most plausible strategy hypothesis.

3.6 Example

Following is a sample student solution in logic programming for the problem *Investment*:

Clause	Type	Implementation
SC1	base	`interest(Amount,_,0,Amount).`
SC2	recursive	`interest(Amount,Rate,Duration,End):-`
		`X is Amount+Amount*Rate/100, Duration>0,`
		`NewDura is Duration-1,interest(X,Rate,NewDura,End).`

According to the algorithm of the error diagnosis, first, the student solution is matched to the generalized solution description of the analytic strategy, then, the tail recursive strategy. This process results in two global mappings, where SP is the solution provided by the student:

$H_{strategy}$={map(Strategy_Analytic, SP)}
$H_{strategy}$={map(Strategy_Tail_Recursive, SP)}

On the solution variant level, the error diagnosis process matches SP against the semantic components of each solution strategy. Since the structural hierarchy of a logic program consists of clause, subgoal, argument/operator, multiplication term, and product factor (where the last two levels exist only if the solution contains mathematical expressions), the matching process is carried out on the corresponding levels of the hierarchy. For instance, the matching between SP and the generalized description of the strategy *tail recursive* (cf. Table 1) on the clause level results in a single mapping H_{clause} which has two entries. The first component (C_i) of each entry represents the expression specified in the generalized description and the second one (SC_i) is provided by the student solution.

H_{clause}={map(C1, SC1), map(C2, SC2)}

On the subgoal level, the subgoals of the student's clause are mapped against the subgoals of the corresponding clause of the generalized description. For example, taking the second element of the mapping H_{clause} above, subgoals of C2 are matched against subgoals of SC2. Considering, for instance, only arithmetic calculation subgoals, matching the two arithmetic calculation subgoals of the student's clause SC2 against two arithmetic calculation subgoals of the generalized description's clause C2 results in two mappings of arithmetic subgoals:

$H_{subgoal}$(cal)= {map(NP is P-1, DecDuration is Duration-1),
map(Ret is NM+R*NM, X is Amount+Amount*Rate/100)}
$H_{subgoal}$(cal)= {map(NP is P-1, X is Amount+Amount*Rate/100),
map(Ret is NM+R*NM, NewDura is Duration-1)}

On the argument/operator level, the arguments of a student's subgoal are matched against the arguments of the corresponding subgoal of the generalized description. E.g., `DecDuration is Duration-1`, an arithmetic subgoal of the student solution, is matched against the subgoal `NP is P-1` of the generalized description: arguments on the left hand-side and on the right hand-side are matched, respectively:

$H_{argument}$={map(NP, DecDuration), map(P-1, Duration-1)}.

Similarly, the hypothesis generation process continues to the product factor level. After local mappings have been generated on the product factor level, their plausibility is evaluated by invoking the constraints of that level. Based on the plausibility score of each local mapping, a set of best mappings is selected and used to multiply the space of local mappings on the next higher level, namely the multiplication terms. Again, each of the local mappings on the multiplication term level is evaluated with respect to its plausibility. The process of evaluating local mappings, choosing the best ones, and extending the space of mappings on the higher level continues until the clause level is reached. At this level, a space of global mappings for each generalised solution description is established.

The algorithm next evaluates the plausibility of these global mappings and determines the one which has the highest score for each generalised solution description. According to the first column of Table 2, the hypothesis that the student has implemented the strategy *tail recursive* is more plausible because it has the highest plausibility score (0.25). The second column of the table shows diagnoses resulting from the evaluation of each hypothesis. The evaluation of the most plausible hypothesis results in the following diagnostic information:

Hint 1 In the 2nd clause, the denominator **100** is superfluous.
Hint 2 In the 2nd clause, a variable is expected instead of **Betrag/100**.

Table 2. Plausibility of hypotheses about the implemented solution strategy

Str.; Score	Weight; Hints
A; 1e-006	0.01; If you intend to define a non-recursion predicate, then it must exist at least a clause of type non-recursive. 0.01; The predicate definition has more base case than required. 0.01; The predicate definition has more recursive case than required.
T; 0.25	0.5; In the 2nd subgoal, the denominator **100** is superfluous. 0.5; In the 2nd clause, a variable is expected instead of **Betrag/100**.

Str. A: Analytic strategy; Str. T: Tail recursive strategy.

4 Evaluation

To test whether the weighted constraint-based model is more capable than the classical CBM approach with respect to its diagnostic capability (of course, a more precise error diagnosis component can be used to give better learning support), we have used the intention analysis capability of a tutoring system for logic programming named INCOM [8]. The purpose of the intention analysis is to determine the rate of student solutions whose solution strategy and components are identified correctly by the system. In literature, this kind of evaluation technique is also noted as *algorithm analysis* [5,10], because the approach of identifying the solution strategy is based on anticipated algorithms for a programming problem.

To compare the diagnostic capability of the weighted constraint-based model with the classical CBM approach, we used two versions of INCOM. The first one applies the weighted constraint-based model (WCBM-INCOM). A modified version of INCOM uses constraints without weight values (CBM-INCOM), and corresponds to a classical CBM tutoring system. Here, the plausibility of hypotheses about the solution strategy intended by the student is computed by summing up the number of constraint violations caused by each hypothesis: $Plausibility_{Add}(H) = |C|$, where C is the set of all constraint violations caused by each hypothesis H. (Note, this formula does not use the weight value of constraints like the formula $Plausibility_{Prod}$).

4.1 Study Design

For the evaluation study, we collected exercises and solutions from past written examinations. The examination candidates had attended a course in logic programming which was offered as a part of the first semester curriculum in Informatics. The following seven tasks have been collected from the written examinations (here, the tasks are described briefly).

1. Access to specific elements within an embedded list;
2. Querying a data base and applying a linear transformation to the result;
3. Modification of all elements of a list subject to a case distinction;
4. Creation of an n-best list from a data base;
5. Computing the sum of all integer elements of a list;
6. Counting the number of elements in an embedded list;
7. Finding the element of an embedded list which has the maximum value for a certain component.

For these problems, 221 student solutions were selected according to the following criteria:

- Any piece of code which satisfies minimal requirements of interpreting it as a Prolog program is considered a solution
- Syntax errors in the solutions are ignored (because during the written examination the students did not have access to a computer)
- Both correct and incorrect solutions are taken into account.

For conducting the intention analysis, we involved a human expert who inspected every student solution manually. Student solutions which could be understood by the human expert, were classified as "*understandable*", the others as "*not understandable*". Then, all "*understandable*" solutions were given as input to the two systems under comparison (WCBM-INCOM and CBM-INCOM), which resulted in two lists of constraint violations – one for each system. The human expert examined the two lists of constraint violations and decided whether the systems analyzed the student solution correctly (in terms of determining the solution variant).

4.2 Results

Table 3 summarizes the statistics of the evaluation. The number of available student solutions per task is indicated in the second column. The third column represents the number of solutions classified as "*not understandable*". The fourth and the fifth column show the absolute and relative amount of solutions which have been analyzed correctly by the WCBM-INCOM system. Similarly, the sixth and the seventh column indicate the absolute and relative number of solutions analyzed correctly by CBM-INCOM.

On average, 90.3% (s.d.=11.0%) of 221 collected student solutions could be analyzed correctly by WCBM-INCOM [8], while CBM-INCOM could achieve correct intention analysis in 35.5% (s.d.=19.9%) of the same solution corpus. In all tasks, the statistics show that the intention analysis capability of WCBM-INCOM is better than the one of CBM-INCOM. In particular, we notice that for Tasks 1-3, the intention analysis capability of CBM-INCOM is higher than 50%, whereas for Tasks 4-7 the intention analysis capability of CBM-INCOM is remarkably worse (Task 6 has the worst intention analysis with only 9.9% correct analysis). This can be explained by the fact that the complexity of Tasks 4-7 is much higher than the one of Tasks 1-3 – here, more possible solution strategies can be (and have been) applied and more implementation variants can be created. The statistics also show that the diagnosis accuracy of the WCBM-INCOM does not show this significant difference between easy and more complex tasks: apparently, adding the constraint weights was sufficient for achieving a much better diagnosis performance.

We next want to illustrate this effect with a concrete example from Task 6 (c.f. Section 4.1) which can be solved by applying either a naive recursive strategy or a tail recursive strategy.

Table 3. Evaluation of the intention analysis

			WCBM-INCOM		CBM-INCOM	
Task	**Solutions**	**Not U.**	**C.A.#**	**C.A.%**	**C.A.#**	**C.A.%**
1	10	0	10	100.0	5	50.0
2	11	0	10	90.9	7	63.6
3	6	1	4	66.7	3	50.0
4	17	1	16	94.1	5	29.4
5	58	2	54	93.1	8	13.8
6	81	0	79	97.5	8	9.9
7	38	2	34	89.5	12	31.6
Sum	221	6	207		48	
Avg.(sd.)				90.3 (11.0)		35.5 (19.9)

Not U.: Not Understandable; C.A.: Correct Analysis.

```
countz(N,L):- L=[], N is 0.
countz(N,L):- L=[Head|Rest], countz(N1, Rest), N is N1+1.
```

The results of error diagnosis executed by WCBM-INCOM and CBM-INCOM for the erroneous student solution above are shown in Table 4. Here, CBM-INCOM was not able to decide on the most plausible hypothesis about the student's solution variant, because two hypotheses of CBM-INCOM (naive recursive and tail recursive) have the same plausibility score (each hypothesis produces five constraint violations). On the contrary, WCBM-INCOM decided on the hypothesis that the student implemented the naive recursive strategy because the plausibility score of this solution strategy (0.064) is higher than the one of the tail recursive strategy (1e-010).

Table 4. Intention analysis for the sample student solution

Strategy	WCBM-INCOM $Plausibility_{Prod}$	CBM-INCOM $Plausibility_{Add}$
Naive recursive	0.064	5
Tail recursive	1e-010	5

In addition to this example, where CBM-INCOM was not able to choose the most plausible hypothesis, there were many cases of wrong intention analysis provided by CBM-INCOM. It even favoured a wrong solution strategy if this had more (but less severe) errors. Here, adding the weights to constraints solved the problem and improved the error diagnosis.

5 Conclusion and Outlook

In this paper, we have argued that the constraint based modelling approach, while useful for ITS development in general, may reach its limitation when it is applied for developing tutoring systems for tasks which have a large solution space. An example for this effect from the domain of programming has been discussed. Additionally, we have introduced a weighted constraint-based model which adopts a probabilistic approach for solving constraint satisfaction problems. In a study, we have shown that this approach is more capable than the pure CBM with respect to error diagnosis: A tutoring system for logic programming based on this extended model was able to analyze the student's intention correctly in 90.3% of 221 student solutions, while a corresponding system which is based on the classical CBM approach was able to achieve a correct intention analysis in only 35.5% of the same corpus of student solutions.

While the examples in this paper were all chosen from logic programming, we believe that the results are valid also beyond this field. Specifically, it is possible to easily apply the weighted constraint-based model to functional programming languages because they are also instances of the declarative programming paradigm. This characteristic, the atemporal nature, of declarative programming languages makes possible to formulate and to check the well-formedness conditions for a program in a static manner. Whether the weighted constraint-based model can also be applied to other declarative domains (e.g., computational modeling) or other programming paradigms (e.g., imperative programming) is a open question which we are currently investigating.

References

1. Dubois, D., Fargier, H., Prade, H.: Possibility Theory in Constraint Satisfaction Problems: Handling priority, preference and uncertainty. J. Applied Intelligence 6, 287–309 (1996)
2. Fargier, H., Lang, J.: Uncertainty in Constraint Satisfaction Problems: a Probabilistic Approach. In: Moral, S., Kruse, R., Clarke, E. (eds.) ECSQARU 1993. LNCS, vol. 747, pp. 97–104. Springer, Heidelberg (1993)
3. Foth, K.: Hybrid Methods of Natural Language Analysis, Shaker, Germany (2007)
4. Freuder, E.C., Wallace, R.J.: Partial Constraint Satisfaction. J. Artificial Intelligence 58, 21–70 (1992)
5. Johnson, W.L.: Understanding and debugging novice programs. J. Artificial Intelligence 42(1), 51–97 (1990)
6. Koedinger, K.R., Corbett, A.: Cognitive Tutors: Technology bringing learning science to the classroom. In: Sawyer, K. (ed.) The Cambridge Handbook of the Learning Sciences, pp. 61–78. Cambridge University Press, Cambridge (2006)
7. Kodaganallur, V., Weitz, R., Rosenthal, D.: An Assessment of Constraint-based Tutors: A Response to Mitrovic and Ohlsson's Critique of "A Comparison of Model-Tracing and Constraint-based Intelligent Tutoring Paradigms". J. of AI in Education 16, 291–321 (2006)
8. Le, N.T., Menzel, W.: Using weighted constraints to diagnose errors in logic programming - The case of an ill-defined domain. J. on Artificial Intelligence in Education 19(4), 382–400 (2009)
9. Le, N.T., Menzel, W., Pinkwart, N.: Considering Ill-Definedness Of Problem Tasks Under The Aspect Of Solution Space. In: 23_{st} Int. Conference of the Florida Artificial Intelligence Research Society (FLAIRS), pp. 534–535. AAAI Press, Menlo Park (2010)
10. Looi, C.K.: Automatic Debugging of Prolog Programs in a Prolog Intelligent Tutoring System. J. of Instructional Science 20, 215–263 (1991)
11. Martin, B.: Intelligent Tutoring Systems: The Practical Implementation Of Constraint-based Modelling. PhD thesis, University of Canterbury (2001)
12. Menzel, W.: Modellbasierte Fehlerdiagnose in Sprachlehrsystemen. Niemeyer, Tübingen (1992)
13. Mitrovic, A., Mayo, M., Suraweera, P., Martin, B.: Constraint-based Tutors: A Success Story. In: 14_{th} International Conference on Industrial and Engineering Applications of AI and Expert Systems, pp. 931–940. Springer, Heidelberg (2001)

14. Mitrovic, A., Suraweera, P., Martin, B., Weerasinghe, A.: DB-Suite: Experiences With Three Intelligent, Web-based Database Tutors. J. of Interactive Learning Research 15(4), 409–432 (2004)
15. Ohlsson, S.: Constraint-based Student Modeling. In: Greer, J.E., McCalla, G.I. (eds.) Student Modelling: The Key to Individualized Knowledge-based Instruction, pp. 167–189. Springer, Berlin (1994)
16. Ohlsson, S., Mitrovic, A.: Constraint-based Knowledge Representation For Individualized Instruction. J. Computer Science and Inf. Systems 3(1), 1–22 (2006)
17. Schiex, T., Cedex, C.T., Fargier, H., Verfaillie, G.: Valued constraint satisfaction problems: Hard and easy problems. In: Joint Conference in AI (1995)
18. Woolf, B.P.: Building Intelligent Interactive Tutors. Morgan Kaufmann, San Francisco (2009)

Evaluation of Social Media Collaboration Using Task-Detection Methods

Johannes Moskaliuk[1], Nicolas Weber[2,3], Hermann Stern[2], Joachim Kimmerle[1], Ulrike Cress[4], and Stefanie Lindstaedt[2,3]

[1] University of Tuebingen, Germany
{j.moskaliuk,j.kimmerle}@iwm-kmrc.de
[2] Know-Center Graz, Austria
[3] Graz University of Technology, Austria
{nweber,hstern,slind}@know-center.at
[4] Knowledge Media Research Center, Tuebingen, Germany
u.cress@iwm-kmrc.de

Abstract. Collaboration using social media is a good way of jointly constructing knowledge. This study aims at better understanding collaborative knowledge construction processes by applying innovative (micro-)task detection approaches. We take a closer look at the interactions of a user with a shared digital artifact by analyzing the captured interaction data. The goal is to identify domain-independent interaction patterns, which can serve as indicators for knowledge development (operationalized as accommodation). We designed an empirical study under laboratory conditions that used our method. The applied task detection approach identified accommodation with a rate of 77.63% without resorting to textual features. This result instantiates an improvement as compared to a previous study in which the text in focus was identified as the feature with best discriminative power. We discuss our hypothesis that our method is independent from the used knowledge domain.

Keywords: collaboration, technology-enhanced learning, task-detection, co-evolution, social media.

1 Learning through Collaboration

Jointly writing a text is a not only a method to share information but also a method to collaboratively develop new knowledge. During their writing activity, users may link their individual ideas to those of others, generate new and innovative ideas, and discuss their own arguments with other users. The creation of a shared, homogenous text engages them to construct shared meaning and build mutual understanding [6]. Emergent knowledge that was not part of the individual knowledge of a single user before may arise during this activity. This yields not only an accumulation of knowledge (whereby the knowledge of many individuals is brought together and made available to others), but also knowledge emergence, i.e. the creation of new knowledge [12], [22].

C. Delgado Kloos et al. (Eds.): EC-TEL 2011, LNCS 6964, pp. 248–259, 2011.

Current social media technologies, such as wikis, enable the following way of collaboration: Individuals collaboratively create a shared digital artifact that is provided through the Internet or a local area network of an organization [27], [31]. This supports individual learning as well as the development of collective knowledge within a community. The shared digital artifact supports learning through collaboration and enables the mutual development of knowledge. The rising success of social media in the context of technology-enhanced learning leads to the question of how to evaluate this collaboration process. How can we measure the collaboration process that takes place within a shared digital artifact?

There are various evaluation methods available which focus on individual learning processes of users employing social media (e.g. [23]) or analyzing the content and structure of the shared digital artifact (e.g. [15]). In addition, it is of high relevance to learning evaluation to develop methods that also measure the collaboration process itself. We are interested in the process of how individual knowledge is integrated into collective knowledge within a shared digital artifact.

This is a relevant research challenge in different contexts. *Organizations and enterprises* are interested in improving their knowledge management strategies. They ask their employees to externalize collective knowledge in shared documents. A method to evaluate the collaboration process could be used to provide feedback about the quality of the externalized knowledge and support the employees to improve their collaboration. *Schools and universities* face the challenge to integrate collaborative learning and teamwork into their curricula, but still need to assess students individually. Information about successful collaboration processes could be helpful to do that, and at the same time ensure a perpetual evaluation and improvement of current learning settings and concepts. In *informal learning settings* (e.g. authors or users of the online-encyclopedia Wikipedia) the collaboration of volunteers could benefit from measurements of current status of the process and the writing product. This could support efficient collaboration and protect against vandalism and disruptions.

In this paper we describe an innovative method to evaluate the knowledge construction processes that take place during the collaboration on a shared digital artifact. Based on current task detection approaches [28, 30, 33] we developed a method to observe the writing process of a user closely by recording the users' interaction with a shared digital artifact. The captured data is used to train and validate a task detection tool and identify user interactions that lead to the development of collective knowledge. We argue that the users' interaction with a shared digital artifact can be used to measure this integration and the development of collective knowledge. We designed an empirical study under laboratory conditions that detects relevant interaction patterns which indicate collaborative knowledge construction. The current study extends our previous work [24] by using a different knowledge domain to examine if our method is independent from the textual features.

In the following we introduce the co-evolution model that highlights from a constructivist point of view the role of shared digital artifacts for learning through collaboration. We describe the process of accommodation as an indicator for the successful integration of individual and collective knowledge. Then, we summarize the task detection approach and argue how it can be applied to evaluate wiki-based collaboration. As a conclusion we present preliminary results of an empirical study and discuss how to implement our method for the evaluation of social media collaboration in different contexts.

2 Co-evolution of Social and Cognitive Systems

From a constructivist point of view, a shared digital artifact like a wiki or other social media tools are external representations of a social and cultural context and represent the knowledge of the corresponding community [18, 36]. Such external representations of knowledge support the reflection of one's own knowledge and may stimulate individual learning processes within a community or an organization. So the interactions with representations may shape individual learning and lead to a development of collective knowledge over time.

This mutual development of individual and collective knowledge is described in the co-evolution model of cognitive and social systems [4, 13]. The model introduces a framework to describe knowledge processes that take place in a wiki. It conceptualizes individual learning and development of collective knowledge as a co-evolution between cognitive and social systems. The co-evolution model is based on Luhmann's system theory [19-21] and considers different relevant systems: the wiki as a social system and the cognitive system of each user. Each system has its specific mode of operation. The social system operates in the mode of communication (which becomes manifest as written text in a digital artifact). The cognitive system operates through individual psychological processes like perception, reasoning, or learning. When people work on a digital artifact, both the social and the cognitive system develop over time and become more and more complex. Individuals externalize their knowledge and contribute it to the wiki, and at the same time they internalize information from the shared digital artifact. So the information in a wiki (that represents the knowledge of a community) evolves in the course of time. At the same time, knowledge in the individual's cognitive system develops. This mutual development of social and cognitive systems is called co-evolution.

Cognitive conflicts that an individual perceives are considered as the key incitement factor of this co-evolution. In the sense of Piaget [25, 26], a cognitive conflict will occur if new information from the environment does not fit to individual knowledge. Such conflicts will motivate individuals to contribute to the wiki or to change their own knowledge structure, in order to establish a balance or equilibrium between their own knowledge and information in the wiki. The model by Cress and Kimmerle [4] specifies the co-evolution process and describes two different processes on the basis of the ideas of Piaget: assimilation and accommodation. Assimilation means interpreting and explaining current experiences by giving them a place in existing schemas. Accommodation means adaptation to the environment in the form of qualitatively changing one's own cognitive schemas. In Piaget's understanding assimilation and accommodation are processes in the cognitive system; the co-evolution model, however, expands Piaget's concepts by describing accommodation and assimilation not only from the perspective of an individual's cognitive system, but also from that of a social system: in the case of assimilation, users make contributions that will not change the basic message and structure of the shared digital artifact, but only add supplementary aspects. Accommodation is also possible in a social system if users contribute their knowledge in such a way that the message is changed and, sometimes, new structures are being created. Accommodation tends to result in some qualitative modification of the artifact, whereas assimilation has primarily to do with quantity, introducing additional arguments or new examples but no fundamental innovation.

To sum up, the co-evolution model is a framework to describe how individuals use digital artifacts to develop knowledge. Knowledge becomes manifest in the digital artifact and is therefore accessible for deeper analysis. So knowledge development can be measured by analyzing the interaction of users with a digital artifact. While assimilation only leads to a quantitative increase of knowledge, accommodation leads to a qualitative knowledge development in the sense of a more sophisticated knowledge structure, balanced argumentation, and innovation.

Thus, we suggest that the process of accommodation is an essential indicator of knowledge development. Emergent effects will usually occur through processes of accommodation in artifacts. This will lead to a higher complexity of the shared digital artifact and, accordingly, to knowledge processes in other people's cognitive systems. We have to focus on the process of accommodation as indicator for the integration of individual and collective knowledge. This integration of individual and collective knowledge (operationalized as accommodation) is an indicator for a higher quality of the collaboratively written text and the development of collective knowledge.

2.1 Analyzing Accommodation as Essential Part of Co-evolution

Analyzing the integration of individual knowledge into the collective knowledge represented by the digital artifact leads to deeper insights into learning through collaboration. We have to analyze the collaboration process in detail to gain a deeper understanding of how the development of knowledge takes place.

This approach enhances current methods of measuring the quality of texts with computational methods, as introduced by several authors. Stvilia et al. [35] assessed the quality of collaboratively created content based on reading scores by counting syllables, words, and sentences [7, 16]. Braun and Schmidt [2] evaluated articles from the online-encyclopedia Wikipedia and used readings scores, presentation format (e.g.: sections, headings, pictures) and number of links to other articles as indicators of text maturity. Other approaches go beyond the surface components of text and extract semantic information from the text by using information retrieval approaches and latent semantic analysis [17]. These methods allow insights into the content or topic of the text and its relevance to readers and their search goals. The Coh-Metrix-tool [9], for example, focuses on the cohesion of text and combines over 200 measurements and analysis algorithms. The cohesion of text may also be used as an indicator for text-quality [8]. All these methods concern texts that were written by single authors. But there are also some methods that consider the specific characteristics of collaboratively written wiki texts and use, apart from text features, also the available information about authors as one of their criteria for assessing text quality.

Other approaches take into account the frequency of editing or reading of wiki articles [38], use the authority of contributors and reviewers of wiki articles as a source to calculate text-quality [11], or analyze phrases in wikis and their development in the course of time (considering also the contribution of different authors and their reputation) as an indicator for text quality

For the current study we do not look at semantic or linguistic features of the text or on contributors and their edits, but only take interactions of users with the shared digital artifact into account. In order to identify the integration of individual and collective knowledge within a shared digital artifact, we observe users' interaction with the shared digital artifact in detail. We use the method of task detection to analyze the writing process. Our task detection approach is described in the following subsection.

3 Task Detection Approach

Context-aware (or sentient) systems are systems that can adapt their operations or behavior to their current usage context, without explicit user intervention. Context awareness increases the usability and effectiveness of a system by taking into account environmental elements (such as time or location), individual and organizational elements (such as a user's identity or position), as well as elements depending on user interaction with the system (such as a pressed button or entered character). The first context-aware systems, developed in the 1990s, mainly focused on providing functionalities that were specific to the user's location [1]. Today's context-aware systems are much more sophisticated and able to integrate complex mechanisms for acquisition and storage of context, abstraction and understanding of context, and adaptation of system behavior based on the context that was recognized by the system.

Context information may be gathered in a variety of different ways, such as applying (physical or virtual) sensors, recording network information and device status, or browsing user profiles and organizational databases. A context model is needed for storing the recorded user context data in a machine readable form. Various context model approaches have been proposed, such as key-value models, markup scheme models, graphical models, object oriented models, logic-based models, or ontology-based models [34]. The ontology-based approach has been advocated as being the most promising one [1, 34], mainly because of its dynamic character, expressiveness and extensibility.

This approach uses and populates an ontology to model the context. Automatic task detection is classically modeled as a machine learning problem, and more precisely as a classification problem. Training instances are constructed and used to classify the current task which users are performing. This method has been used to recognize Web based tasks [10], tasks within emails [33] or tasks from the complete desktop of a computer user [28, 32, 33]. Detecting the users' tasks makes it possible to provide personalized and relevant support to the users [3, 5]. For the current study, we adopt an ontology-based task detection approach that has already proven to be applicable to routine and knowledge-intensive computer desktop tasks [28, 29]. We consider accommodation as specific form of refined editing tasks (transformations of the wiki text) which we would like to detect. This allows us to measure knowledge maturing by analyzing the interactions of users with a digital artifact. We designed a study in which participants had to work with a wiki and add new information. Experts then rated each text transformation carried out by the the participants to evaluate the amount of accommodation created by doing this transformation. These expert ratings were used for labeling the recorded user interaction data for the task detection evaluations. In order to identify features for accommodation transformations, we applied the ontology-based task detection approach for the selection of (quantitative) interaction-based features in order to automatically detect accommodating activities as a maturity indicator. In a previous study using the same method [24] our ontology-based task detection approach yielded an identification rate of 79.12%. Based on the ratings of two experts, we were able to distinguish between three classes of accommodation (high, medium, low). We could show that it was indeed possible to detect accommodation during the writing process, using only a limited amount of contextual information.

But the results also showed that the textual content of the transformations plays an essential role for the detection of accommodation. The best performing feature was the *content in focus feature*. This feature was constructed based on the text the user had interacted with (represented as a term vector from which all stop words were removed). In the current study we aim to show, that our method also works independently from the textual content of the interaction. We run an experiment under laboratory conditions but use a different knowledge domain. The next section will describe this study in detail.

4 Study: Integrating Individual Knowledge into a Shared Artifact

4.1 Research Setting

We designed a research setting in which participants had to work with a wiki and add new information. This research setting followed closely the setting of our previous study [24] in which participants were asked to integrate pro and contra arguments on the risk and danger of violent computer games. The study presented in this paper focused on the current debate on minarets and other religious buildings in German-speaking countries. The wiki and the additional information were presented on the participants' computers. In order to induce cognitive conflicts, the information presented in the wiki and the additional information contradicted each other. This was supposed to provoke accommodation processes and was therefore considered as an ideal condition for constructing the required data set. We used "pros and cons" of a contrary topic and the participants' task was to complete a given text of a wiki (initial state), with the aim that a resulting article should be a scientifically balanced article about the specific topic. Two windows were shown during the study: a wiki page and a page with additional information. Those two windows were accessible through two tabulators at the top of the page. The wiki page (wiki-tab) presented a text that initially was biased toward one position. Here, only arguments for the one position of the topic were presented ("The construction of new minarets should not be allowed."). The page with additional information (info-tab) contained ten different arguments that were arguing towards the other direction, invalidating arguments in the wiki text or explaining why these were one-sided or incorrect ("Religious buildings support integration of foreigners."). Participants could copy, paste, delete, and edit the wiki freely or type in new text. The participants were 10 graduate students from Germany, their mean age was 25,5 (SD = 3.38). 7 of these were women, 3 were men.

Participants had a total of 50 minutes to edit the wiki. They were instructed to save their changes after each set of activities that belonged together (referred to as transformations). This procedure led to a series of single editing tasks. We asked two experts to rate each single transformation on a five-point Likert scale ranging from low to high accommodation. These rating were used to build three classes of accommodation: low, medium, and high.

4.2 Automated Detection of Accommodation

Considering micro-events such as mouse movements and keystrokes enables us to understand and detect usage patterns characteristic for accommodation. Due to technical limitations, common online collaboration systems like wikis do not allow tracking fine-granular user events like single edits or copy and paste activities. Instead, in wiki systems, user activities can only be reproduced from article history, which contains an aggregation of user activities in the article revisions. In order to observe fine-granular user interactions during the editing process for the experiment we developed a wiki-like user interface that provides the ability to track user generated events.

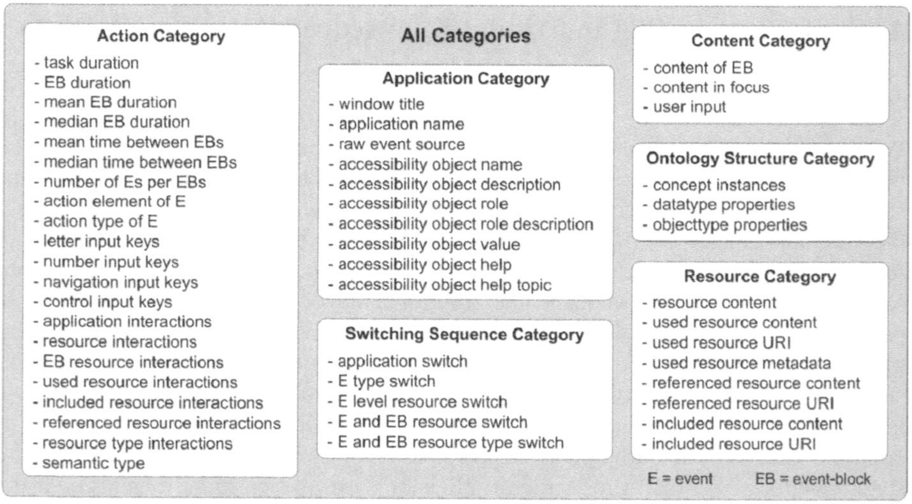

Fig. 1. Feature categories used in the experiment [29]

Detection of accommodation is based on data from the experiment. Therefore concepts and their attributes have been extracted to create the semantic dataset. The interaction of users with the system is referred to as *user interaction context* which is represented by the instantiation of the *User Interaction Context Ontology (UICO) [29]*. From a top-level perspective, the concepts of this ontology can be grouped into five different feature categories (see Figure 1): (i) action category, (ii) resource category, (iii) user category, (iv) switching sequence category, (v) application category, and (vi) ontology structure category. The **action category** contains concepts that are related to the user activities aggregated in different granularity levels. The event level contains atomic user events like keystrokes or mouse movements. These are aggregated to event-blocks, groups of uninterrupted events belonging to the same data artifact. Besides the activities, also the activity attributes are part of the model. This includes the temporal component, duration of activities and time between activities as well as the type of resource interaction. The **application category** contains metadata about used applications, like window title and the name of the application. Since users in both experiments were limited to one application, these features have been only used as static features in the task detection approach. The **content category** consists of textual content which is related to the activities and the content which was focused by the user during the activities. Metadata about the resources is provided by concepts in the **resource category**. One important feature in task mining is the switching behavior between resources and applications. This information is stored in the **switching sequence category**. Overall the model contains 88 concepts and 272 properties, and is modeled in OWL-DL (the Ontology Web Language, a W3C standard). The **ontology structure category contains** metadata about this model.

For data analysis, the properties are converted into attributes using the machine learning toolkit Weka [37]. These attributes serve as features feeding the classification algorithms. We use standard classification algorithms in combination with different feature sets. Among the classifiers are the Naive Bayes, the Linear Support Vector Machine (SVM) with cost parameter settings c = $\{2^{-5}, 2^{-}, 2^{-1}, 2^{}, 2^{}, 2^{}, 2^{}, 2^{}, 2^{10}\}$, the J48 decision tree and the k-Nearest Neighbor (KNN-k) algorithm with k = {1, 5, 10, 35} settings. In order to identify the best performing feature set for classifying accommodation, different sets of features are tested. Each feature is evaluated as single feature. Groups of features are clustered into: all features of one category, the best performing features in groups of 2,3,4,5 features.

More detailed information on the task detection approach described can be found in our previous study [24]. We used the same interface, and algorithms and the input parameter groups were also identical.

4.3 Preliminary Results

During the study, we recorded a dataset of 102 editing tasks. Each participant saved on average 10.2 ($SD = 3.55$, min = 5, max = 17) editing tasks. Correlation between the two independent experts for each transformation was $r(102)=.87$, $p<.01$.We regard this as evidence for the validity of the expert ratings and as a hint that it is possible to evaluate single transformations on the basis of the accommodation concept.

The best performing feature category for detecting accommodation was the action category using a J48 classifier. This category provided an accuracy of 73.72% (precision 0.77, recall 0.55). The content category which performed best in our previous study [24] reached only 68.08% (precision 0.73, recall 0.52) applying a SVM-C classifier. Considering single features, interaction based features performed better than content based features as well. The best single feature for detecting accommodation were *resource interaction* with an accuracy of 77.63% (precision

0.75, recall 0.55) and *used resource interaction* 66.15% (precision 0.77, recall 0.61) with a J48 classifier. Both features provide information about the type of interaction events between users and resources. The top scoring single feature in the previous experiment [24] *content in focus* reached an accuracy of 61.53% (precision 0.68, recall 47) with SVM-C classification.

4.4 Discussion

Our previous experiment [24] showed that the best performing features for detecting accommodation were related to *content*. Best feature categories (content category) as well as best single features were based on the textual content of the experiment. The best performing single features were *content in focus* and *user input*. This implies that the performance of the features was highly depending on domain specific text patterns. In the present experiment, the content domain was changed. The results show that the content related feature is not the best performing one (as in the previous experiment). We assume that different domains are more sensitive for content related features due to a special domain specific wording. The best performing features in the present experiment in the feature category as well as single features were related to *user activity*. Based on the user's interactions with the resource (writing/editing wiki page, reading/copying additional information), we detected accommodation with an accuracy of 77%. This implies that content artifacts characteristic for accommodation activities are mainly domain dependent: Analysis of the experiment data showed that some words are characteristic for accommodation in the tested domain. Hence, the content based feature was evaluated as highly content dependent. The good performance of activity based features indicates that accommodation interaction patterns exist. Independent from the content domain, some of the features performed well in both experiments. *Used resource interaction* and *application interaction* performed with accuracy around 66% in both experiments. Both features seem to be stable independent from the content domain.

5 Outlook

Within this second study we demonstrate the benefit of using task detection methods to detect accommodation tasks as indicator for the integration of individual and collective knowledge. The preliminary results support our hypothesis that we will be able to develop a method to automatically analyze users' interactions with social software. This is a crucial step for the development of enhanced methods for evaluating learning trough collaboration. As a next step, we will analyze the results in more detail to gain further insights, especially regarding the kind of interaction patterns, which lead to accommodation. We also will identify the characteristic words for accommodation in the tested domain and compare them with the results of our previous experiment. This is necessary to improve our method and ensure that it is independent from (domain related) textual features.

Our study leads to a better understanding of collaborative knowledge construction processes by taking a closer look at the interaction of a user with a shared digital artifact. The close capturing of the writing process is an innovative method to evaluate learning trough collaboration. The results could be applied to support the collaboration process by using the findings of the task detection as online feedback. Information regarding the current status of the integration of individual and collective knowledge could support users to write good texts and contribute to the mutual development of knowledge. In following studies we will use our method with larger data sets in the field, e.g. real users of an enterprise wiki or of a course wiki in an educational context. Therefore we plan to implement our tool as a MediaWiki extension. This will lead to a large amount of data we will use to validate and specify the identified domain-independent interaction patterns which can serve as indicators for knowledge development.

References

1. Baldauf, M., Dustdar, S., Rosenberg, F.: A survey on context-aware systems. International Journal of Ad Hoc and Ubiquitous Computing 2(4), 263–277 (2007)
2. Braun, S., Schmidt, A.: Wikis as a technology fostering knowledge maturing: What we can learn from wikipedia. In: Proceedings, 7th International Conference on Knowledge Management (IKNOW 2007), Special Track on Integrating Working and Learning in Business (IWL), pp. 321–329 (2007)
3. Coutaz, J., Crowley, J.L., Dobson, S., Garlan, D.: Context is key. Communications of the ACM 48(3), 49–53 (2005)
4. Cress, U., Kimmerle, J.: A systemic and cognitive view on collaborative knowledge building with wikis. International Journal of Computer-Supported Collaborative Learning 3(2), 105–122 (2008)
5. Dey, A.K., Abowd, G.D., Salber, D.: A conceptual framework and a toolkit for supporting the rapid prototyping of context-aware applications. Human-Computer Interaction 16(2), 97–166 (2001)
6. Erkens, G., Jaspers, J., Prangsma, M., Kanselaar, G.: Coordination processes in computer supported collaborative writing. Computers in Human Behavior 21(3), 463–486 (2005)
7. Flesch, R.: A new readability yardstick. Journal of Applied Psychology 32(3), 221–233 (1948)
8. Graesser, A.C., McNamara, D.S., Louwerse, M.M.: What do readers need to learn in order to process coherence relations in narrative and expository text. In: Sweet, A.P., Snow, C.E. (eds.) Rethinking Reading Comprehension, pp. 82–98. Guilford Publications, New York (2003)
9. Graesser, A.C., McNamara, D.S., Louwerse, M.M., Cai, Z.: Coh-Metrix: Analysis of text on cohesion and language. Behavior Research Methods, Instruments, & Computers 36, 193–202 (2004)
10. Gutschmidt, A., Cap, C.H., Nerdinger, F.W.: Paving the path to automatic user task identification. In: Proceedings, IUI Workshop on Common Sense Knowledge and Goal-Oriented Interfaces (2008)
11. Hu, M., Lim, E.P., Sun, A., Lauw, H.W., Vuong, B.Q.: Measuring article quality in wikipedia: models and evaluation. In: Proceedings, 16th ACM Conference on Conference on Information and Knowledge Management, pp. 243–252 (2007)
12. Johnson, S.: Emergence: The Connected Lives of Ants, Brains. Cities, and Software. Scribner, New York (2001)

13. Kimmerle, J., Cress, U., Held, C.: The interplay between individual and collective knowledge: Technologies for organisational learning and knowledge building. Knowledge Management Research & Practice 8, 33–44 (2010)
14. Kimmerle, J., Moskaliuk, J., Cress, U.: Using Wikis for Learning and Knowledge Building: Results of an Experimental Study. Journal of Educational Technology & Society (in press)
15. Kimmerle, J., Moskaliuk, J., Harrer, A., Cress, U.: Visualizing Co-evolution of individual and collective knowledge. Information, Communication & Society, 1099–1121 (2010)
16. Klare, G.R.: Assessing Readability. Reading Research Quarterly 10(1), 62–102 (1974)
17. Landauer, T.K., Dumais, S.T.: A solution to Plato's problem: The latent semantic analysis theory of acquisition, induction, and representation of knowledge. Psychological Review 104(2), 211–240 (1997)
18. Lave, J.: Situating learning in communities of practice. In: Resnick, L., Levine, J., Teasley, S. (eds.) Perspectives on socially shared cognition, pp. 63–82. APA, Washington, D.C (1991)
19. Luhmann, N.: System as Difference. Organization 13(1), 37–57 (2006)
20. Luhmann, N.: Social systems. Stanford University Press, Stanford (1995)
21. Luhmann, N.: The Autopoiesis of Social Systems. In: Geyer, F., van der Zouwen, J. (eds.) Sociocybernetic Paradoxes: Observation, Control and Evolution of Self-Steering Systems, pp. 172–192. Sage Publications, London (1986)
22. Moskaliuk, J., Kimmerle, J.: Using wikis for organizational learning: Functional and psycho-social principles. Development and Learning in Organizations 4(23), 21–24 (2009)
23. Moskaliuk, J., Kimmerle, J., Cress, U.: Wiki-supported learning and knowledge building: Effects of incongruity between knowledge and information. Journal of Computer Assisted Learning 25(6), 549–561 (2009)
24. Moskaliuk, J., Rath, A., Devaurs, D., Weber, N., Lindstaedt, S., Kimmerle, J., Cress, U.: Automatic detection of accommodation steps as an indicator of knowledge maturing. Interacting with Computers 23(3), 247–255 (2011)
25. Piaget, J.: Problems of equilibration. In: Appel, M.H., Goldberg, L.S. (eds.) Topics in Cognitive Development, pp. 3–14. Plenum, New York (1977)
26. Piaget, J.: The development of thought: equilibration of cognitive structures. Viking Press, New York (1977)
27. Raitman, R., Augar, N., Zhou, W.: Employing Wikis for Online Collaboration in the E-Learning Environment: Case Study (2005)
28. Rath, A.S., Devaurs, D., Lindstaedt, S.N.: Detecting real user tasks by training on laboratory contextual attention metadata. In: Exploitation User Attention Metadata (EUAM 2009). Lecture Notes in Informatics (2009)
29. Rath, A.S., Devaurs, D., Lindstaedt, S.N.: UICO: an ontology-based user interaction context model for automatic task detection on the computer desktop. In: Proceedings, 1st Workshop on Context, Information and Ontologies, p. 8 (2009)
30. Rath, A.S., Weber, N., Kröll, M., Granitzer, M., Dietzel, O., Lindstaedt, S.N.: Context-aware knowledge services. Personal Information Management: PIM, 5–6 (2008)
31. Reinhold, S.: WikiTrails: Augmenting Wiki Structure for Collaborative, Interdisciplinary Learning. In: Proceedings, 2006 international symposium on Wikis (2006)
32. Shen, J., Irvine, J., Bao, X., Goodman, M., Kolibaba, S., Tran, A., Carl, F., Kirschner, B., Stumpf, S., Dietterich, T.G.: Detecting and correcting user activity switches: Algorithms and interfaces. In: Proceedings, 13th International Conference on Intelligent User Interfaces, pp. 117–126 (2009)

33. Shen, J., Li, L., Dietterich, T.G., Herlocker, J.L.: A hybrid learning system for recognizing user tasks from desktop activities and email messages. In: Proceedings, 11th International Conference on Intelligent User Interfaces, pp. 86–92 (2006)
34. Strang, T., Linnhoff-Popien, C.: A context modeling survey. In: Workshop on Advanced Context Modelling, Reasoning and Management, UbiComp, pp. 1–8 (2004)
35. Stvilia, B., Twidale, M.B., Smith, L.C., Gasser, L.: Assessing information quality of a community-based encyclopedia. In: Proceedings, International Conference on Information Quality, pp. 442–454 (2005)
36. Vygotsky, L.S., Cole, M., John-Steiner, V., Scibner, S., Souberman, E.: Mind in society: The developmentof higher psychological processes. Harvard University Press, Cambridge (1978)
37. Witten, I.H., Frank, E.: Data Mining: Practical machine learning tools and techniques. Morgan Kaufmann, San Francisco (2005)
38. Wöhner, T., Peters, R.: Assessing the Quality of Wikipedia Articles with Lifecycle Based Metrics. In: Proceedings, 5th International Symposium on Wikis and Open Collaboration (2009)

Framework for Contextualized Learning Ecosystems

Mario Muñoz-Organero[1], Gustavo A. Ramírez[2], Pedro J. Muñoz-Merino[1], and Carlos Delgado Kloos[1]

[1] Universidad Carlos III de Madrid
Av. Universidad 30 E-29811 Leganés Madrid (Spain)
{munozm,pedmume,cdk}@it.uc3m.es
[2] Universidad del Cauca Popayán Colombia
gusramir@gmail.com

Abstract. Using mobile personal devices to interact with pervasive smart learning objects and services that create contextualized learning ecosystems can enhance both the learning outcomes and the motivational states of students. This paper defines and analyzes several pervasive learning ecosystems in which students at the Carlos III University of Madrid interact with contextualized learning objects and services. The technology defining the contextualized learning environments is first introduced and later used in two user experiments. These experiments provide data both about the learning outcomes for students after interacting with smart learning objects and services and about the motivational impact that the use of these technologies have on them.

Keywords: Learning ecosystems, contextualized learning in an Internet of Things, Learning outcomes, student motivation enhancement, user experiments, RFID, NFC, mobile learning.

1 Introduction

Ubiquitous computing embeds computational capabilities into contextualized objects creating an invisible part of the fabric of everyday life [1]. Providing communication and computational capabilities to smart objects adds a new dimension to an any-time any-where and for any-body pervasive computing environment into an Internet of Things designed to communicate to and from any-thing. "The Internet of Things will enable forms of collaboration and communication between people and things, and between things themselves, hitherto unknown and unimagined" [2]. Human users can make use of mobile personal devices to interact with such an Internet of things [3]. Depending on the purpose of the user when interacting with an Internet of Things, several ecosystems can be defined [4]. Learning ecosystems in an Internet of Things enable anyone to learn at anytime and anywhere [1] by interacting with contextualized smart objects.

The use of mobile devices in learning activities as an interaction mechanism with contextualized pervasive learning ecosystems can enhance the shift from pure instructor centered classroom teaching to constructivist learner centered educational settings [5] in which students interact with smart learning objects and services in a

C. Delgado Kloos et al. (Eds.): EC-TEL 2011, LNCS 6964, pp. 260–270, 2011.

contextualized environment. Several experiments during the last years have concentrated on studying how the use of mobile phones and PDAs in educational settings can enhance the learning process and learning outcomes. A context-aware language-learning support system using PDAs, GPS and RFID tags is described in [6]. The paper in [7] proposes a context-aware ubiquitous English learning system in a campus environment which can detect the learning location by wireless positioning techniques and retrieve specific learning content based on location information to individual learners through wireless networks. An environment for learning with educational resources based on RFID and ubiquitous computing technologies is described in [8]. A similar experiment which uses a context-aware ubiquitous learning environment which utilizes mobile devices, sensors and wireless networks to conduct situated learning is described in [9].

Among the main characteristics normally present in experiments using mobile devices in learning settings three of them are of particular importance for the scope of this paper: their use for exploratory learning, in situated environments, for problem solving. In exploratory learning, the students investigate a system on their own, often in pursuit of a goal. Exploratory learning [10] is a constructivist instructional approach, wherein the learners are encouraged to explore and to experiment on their own. Situated learning was defined by Brown et al. as "embedded within and inseparable from participating in a system of activity deeply determined by a particular physical and cultural setting" [11]. Problem based learning (PBL) originates form education in medicine [12] and consists of the manipulation of problematic situations, comprising the appraisal of the problem, creation of a problem space, the selection of goals and the deployment and monitoring of cognitive structures to reach those goals [13].

Among the main objectives targeted when using mobile devices in learning experiments two of them are of especial importance for the scope of this paper: improving the learning outcomes and enhancing the motivation of students. Liu et al. [8] evaluated the impact of using mobile devices in a situated learning environment on the grades of the students. Moura and Carvalho [14] studied if the use of mobile phones and podcasts encouraged the students to learn. The activity theory was used in [15] for designing ubiquitous learning scenarios showing that learners become motivated and engage in active pursuit and construction of knowledge when using ubiquitous technologies.

This paper defines and analyzes several pervasive learning ecosystems in which students at the Carlos III University of Madrid interact with contextualized learning objects and services. The technology defining the contextualized learning environments is first introduced and later used in two user experiments. These experiments provide data both about the learning outcomes for students after interacting with smart learning objects and services and about the motivational impact that the use of these technologies have on them.

2 Defining Contextualized Pervasive Learning Ecosystems

The vision of the Internet of Things extends the connectivity to the Internet into our everyday lives through the use of wireless technologies and identification mechanisms such as RFID which generate technology enhanced pervasive ecosystems [4]. This section presents an analysis and construction proposal for a particular case of such ecosystems: the contextualized pervasive ecosystems for learning.

The definition of the concept of the contextualized learning ecosystem should incorporate together the use of three technology enhanced components: pervasive learning objects, pervasive learning services and proactive learning environments. Combining together technology enablers such as RFID tags, mobile devices and pervasive service items, learning environments can be created in which users interact with pieces of knowledge embedded in communicating smart objects and contextualized services. Using appropriate technologies such as NFC (Near Field Communications) [16] integrated in user devices such as smart-phones, the users can consume the information provided by pervasive (either passive or active) learning objects. The process in which the user consumes this ambient knowledge can be interactive either by using the communication link with the pervasive learning object or by using an external representation of such object in the Internet [17]. The second component of a contextualized learning ecosystem, the learning service, provides additional learning capabilities such as projection rooms, printing services or collaborative information sharing to the learning environment. The assimilation of concepts and the acquisition of practical competencies can be simplified by the use of such services. Finally, the third component in a pervasive learning ecosystem should also allow accidental learning to take place in a user friendly manner. Proactive learning environments should be used to provide smart, contextualized and user adapted learning information when the user needs it without requiring the user intervention.

This section presents some technological tools developed inside the Gradient group at the Carlos III University of Madrid to define a contextualized pervasive learning ecosystem categorized as pervasive learning objects, pervasive learning services and proactive learning environments.

2.1 Pervasive Learning Objects

Pervasive learning objects can be either physical things and tools (such as the different cards in a telecommunication exchange or a working engine inside a car) or conceptual representations of them (such as a panel or a poster displaying how to send a packet over an IP network). Pervasive learning objects in a contextualized situated environment should provide multimedia information to nearby users by means of a wireless communication link such as NFC or Bluetooth (remote access can also be provided to such objects over the Internet such as in remote labs [18]). Figure 1a captures a user interacting with a telecommunication exchange using an NFC enabled mobile phone (Nokia 6131 NFC). Multimedia information is obtained from NFC tags in different parts of the exchange either directly or by following a link to an external object in the Internet (which can be cached inside the mobile device for efficiency). Figure 1b captures a virtual recreation of a physical environment using an NFC enhanced panel presenting information about how to configure certain services in an IP network. A user can interact with the panel using an NFC compatible mobile device.

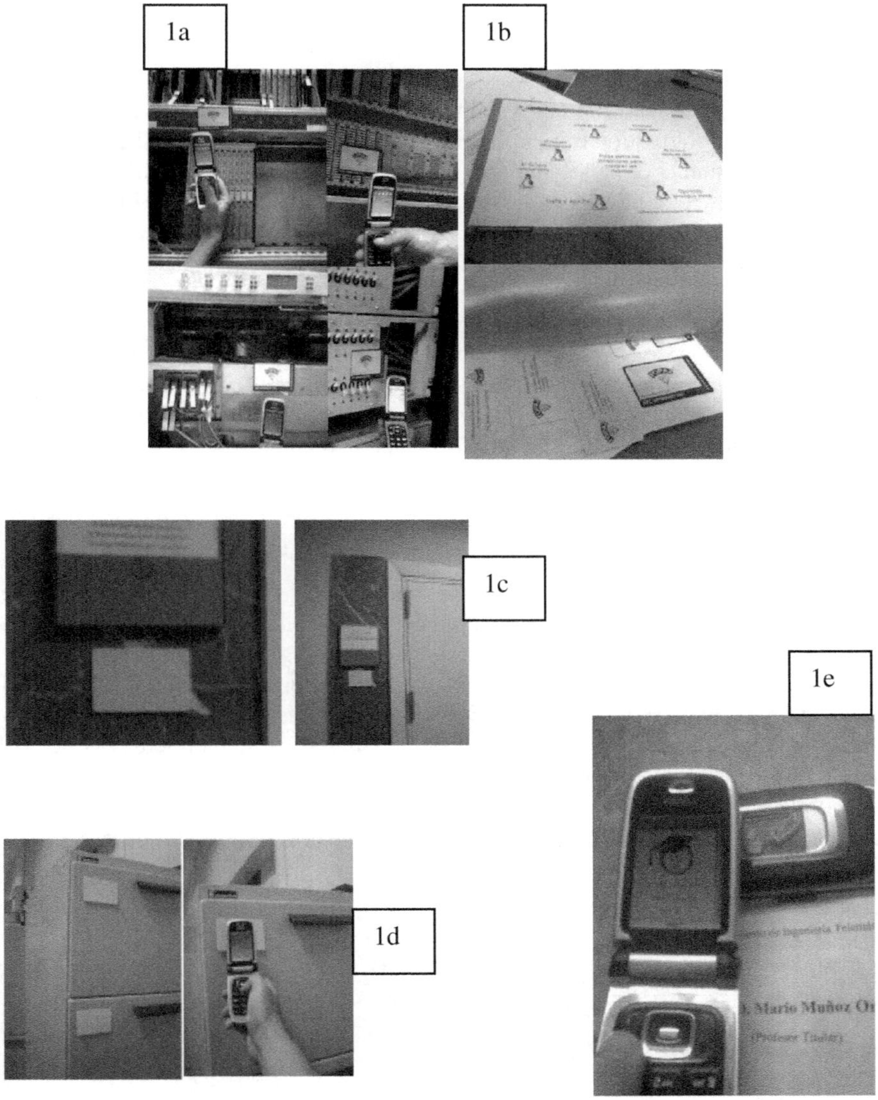

Fig. 1. Pervasive learning objects. 1a shows a user interacting with NFC tags in a telecommunication exchange. 1b shows the front and back of a panel containing NFC tags to interact with a mobile user. 1c presents a touching note associated to an office. 1d shows a touching cabinet and 1e presents a bidirectional learning object using HTTP over NFC.

Any existing object in a learning environment can be tagged with information. Figures 1c and 1d capture two examples more [19]. Figure 1c represents a touching note associated to any office or lecture room in a university. A student can touch the note to get information about the location of a particular professor or about the activity been carried out in a particular lecture room at a particular time. The touching note also allows the student to leave a message back to the professor if required. Figure 1d captures a touching cabinet in which information is provided about the contents of such objects or about who has interacted with them.

A contextualized learning ecosystem can also incorporate active learning objects augmented with bidirectional input/output facilities. Students should be able not only consume the contents of smart learning objects but also to interact with them. Several technologies can be applied for implementing embedded interactions in an Internet of Things [20]. Figure 1e captures the interaction of an NFC enabled mobile phone running our implementation of an HTTP browser [21] accessing the information on an HTTP server over a bidirectional NFC channel. Our implementation of the HTTP server [22] is executed in a similar NFC enabled mobile device providing contextualized access to students touching on it. Figure 1e implements an examination platform deployed at the door of the office of one of the authors of this paper. The student receives a short, single choice type exam that after completion can be uploaded to the examination platform (the NFC enabled phone at the door) receiving instant feedback about the answers.

2.2 Pervasive Learning Services

A second family of components in contextualized pervasive learning ecosystems is constituted by pervasive learning services offered to nearby learners. Services embedded in smart pieces of the contextualized learning environment should provide instant access to learning facilitators such as printers, projection panels or interactive platforms for collaborative editing and document processing. Figure 2a captures one of such services: a projection panel controlled using an NFC mobile device [23]. The back part of the projection panel contains an array of NFC tags that are detected by the mobile phone and sent to the PC controlling the projection of the information. The reading of a tag in the panel by the mobile device is mapped to a click of a mouse at the appropriate location.

Fig. 2. Pervasive learning services. 1a presents a projection panel controlled using an NFC mobile device. 2b shows an NFC enabled printing service.

Figure 2b captures our implementation of a printing service. A student can print an image, a presentation of slides or a pdf document simply by selecting it on his or her mobile phone and by touching an NFC tag on the printer with it. This tag contains the Bluetooth address of a controlling computer to which to send the resource to be printed. The controlling computing opens that resource with a reading application and sends it to printing queue of the printer. A similar interaction interface has also been implemented for controlling the image projected in the panel in figure 2a so that a learner can share and discuss information with other nearby students.

Other examples of services that we have implemented for the creation of contextualized pervasive learning ecosystems are a document sharing application controlling a PC hosted document using NFC and Bluetooth enabled mobile devices and a "touchable beamer" projecting an image simply by touching it with an NFC mobile device.

2.3 Proactive Learning Environments

Contextualized pervasive learning ecosystems should facilitate and enhance both deliberate and accidental learning. Pervasive learning objects and services described in previous subsections are mainly targeted to a learning user having the control of the learning interaction process in a deliberate manner (the learning user is normally responsible for starting the interaction process). Proactive learning environments, on the other hand, should be used to provide smart, contextualized and user adapted learning information when the user needs it without requiring the user intervention in an accidental manner. Proactive parts in a pervasive learning ecosystem should know what the learning user needs by inferring what the user is, has and does and provide contextualized and relevant information to this user. [20] presents some examples of proactive environments in an Internet of Things such as a kitchen. We have concentrated on the creation of proactive learning ecosystems. This section describes two examples.

A first example of a proactive learning environment that we have implemented is a smart gate at the entrance of the Torres Quevedo building inside the Carlos III University of Madrid campus in Leganes (Madrid, Spain). A laptop with the profile of some students (the group of the experiment) uses Bluetooth to detect such students when they go inside that building. Depending on the subjects that each student is taking the system sends relevant information to that student using OBEX object transfer over bluetooth.

A second example is a proactive application for sending learning pills to interested students in class (later described as a case study in this paper). As in the previous example, OBEX object transfer over Bluetooth is used to automatically send information to students. In this case, the information sent is a set of complementary exercises to reinforce what is being taught by the professor at this particular moment in class.

3 Case Studies

This section captures two case studies that we have carried out in order to assess both the learning outcomes for students after interacting with smart learning objects and services and the motivational impact that the use of pervasive learning technologies have on them.

3.1 Interactive Panels

An interactive panel such as the one presented in Figure 1b has been used to simulate a server room running some network services in Linux. Students attending the "advanced telematic applications" course at the Carlos III University of Madrid (dedicated to teach third year Telecommunication Engineering students how to configure IP network services such as DNS, NFS, NIS, SMTP and HTTP) were divided into control and experiment groups. The group of the experiment used an NFC panel such as the one in figure 1b while the control group attended a traditional face-to-face instructor-led class. Our objective was to show if using an NFC panel alone a student is able to obtain a similar learning gain as in a traditional face-to-face instructor-led class (but with no limitations in when, where and how fast to learn). In order to have enough data for statistical analysis, two different experiments were carried out.

In the first experiment, the 31 students attending the class were randomly divided into two groups, one of 10 students (the group of the experiment) and the other of 21 students (the control group). The control group received a normal lecture about the main characteristics of some services in a TCP/IP network. The session took place in a classroom with slides showing a diagram containing multiple servers as a simulated servers' room and was explained by the professor of the course. The group of the experiment was moved to another room with NFC mobile phones. Students interacted with NFC panels as presented in Figure 1b simulating the interaction with the physical servers in the servers' room. The learning objects used were based on video and audio and contained the same information that the professor gave to the control group in class. The students were able to control the learning experience by exploring the panels at their own pace, replaying the contents if needed. No synchronous communication channel was provided to allow students to ask questions to an online instructor.

At the beginning of the session, each student of both groups answered a pre-test with questions about the contents of the experiment. The objective of this pre-test was to measure any prior knowledge that students may have had so that the learning gain could be measured at the end of the experiment. The test consisted of 7 questions that were rated on a scale between 0 and 7. Once the session finished, both groups were asked to answer a post-test containing again 7 questions to assess the learning gain of the process in both groups.

In the second experiment, the 24 students attending the class were again randomly divided into two groups, one of 10 students (the group of the experiment) and the other of 14 students (the control group) taking into account that students in the group of the experiment in the first experiment were not again in the group of the experiment in this second experiment. The methodology was the same. The learning content described the details of the "inetd" service, its configuration files in a Linux system and its component architecture.

The learning gains for the two groups in the two experiments are captured in Table 1 (represented by the difference between pre-test and post-test). The learning gains for the group of the experiment were slightly worse in the first experiment but slightly better in the second one.

In order to assess if the learning gains of the group of control are similar to those of the group of the experiment or not an Anova test was performed to the learning gains for the 4 samples of students (including the two groups in the two experiments). The F value obtained is 1.37 which is smaller than the critical value 2.79 and the p-value obtained is 0.26 which is bigger than the critical value of 0.05. Therefore it can be assumed that the differences in the learning gains are due to random factors.

Table 1. Average increase of results by experiment

Group	First experiment	Second experiment
Group of Experiment	1.9	3.22
Control Group	2.43	2.93

Some intrinsic motivation related factors were also evaluated in these experiments using a questionnaire at the end of them. The questions answered by students were (using a Likert-type scale):

- How would you define your intrinsic motivation towards the use of mobile devices in class? (1. Very small, 2. Small, 3 Neither small nor big, 4. Big, 5. Very big)
- Comparing learning with mobile devices and traditional instructor based classes, what do you enjoy most? (Learning with mobile devices: 1. Much less, 2. Less, 3 The same, 4. More, 5. Much more)
- Would you be willing to do a second experiment like this in the future? (1. Strongly disagree, 2. Disagree, 3 Neither agree or disagree, 4. Agree, 5. Strongly Agree)

Table 2 captures the answers of students. Students recognized that they were motivated by the use of mobile devices in class. Similar results are presented in [14] for the use of mobile phones and podcasts. The use of mobile devices can be used therefore to increase the motivational state of students. Students also reported that they slightly prefer learning with mobile devices instead of learning in a traditional environment. Finally, students also said that they would enjoy doing a similar experiment a second time.

Table 2. Answer of the students to motivation questions

Questions	1	2	3	4	5	Mean
1	0	0.08	0.34	0.5	0.08	3.58
2	0	0.17	0.58	0.25	0	3.08
3	0	0.08	0.17	0.5	0.25	3.92

3.2 Proactive Learning Pills

A second case study has been deployed and evaluated implementing a proactive application for sending learning pills to interested students in class as described in section 2.3. A set of complementary exercises was sent to the mobile phones of 38 students via Bluetooth to reinforce what was taught by the professor in class. A total of 6 learning pills were sent in 6 different classes to students attending the "advanced telematic applications" course at the Carlos III University of Madrid (using a different group of students than in the previous case study). As in the previous case study, both learning gains and motivational impacts have been assessed.

In order to validate if the introduction of proactive learning pills in class promotes an increase in learning gains of students of the course (in a scale from 0 to 10), the group of 60 students was divided in a control group (those students with no o little use of the learning pills) and the group of the experiment (those students consuming at least half of the pills). The learning gains were assessed using a pre-test at the beginning of the course and a post-test at the end. The results are captured in table 3. The results clearly show that students regularly consuming the learning pills obtained a bigger increase in their learning outcomes. This increase is also influenced by other factors that were also present in the experiment such as the fact that students receiving the pills were also physically attending the lectures where many of the students that did not receive the pills were not.

Table 3. learning gains for students consuming at least half of the pills and the rest of students

Group	Learning gain
Group of Experiment	5.4
Control Group	1.6

The motivational impact that the use of the learning pills had on students was assessed using a questionnaire. The answers of the students to the question: "Does the use of mobile phones for receiving learning pills increase your motivation for the subject?" using a Likert-type scale (1. Not at all, 2. Very Slightly, 3. Somehow, 4. Significantly, 5. Very much) is captured in table 4. The results are divided into two groups: students consuming at least half of the pills (frequent users) and the rest of the students (sporadic users). Although there is a bigger impact on students frequently consuming the pills, event sporadic users of the pills considered more than significant the motivational impact that the learning pills had on them.

Table 4. motivational impact for students consuming at least half of the pills and the rest of students

Group	Motivational impact
Frequent users	4.18
Sporadic users	4.07

4 Conclusions

This paper has studied and analyzed the basic components of pervasive contextualized learning ecosystems defining and implementing some learning objects, services and proactive environments. The paper has also presented how the use of pervasive learning smart objects and services can create contextualized learning ecosystems that enhance both learning outcomes and motivational states of students interacting with them using mobile personal devices. This paper has presented the results of two case studies that implemented pervasive learning ecosystems in which students at the Carlos III University of Madrid interacted with contextualized learning objects and services. These experiments have provided data both about the learning outcomes for students after interacting with smart learning objects and services and about the motivational impact that the use of these technologies have on them.

Acknowledgements. The research leading to these results has been partially funded by the ARTEMISA project TIN2009-14378-C02-02 within the Spanish "Plan Nacional de I+D+I", the Madrid regional community projects S2009/TIC-1650 and CCG10-UC3M/TIC-4992 and the SOLITE CYTED Program 508AC0341. Thanks to INNOVISION for providing the NFC tags for this experiment. Gustavo Ramirez-Gonzalez is funded by the EU Programme Alban, scholarship number E06D101768CO and by the Universidad del Cauca.

References

1. Sakamura, K., Koshizuka, N.: Ubiquitous computing technologies for ubiquitous learning. In: IEEE International Workshop on Wireless and Mobile Technologies in Education, WMTE 2005, November 28-30, pp. 11–20 (2005)
2. International Telecommunication Union ITU, ITU Internet Reports 2005: The Internet of Things (2005)
3. Wiechert, T., Thiesse, F., Michahelles, F., Schmitt, P., Fleisch, E.: Connecting Mobile Phones to the Internet of Things: A Discussion of Compatibility Issues between EPC and NFC. In: Americas Conference on Information Systems (AMCIS), Keystone, Colorado, USA (2007)
4. Welbourne, E., Battle, L., Cole, G., Gould, K., Rector, K., Raymer, S., Balazinska, M., Borriello, G.: Building the Internet of Things Using RFID: The RFID Ecosystem Experience. IEEE Internet Computing 13(3), 48–55 (2009)
5. Holzinger, A., Nischelwitzer, A., Meisenberger, M.: Mobile Phones as a Challenge for m-Learning: Examples for Mobile Interactive Learning Objects (MILOs). In: IEEE International Conference on Pervasive Computing and Communications Workshops, Third IEEE International Conference on Pervasive Computing and Communications Workshops, Hawai, pp. 307–311 (2005)
6. Ogata, H.: Computer Supported Ubiquitous Learning: Augmenting Learning Experiences in the Real World. In: Fifth IEEE International Conference on Wireless, Mobile, and Ubiquitous Technology in Education (wmute 2008), Beijing, China, pp. 3–10 (2008)
7. Hsieh, H.C., Chen, C.M., Hong, C.M.: Context-Aware Ubiquitous English Learning in a Campus Environment. In: 7th IEEE International Conference on Advanced Learning Technologies (ICALT 2007), Niigata, Japan, pp. 351–353 (2007)

8. Liu, T.Y., Chu, Y.L., Tan, T.H., Chang, C.C.: RFID-based Ubiquitous Learning Environment for Outdoor Learning. In: 7th IEEE International Conference on Advanced Learning Technologies (ICALT 2007), Niigata, Japan, pp. 675–677 (2007)
9. Wu, T.T., Yang, T.C., Hwang, G.J., Chu, H.C.: Conducting Situated Learning in a Context-Aware Ubiquitous Learning Environment. In: Fifth IEEE International Conference on Wireless, Mobile, and Ubiquitous Technology in Education, Beijing, China, pp. 82–85 (2008)
10. Bruner, J.S.: The Act of Discovery. Harvard Educational Review 31, 21–32 (1961)
11. Brown, J.S., Collins, A., Duguid: Situated Cognition and the Culture of Learning. Educational Researcher 18(1), 32–42 (1989)
12. Barrows, H.S., Tamblyn, R.M.: An evaluation of problem-based learning in small groups utilizing a simulated patient. Journal of Medical Education 51, 52–54 (1976)
13. Newell, A., Simon, H.A.: Human Problem Solving. Prentice-Hall, Englewood Cliffs (1972)
14. Moura, A., Carvalho, A.A.: Mobile Learning: Teaching and Learning with Mobile Phones and Podcasts. In: Eighth IEEE International Conference on Advanced Learning Technologies, Santander, Spain, pp. 631–633 (2008)
15. An, J.: Activity Theory for Designing Ubiquitous Learning Scenarios. In: Innovative Techniques in Instruction Technology, E-learning, E-assessment, and Education, pp. 205–209. Springer, Netherlands (2008)
16. Ortiz Jr., S.: Is near-field communication close to success? IEEE Computer Publication, Los Alamitos (2006)
17. Yan, L., Zhang, Y., Yang, L.T., Ning, H., Zhang, Y.: The Internet of Things From RFID to the Next-Generation Pervasive Networked Systems. Auerbach Publication (March 2008)
18. Garcia-Zubia, J., Lopez-de-Ipia, D., Ordua, P.: Mobile Devices and Remote Labs in Engineering Education. In: Eighth IEEE International Conference on Advanced Learning Technologies, ICALT 2008, pp. 620–622 (2008)
19. Gonzalez, G.R., Organero, M.M., Kloos, C.D.: Early Infrastructure of an Internet of Things in Spaces for Learning. In: Eighth IEEE International Conference on Advanced Learning Technologies, ICALT 2008, July 1-5, pp. 381–383 (2008)
20. Kranz, M., Schmidt, A., Holleis, P.: Embedded Interaction - Interacting with the Internet of Things. In: IEEE Internet Computing Online (December 2009) (accepted for future publication), doi:10.1109/MIC.2009.141
21. Muñoz-Organero, M., Muñoz-Merino, P.J., Delgado Kloos, C.: Enabling M-Learning Applications In MIDP Devices: Specification And Development Of An HTML To MIDP Automatic Converter. In: International Technology, Education and Development Conference, Valencia (March 2007)
22. Muñoz-Organero, M., Delgado Kloos, C.: Web-enabled middleware for mobile devices. In: International Wireless Applications and Computing 2007 Conference, Lisbon, Portugal, July 6-8 (2007)
23. Ramírez, G., Organero, M.M., Kloos, C.D., Astaiza, Á.C.: Exploring NFC Interactive Panel. In: The Fifth Annual International Conference on Mobile and Ubiquitous Systems: Computing, Networking and Services (MobiQuitous 2008), Dublin, Ireland, July 21-25 (2008)

On the Use of Learning Object Metadata: The GLOBE Experience

Xavier Ochoa[1], Joris Klerkx[2], Bram Vandeputte[2], and Erik Duval[2]

[1] Information Technology Center, Escuela Superior Politecnica del Litoral, Guayaquil, Ecuador
xavier@cti.espol.edu.ec
[2] Dept. Computerwetenschappen, Katholieke Universiteit Leuven, Celestijnenlaan 200A, B-3001, Heverlee, Belgium
{Joris.Klerkx,Bram.Vandeputte,Erik.Duval}@cs.kuleuven.be

Abstract. Since IEEE LTSC LOM was published in 2002, it is one of the widest adopted standard for the description of educational resources. The GLOBE (Global Learning Objects Brokered Exchange) alliance enables share and reuse between several Learning Object Repositories worldwide. Being the largest and more diverse collection of Learning Object Metadata, it is an ideal place to perform an analysis of the actual use of the LOM standard in the real world. This paper presents an in-depth analysis of the use and quality of 630.317 metadata instances.

Keywords: Interoperability, Learning Object Metadata.

1 Introduction

The Learning Object Metadata standard for describing learning resources has been published in 2002. It is based on early metadata schemes that were developed by the ARIADNE Foundation [1] and the IMS Global Learning Consortium [2]. Main goal of LOM is "to facilitate search, evaluation, acquisition, and use of learning objects, for instance by learners, instructors or automated software processes" [3] through the description of a common set of metadata elements. LOM proposes around 50 different metadata elements grouped into nine categories: General, Lifecycle, Meta-Metadata, Technical, Educational, Rights, Relation, Annotation and Classification.

As important as the LOM standard is for sharing and reusing learning materials, very little research has been performed on the actual use of the standard in real-world situations. To our knowledge, only three studies have been performed on the use of LOM. Neven and Duval [4] made a first survey of LOM based repositories, but only considered the use of qualitative criteria about the repository infrastructure and did not consider the actual use of metadata. Najjar and Duval [5] performed a quantitative study of the use of metadata for indexing objects in ARIADNE. While the original ARIADNE format was similar to LOM, it was simpler, lacked several key elements (e.g. description and learning resource type) and had different ways to store the information (i.e. the

C. Delgado Kloos et al. (Eds.): EC-TEL 2011, LNCS 6964, pp. 271–284, 2011.

fixed Semantic Section in ARIADNE versus the flexible Classification category in LOM). Another effort to characterize LOM usage was performed by Friesen [6], who collected samples from several repositories worldwide to study their usage of LOM. However, this survey has deep methodological flaws (e.g the metadata was not randomly selected, but hand-picked in each repository) and was based on a very small sample that weakens the validity of the conclusions drawn by the author. After this last work, there has been a large silence in the research community about the actual use of LOM.

This lack of research has not stopped the wide adoption of LOM as the way in which most Learning Object Repositories (LORs) represent their metadata internally, or the way in which they interchange their metadata with search facilities and other repositories. The largest organization that uses LOM as a common medium to enable the sharing of learning materials is GLOBE (Global Learning Objects Brokered Exchange). Jointly, nearly 1,2 million learning objects are shared inside the GLOBE alliance. Being the largest and more diverse collection of Learning Object Metadata, GLOBE is an ideal place to perform an analysis of the actual use of the LOM standard in the real world. This paper presents a large-scale study of the use and quality of more than 50% (630 317) of LOM instances in GLOBE.

We start this paper in Section 2 with information about the collection of data. Section 3 presents basic statistics about the metadata. A study how the various LOM elements and their values have changed over time the presence of new data elements is presented in Section 4. Section 5 performs a deeper study on the (mis-)use of LOM and the actual information that is contained in the metadata. We conclude in Section 6 with lessons learned and future work.

2 Data Collection

To obtain a representative sample of LOM metadata, approximately half of the instances of GLOBE has been obtained by using the OAI-PMH harvester tool developed in ARIADNE. In total, 630 317 LOM metadata instances were harvested from those GLOBE repositories that provide a metadata harvesting service based on the OAI-PMH protocol [7]. All data, code and figures used in this work can be downloaded[1]. The obtained metadata have a high degree of diversity which is the result of different methodologies used to create and aggregate metadata in the different repositories. For example, KOCW, LACLO and OUJ expose a single collection of metadata instances with a common provenance. Other GLOBE repositories expose the result of the aggregation of several metadata collections that have different provenance. This is the case of ARIADNE, where only a 5% of its objects has been produced by its members and the remaining 95% has been collected from other repositories worldwide. LRE, OER and LORNET also aggregate metadata instances from different provenances, but those are altered and enriched after-collection to follow a repository-wide standard. Another example of diversity is the methodology used to create the metadata. While most

[1] http://ariadne.cs.kuleuven.be/lomi/index.php/LOM_in_GLOBE

Table 1. LOM Repositories studied

Repository	Instances	Provenance	Creation
ARIADNE Foundation	374 857	Aggregated	Mostly Manual
Learning Resource Exchange (LRE)	169 736	Enriched	Manual
Community on Learning Objects (LACLO)	49 943	Single	Automatic
OER Commons (OER)	25 794	Enriched	Manual
Korean OCW (KOCW)	7 183	Single	Manual
LO Repository Network (LORNET)	1 804	Enriched	Manual
Open University Japan (OUJ)	1 000	Single	Manual

metadata are created manually, some repositories, like ARIADNE, have a non-negligible amount of instances that were produced by semi-automatic metadata generators [8]. An extreme case is LACLO which is exclusively composed of automatically generated metadata. While this diversity makes it harder to analyze the metadata instances as a whole, it is unavoidable in a real-world scenario and it makes the result of our analysis richer and more relevant. Table 1 lists the details from those repositories and the number of instances obtained from them.

3 Basic Statistical Analysis

3.1 LOM Size

The first useful analysis that can be conducted over the collected data is to determine the average size of a LOM instance. The size of the instance is determined by the binding in which it is represented. In this study, we used the XML binding of LOM. The arithmetic mean of the size of the XML files is 4,25 Kb. However, the distribution of the sizes does not follow a normal distribution. The size distribution for each repository, as well as for the aggregated set, presents a right tail (positive skewness). As suggested in literature [9], this distribution can be approximated (but not exactly fitted) by a log-normal distribution, meaning that if the logarithm of the file size is taken, the obtained values follow approximately a normal distribution. An ANOVA analysis of the size of the files in the different repositories shows that there is no major difference in the size of the LOM instances between the GLOBE repositories. The average size and the distribution of values could therefore be used to model the space requirements for LORs and the capacity needed to interchange those instances over the network.

3.2 LOM Data Elements Usage

Our second analysis is a frequency analysis of the data elements in LOM. For this analysis, only top-level fields were counted. For example, in the XML binding of LOM, the field *General.Structure* has two subfields: *General.Structure.source* and *General.Structure.value*. In this case, only the number of appearances of *General.Structure* is counted. The justification for this is that not all repositories have the *source* subfield and, in most cases, the number of appearances of

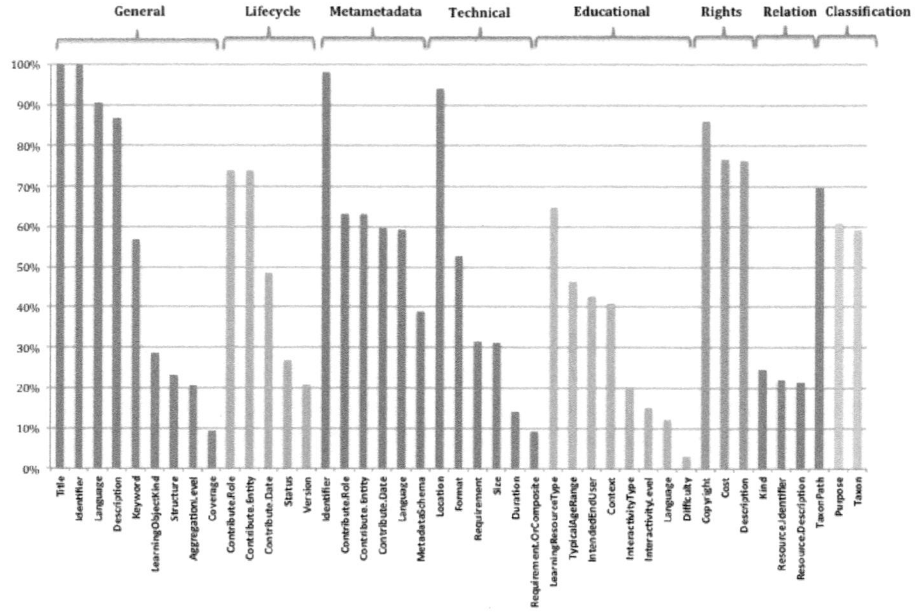

Fig. 1. Percentage of Usage of Different LOM Data Elements in GLOBE

top-level fields is equal to the number of appearances of the *value* subfield. If the data element is filled more than once, it is counted only once. Figure 1 shows the percentage of metadata that have a value for the different data elements of LOM in GLOBE.

Main finding of this study is that only a small fraction of the LOM standard is frequently used to describe the learning objects. Only 20 of the 50 data elements are used more than 60% of the time. Also, 16 data elements are used less than 10% of the time. At first sight, this result seems to corroborate the Friesen study [6] conclusion that the added complexity of LOM is not used in the real-world. However, a similar study performed by Wand [10] over almost 1 million metadata instances from the Open Archives Initiative (OAI) that uses the much simpler Dublin Core (DC) [11] metadata schema, found that of 15 DC data elements, five (creator, identifier, title, date, and type) are used 71% of the time and the five least used elements (language, format, relation, contributor and source) are used less than 6% of the time. Contrasted with this last study, it seems that LOM, while being more complex, helps to collect proportionally more information than a simpler and more general schema such as DC. The inequality of usage of the different data elements seems to be something inherent to the creation of metadata. This inequality deserves further research, through a comparative analysis of the use in different metadata standards.

Considering that LOM is specifically designed to describe educational material, it is important to review the use of the Educational section. 4 out of 11 educational data elements (Learning Resource Type, Intended End User Role, Typical Age Range and Context) are used more than 40% of the time, 3 elements

TOTAL	LACLO	ARIADNE	LORNET	LRE	OER	KOCW	OUJ	Path
0.69	0.96	0.51	0.84	0.97	1.0	0.0	1.0	educational
0.54	0.83	0.39	0.75	0.76	0.84	0.0	0.83	educational.context
0.03	0.13	0.03	0.11	0.01	0.0	0.0	0.07	educational.description
0.03	0.0	0.03	0.0	0.05	0.0	0.0	0.0	educational.difficulty
0.57	0.83	0.36	0.37	1.08	0.0	0.0	0.0	educational.intendedenduserrole
0.15	0.0	0.24	0.01	0.03	0.0	0.0	0.0	educational.interactivitylevel
0.2	0.0	0.33	0.02	0.02	0.0	0.0	0.0	educational.interactivitytype
0.13	0.0	0.09	0.02	0.21	0.32	0.0	0.0	educational.language
0.73	0.96	0.55	0.71	1.01	1.26	0.0	0.0	educational.learningresourcetype
0.02	0.0	0.03	0.0	0.0	0.0	0.0	0.0	educational.semanticdensity
0.56	0.83	0.38	0.01	0.8	1.1	0.0	0.0	educational.typicalagerange
0.02	0.0	0.01	0.1	0.04	0.0	0.0	0.0	educational.typicallearningtime

Fig. 2. Heat map table comparing the usage of educational elements

(Language, Interactivity Level and Interactivity Type) are used more than 10% but less than 20%, and the 4 remaining elements (Description, Difficulty, Semantic Density and Typical Learning Time) are used less than 10% of the time. While these values are lower than desirable, they provide proof that LOM is used in the real-world to capture educational information about digital objects. Current research on automatic generation and enrichment of LOM that is based on the context and use of the resource [8] seems to be a way to increase the completeness of the Educational section. To easily find differences in the usage of the educational data elements across the different providers, a heat map view is presented in Figure 2. It presents the relative frequency of the use of a given data element in each studied repository. Values higher than 1 represent that the described element is commonly used more than once in each instance. The first column is the average relative frequency over all the repositories. At this level, it is easy to see the large diversity with respect to the use of the Educational section of LOM that was hidden in the previous general analysis. There are two different repositories, LRE (manually enriched metadata) and LACLO (automatically generated metadata), that use several educational data elements in more than 70% of the cases. Other repositories, such as KOCW and OUJ provide almost no information in this section. ARIADNE, being in itself a collection of several other repositories, mirrors very closely the results of GLOBE as a whole. This suggests, by an admittedly unsafe extrapolation, that the use of the Educational section found in the studied set would be similar to the one existent in the theoretical universe of LOM instances.

3.3 LOM Vocabulary Usage

Vocabulary elements are data elements that can only be filled with a limited set of values established by the metadata standard. The main goal of these vocabulary elements is to provide a higher level of semantic interoperability [3]. In this analysis, the distribution of permitted values for the vocabulary elements has been obtained by parsing the XML instances in a similar way than in our previous studies. Based on the distribution of their values, it was easy to identify at least 3 main groups of vocabulary elements. The Heavy-Tailed group is

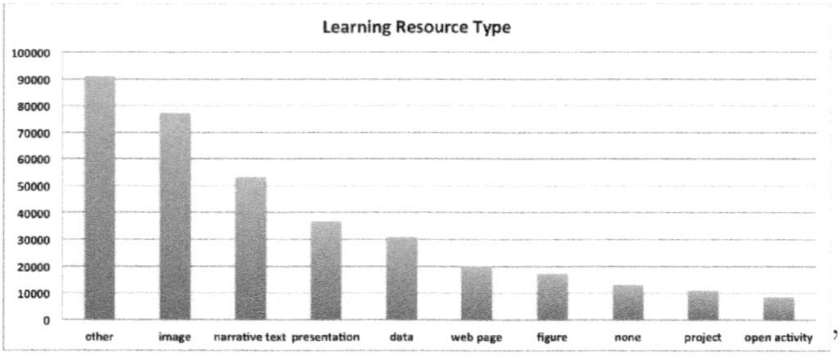

(a) Heavy-Tailed

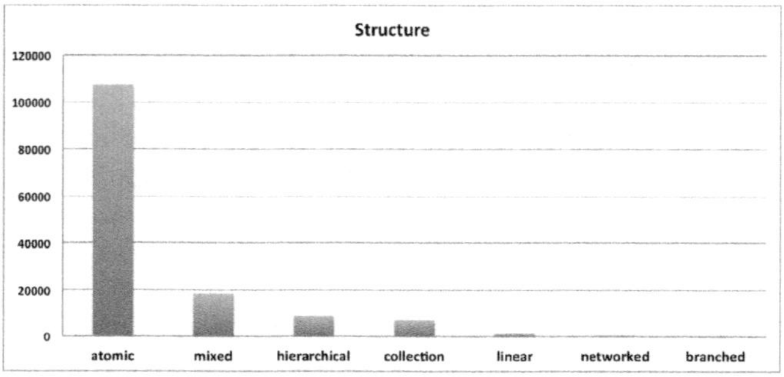

(b) Light-Tailed

Fig. 3. Different Vocabulary Elements Groups based on their Distribution

characterized by two or three values that are heavily used (more than 25% of time), followed by several values with a lower, but still significant usage between 5% and 25%. An example of this distribution can be seen in Figure 3a. The Language (of the resource and the metadata), Format, Resource Type and Context elements are part of the Heavy-Tailed group. The Light-Tailed group is characterized by a dominant value (usage of more than 70% of the time), followed by few (2 or 3) values that are used more rarely (from 2% to 10% of the time). An example of this distribution can be seen in Figure 3b. Structure, Intended User Role, Interactive Type and Status seem to be part of the Light-Tailed group. The third group, called Symmetrical group, contains all vocabulary elements that are based on a scale. In this group, the central value of the scale is by far the most common value (60% to 70%). Difficulty, Interactivity Level, Semantic Density and Aggregation Level are part of this group. Other vocabulary fields, such as Cost, Copyright, Classification Purpose, Contributor Role, etc. lie between these groups.

One possible explanation for these field distributions involves the objectivity of the information and the use of default values. The elements in the Heavy-Tailed group, contains information that could be easily obtained from the object or the context where it is used. For example, it is easy for the indexer to determine the

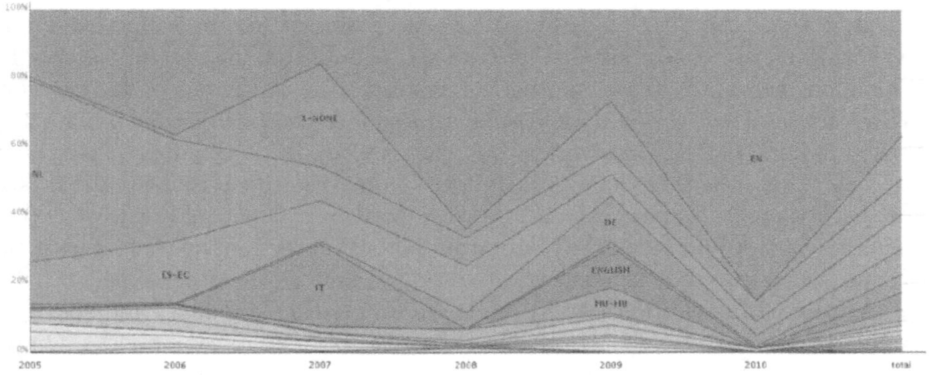

Fig. 4. Change over time of the values for general.language

language of the element, specifically because it is most probably in a language that he knows. On the other hand, data elements in the Light-Tailed group, seem to be more difficult to determine objectively (such as the Structure of the object) or there is safe default value for most objects (for example, it is safe to assume that the Intended End User is the learner). The existence of a safe default (usually the middle value) seems to be also a possible explanation for the value distribution in the elements in the Symmetrical group. Although there is anecdotal evidence to support this interpretation, further quantitative and qualitative studies of the indexation process are needed in order to provide a strong explanation.

4 LOM Evolution and Expansion

The previous section showed basic statistics about the GLOBE metadata. This section presents the use of LOM data elements through time to find existing trends in metadata indexation. To investigate this, only those metadata instances with value for *lifecycle.contribution* date have been considered. "Many Eyes", a set of data visualization tools provided by IBM, has been used to create stack graphs for the values of the data elements which is meant for visualizing the total change over time of a group of quantities. Each ribbon of color in the stack graph represents a data item changing over time. Figure 4 shows the change over time of the values for *general.language*. This figure shows a number of trends:

In 2005, more than 50% of investigated metadata instances were in dutch. After 2005, a decrease is noticed by 40%. This raises the question why so many instances were indexed in dutch around 2005. One explanation could be the uptake of sharing educational resources by several institutions such as ARIADNE and Klascement in Flanders and the Netherlands around that time. Following this reasoning, organizations from Italy started to share their resources in 2007. English is the most used language for educational resources which is obvious.

Note however, that besides "en" also "english" is added as a value. Looking back to the data, "eng" was found for a minority of resources. Developers of applications that make use of this data set, should thus be careful to consider this interoperability issue. Another interesting trend is the use of the category relationship. There are several possible values for a relationship type provided by the standard such as "has worked on", "is required by", "requires", "is version of", etc. However, there are only two types in use that both take up about 50% of the instances: "is part of" and "is referenced by". One could at least expect that indexers would index the relationship in both directions so that at least the types "has part" and "references" are used as much as the other types.

Application profiles (APs) are based on existing standards and specifications [12] but can add elements if needed for the purpose of the provider. One of our goals was to list elements used by the GLOBE providers that are not in the LOM standard but were added by extending the standard in one or more application profiles. The following 3 elements were found to be the most significant additions in the GLOBE metadata set:

- *general.learningobjectkind* is added in the MACE Application Profile to distinguish between a real world object (i.e. a building) or a media object (digital resource) that e.g. describes that real world object. As this is a mandatory field in the AP, 34.532 metadata instances were found in the investigated set that made use of this field.
- *technical.geolocation* is also added in the MACE AP to be able to add geo-coordinates in the case of a real world object such as a building. This field enabled the creation of typical mobile applications where users walk through a city and get extra information about the buildings in their neighborhood. However, a mismatch was found between the number of geo-locations (7 369) and the number of real world objects (11 622).
- *general.subtitle* is added by the AP of KOCW. It has a value in 73% of their metadata instances. In LOM, only 1 title is allowed with one or more translations of that title which is obviously not enough in the case of KOCW.

5 Metadata Quality Metrics

5.1 LOM Compliancy

As the amount of metadata instances grows, so does the need for automatic validation against agreed rules of conformance, based on an application profile for example. An open source tool, the ARIADNE validation service, has been developed for this purpose. On the GLOBE metadata set, the conformance against the IEEE LOM loose and LOM strict bindings have been checked. LOM loose mostly focuses on validating the structure of the elements in the LOM namespace, checking for patterns of datatypes like date stamps, and the peculiarities of the vCard structure. The vCard standard is used to represent information about an individual, as defined by the IETF proposed standard RFC 2426[2].

[2] http://www.ietf.org/rfc/rfc2426.txt

LOM Strict has the same set of rules as LOM loose, with additional constraints on the used vocabularies. Furthermore, LOM strict does not allow to extend the structure with new elements. Figure 5 shows a summary of the validation error rate in total and split down per provider. The overall error rate against LOM Loose is 30,1%, which means that 30% of the records can cause issues when processing these with tools that assume LOM compliancy. Looking at the error rate per provider, it can be seen that OUJ, LORNET and KOCW have an error rate of over 80%. Also striking is the error rate of the LOM strict validation. Only 4% of all records are fully LOM Strict compliant. These numbers are explored in more detail further down this section. The validation errors can be split up in the following categories:

- Pattern matching errors : Certain datatypes have very specific syntax constraints. These are checked with pattern matching rules. All kinds of pattern matching errors are put in this category. An example mistake in this category is a wrongly denoted timezone in a date stamp. For example, "GMT-5" should be written as "-05:00".
- vCard errors : This category contains all errors concerning the vCard standard. This datatype is used in the lifecycle.contribute.entity and metaMetadata.contribute.entity elements.
- LOM extension errors : When extending LOM in an application profile, elements can be added using a separate namespace and different vocabularies can be used. However they can not be used inside the same LOM namespace, otherwise they are seen as a validation error. Furthermore, as said earlier, LOM strict imposes a fixed set of allowed values for all elements with a vocabulary datatype. Thus, all other values are considered invalid when validating against LOM strict.
- Various errors : This contains all other types of errors. Mostly these are structural errors, e.g. putting a child element in a parent element must only contain plain text.

A more in depth look at the validation errors against LOM loose is shown in Figure 6. The most prominent error is a vCard error, where the vCard is missing 1 of 3 mandatory elements (i.e. N, FN or VERSION). It represents 68% of all LOM loose validation errors. vCard errors in general are the most common

		LACLO	OUJ	ARIADNE	LORNET	LRE	OER	KOCW	TOTAL
	total records	49943	1000	374857	1804	169736	25794	7183	630317
LOMLoose	invalid records	145	878	165616	1790	7191	8293	5794	189707
	error rate	0,3%	87,8%	44,2%	99,2%	4,2%	32,2%	80,7%	30,1%
LOMStrict	invalid records	49922	990	351508	1800	169736	25794	7183	606933
	error rate	100%	99%	93,8%	99,8%	100%	100%	100%	96%

Fig. 5. A summary of error rates in GLOBE against LOM loose and LOM strict

	LACLO	OUJ	ARIADNE	LORNET	LRE	OER	KOCW	TOTAL
datetime	0,0%	0,0%	9,6%	0,0%	0,0%	100,0%	32,2%	13,2%
duration	0,0%	0,0%	0,7%	0,0%	0,0%	0,0%	0,0%	0,6%
language	2,1%	0,0%	2,2%	0,0%	6,4%	0,0%	0,0%	2,2%
other	11,0%	0,0%	0,0%	0,0%	0,0%	0,0%	0,0%	0,0%
declaration	0,0%	0,0%	0,0%	0,0%	59,6%	0,0%	0,0%	2,8%
no mandatory	0,0%	0,0%	77,0%	100,0%	20,3%	0,0%	6,6%	68,4%
no value	86,9%	0,0%	0,0%	0,0%	0,0%	0,0%	0,0%	0,1%
other	0,0%	0,0%	0,1%	0,0%	0,0%	0,0%	0,0%	0,1%
wrong namespace	0,0%	0,0%	0,0%	0,0%	0,0%	0,0%	61,2%	2,4%
other	0,0%	100,0%	10,4%	0,0%	13,6%	0,0%	0,0%	10,2%

Fig. 6. Validation results against LOM loose. Blue represents pattern matching errors, green, vCard errors, red, LOM extension errors and orange, other errors.

	LACLO	OUJ	ARIADNE	LORNET	LRE	OER	KOCW	TOTAL
datetime	0,0%	0,0%	1,1%	0,0%	0,0%	4,5%	7,4%	0,7%
duration	0,0%	0,0%	0,1%	0,0%	0,0%	0,0%	0,0%	0,0%
language	0,0%	0,0%	0,2%	0,0%	0,0%	0,0%	0,0%	0,1%
other	0,0%	0,0%	0,0%	0,0%	0,0%	0,0%	0,0%	0,0%
declaration	0,0%	0,0%	0,0%	0,0%	0,3%	0,0%	0,0%	0,1%
no mandatory	0,0%	0,0%	8,5%	22,2%	0,1%	0,0%	1,5%	3,5%
no value	0,1%	0,0%	0,0%	0,0%	0,0%	0,0%	0,0%	0,0%
other	0,0%	0,0%	0,0%	0,0%	0,0%	0,0%	0,0%	0,0%
source vocabulary	45,4%	39,5%	30,9%	27,8%	75,0%	52,4%	62,9%	55,6%
value vocabulary	54,5%	0,7%	46,2%	27,8%	24,5%	43,1%	14,2%	34,5%
wrong namespace	0,0%	0,0%	0,0%	0,0%	0,0%	0,0%	14,0%	0,1%
other	0,0%	59,8%	13,1%	22,1%	0,1%	0,0%	0,0%	5,3%

Fig. 7. Validation results against LOM strict. Blue represents pattern matching errors, green, vCard errors, red, LOM extension errors and orange, other errors.

errors, spread across almost all providers. In 4 out of 7 providers, vCard errors represent at least 60% of all errors. Only 2 providers have no vCard errors at all. Further analysis has shown that 79% of the invalid records contain at least one vCard error. Solving only this vCard error would turn at least 68% of the invalid records into valid records. This indicates that vCards have been wrongly used across all providers. When analyzing the error type of the providers with an error rate of over 80%, it was found that 100% of the errors of LORNET are about a missing mandatory element in a vCard. Looking at OUJ's errors reveals that all of these are structural errors, by putting a "string" or a "entry" element inside an invalid parent element. Finally, validation errors of KOCW are largely due to extending LOM in the wrong way, by adding a new element without using a new namespace for this element. When examining the validation errors against LOM strict, as shown in Figure 7, a completely different picture can be seen. Here, 90% (55,6% source vocabulary and 34,5% value vocabulary errors) of the validation errors are caused by the "LOM extension" category. Hence it is also no surprise that for 6 of the 7 providers, this same category contains more than 50% of the errors. By looking at a number of randomly hand-picked errors, it becomes clear that the larger the amount of invalid records for one error, the bigger the chance that the error is more syntactical in nature. A possible explanation could be that these errors are produced by (semi-)automatically generated metadata or by programming bugs in client tools. To conclude this section, we can safely say that the vCard standard is very error prone, and not very portable. One of the main reasons is its syntax. It relies on new line characters, it is heading space sensitive, etc. There are also many DateTime formatting issues. One cause is an inconsistency in the XML binding of LOM and the ISO 8601:2000 standard it claims to adhere to. When a TZD (TimeZone Designator) is present, the digits representing a decimal fraction of a second become mandatory in LOM, whereas this should actually not be the case. Finally, a major problem with metadata validation is that providers are usually not aware of these errors, and often they do not use validators to check their own compliancy. People are also not aware of how to properly extend LOM, and they do not understand properly how xml namespaces work.

5.2 Information Content

One of the main purposes of any metadata schema is to carry information about the object that is described. One method that has been extensively used to determine the informativeness of a metadata instance is to measure its completeness; the presence or absence of a value for the different data elements of the metadata schema. However, the mere existence of a value in 1 of the data elements does not necessarily mean that this value is carrying information. For example, if all metadata instances in the repository have the same title, that data is useless to find and select a particular object within that repository. On the other hand, if values for a data element are well distributed among the metadata instances, the information carried by that element is maximized and that field could be easily used to filter and select elements from the repository. Shannon was the first to propose variety or disorder as a measurement of information content [13]. He proposed "entropy" as a measurement of information contained in a message. This subsection will apply different measurements to measure entropy of values contained in various types of data elements of LOM in order to determine the amount of information they carry.

Vocabulary Data Elements. When a variable can take only a determined set of values, the information content of that variable can be determined by its entropy. The entropy is equal to the sum of the probability of all given values, multiplied by the negative of the logarithm of the value probability. If the logarithm used is base-2, entropy is measured in bits. This concept is applied to determine the entropy of the vocabulary data elements [14]. For each of the studied repositories, entropy values have been calculated for the 14 vocabulary data elements presented in Table 2. If the repository does not have any value for that element it is marked with an X. This entropy value depends on the variety of the distribution of values in the data elements. For example, in LACLO, all instances have "es-EC" as the value for "Language". The calculated entropy for this case is 0 bits, as there is no variety in the data. On the other hand, LRE has materials from various european countries which makes its content highly multi-lingual. As expected, entropy value for "Language" in LRE is the highest in GLOBE (3,43 bits). The entropy value also depends on the number of choices for each element. For example, the maximum entropy for the element Cost, that could only take two values (Yes or No) is 1 bit, while the maximum entropy for the element Difficulty that could take five values is 2,32 bits. Because of this, the value of entropy is only comparable between elements with the same number of possible values.

It can be easily deduced from Table 2 that LRE has the most diverse information in their vocabulary elements. One explanation for this observation is the process of metadata enrichment that has been performed over most metadata [15] in that repository. On the other hand, LACLO instances carry very little information because all the values for an element are the same. The lack of sophistication of the automatic algorithm used to generate the metadata of this repository (for example adding the Aggregation Level atomic to all its instances) is responsible for these results. However, LACLO has the highest entropy value for "Format", which indicates that the automatic metadata generator seems to be better equipped than manual indexers to obtain the MIME type of digital objects. The main conclusion that can be obtained from this analysis is that quality control processes, such as the enrichment conducted in the LRE repository, helps to improve the amount of information contained in the metadata instance.

Free Text Data Elements. When data elements can be filled with free text, traditional entropy measurement is not possible as the amount of values that the text could take is infinite. However, the entropy formula could be adapted to count the information carried by each word in the free text elements, divided by the number of words used. In this way, a value of entropy can be calculated for each instances based on the text that they contain. The Relative Entropy metric provides such adaptation [16]. Figure 8 presents the distribution of the Relative Entropy metric for each repository. In general, all repositories have a main peak between 6 and 12 bits per word. Repositories that are the aggregation of other repositories present one or more lower peaks at different values (e.g. at 6 and 17 for LRE). While in this study, the enriched LRE again contains more information per word in their text fields, the automatically generated instances from LACLO, that took their free text from the text available in the context where the object was published, seems to contain a similar amount of information than other repositories with manually generated metadata such as LORNET, OUJ and OER.

Table 2. Entropy value for the Vocabulary Data Elements in each Repository

	ARIADNE	LRE	LACLO	OER	KOCW	LORNET	OUJ
Language	2,22	3,43	0	0,52	1,38	1,10	0,70
Structure	1,65	0,60	X	X	X	1,39	X
Aggregation Level	1,09	1,85	0	X	X	1,30	0,21
Status	0,99	0	X	X	X	1,09	X
Format	2,72	X	3,04	2,10	0,07	2,70	2,27
Interactivity Type	0,71	1,55	X	X	X	1,02	X
Learning Resource Type	2,87	3,04	1,76	3,61	0	1,05	X
Interactivity Level	1,18	2,02	X	X	X	1,49	X
Semantic Density	1,51	X	X	X	X	0	X
Intended End-User Role	1,56	1,91	X	X	0	2,42	X
Context	2,62	2,23	X	1,43	0	2,55	0,14
Difficulty	1,24	1,71	X	X	X	0	X
Cost	0,12	0,01	0	0	0	0,69	0
Copyright	0,82	0,55	0	0,30	0	0,94	X

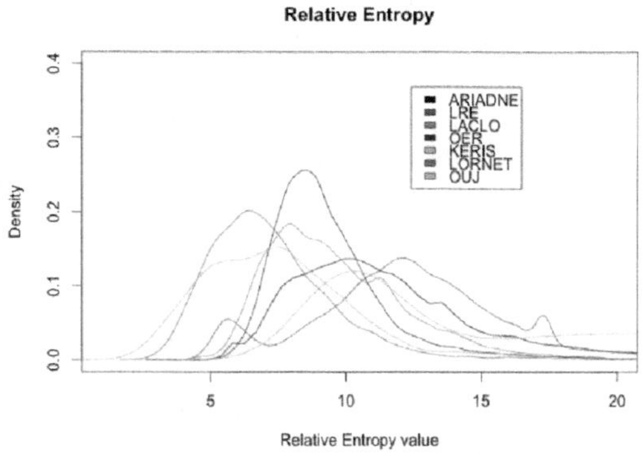

Fig. 8. Distribution of the Relative Entropy (RE) metric

6 Conclusions and Future Work

The main conclusions of this work can be summarized in the following points:

- Out of 50 LOM elements, 20 are used consistently in GLOBE. This is in itself more metadata than traditionally collected metadata in simpler schemas such as DC. Results of this analysis can be used to improve search facilities, focusing on data that is actually contained in the metadata.
- 4 out of 11 educational elements are used in average. However, their actual use is repository/community dependent. The LOM Working group and repository managers should analyze the reasons behind different levels of adoption.
- 3 main extended data elements have been found. These can be a starting point for a LOM Working group discussion about the merit of inclusion of new elements in the next version of the standard. A more exhaustive analysis of Application Profiles should be performed to get a broader understanding of the need for additional elements.
- Solving the vCard related errors would turn at least 68% of the LOM loose invalid records into valid records. Also, in GLOBE it does not make sense to adhere to the LOM strict binding, as only 4% of all records in GLOBE are fully compliant with LOM strict.
- The information content of a record is strongly related with both the quality management process implemented by the repositories and the inherent capabilities of different types of metadata generation.

This work also rises new questions that require further research in order to be answered:

- What is the mechanism behind the usage level of different metadata elements in a given community?
- The distribution of values in the fields follows patterns. What are the forces that shape those distributions?

It is the perception of the authors of this work that this kind of studies should be made an integral part of the development of LOM (or any other metadata schema) solution. Only by analyzing actual use of different schemas in real life settings, decisions can be taken to improve or adapt those schemas so that teachers and learners can really benefit from them.

References

1. Duval, E., Forte, E., Cardinaels, K., Verhoeven, B., Van Durm, R., Hendrikx, K., Forte, M., Ebel, N., Macowicz, M., Warkentyne, K., et al.: The Ariadne knowledge pool system. Communications of the ACM 44(5), 72–78 (2001)
2. IMS-GLC: IMS Learning Resource Metadata specification v.1.2 (2001), `http://www.imsglobal.org/metadata/`
3. IEEE: IEEE 1484.12.1 Standard: Learning Object Metadata (2002), `http://ltsc.ieee.org/wg12/par1484-12-1.html`
4. Neven, F., Duval, E.: Reusable learning objects: a survey of LOM-based repositories. In: Proceedings of the tenth ACM international conference on Multimedia, MULTIMEDIA 2002, pp. 291–294. ACM, New York (2002)
5. Najjar, J., Ternier, S., Duval, E.: The actual use of metadata in ARIADNE: an empirical analysis. In: Duval, E. (ed.) Proceedings of the 3rd Annual ARIADNE Conference, pp. 1–6. ARIADNE Foundation (2003)
6. Friesen, N.: The international learning object metadata survey. The International Review of Research in Open and Distance Learning 5(3) (2004)
7. Van de Sompel, H., Nelson, M., Lagoze, C., Warner, S.: Resource Harvesting within the OAI-PMH Framework. D-Lib Magazine 10(12), 1082–9873 (2004)
8. Meire, M., Ochoa, X., Duval, E.: Samgi: Automatic metadata generation v2.0. In: Seale, C.M.J. (ed.) Proceedings of the ED-MEDIA 2007 World Conference on Educational Multimedia, Hypermedia and Telecommunications, pp. 1195–1204. AACE, Chesapeake (2007)
9. Downey, A.B.: The structural cause of file size distributions. In: Proceedings of IEEE Ninth International Symposium on Modeling, Analysis and Simulation of Computer and Telecommunication Systems, pp. 361–370 (2001)
10. Ward, J.: Unqualified Dublin Core usage in OAI-PMH data providers. OCLC Systems & Services 20(1), 40–47 (2004)
11. Weibel, S.: The Dublin Core: a simple content description model for electronic resources. Bulletin of the American Society for Information Science and Technology 24(1), 9–11 (1997)
12. Heery, R., Patel, M.: Application profiles: mixing and matching metadata schemas. Ariadne 25, 27–31 (2000)
13. Shannon, C.E.: A mathematical theory of communication. ACM SIGMOBILE Mobile Computing and Communications Review 5(1), 3–55 (2001)
14. Ochoa, X., Duval, E.: Automatic evaluation of metadata quality in digital libraries. International Journal of Digital Libraries 10(2), 67–91 (2009)
15. Massart, D.: Towards a pan-European learning resource exchange infrastructure. In: Next Generation Information Technologies and Systems, pp. 121–132 (2009)
16. Stvilia, B.: Measuring information quality. PhD thesis, University of Illinois (2006)

GLUE!-PS: A Multi-language Architecture and Data Model to Deploy TEL Designs to Multiple Learning Environments

Luis Pablo Prieto, Juan Ignacio Asensio-Pérez, Yannis Dimitriadis, Eduardo Gómez-Sánchez, and Juan Alberto Muñoz-Cristóbal

GSIC-EMIC Group, University of Valladolid, Spain
lprisan@gsic.uva.es, juaase@tel.uva.es, yannis@tel.uva.es, edugom@tel.uva.es, juanmunoz@gsic.uva.es

Abstract. The complexity of orchestrating TEL scenarios prompts for a careful learning design by practitioners. Currently, teachers can use variety of tools and languages to express their designs, but they are unlikely to be supported in deploying such designs in the learning environment of their choice. This paper describes a multi-tier architecture and data model to support the deployment of learning designs, expressed in multiple languages, to different learning platforms. The proposal strives to be sustainable in authentic scenarios by minimising both software development costs and changes to current installations. The architecture and data model are theoretically validated through the transformation of a well-known learning scenario from several design languages to different learning platforms, preserving the design's essential characteristics.

Keywords: learning design, deployment, instantiation, learning environments.

1 Introduction

As technology-enhanced learning (TEL) research outcomes permeate our educational system, several trends can be recognized. From a technological standpoint, we can see a proliferation of learning environments being used by teachers and institutions to cater for their increasing needs in blended and distance courses. These environments include virtual learning environments (VLEs, such as Moodle[1] or .LRN[2]), but also other platforms, especially "web 2.0" tools (blogs, wikis, social networks...) which are now often used with educational purposes [3]. From a pedagogical perspective, purely instructional, content-based approaches are being abandoned in favor of multiple other approaches (collaborative learning, inquiry-based learning, etc.). The complexity of coordinating all these factors (sometimes referred to as "orchestrating learning" [5,18]) prompts for a careful

[1] http://moodle.org
[2] http://dotlrn.org

C. Delgado Kloos et al. (Eds.): EC-TEL 2011, LNCS 6964, pp. 285–298, 2011.

preparation of the learning activities to be performed by students. This preparation has been extensively studied in the field of learning design (LD) [14].

Despite the efforts of LD research in proposing methods, languages and tools to structure and plan learning experiences, their outcomes have not been widely adopted by teachers. This lack of adoption can be understood if we consider how those design tools are connected to the learning platforms that are normally used for their enactment: some LD tools generate formal designs (also known as *scripts*) which cannot be deployed easily in any learning environment (e.g. CompendiumLD [2]); other LD tools generate formal designs following the IMS-LD specification [10], which is not natively supported by most major learning platforms; finally, there exist LD tools associated to *one* learning environment, which allow for the deployment of the design only into that execution platform (e.g. LAMS[3]). The fact that teachers often intend to enact their learning scenarios in VLEs chosen and provided by their institutions, or in web 2.0 platforms (which were not designed for education, and do not follow any educational standards), highlights the need for a way of *supporting the deployment of learning designs, expressed in a number of LD languages, into a variety of learning environments.*

This paper proposes a compromise solution to the problem of deploying learning designs to multiple learning environments (section 2). The proposal tries to achieve the ideal of "design once, deploy anywhere" while maintaining a low barrier of entry for the involved stakeholders (teachers, institutions, software developers and integrators), by *not* requiring them to use a specific LD language or learning environment and minimising software development costs. Our proposal (section 3) consists of a multi-tier service-based architecture and a data model to capture the main learning design aspects that are deployable in current learning environments. As a theoretical validation of the proposal (section 4), we have taken a well-known learning design scenario and we have performed its translation for deployment, from its expression in several LD languages to different possible learning environments maintaining the scenario's essential qualities.

2 Deploying Learning Designs in a Fragmented Learning Environment Landscape

Although the preparation of the activities to be performed by learners is as old as teaching itself, the concept of learning design (normally related to e-learning systems and educational technologies) is relatively recent [13]. Multiple methods, tools and languages have been proposed by learning design researchers to aid practitioners in creating and describing learning activities, as tools for reflection and for their reusability by practitioners in other contexts [14]. This objective can also be formulated as "design once, deploy anywhere". With respect to this objective, current support for LD can take various forms:

Non-deployable LD Tools. These are tools and methods to support learning design, mainly intended for personal use and reflection, but also to com-

[3] `http://lamsfoundation.org`

municate pedagogic practices (e.g. CompendiumLD [2], Phoebe[4]). Although many of these tools are computer-based and express learning designs in some formal, computerized format, they are not intended for the automatic deployment of the designed activities in any execution environment.

IMS-LD Tools. The IMS Global Consortium proposed the Learning Design specification (IMS-LD, [10]) as the standard for describing and exchanging reusable learning designs. This standards-based approach assumed that, once the specification was in place, developers of learning environments (such as VLEs) would implement or integrate IMS-LD players in their software. There was a considerable uptake in the researcher community about IMS-LD, in the form of authoring tools (e.g. Reload[5] or Collage[6]) and a few execution engines and players (such as CopperCore[7]). Still, none of the major learning environment developers has adhered to the specification[8], whose development has now stalled, and is being criticized for its complexity and the difficulty in implementing compliant execution environments [17].

Integrated LD and Execution Environments. Other approaches to LD propose an integrated environment that allows for the specification, deployment and execution of learning designs using an integrated environment. These integrated environments include moderately successful VLEs like LAMS[9], as well as more research-oriented prototypes such as the ones developed by the SCY project [15], or the one based on the LDL language [8]. Although these systems make it relatively easy to deploy and execute a learning design, they are "walled gardens", requiring practitioners to adopt both the learning design methodology and the execution environment that they propose.

As mentioned before, to this fragmented LD landscape we have to add the current multiplicity of learning platforms, including not only VLEs, but also web 2.0 tools like wikis and social networks, as well as customized platforms based on generic web systems like Elgg[10] or Drupal[11]. In this context, learning design and deployment features are only one of the many factors that institutions take into account when choosing learning platforms to be used by teachers and other practitioners. As a result, currently, a practitioner has very slim chances of her institution having chosen one of the few LD-enabled learning environments.

Taking into account the aforementioned kinds of support for learning design, practitioners today have only two options. One is to switch to a LD-enabled learning environment, such as LAMS or .LRN with the GRAIL extension, and accept the learning design methods and execution support that they offer. However,

---

[4] `http://www.phoebe.ox.ac.uk`
[5] `http://www.reload.ac.uk/ldesign.html`
[6] `http://www.gsic.uva.es/collage`
[7] `http://coppercore.sourceforge.net/`
[8] With the exception of an optional .LRN extension (the GRAIL player [7]) to execute IMS-LD designs.
[9] `http://lamsfoundation.org`
[10] `http://www.elgg.org/`
[11] `http://drupal.org/`

in most cases this is an institution-wide decision, making such a switch unlikely to happen. The remaining alternative is to convert the concepts in the learning design to the practitioner's learning environment of choice *manually*, and then deploy them by hand using the learning environment's authoring features (if any). This is a time-consuming and error-prone process (sp. for complex designs), which highlights the *need for a multi-language solution for the deployment of learning designs in a variety of learning environments.*

There is another problem associated with the deployment of learning designs in this fragmented landscape, which is often overlooked: that of providing instantiation information. Learning designs are de-contextualized models of pedagogic methods (to make them reusable in other contexts), which have to be *re-contextualized* by the teachers using them (e.g. by stating how many groups there will be, who will be in them, or which instances of a tool will be used by each group). This process is called operationalisation or *instantiation* of the design. The form that this instantiation takes is dependent both on the LD language and the target learning environment used. Thus, while the instantiation process is rather straightforward in integrated LD systems such as LAMS, it has to be performed in an ad-hoc way for other combinations of LD language and learning platforms. Although specialized tools for this purpose in specific LD contexts exist [20], *a way of adding instantiation information to designs, applicable to the wide range of possible LD and LE combinations*, is highly desirable.

3 GLUE!-PS

We propose a general solution to the problem of deploying learning designs into different learning environments. Our approach tries to find a compromise between the expressiveness of LD languages and the capabilities of current learning environments. Taking into account the current fragmented state of both the learning design and learning environment landscapes, as well as past experiences (such as IMS-LD not getting widely adopted), our main design principle has been to *maximise community acceptance*, by lowering the barrier of entry for all involved stakeholders (teachers and learners, institutions, learning environment and other software developers). Thus, our solution should be able to integrate as many LD tools/languages and learning environments as possible, while avoiding modifications to the source code of existing systems, minimising the average development effort, and allowing institutions that have opted for specific LD tools or learning environments to keep their choices. Moreover, the solution should also provide a more time-efficient deployment alternative for practitioners.

3.1 GLUE!-PS Architecture

Following the aforementioned design principles, we propose a service-oriented architecture following an *Adapter* pattern [16]. The architecture, which can be seen in Figure 1, has a central component called Group Learning Unified Environment – Pedagogical Scripting (GLUE!-PS) and two sets of adapters, one

(LD adapters) for the different learning design languages available (IMS-LD, LDL, CompendiumLD, etc.) and another set (LE adapters) for the different target learning environments (e.g. Moodle, LAMS or wikis). Coarsely speaking, the architecture would translate from the learning designs in the different formats to a common data format native to the GLUE!-PS using the different LD adapters. Then, the design information in this data format is completed with instantiation information about the specific context provided by the teacher (e.g. number and composition of groups, specific tools to be used) and finally translated automatically to the concepts and data format of the target learning environment using the LE adapters, which deploy the instantiated design to the target environment.

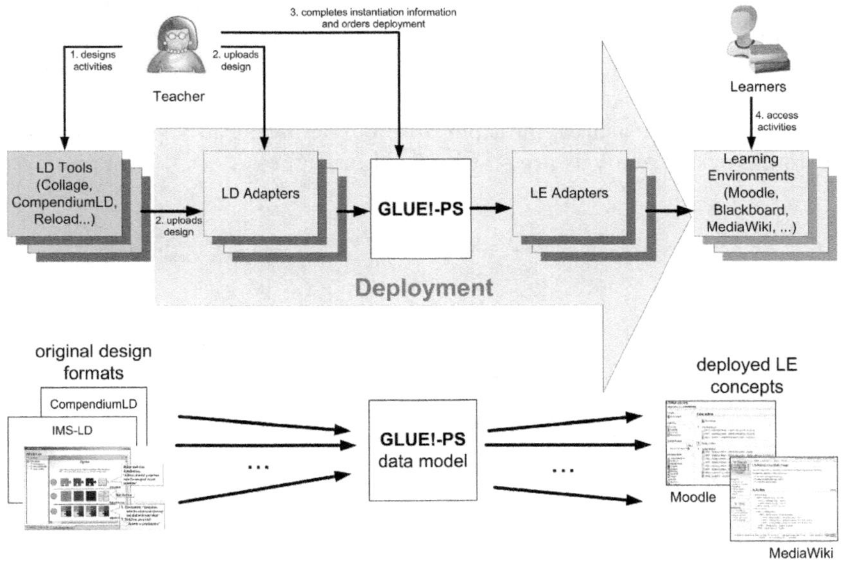

Fig. 1. GLUE!-PS simplified architecture (top) and data model view (bottom)

This kind of service-oriented architecture poses several well-known advantages: a) it allows third parties (authoring tool developers, VLE providers and other interested parties like institutions) to develop the adapters, as long as they comply with the GLUE!-PS service interface; b) it allows implementors of the different components to use any desired technology internally; c) using the *Adapter* pattern allows for extension of the architecture to new languages and environments, maximising the code reuse among adaptations and reducing the average development effort by the developers (i.e. once an LD adapter is implemented, that LD language can be used with all the supported learning environments). Moreover, the architecture also complies with aforementioned design principles such as not requiring modifications to either design tools or existing learning environments. Finally, it is worth mentioning that, due to the existence of a central element (the GLUE!-PS service), a way is provided to introduce instantiation information in a unified way, regardless of the original LD language or the target learning environment (see section 2 above).

3.2 GLUE!-PS Data Model

The development effort needed to implement the LD and LE adapters above (i.e. the barrier of entry for software developers) depends mostly on the complexity of the GLUE!-PS service contract and the underlying data model of such service. Since this data model has to integrate languages and concepts with opposing goals (e.g. de-contextualizing for reuse in LD languages, contextualizing for execution in learning environments), we have analyzed and compared several languages and sets of concepts from both sides, in order to determine the most commonly supported *deployable characteristics of computerized learning designs*. These traits have been extracted from the review of several scripting conceptual frameworks available in literature [12,11,21], and include aspects like the presence of sequencing features, multiple roles participating in a single activity, etc. The results of our analysis are summarized in Figure 2.

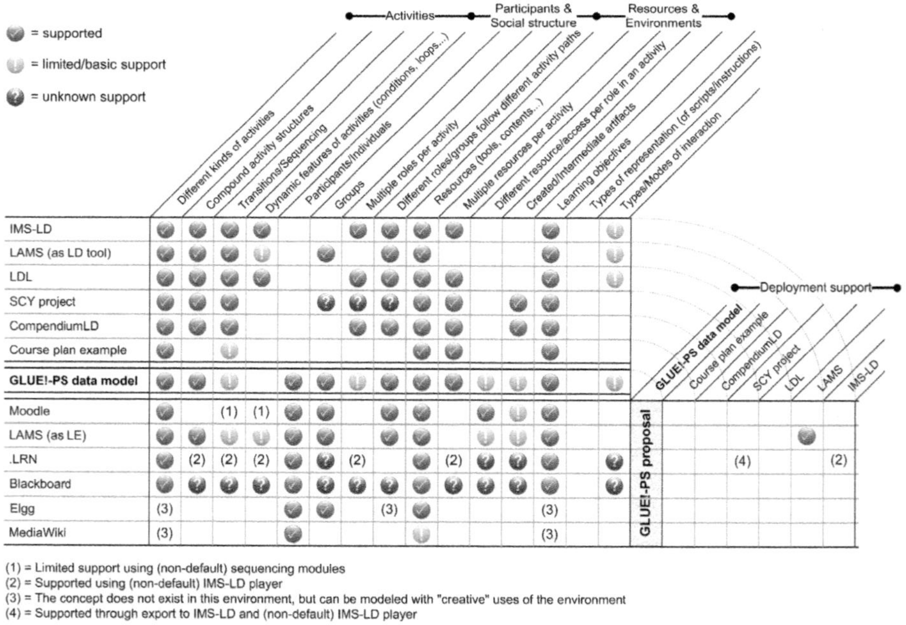

Fig. 2. Summary of the main scripting characteristics supported by learning design languages and learning environments, extracted from [12,11,21]

Our analysis of LD languages (upper rows in Figure 2) included the aforementioned IMS-LD [10], plus LAMS [4], LDL [8] and the SCY project design approach [15]. In order to get a wider view and include current LD practice, we also included two languages not intended for automated deployment, such as the language of CompendiumLD [2] and the implicit language used in course plans in an authentic setting (at University of Valladolid). As we can see, they are all activity-based[12], and provide some kind of structuring of the learning activities. Often, they also model learning objectives and resources, as well as the social structure of the design, often in the form of roles (but not individuals).

When analysing learning environments (lower rows in Figure 2), we chose several of the main players in the VLE arena (like Moodle or .LRN), but also other web 2.0 tools that are being used as learning environments, such as the MediaWiki[13] wiki engine or the Elgg[14] social networking software. These environments always recognize individual participants, but invariably have poorer activity sequencing and dynamics (if any) than those offered in LD languages. Many of them define groups of participants (but not functional roles in activities), and all provide modeling of resources. We have also represented the current LD deployment support in the different environments, which is currently very scarce – precisely the gap that the GLUE!-PS proposal intends to cover.

With the results of this analysis, we have constructed a data model that represents the scripting properties available in LD languages, to the extent that learning environments support them, pondering at the same time the relative acceptance of the different conceptual sets (e.g. the concepts of Moodle or IMS-LD had more weight in our design decisions than those of LAMS or LDL). The proposed model appears in Figure 3: each *learning design* is composed of a number of *activities*, which can be structured in an arbitrarily deep tree structure, and can be sequenced (even if just for presentation purposes). Each activity can be performed by different functional *roles*, and is mediated by one or more *resources* (which can be static *objects* in the sense of its location being known at design time, or *tools* that can be instantiated so that each group accesses a different instance of the tool, e.g. a shared drawing board for each learner team). A learning design can be deployed, and each *deploy* is a contextualization of that learning design for a specific *learning-environment*. In such a deploy, the concrete *participants* (i.e. users in that learning environment) that will take part on the activities have to be specified. For each activity in the design, a number of *instanced-activities* will be created, one per *group*. Finally, each group performing an activity that requires the use of (instantiable) tools will be assigned an *instanced-tool* so that they can work independently from other groups if needed. The scripting properties of this model can also be seen in Figure 2 (middle row).

It is important to note that this data model does not intend to be "yet another" LD language or conceptual framework. Rather, it strives to represent *deployable characteristics of learning designs*, taking into account the practitioners' desire for contextualization, since they "are more interested in effectively and

[12] Even if LDL puts more emphasis on single interactions, the notion of activity exists and is a basic organizing concept [8].

[13] `http://mediawiki.org`

[14] `http://www.elgg.org`

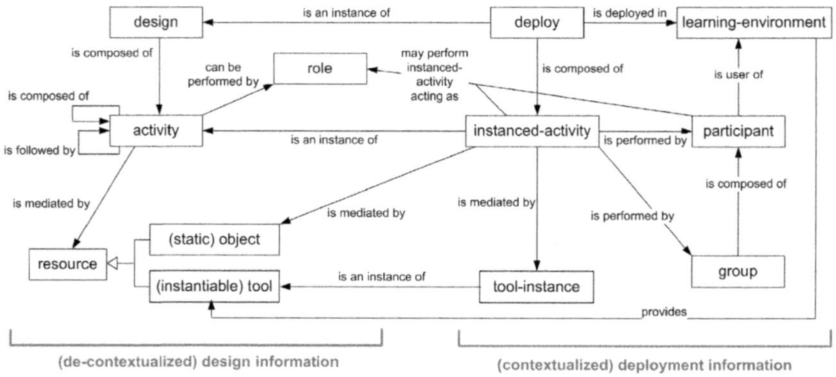

Fig. 3. GLUE!-PS data model

efficiently bringing these proposals in the real classroom with certain guarantees for sustainability and scalability" [21].

4 Validating GLUE!-PS: The Planet Game Scenario

As a first theoretical validation of the proposed infrastructure and its data model, we have simulated (manually) the process proposed in the previous section, to deploy a well-known learning scenario expressed in several LD languages, to various learning environments not supported originally. Then, we analysed the original scripts and their deployed counterparts, to ascertain *whether the original scenario's essential pedagogical qualities* (which some authors have termed "intrinsic constraints" [6]) *were maintained or not* throughout the process.

The *Planet Game scenario* [19] is a game-based collaborative learning scenario which is part of a real lifelong learning situation in Astronomy. Its main goal is for learners to acquire knowledge about the planets in the Solar System, classifying them with respect to their distance from the Sun. In the scenario, learners are grouped into two teams, and each team is given part of the needed information, so that they must collaborate to succeed (although they also compete to be the winners of the game). The scenario is structured as follows: Team A is given expert interviews which contain some planets' properties from which they can deduce the planets' order, but not their names. Team B is given another set of clues that contain the planets' names and a few properties so that the order cannot be deduced from them alone. In order to collaborate, each team will have a private chat room, and all the participants can communicate through a forum. The teacher may add new clues to the interviews of both teams, and see how the activity progresses through the forum. When the teacher deems so, the collaborative part ends, and each learner takes an individual questionnaire to assess their knowledge about the planets' distance from the Sun. In light of the results, the teacher finally announces who is the winner of the game.

This scenario has been used by the learning design community as a benchmark to compare and explore the advantages and disadvantages of different languages and approaches. It was the object of a LD languages workshop in 2006 and a followup special issue at the Journal of Interactive Media in Education[15] in 2008. In this process, the scenario has been modeled using different LD languages and approaches (e.g. various forms of IMS-LD, LAMS, LDL), including also the operationalisation and execution of the generated computer-specified scripts in the supported learning environments (CopperCore, LAMS, LDI, etc.).

To explore the limits of the deployment of the Planet Game scenario using GLUE!-PS, four translations are described, trying to cover both commonplace and extreme cases. Two original LD formalizations have been considered: the IMS-LD script described in [9] (an example of the most common LD language) and the LDL model detailed in [8] (an extreme example, since LDL is a more expressive language in terms of collaborative learning). As target learning environments, we have chosen Moodle (currently the most widely used VLE) and MediaWiki (as a extreme case, due to its limited feature set, see Figure 2).

The full description of the four mentioned translations, and the decisions that are to be taken by the LD and LE adapters to translate from the original LD language to the target learning environment, would exceed the scope of this paper. Thus, we will concentrate only on two concrete aspects of the most commonplace (IMS-LD to Moodle) and most extreme cases (LDL to MediaWiki), as a sampler of how the translations were conducted and which are their limitations[16].

The transformation of the concepts of *roles and groups*, from the IMS-LD formalization [9] to its deployment in Moodle (using Moodle's course backup data format), is depicted in Figure 4. The two teams in the Planet Game scenario are modeled in IMS-LD as roles of the "jigsaw group" type. The IMS-LD adapter would then transform those roles to *role* entities in the GLUE!-PS data model. However, nowhere in the original script it is specified that there are two teams, or which learners are in each team (to increase reusability, e.g. in the case that we want three teams). It is the teacher who has to complete this "instantiation data", defining how many *group* entities should exist. In the original specification of the scenario in IMS-LD this kind of information would have been completed manually by the teacher in the execution environment's (e.g. CopperCore's) administrative interface, and is done in an ad-hoc way for each execution environment. With GLUE!-PS, once the instantiation information is in place, the conversion from the GLUE!-PS data model to the Moodle concepts of "group" and "grouping"[17] is easily automatable (even if it is time-consuming and error-prone to do it manually), and it is introduced at a unified point and in a unified manner, no matter what the target learning environment is.

Let us now take a look to the translation of the concepts related to *activities*, in the extreme case of deploying the LDL formalization of Planet Game [8] in

[15] `http://jime.open.ac.uk/`

[16] Complete diagrams and data from the translations are available at `http://www.gsic.uva.es/~lprisan/Prieto2011-ECTEL-AdditionalData.zip`

[17] Optional feature in Moodle in order to have activity separation between groups.

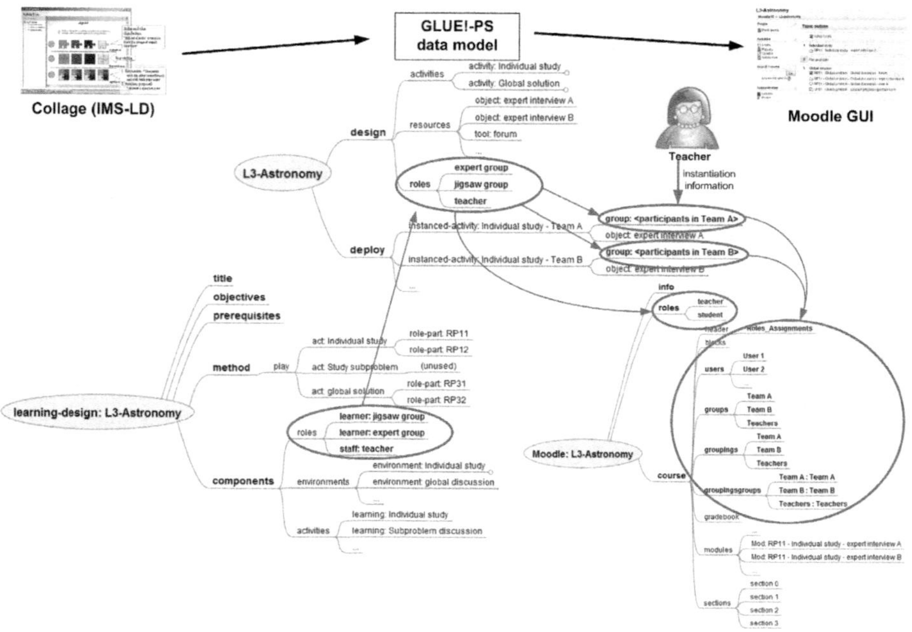

Fig. 4. Converting roles and groups in GLUE!-PS: from IMS-LD to Moodle through the GLUE!-PS data model

MediaWiki. LDL allows for more detail than IMS-LD, and allows for a high degree of expressiveness regarding collaborative activities. Albeit it supports the concept of activity, its central emphasis is in a more fine-grained element: the interaction (i.e. for each activity, several interactions among participants are defined). We have made the LDL adapter to convert interactions and activities to a hierarchy of activities. The resulting activity structure is much more verbose than the one coming from the IMS-LD specification (making the instantiation process more time-consuming). The activity tree is then converted to wiki pages by the MediaWiki adapter. However, there is an important limitation in the deployment of the design in MediaWiki: due to its default open access policy, it is very difficult to implement a hard separation of resources among teams (e.g. nothing precludes team A from accessing team B's expert interview), making the game's success dependent on the participants' good will. This is a general consideration of our approach: *the basic nature of the learning environment will not be modified*, and so the teachers should consider whether the selected learning environment is adequate for the design or not (e.g. MediaWiki may not be the best option for a scenario that requires separation of who sees what).

Figure 5 summarises the analysis of the Planet Game scenario described in [19], and how those properties were preserved throughout the different translations. The script aspects have been classified as intrinsic (essential for the pedagogic intent) or extrinsic (due to implementation restrictions, such as the (supported icon) = supported

(limited icon) = limited/basic support

	Intrinsic characteristics: Different teams have access to different clues	Environment for inter-team collaboration (forum)	Separated collaboration environment for teams (chat)	Teacher can intervene in the inter-team collaboration	Teacher can add clues	Teacher decides when the collaboration finishes	At the end, individual knowledge is assessed (quiz)	Teacher decides the winner based on the assessment	Extrinsic characteristics: Choice of tools for activities (forum, chat, quiz)	The learning activity can be observed	Teacher can modify the activity based on the observation	Activities are traced	Traces can be easily used by teacher	Reusability of structure for a different topic	Adaptability of the activity structure	Automated script workflow
Collage IMS-LD + Gridcole	supported	supported	supported	supported	supported	limited	supported	supported	supported	supported		limited		supported		supported
IMS-LD → GLUE!-PS	supported	supported	supported	supported	supported	(1)	supported	supported	supported	(1)	(1)	(1)	(1)	supported	supported	(1)
IMS-LD → GLUE!-PS → Moodle	supported	supported	supported	supported	supported	limited	supported	supported	supported	supported	supported	limited		supported	supported	limited
IMS-LD → GLUE!-PS → MediaWiki	(2)	supported	(2)	supported	supported	limited	supported	supported	(3)	supported	supported	limited	limited		supported	
LDL + LDI	supported	supported	supported	supported	supported	supported	supported	supported	supported	supported	supported	supported	supported	supported	limited	supported
LDL → GLUE!-PS	supported	supported	supported	supported	supported	(1)	supported	supported	supported	(1)	(1)	(1)	(1)	supported	supported	(1)
LDL → GLUE!-PS → Moodle	supported	supported	supported	supported	supported	limited	supported	supported	supported	supported	supported	limited		supported	limited	limited
LDL → GLUE!-PS → MediaWiki	(2)	supported	(2)	supported	supported	limited	supported	supported	(3)	supported	supported	limited	limited		supported	

(1) = Not applicable in absence of an execution environment
(2) = Not supported due to MediaWiki's open policies, but could be supported by accessing GLUE!-PS during runtime (see text)
(3) = Not supported by default, but could be implemented with MediaWiki extensions or links to external tools

Fig. 5. Summary of the intrinsic and extrinsic characteristics of the Planet Game script extracted from [19], and how they are preserved throughout the translations

specific technologies used), following the categorisation from [6]. As we can see, the Moodle deployment preserves all intrinsic characteristics, and most extrinsic ones, regardless of the original LD language. The MediaWiki deployment preserves most aspects, although it breaches the "separation of teams" constraint, due to its lack of group-based access control. Nevertheless, several possible solutions for this learning environment limitation exist, from integrating advice mechanisms into GLUE!-PS (so as to warn teachers of these breaches in the script principles), to accessing the activities through GLUE!-PS during runtime, so that the group-based access control is enforced by this service.

5 Discussion

The proposed architecture and data model is capable of deploying TEL designs, *in a variety of LD languages*, to *multiple execution infrastructures*, while *preserving the intrinsic characteristics of the original script* (as long as the target learning environment supports them). Also, due to the use of service orientation and adapters in the architecture, this can be achieved for a large number of combinations of LD languages and learning environments, *at a reduced development cost, with minimal modifications to installed infrastructures* (LD tools, VLEs, etc.). By providing a time-efficient way of deploying learning designs in existing learning environments, this system can greatly help in the orchestration of complex TEL scenarios. However, this approach is not without limitations.

The comparison of the LDL and IMS-LD examples above show a *lack of unicity* in the resulting deployments of a learning scenario (even for a same target learning environment). This variability not only depends on the available concepts in the source LD language (and the fact that designs were conceptualized by different designers), but also on the decisions taken by the implementors of the different LD and LE adaptrs (e.g. converting each interaction in LDL to one GLUE!-PS activity vs transforming it to instructions in the description of a high-level GLUE!-PS activity). These mappings are very subjective, and largely depend on the practitioners' pedagogical beliefs. Indeed, the service-oriented nature of the proposed architecture favors the appearance of multiple implementations of the adapters, rather than enforcing a single view of how such transformations should be done. Thus, practitioners can potentially choose, among different implementations of an adapter, the one that better suits their needs.

Still, the most important limitation of our approach is the *loss of expressiveness* that is bound to appear in the two translations involved (first from LD language to the GLUE!-PS *lingua franca*, and then to the learning environment concepts). As we have mentioned, the proposed data model is expressive enough to capture most essential structural elements of a script (e.g. activities, resources, groups, roles, sequencing), to the extent that the learning environments themselves support them. In this sense, the biggest losses are in the area of complex sequencing and other dynamic properties of activities (conditions, loops and features related to advanced activity flow automation). Given that the implementation of those features varies wildly between one LD language and another, and that few learning environments support them anyway, we have chosen to leave most of these features out of our data model. Thus, even if the loss in translations will be higher for complex designs that exploit the full capabilities of a highly expressive LD language, we contend that the support given by our proposal matches the support given by current learning environments, with a comparatively low development effort[18]. Furthermore, the provision of a single point for adding instantiation data (the GLUE!-PS service) would allow teachers to check the translated script model so that such loss can be remedied.

Another feature of this approach that should not go unnoticed is its *extensibility*, both on the side of LD languages, as well as in the case of new, emergent learning environments (e.g. in case a new VLE gains wide adoption). By not tying ourselves to a concrete LD specification or implementation of execution environment, we hope to remain adaptable to the ever-changing evolution of TEL practice. Moreover, this approach also leaves the door open to integrate current TEL design practices into the "LD lifecycle": for example, we could develop an adapter that takes teachers' lesson plans (which teachers already develop for administrative and personal purposes) and deploy them in their institution's learning environment (in fact, one such language was included in the LD language analysis for our proposal, in section 3). This would widen the adoption of learning design in TEL, while helping teachers save time and effort.

[18] As a first-level approximation and example, IMS-LD support in Moodle was considered for months by an external group of researchers and teachers using Moodle (see [1]), and finally dismissed as not worth the effort. On the other hand, developing an IMS-LD adaptor and a Moodle adaptor for the GLUE!-PS architecture has taken 45 days for two developers working part-time on the project.

In summary, we have presented the architecture and data model for a technological infrastructure to deploy learning designs to a variety of learning environments, which currently is a time-consuming and error-prone task for the average TEL practitioner that has to perform it manually. Our proposal has the advantage of requiring a comparatively small development effort, as well as minimising the impact to currently deployed infrastructures. We believe that the implementation of this kind of architecture has the potential of making the use of solid learning design practice more widespread in real-world situations. Apart from the implementation of prototypes that reify this proposal, our current and future work in this direction is aimed at investigating the advantages of using the GLUE!-PS infrastructure also during runtime, to help in the orchestration of the learning activities, and complement under-featured learning environments.

Acknowledgements. This research has been partially funded by the Spanish Ministry of Science and Innovation Projects TIN2008-03-23 and IPT-430000-2010-054, and the Autonomous Government of Castilla and León, Spain, project VA293A11-2.

References

1. Berggren, A., Burgos, D., Fontana, J.M., Hinkelman, D., Hung, V., Hursh, A., Tielemans, G.: Practical and pedagogical issues for teacher adoption of IMS Learning Design standards in Moodle LMS. Journal of Interactive Media in Education 2005(2) (2005)
2. Brasher, A., Conole, G., Cross, S., Weller, M., Clark, P., White, J.: CompendiumLD – a tool for effective, efficient and creative learning design. In: Proceedings of the 2008 European LAMS Conference (2008)
3. Conole, G., Alevizou, P.: A literature review of the use of web 2.0 tools in higher education. Report commissioned by the Higher Education Academy, The Open University (August 2010)
4. Dalziel, J.: Implementing learning design: The learning activity management system. In: Proceedings of the 20th Annual Conference of the Australasian Society for Computers in Learning in Tertiary Education, ASCILITE (2003)
5. Dillenbourg, P., Järvelä, S., Fischer, F.: The evolution of research in computer-supported collaborative learning: from design to orchestration. In: Balacheff, N., Ludvigsen, S., de Jong, T., Lazonder, A., Barnes, S. (eds.) Technology-Enhanced Learning: Principles and Products, pp. 3–19. Springer, Heidelberg (2009)
6. Dillenbourg, P., Tchounikine, P.: Flexibility in macro CSCL scripts. Journal of Computer-Assisted Learning 23(1), 1–13 (2007)
7. Escobedo, J., de la Fuente, L., Gutiérrez, S., Pardo, A., Delgado-Kloos, C.: Implementation of a learning design run-time environment for the.LRN learning management system. Journal of Interactive Media in Education (JIME) (July 2007), `http://jime.open.ac.uk/2007/07/escobedo-2007-07-paper.html`
8. Ferraris, C., Martel, C., Vignollet, L.: Modelling the "Planet Game" case study with LDL and implementing it with LDI. Journal of Interactive Media in Education (JIME) Special issue on Comparing Educational Modelling Languages on the'Planet Game' Case Study (December 2008), `http://jime.open.ac.uk/jime/article/view/2008-20`

9. Hernández-Leo, D., Villasclaras-Fernández, E.D., Asensio-Pérez, J.I., Dimitriadis, Y., Bote-Lorenzo, M., Jorrín-Abellán, I.: The added value of implementing the planet game scenario with Collage and Gridcole. Journal of Interactive Media in Education (JIME) Special issue on Comparing Educational Modelling Languages on the'Planet Game' Case Study (December 2008), `http://jime.open.ac.uk/jime/article/view/2008-22`
10. IMS Global Learning Consortium: IMS Learning Design v1.0 final specification (2003), `http://www.imsglobal.org/learningdesign/`
11. Kobbe, L., Weinberger, A., Dillenbourg, P., Harrer, A., Hämäläinen, R., Häkkinen, P., Fischer, F.: Specifying computer-supported collaboration scripts. International Journal of Computer-Supported Collaborative Learning 2(2/3), 211–224 (2007)
12. Kollar, I., Fischer, F., Hesse, F.: Collaboration scripts – a conceptual analysis. Educational Psychology Review 18, 159–185 (2006)
13. Koper, R.: Current research in learning design. Educational Technology & Society 9(1), 13–22 (2006)
14. Koper, R., Tattersall, C.: Learning Design: a Handbook on Modelling and Delivering Networked Education and Training. Springer, Heidelberg (2005)
15. Lejeune, A., Ney, M., Weinberger, A., Pedaste, M., Bollen, L., Hovardas, T., Hoppe, U., de Jong, T.: Learning activity spaces: Towards flexibility in learning design? In: Proceedings of the Ninth IEEE International Conference on Advanced Learning Technologies, ICALT 2009 (2009)
16. Monday, P.B.: Web Service Patterns: Java Edition, ch. 4: Exploring the Architecture Adapter Pattern. Apress (2003)
17. Neumann, S., Klebl, L., Griffiths, D., Hernández-Leo, D., de la Fuente-Valentín, L., Hummel, H., Brouns, F., Derntl, M., Oberhuemer, P.: Report of the results of an IMS Learning Design expert workshop. International Journal of Emerging Technologies in Learning 5(1), 58–72 (2010), at `http://dspace.ou.nl/bitstream/1820/2154/2/iJET_onlineaccess_version.pdf`
18. Prieto, L.P., Dimitriadis, Y., Villagrá-Sobrino, S., Jorrín-Abellán, I.M., Martínez-Monés, A.: Orchestrating CSCL in primary classrooms: One vision of orchestration and the role of routines. Paper Accepted at the 9th International Conference on Computer-Supported Collaborative Learning, CSCL 2011 (2011)
19. Vignollet, L., Martel, C., Ferraris, C.: Description of the 'Planet Game' case study and guidelines to the authors. Journal of Interactive Media in Education (JIME) Special issue on Comparing Educational Modelling Languages on the'Planet Game' Case Study (December 2008), `http://jime.open.ac.uk/jime/article/view/2008-18`
20. Villasclaras-Fernández, E., Hernández-Gonzalo, J., Hernández-Leo, D., Asensio-Pérez, J., Dimitriadis, Y., Martínez-Monés, A.: InstanceCollage: A tool for the particularization of collaborative IMS-LD scripts. Educational Technology & Society 12(4), 56–70 (2009)
21. Weinberger, A., Kollar, I., Dimitriadis, Y., Mäkitalo-Siegl, K., Fischer, F.: Computer-supported collaboration scripts: Perspectives from educational psychology and computer science. In: Balacheff, N., Ludvigsen, S., de Jong, T., Lazonder, A., Barnes, S. (eds.) Technology-Enhanced Learning: Principles and Products. Amsterdam University Press (2009)

Automatic Assessment of Collaborative Chat Conversations with PolyCAFe

Traian Rebedea[1], Mihai Dascalu[1], Stefan Trausan-Matu[1], Gillian Armitt[2], and Costin Chiru[1]

[1] University "Politehnica" of Bucharest, Computer Science Department, 313 Splaiul Independentei, 060042 Bucharest, Romania
[2] University of Manchester, Higher Cambridge Street, Manchester, M15 6FH, UK
{traian.rebedea,mihai.dascalu,stefan.trausan, costin.chiru}@cs.pub.ro, gillian.armitt@manchester.ac.uk

Abstract. The wider acceptance and usage of instant messaging (chat) represents one of the consequences of undertaking Computer-Supported Collaborative Learning (CSCL) practices in formal education settings. However, the difficulty of analyzing these textual artifacts of learners in order to offer them feedback represents a serious problem in further extending the usage of chat conversations. PolyCAFe is a system that was designed to support the tutors and to provide automatic feedback for the learners engaged in collaborative chat conversations and discussion forums. The architecture of the system is presented by focusing on two key components: the assessment of the utterances and of the collaborative discourse. PolyCAFe's effectiveness has been proved in a validation experiment with students and tutors from a University course. The main findings from this trial, together with the conclusions of domain experts verifying the accuracy of the assessment provided by PolyCAFe, are also analyzed and commented in detail.

Keywords: CSCL, assessment, verification, feedback, chat conversations, dialogism.

1 Introduction

Chat conversations and other technologies that provide online textual interactions between learners have been widely used in practice for more than a decade, especially in informal contexts. However, with the advent and rise of Computer-Supported Collaborative Learning (CSCL), these technologies have been brought by teachers to more formal education environments, including schools and universities [1]. In the particular case of chat conversations, the typical learning scenario is to divide the students in small groups (usually, of 4-10) that have to solve a problem or participate in a debate in order to reach a solution or a conclusion within the entire group. However, one of the main problems of using these technologies in formal education is the difficulty of analyzing this kind of textual artifacts in order to provide feedback and guidance to the participants of the conversations. For this reason, when chat

C. Delgado Kloos et al. (Eds.): EC-TEL 2011, LNCS 6964, pp. 299–312, 2011.

conversations are used as part of learning tasks, the students are rarely offered feedback and assessed for their participation in the conversation.

In order to address these issues, there has been a lot of research on how to support the tutors that analyze online textual interactions of students, including chat conversations. Some of them present a general analysis at the conversation level, such as topic detection and extraction in chats [2] or concept formation in a group discussion [3]. Other studies were concentrated on summarizing the chat conversation [4] in order to provide a brief overview of the entire discussion. Meanwhile, feedback for each utterance/posting in a discussion has also been considered important, for example in the analysis of argumentation and transactivity in each posting [5] or classifying utterances based on the degree of concept coverage and social score [6].

However, despite the extensive study of methods for analyzing chat conversations, both at utterance level and at the level of the whole discussion, there have been no other attempts, except the software presented in this paper, of integrating these techniques in order to provide a complete analysis tool. Thus, PolyCAFe [7] is the only tool that has been designed to provide detailed feedback for the learners engaged in CSCL chats and to support the tutors in analyzing the conversations of their students. It uses a novel theory of discourse [8, 9] that is developed starting from Bahktin's dialogic theory [10] and integrates current technologies available in Natural Language Processing (NLP) and in Social Network Analysis (SNA). Its main objectives are to provide an immediate feedback to the learners that allows them to reflect on their collaboration and corresponding coverage of the topics of the chat conversation on one hand, and to reduce the amount of time the tutors need in order to provide a manual feedback to the students, on the other. The only alternative for PolyCAFe seems to be the recently advances in conversational tutors [11] that can provide a detailed analysis of the chat, but also influence the conversation.

The remainder of this paper presents the main functionality of PolyCAFe and the results of a validation experiment, together with the verification of its accuracy. Thus, section 2 presents the basic theoretical framework used for designing the system, while section 3 focuses on the architecture, functionality and technologies used by PolyCAFe. The setup and main findings of a validation experiment run in an academic context with 35 students are described in section 4. In order to demonstrate the accuracy of the system, two verification experiments with domain experts are presented in section 5. The paper ends with conclusions and observations related to the transferability of the system in other educational settings.

2 Theoretical Framework

Discourse is viewed in NLP as being differentiated in monologues and dialogues. One-way, speaker-listener directed models of communication are considered in monologues [12]. The usual way of analyzing discourse in this case is the segmentation of text, the search for different relationships among segments, the measurement of coherence and obtaining some discourse abstractions like co-references or summaries. Coherence is a key issue in this enterprise and is considered as granted, in different degrees when analyzing texts.

Probably the most known discourse theories belong to Hobbs [13], Grosz [14] or Mann and Thompson [15]. Jerry Hobbs' theory is based on considering semantic coherence relations and on using abduction inferences in formal logic [13]. Rhetorical Structure Theory (RST) [15] identifies hierarchical rhetorical structures built using a limited set of rhetorical schemas (patterns) like antithesis, elaboration etc. *Coherence* is obtained in the centering theory [14] at both local and global levels, centered on two different aspects: intentional and attentional states, which together with the linguistic structure of an utterance sequence, form a tripartite organization. An intentional structure should be present in each discourse, assuring that discourse is rational. This structure is built from intentions (purposes) and, sometimes, from beliefs of the author of the discourse (or of each participant in a conversation) and from relationships among linguistic segments [14].

In dialogues usually a phone-like (or face-to-face) type of conversation is considered. Typically, speech acts, dialog acts or adjacency pairs [12] are the units of analysis. Even if there are attempts to analyze conversations with multiple participants using, for example, transacts [4, 5], this approach is also based on a two interlocutors' model.

Starting from Bakhtin's dialogic theory [10], the theoretical framing for PolyCAFe is based on discovering the evolution of voices in a chat conversation. Thus, any utterance becomes a voice in the discourse that may be related to previous utterances by means of explicit (specifically defined by the user within the conversation interface) and, more often, implicit links. Thus, one of the main components of PolyCAFe is to capture these implicit references between utterances in order to determine voices that are related. This task is very difficult, but several links can be discovered using existing technologies from Natural Language Processing (NLP), like repetitions, lexical and semantic similarity, co-references and adjacency pairs (e.g. question-answer).

Starting from Bakhtin's [10] perspective of discourse analysis, each *thread* may be considered as a voice which becomes more or less powerful than the others and which influences the others. Among chat voices there are sequential and transversal relations, highlighting a specific point of view in a counterpointal way, as mentioned in our previous work [16]. The co-occurrence of several voices which enter in dialogue is a phenomenon considered by Bakhtin to be universal, present in any text, not only in conversations: 'Life by its very nature is dialogic … when dialogue ends, everything ends' [10]. He moves the focus of analysis from sentences to utterances in an extended way, in which even an essay contains utterances and is, at its turn, an utterance. Moreover, each utterance is filled with 'overtones' that contain the echoes and influence of other previous utterances.

A voice is generated by an utterance with effects (echoes) on the subsequent utterances via explicit and implicit links. Moreover, by the simple fact that they co-occur, voices are permanently *interacting*, *overlapping* and *inter-animating*, entering in competition, and generating multivocality in any conversation. The ideal situation of a successful conversation or a coherent discourse is achieved when the voices are entering inter-animation patterns based on the discussion threads they are part of.

Moreover, in order to thoroughly analyze chats and forums our approach integrates TF-IDF [2, 17], Latent Semantic Analysis (LSA) [3, 17], Social Network Analysis

[3], WordNet (wordnet.princeton.edu) [2, 3], dialog acts [18] and other complex relations [5], some of which are presented in detail in the following section.

3 Architecture and Technologies

This section addresses the functional architecture behind PolyCAFe and presents in detail the process of evaluating utterances and collaboration throughout the entire conversation. Further details on the technologies and algorithms used for PolyCAFe can be found in [19].

3.1 Multilayered Architecture

Starting from the initial architecture of the system proposed in [9], we have devised a refined model that highlights all the functionalities and underlying technologies, also offering a deep insight in the theoretical framework behind PolyCAFe. This multi-layered model is presented in Fig.1, where similar technologies are grouped together and denote the functionality of the layer.

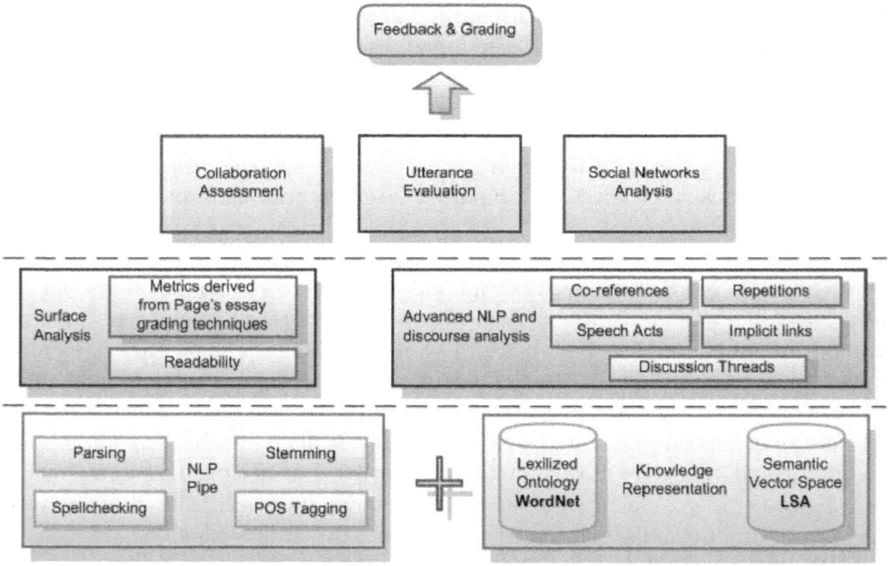

Fig. 1. The detailed architecture of PolyCAFe: similar functionalities and the underlying technologies are grouped in layers

The *first layer* is made up of two different components: the NLP pipe and the representation of the domain. The **NLP pipe** is composed of traditional natural language processing (NLP) tools such as spellchecking, stop words removal, stemming, probabilistic parsing and log-linear POS tagging. The modeling of the domain of the conversation is centered on **knowledge representation** based on a linguistic ontology (WordNet, in our case) and vector space models derived from

Latent Semantic Analysis (LSA). All functionalities and technologies integrated in this level are widely available and form the starting point of the analysis.

On the other hand, the *second layer* enriches the initial context of the discussion by providing both surface analysis and advanced discourse analysis techniques. **Surface analysis** consists of a set of metrics derived from Page's essay grading studies [6, 20] and readability assessment [6]. The **discourse analysis** layer is used for constructing the utterance (or conversation) graph from explicit links (defined from the user interface of the ConcertChat system used in the validation experiment) and implicit links automatically identified using repetitions, semantic similarities, co-references and adjacency pairs (e.g. question-response, claim-agreement/disagreement) [9]. Discussion threads, speech acts and argumentation acts [9] are also detected and integrated in the next steps of the evaluation process in order to provide assessment and enriched feedback.

Finally, **the third layer** addresses the core elements of our feedback system, namely collaboration and utterance assessment, combined with social network analysis specific metrics for obtaining a holistic perspective of each participant's involvement and knowledge within the conversation. Social networks analysis (SNA) plays an important role in current evaluation by integrating a quantitative perspective based on surface metrics and the qualitative one derived from the utterance assessment process described in the following subsection.

At this level, SNA models best the involvement of each participant in the discussion, whereas the utterance grading process grants the possibility of evaluating the actual knowledge, meaning and the impact of each intervention within the conversation. In our current approach, various metrics are applied on both the utterance graph and the participants' graph depicting the interactions between users: degree (indegree, outdegree), centrality (closeness centrality, graph centrality, eigen-values) and user ranking based on the PageRank algorithm [21]. Besides the evaluation of each utterance, the assessment of collaborative discourse (or collaboration in the conversation) is also a key component of our analysis and is described in detail later in this section. As can be easily observed, the scoring and the feedback provided by PolyCAFe use both content/knowledge-related features and involvement/collaboration criteria because both of these perspectives are important within a good collaborative conversation.

3.2 Utterance Assessment

The utterance scoring process intertwines 3 different perspectives: a *quantitative* one inspired from traditional information retrieval, a *qualitative* dimension based on Latent Semantic Analysis and a *social perspective* derived from social network analysis applied on the utterance graph (see Fig. 2).

The *quantitative dimension* assesses at surface level the impact of an utterance and is derived from information retrieval techniques in the sense that term frequency expressed as a logarithm function is used instead of the actual number of occurrences of stems. The purpose of this perspective is to reward the involvement of each participant in terms of written communication by taking into consideration the length in characters of each processed word (after spellchecking, stop words elimination and stemming are applied on the initial text).

On the other hand, the *qualitative perspective* grants a more cognitive insight of the actual importance of each utterance and covers 4 different aspects with regard to the utterance graph and the corresponding discussion threads. All these factors can be seen as a bonus from the initial quantitative mark and define a clear perspective of the influence, contextual importance and knowledge withheld by each utterance.

Thread coherence is focused on evaluating inner cohesion of the discourse and continuity within a discussion thread. Therefore, it represents the percentage of links, starting from a specific utterance, which share a (lexical and semantic) similarity above a given threshold determined experimentally through multiple runs.

Future impact highlights the impact of the current utterance onto future interventions in the discussion, with regard to utterance inter-animation within discussion threads. Due to the high correlation between the semantic similarity of two utterances and the impact of one utterance onto another in the discourse, this factor is obtained by summing up all cosine similarities exceeding a predefined threshold between the current utterance and all further interlinked utterances. Another aspect that needs to be addressed is *relevance* with regards to the entire discussion, measured in terms of similarity between current utterance and the vector of the entire chat.

Due to the nature of the discussions in a predefined academic setup, *completeness* is seen as the coverage of key topics defined for each conversation by the teacher, and it also plays an important role in the overall grading process. If no topics are manually set before commencing the discussion by the teacher, the system can automatically determine the central keywords within the discussion by means of LSA and computes this factor with regards to that list. Otherwise, the central keywords or concepts are those selected by the teacher before analyzing the conversations.

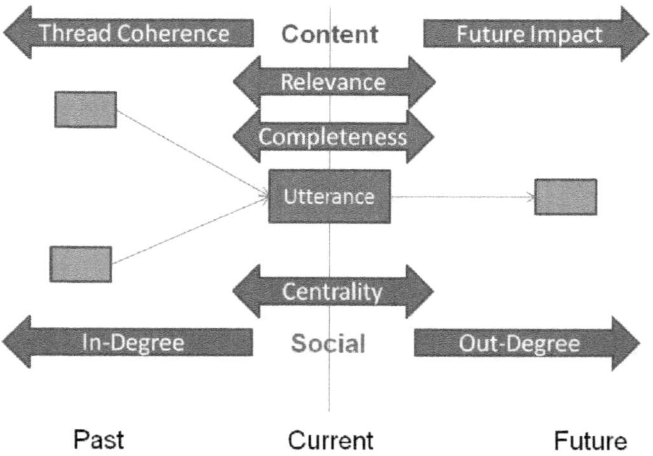

Fig. 2. The main social and content-oriented components of the utterance assessment process

From another perspective, *social network analysis* is performed on the utterance graph, granting further insight into the importance of each unit of analysis within its corresponding discussion threads. This enables a deep understanding of the discourse in term of utterance intertwining based on explicit and implicit links [19].

The final score of each utterance is obtained by multiplying all previous factors from the simplest (the qualitative mark), to the later (qualitative and social dimensions). This approach grants a deep understanding of the importance of each utterance within the discourse structure by emphasizing on the impact, the coherence and the relevance of each unit of analysis within the overall conversation [19].

3.3 Collaboration Assessment

Starting from the perspective that knowledge advancement can be achieved as a community rather than as individual achievement [22], the concept of *gain* transposes in our computational approach both personal and collaborative knowledge building. Gain expresses the cognitive input of each utterance in the overall discourse by taking into consideration its punctual importance (the previously assigned mark) and its future impact.

The natural way of achieving the split between inner and collaborative knowledge building is by considering the interlocutors within a discussion thread. Therefore, if interlinked (explicitly or implicitly) utterances have the same speaker we consider that information is derived from self-experience and by continuing internal thoughts. On the other hand, if interlinked utterances have different participants, collaborative gain is identified in the interchange of ideas and knowledge between the participants. The degree of collaboration within the conversation, which can be also seen as the extent of a collaborative discourse, assesses how well these ideas inter-animate each other and how collaborative gain supports the discussion by integrating multiple perspectives of different participants.

To sum up, gain concentrates the mark and the gain built within all previous interlinked utterances in the current one, filtered by means of semantic similarity measured by LSA.

This approximation of knowledge building as gain clearly expresses the idea of interchanging concepts, thoughts and knowledge between participants. From this perspective, each utterance has its own static mark as a symbol of its importance and relevance within the discourse, whereas collaboration is built constantly within discussion threads by inter-animating interventions between participants.

4 Validation Experiment and Results

After the first validation pilot of using PolyCAFe showed that the system was efficient and effective for both learners and tutors [7, 23], a new validation experiment has been undertaken to further study the effects of using an improved version of the system with a larger group of students. The new experiment was integrated as a learning task and assignment for a group of senior year undergraduate students studying Human-Computer Interaction (HCI). A total of 35 students have been engaged in the experiment for several weeks: 25 students were part of the experimental group and 10 students were assigned to the control group. The only difference between the experimental and control group is that the latter did not receive any feedback from PolyCAFe, but only from the tutors. The experiment was structured as follows: the learners were divided into groups of 5 students, thus having 5 experimental and 2 control groups, and were given two successive chat assignments related to HCI to debate using ConcertChat. The experimental group was asked to use PolyCAFe to get feedback for the first assignment, while the control group did not use the system for this assignment. The use of PolyCAFe for the second assignment was not mandatory, so the learners had an option to use PolyCAFe only if they considered it would be useful for them. The two topics for the assignments were:

- "A debate about the best collaboration tool for the web: chat, blog, wiki, forums and Google Wave. Each student shall choose one of the 5 tools and shall present its advantages and the disadvantages of the other tools. Thus, you will act as a "*sales person*" for your tool and try to *convince* the others that you have the best offer. You must also defend your product whenever possible and criticize the other products if needed."
- "Each of you has studied one of the following technologies that are considered by your company: chat, blog, wiki, forums and Google Wave. *Engage into a collaborative discussion in order to decide for which activities it is indicated to use each technology*. You should give the best advice for the technology that you support and convince the others to use it. The result should be a plan of using these technologies to have the best outcomes for your company. "

The tutors had to provide manual feedback to each of the students involved in the chat conversations for the first assignment. Each tutor assessed at least one conversation without using PolyCAFe and one conversation using PolyCAFe. However, no manual feedback has been provided for the second assignment, only the feedback provided by PolyCAFe.

At the end of the validation experiment, all the students and tutors had to answer a questionnaire and also participated in focus groups and interviews. The results of the experiment have been devised into several validation topics (VT): tutor efficiency, quality and consistency of the automatic feedback, making the educational process transparent, quality of educational output, motivation for learning, etc. All these topics have been validated conditionally or with minor qualifications, but the results for only three of them are reported in the remainder of this section:

- VT1: Tutors/facilitators spend less time preparing feedback for learners compared with traditional means (tutor efficiency);
- VT2: Learners perceive that the feedback received from the system contributes to informing their study activities (quality and consistency of automatic feedback);
- VT3: Learner performance in online discussions is improved in the areas of content coverage and collaboration when using PolyCAFe (quality of educational output).

4.1 Tutor Efficiency

In order to validate the topic VT1: "Tutors/facilitators spend less time preparing feedback for learners compared with traditional means" related to tutor efficiency, several methods have been used: measurements, questionnaires and the answers to the interviews. The complete set of statements related to this validation topic is presented in Table 1, together with the mean, standard deviation and agreement percentage for each statement. The results are on a Likert scale from 1 to 5, with 5 = strongly agree and they show a good consensus of the 6 tutors with regard to the efficiency of using PolyCAFe, with averages over 4.5 and agreement factors of over 83% for all statements.

Table 1. Results for the validation statements related to VT1

Questionnaire no. & statement	Mean	Standard deviation	% Agree / Strongly agree
7. It takes less time to complete my teaching tasks using PolyCAFe than without the system.	4.5	0.84	83%
8. Using PolyCAFe enables me to work more quickly than without the system.	4.5	0.84	83%
9. I do not wait too long before receiving the requested information.	4.8	0.41	100%
10. PolyCAFe provides me with the requested information when I require it (i.e. at the right time in my work activities).	4.7	0.52	100%
36. I find the information needed to write the feedback for the learners more quickly using PolyCAFe than without it.	4.7	0.52	100%

Besides the questionnaires, time measurements for preparing the feedback by the tutors were also used. Thus, each chat conversation has been analyzed and provided feedback for by 4 tutors: 2 using PolyCAFe and 2 not using the system. This data has been compared for all the 7 chats resulted for the first assignment. The average time needed to prepare feedback without PolyCAFe was *84 minutes*, with a standard deviation of 15 minutes, while the average time required for the feedback with PolyCAFe was of *55 minutes* with a standard deviation of 20 minutes. These results show that the average time saved for a chat conversation is (84 – 55) / 84 = 35%, resulting in almost half an hour less time required per chat. However, as the standard deviation has increased, it also demonstrates that not all the tutors have managed to use the software efficiently.

4.2 Quality and Consistency of the Automatic Feedback

The validation topic "Learners perceive that the feedback received from the system contributes to informing their study activities" has been validated using questionnaires for the experimental group of 25 students, plus system logging. Table 2 presents the results for the validation statements that were linked to this topic, that denote agreement factors between 60-80% and means between 3.7-4.0.

The system logging utilities that monitored the access of the students to PolyCAFe have shown that for the whole period of the validation experiment there have been 285 visits to PolyCAFe and 1447 page-views, that results in more than 40 page-views in average per student. Therefore, the students have been actively using the system in order to reflect on their activity in the chat conversation.

Table 2. Results for the validation statements related to VT2

Questionnaire no. & statement	Mean	Standard deviation	% Agree / Strongly agree
6. The information the system provides me is accurate enough for helping me perform my learning tasks.	3.7	0.52	60%
31. PolyCAFe provides feedback that is relevant to my study activities.	3.9	0.91	72%
32. PolyCAFe provides feedback that is useful to my study activities.	3.8	0.85	72%
33. PolyCAFe's feedback is sufficiently accurate to inform my study activities.	3.8	0.88	64%
34. I trust PolyCAFe to provide helpful feedback.	4.0	0.87	80%

4.3 Quality of the Educational Output

The last validation topic covered in this section is VT3: "Learner performance in online discussions is improved in the areas of content coverage and collaboration when using PolyCAFe" was validated using measurements computed by PolyCAFe for the second chat assignment: the most important concepts in the conversation and their score, plus the average grade for a utterance in a chat conversation. Moreover, the density of implicit and explicit links between utterances was also a factor to account for collaboration. The results presented in Table 3 show a better average score for the 5 chats of the experimental group compared to the one of the 2 chats in the control group. However, only for the estimation of collaboration measured in the density of links (average number of links/utterance) and for the average score of a utterance there was found a substantial increase between the two groups.

Table 3. Results for the measurements of the differences between the chats for the second assignment between the experimental and the control group

Criteria	Experimental group for the second assignment	Control group for the second assignment	Improvement over control group
Average score for a chat conversation (collaboration + content)	6.80	6.37	(6.80-6.37)/6.37 = 6.8%
Average importance of the most important 20 concepts	0.194	0.192	1.2%
Average number of utterances	351	338	(351-338)/338 = 3.8%
Average density of (implicit and explicit) links between utterances	1.12	0.87	(1.12-0.87)/0.87 = 29%

5 Verification Results

Several verification experiments were also undertaken in order to assess the accuracy of the results provided by PolyCAFe. However, in this section only the ones that are relevant to the functionalities described in section 3 are presented: ranking of the participants and utterance assessment.

5.1 Ranking of the Participants

For all the chats of the first assignment, the tutors had to rank the participants according to the importance they had in the conversation, taken into account both the content of their utterances and their involvement and degree of collaboration with the other participants. Therefore, for each participant to a conversation a rank from 1 to 5 was assigned by each tutor. For experimental and pedagogical reasons, each student was also asked to rank the other participants in his conversation using scores from 1 to 4. These results were then compared with the automatic ranking of the participants provided by PolyCAFe that takes into account the average score of an utterance for each participant, plus SNA metrics computed on the graph of the participants (e.g. centrality and PageRank). An example of rankings received from students, tutors and the system for a single conversation is presented in Table 4.

Table 4. Example of ranking the participants in a chat conversation: by their colleagues, the tutors and the system

Rank	Student 1	Student 2	Student 3	Student 4	Student 5
Student 1	-	2	3	1	4
Student 2	2	-	3	1	4
Student 3	2	3	-	1	4
Student 4	1	2	3	-	4
Student 5	1	2	4	3	-
Student average	2	3	4	1	5
Tutor 1	4	1	5	2	3
Tutor 2	4	2	5	1	3
Tutor average	4	1-2	5	1-2	3
PolyCAFe	4	2	5	1	3

The results of this experiment show that the system achieves a 66% precision and an 84% correlation with the average student scores. Moreover, the average ranking distance between the students and the system is only 0.43 ranks per participant. However, *the most interesting result is that the system achieves a much higher precision and correlation with the average tutor score: 77% and 94% respectively*. In this case the average difference between the tutor score and the system score is of only 0.23 ranks per participant. The fact that the system achieves a better correlation with the tutors' score rather than the students' one might show: (1) the system evaluated correctly the participants to a collaborative chat conversation, and (2) the students do not have the same assessment abilities as tutors. These results are summarized in Table 5.

Table 5. Comparison overview of the average rankings provided by the tutors, the students and the system

Rankings compared	Correlation	Precision	Average distance
Tutors – System	0.94	77%	0.23
Students – System	0.84	66%	0.43
Tutors – Students	0.84	71%	0.40

The most important conclusion of this experiment is that although the average rankings of the students, the tutors and the system are quite well correlated one with the other, on an individual basis the correlation between different tutors (or students) drops dramatically. For example, the average inter-tutor correlation for the same set of conversations is just 56%, while the average precision of the rankings by two different tutors is of just 40%. Moreover, the mean distance between the ranks provided by two tutors for the same participant rises to 0.84. All these results encourage us to say that although the grading/evaluation/ranking criteria of tutors are not the same, *the system is well correlated with the average rank provided by different tutors, thus being more objective*. Even more important is that *the correlation between the system and the tutor average is better than the average inter-tutor correlation (77% as compared to just 56%)*.

5.2 Utterance Assessment Verification

The second verification experiment involved four tutors who have manually annotated each utterance from three distinct chat conversations from the first assignment with scores from 1 to 4. In order to be able to compute the inter-rater agreement, one chat conversation was annotated by two tutors. The results are presented in Table 6 and they show that *the system has an average correlation to the tutor average just several point below the inter-tutor correlation*. Moreover, the average precision of the automatic scoring is 51%.

Table 6. Overview of the utterance assessment and scoring of PolyCAFe compared to human experts

Chat number	Number of utterances	Tutor 1 – Tutor 2 correlation	Tutor 1 – PolyCAFe correlation	Tutor 2 – PolyCAFe correlation	Tutor average – PolyCAFe correlation
Chat no. 1	331	0.61	0.60	0.51	0.57
Chat no. 2	277	N/A	0.55	N/A	0.55
Chat no. 3	321	N/A	0.52	N/A	0.52

The average error in absolute value between the scores provided by PolyCAFe and the one of the tutors is of 0.63 units per utterance and the average (non-absolute) error is of -0.35 units per utterance, therefore, on average, the system underestimates the score of an utterance. It should be taken into consideration that this task is much more difficult than the previous one, as the classes are not disjoint and the baseline precision of selecting a random score for an utterance is of just 25%.

6 Conclusions

The system presented in this paper was designed particularly for providing feedback to learners using chat conversations for collaborative assignments. PolyCAFe is employing NLP and SNA technologies in order to implement a novel theory of discourse based on detecting the discussion threads of a conversation, based on the explicit and implicit linkages between the different postings. Several indicators and textual feedback are presented to the learners in order to help them reflect on their learning activities. Moreover, the system is also used by tutors to reduce the time needed to provide manual feedback to their students.

The validation experiment run in a formal education context has shown the acceptance of PolyCAFe by both learners and tutors. Moreover, the first results demonstrate that the feedback provided by PolyCAFe has positive effects to the subsequent collaborative chats of the learners, mostly related to the degree of collaborative discourse than to the actual content of the conversations.

However, in order to use the system in other contexts several points should be taken into consideration. The system was developed to work for English and language transferability implies integrating several linguistic tools in the current framework for each new language (e.g. NLP pipe, linguistic ontology, adjacency pairs and other linguistic patterns). In order to use the system for another domain, the only required resource is a latent semantic space or a corpus in order to train this space that is relevant to the actual assignment and the system has already been transferred to a new domain: Medicine. The last aspect is related to learning task transferability: the system is suited especially for tasks where the students are encouraged to engage into a collaborative discourse and that are not scripted.

Acknowledgments. The research presented in this paper was partially performed under the FP7 ICT STREP project LTfLL (http://www.ltfll-project.org/). The authors would like to thank the students and tutors that participated in the validation and verification experiments.

References

1. Stahl, G.: Group cognition: Computer support for building collaborative knowledge. Cambridge. MIT Press, Cambridge (2006)
2. Adams, P.H., Martell, C.H.: Topic detection and extraction in chat. In: Proceedings of the 2008 IEEE International Conference on Semantic Computing, pp. 581–588 (2008)
3. Dong, A.: Concept formation as knowledge accumulation: A computational linguistics study. Artif. Intell. Eng. Des. Anal. Manuf. 20(1), 35–53 (2006)
4. Joshi, M., Rosé, C.P.: Using transactivity in conversation summarization in educational dialog. In: Proceedings of the SLaTE Workshop on Speech and Language Technology in Education (2007)
5. Rose, C.P., Wang, Y.C., Cui, Y., Arguello, J., Stegmann, K., Weinberger, A., Fischer, F.: Analyzing Collaborative Learning Processes Automatically: Exploiting the Advances of Computational Linguistics in Computer-Supported Collaborative Learning. International Journal of Computer Supported Collaborative Learning (2007)
6. Dascalu, M., Trausan-Matu, S., Dessus, P.: Utterances Assessment in Chat Conversations. Research in Computing Science, Special Issue on Natural Language Processing and its Applications 46, 323–334 (2010)

7. Rebedea, T., Dascalu, M., Trausan-Matu, S., Banica, D., Gartner, A., Chiru, C., Mihaila, D.: Overview and preliminary results of using polyCAFe for collaboration analysis and feedback generation. In: Wolpers, M., Kirschner, P.A., Scheffel, M., Lindstaedt, S., Dimitrova, V. (eds.) EC-TEL 2010. LNCS, vol. 6383, pp. 420–425. Springer, Heidelberg (2010)
8. Trausan-Matu, S., Stahl, G., Sarmiento, J.: Supporting Polyphonic Collaborative Learning. E-service Journal 6(1), 58–74 (2007)
9. Trausan-Matu, S., Rebedea, T.: A Polyphonic Model and System for Inter-animation Analysis in Chat Conversations with Multiple Participants. In: Gelbukh, A. (ed.) CICLing 2010. LNCS, vol. 6008, pp. 354–363. Springer, Heidelberg (2010)
10. Bakhtin, M.: Problems of Dostoevsky's poetics (C. Emerson, Trans.). University of Minnesota Press, Minneapolis (1984)
11. Kumar, R., Ai, H., Beuth, J.L., Rosé, C.P.: Socially Capable Conversational Tutors Can Be Effective in Collaborative Learning Situations. In: Aleven, V., Kay, J., Mostow, J. (eds.) ITS 2010. LNCS, vol. 6094, pp. 156–164. Springer, Heidelberg (2010)
12. Jurafsky, D., Martin, J.H.: Speech and Language Processing, 2nd edn. An Introduction to Natural Language Processing, Computational Linguistics, and Speech Recognition. Pearson Prentice Hall, London (2009)
13. Hobbs, J.: On the coherence and structure of discourse. Technical Report CSLI-85-37, Center for the Study of Language and Information, Stanford University (1985)
14. Grosz, B.J., Joshi, A.K., Weinstein, S.: Centering: A framework for modeling the local coherence of discourse. Computational Linguistics 21(2), 203–225 (1995)
15. Mann, W.C., Thompson, S.A.: Rhetorical structure theory: A theory of text organization. Tech. rep. RS-87-190, Information Sciences Institute (1987)
16. Trausan-Matu, S., Rebedea, T.: Polyphonic Inter-Animation of Voices in VMT. In: Stahl, G. (ed.) Studying Virtual Math Teams, pp. 451–473. Springer US, Boston (2009)
17. Schmidt, A.P., Stone, T.K.M.: Detection of topic change in IRC chat logs, `http://www.trevorstone.org/school/ircsegmentation.pdf`
18. Kontostathis, A., Edwards, L., Bayzick, J., McGhee, I., Leatherman, A., Moore, K.: Comparison of Rule-based to Human Analysis of Chat Logs. In: 1st International Workshop on Mining Social Media Programme, Conferencia de la Asociación Española para la Inteligencia Artificial (2009)
19. Dascalu, M., Rebedea, T., Trausan-Matu, S.: A Deep Insight in Chat Analysis: Collaboration, Evolution and Evaluation, Summarization and Search. In: Dicheva, D., Dochev, D. (eds.) AIMSA 2010. LNCS (LNAI), vol. 6304, pp. 191–200. Springer, Heidelberg (2010)
20. Wresch, W.: The Imminence of Grading Essays by Computer–25 Years Later. Computers and Composition 10(2), 45–58 (1993), `http://computersandcomposition.osu.edu/archives/v10/10_2_html/10_2_5_Wresch.html`
21. Brin, S., Page, L.: The Anatomy of a Large-Scale Hypertextual Web Search Engine. In: Seventh International World-Wide Web Conference, Brisbane, April 14-18 (1998)
22. Scardamalia, M.: Collective cognitive responsibility for the advancement of knowledge. In: Smith, B. (ed.) Liberal Education in Knowledge Society, pp. 67–98. Open Court, Chicago (2002)
23. Dascalu, M., Rebedea, T., Trausan-Matu, S., Armitt, G.: PolyCAFe: Collaboration and Utterance Assessment for Online CSCL Conversations. Accepted at the 9th International Conference on CSCL, CSCL 2011 (2011)

Monitoring Pattern-Based CSCL Scripts: A Case Study

María Jesús Rodríguez-Triana, Alejandra Martínez-Monés,
Juan Ignacio Asensio-Pérez, Iván Manuel Jorrín-Abellán, and Yannis Dimitriadis

GSIC-EMIC, University of Valladolid, Spain
{chus@gsic,amartine@infor,juaase@tel,ivanjo@pdg,yannis@tel}.uva.es
http://www.gsic.uva.es

Abstract. Two strategies have been proposed in CSCL to foster effective collaboration: structuring the learning scenario by means of collaboration scripts and monitoring interactions among participants in order to detect and regulate potential deviations from the initial plan. In order to help teachers in this endeavor, we propose to combine these two approaches by means of a process where design takes into account the especial requirements posed by monitoring, and monitoring is informed by the characteristics of the scripts that must be met to achieve the learning goals. These desired features are obtained from the constraints defined by the collaborative-learning flow patterns on which the scripts are based. The result is an automated and higher-level view about the evolution of the learning process, integrating the data gathered from the different tools. This paper also presents a case study based on an authentic experience in higher education where these ideas were tried out.

Keywords: CSCL, scripting, monitoring, Collaborative Learning Flow Patterns.

1 Introduction

Computer-Supported Collaborative Learning (CSCL) scenarios are a clear example within TEL where Information and Communication Technologies (ICT) are employed in order to improve learning and teaching [17] [21]. Since in these scenarios learning occurs largely through interactions between participants, promoting effective collaboration is one of the main concerns [3]. *Scripting* and *monitoring* are two techniques long discussed in the research community aimed to foster effective collaborative learning interactions [13]. On the one hand, CSCL scripting structures the learning scenario and provides students with a set of instructions that guide potentially fruitful collaboration; on the other hand, monitoring the collaboration facilitates the intervention in order to redirect the group work in a more productive direction.

Within the wide range of CSCL scripting techniques, we can find the category of *macro-scripts* [7] (scripts hereafter), which mainly describes how groups (and individual roles) should perform a set of interrelated activities [4]. Such descriptions can even be made computer-interpretable by means of Educational Modeling Languages such as IMS Learning Design [12]. Since modeling potentially effective CSCL scripts is a difficult task, the use of patterns that reflect best practices in structuring collaborative learning has proved to be helpful [6] [19] [23]. For instance, *Collaborative Learning Flow*

C. Delgado Kloos et al. (Eds.): EC-TEL 2011, LNCS 6964, pp. 313–326, 2011.

Patterns (CLFPs) are a particular type of CSCL scripting patterns that describe well-accepted ways of arranging activities in collaborative learning scenarios [11]. Among other elements, CLFPs describe what types of interaction among learners and teachers should occur within activities.

Even though design is one important factor in building effective collaborative learning [8], having a "plan" is not enough to ensure that desired collaboration occurs: eventualities and contingencies require modifications of the course of the learning situation, leading teachers to adapt the original design [3]. At this point, monitoring the learning scenario and comparing its actual and expected states may provide useful information that the teacher can use to regulate collaboration [20].

In these scenarios collaboration takes place, at least in part, through computers. Analyzing these computer-mediated interactions may help to better understand collaboration [20]. This has been a strong trend in CSCL in the last years, but it has mostly focused on detailed interaction-analysis approaches aimed at letting researchers gain insight into the collaborative processes [14]. However, teachers need a more abstract idea of collaboration, easier to interpret so that they can react on time if needed. Additionally, the use of diverse ICT tools in the classroom would oblige teachers to integrate data from all the tools and environments where the learning process takes place. This is time-consuming and error-prone, if not unfeasible.

This paper proposes a method to get an automated and higher level view about the evolution of the learning process by combining the two mentioned approaches, i.e., structuring by means of pattern based scripting (to be more precise, supported by CLFPs), and monitoring collaboration. We propose to use the knowledge about the expected collaboration provided by the script, together with the data provided by the different sources in the technological infrastructure used for the enactment (VLE, Web 2.0 tools, etc.). We hypothesize that monitoring at a coarse level using these data is enough to provide teachers with relevant information about the current state of the learning process, helping them to react in time to redirect the course of the learning situation.

We present a pilot study based on data from an experience in higher education where these ideas were tried out. While the students collaborated following a script, we analyzed the data obtained from log-files and compared them with the expected collaboration as defined by the script. With these results, we provided teachers with visualizations of the process that helped them to understand and detect situations where actions needed to be taken (e.g. solving technical problems, extending the deadlines or modifying the distribution of the groups). The results were positive and set the lines for future research plans.

The remaining of this paper is structured as follows: section 2 introduces the proposal; section 3 is devoted to explain the case study (description, life-cycle and findings); finally conclusions and future work are summarized in section 4.

2 Combining Scripting and Monitoring to Enhance Collaboration

As mentioned before, the work presented in this paper combines both scripting and monitoring during the whole CSCL learning lifecycle to enhance collaboration. This section describes the role that scripting and monitoring play in the proposal. The ideas are illustrated by means of an example.

2.1 The Approach throughout the CSCL Learning Lifecycle

The process proposed in this work spans different phases of the CSCL learning life-cycle [9] (see Figure 1): the preparation of the learning scenario (design); the deployment/contextualization of the designed activities to address the concrete tool instances, participants and groups that will participate in their execution (particularization and instantiation); the execution of the activities themselves (enactment) and, eventually, the revision and refinement of those activities (evaluation).

First of all, the **design** must take into account the especial requirements posed by monitoring, in a similar way to existing proposals in which other tasks, such as assessment or evaluation are considered in advance [18]. As a consequence, some decisions must be taken explicitly, together with the rest of the learning process before the learning situation is implemented: defining how collaboration is expected to happen, choosing suitable tools for these collaboration purposes, identifying the key collaborative aspects and planning how they will be monitored. The result of this phase is what we have called the *monitorable collaboration schema*. It contains pointers to those phases of the script that have to occur in order to obtain its expected benefits.

During **enactment**, *actual monitoring results* are obtained from those sources that offer relevant information for collaboration analysis purposes (e.g., looking for evidence of expected interaction during a discussion activity in the logging data of a chat tool).

Finally, during **evaluation**, the *actual monitoring results* are contrasted with the *monitorable collaboration schema*, obtaining graphical visualizations of the progress of the learning situation, with the aim of helping teachers to detect potentially critical situations where actions must be taken.

The key issue in this process lies in the identification of the script constraints since this information guides the whole process. Next section deals with this issue in depth.

2.2 Pattern-Based Guidance of the Monitoring of the Collaborative Scripts

The design phase of the process described in the previous section is to be informed by an analysis of the collaboration scripts. During this phase, authoring tools will guide designers to identify those aspects of the script that must be accomplished in order to achieve the learning goals. Several authors [5] [2] [15] have analyzed scripts in order to identify which features are modifiable (extrinsic constraints) and which ones have to be accomplished in order to maintain their pedagogical intentions (intrinsic constraints). In this section, we focus on determining which collaborative interactions should take place in order not to compromise the fulfillment of the intrinsic constraints of the script.

As mentioned before, the scripts considered in this work are based on a especific type of patterns, the CLFPs. For this reason, the requirements that the script must satisfy in order to achieve the learning objectives derive from the ***pattern constraints***. These *constraints* are the conditions that should be met in a learning situation to be considered an implementation of such a pattern. Attending to the classification of the structural components of the scripts [3] [16], these constraints are represented by the group formation policies, the sequence of phases, the expected interactions, etc. The rest of the constraints of a script are the result of the ***particularization and instantiation*** of the pattern to each specific context [24], so they are modeled by means of elements like concrete tasks, resources, tools, participants, group composition, etc.

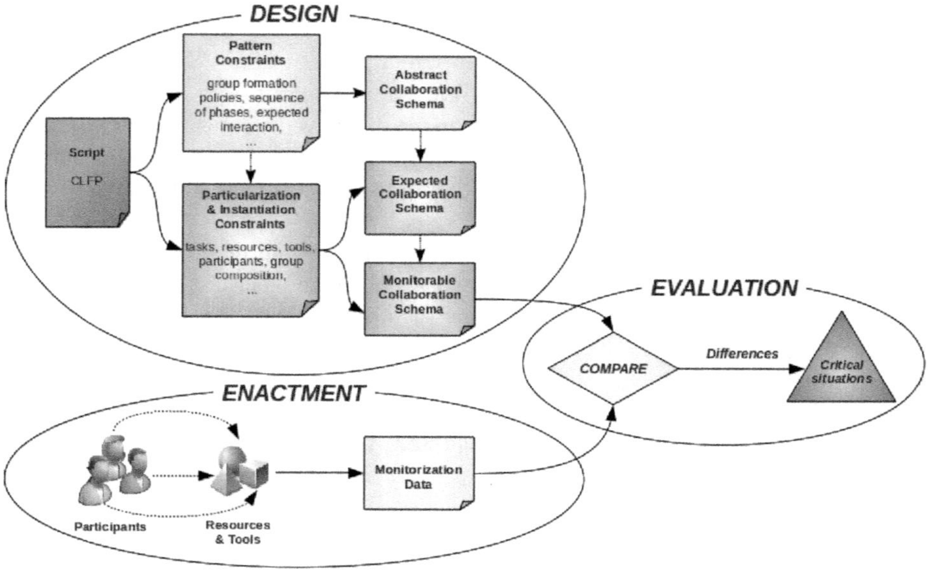

Fig. 1. Overview of the approach: associated tasks and expected results

The determination of the script constraints has significant implications for the monitoring of the learning situation. First, by means of the pattern constraints it is possible to extract the heuristics that define the expected sequence of phases, as well as some collaboration guidelines and the rules for structuring groups. These heuristics constitute what we have called an *abstract collaboration schema*. Secondly, particularization and instantiation constraints provide the description of the context and the details about how such collaboration should happen (face to face, by means of a collaborative tool or a shared resource, synchronous or asynchronous...). This information is used to build the *expected collaboration schema*. Additionally, it is possible to establish which collaboration may be computationally monitored depending on the expected interaction and the computer support (e.g. which resources or tools provide history data), obtaining with it the *monitorable collaboration schema*. During the enactment of the learning situation, the monitorable collaboration schema guides the data gathering, focusing the process on obtaining evidence about the expected collaboration.

Summarizing, by means of the analysis of the CLFPs, we expect to support CSCL practitioners during the design process in the decision-making and the identification of the key aspects of collaboration. The analysis of two CLFPs, *Brainstorming* and *Pyramid* [10], is presented in sections 2.3 and 3.2. The result of this process, the monitorable collaboration schema, will guide the data gathering during the enactment of the learning situation, focusing on monitoring those key aspects of the script.

2.3 An Illustrative Example

This section illustrates the proposal with an example. Let us suppose that a teacher is designing a collaborative learning situation that implements a *Brainstorming* CLFP

[10]. In this activity, students, working *in pairs*, have to collect the relevant keywords related to a lesson, putting them altogether in a list. Then, they have to review the list and draw a concept map using the terms included in the list. The activity lasts one week, combining *face-to-face* sessions with *distance* work mediated by several ICT tools. Participants will have at their disposal a shared document (*Google Document*[1]), to elaborate the list, and a shared board (*Dabbleboard*[2]), to draw the concept map.

This description of the learning situation contains some issues that are key to achieve the learning objectives: there are two phases, the brainstorming and the review, and intra-group collaboration is expected in both of them. Such issues, are represented by the *abstract collaboration schema* depicted in Figure 2 (step 1).

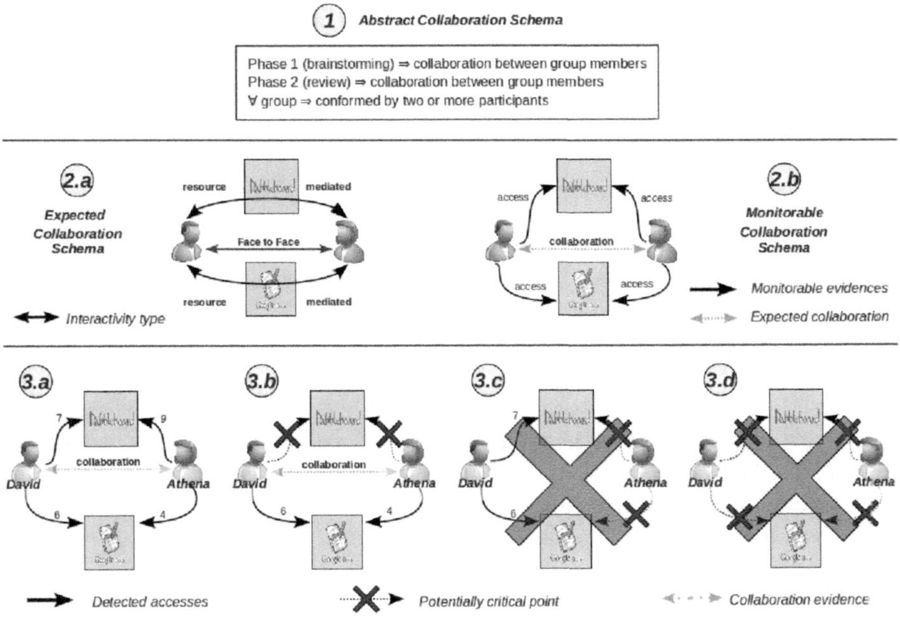

Fig. 2. Collaboration schemas and potential monitorized scenarios

Taking into account that the expected collaboration from students during the learning situation is to take place both face to face and through the shared resources (GoogleDocs and Dabbleboard instances), the heuristics obtained in the first step appear particularized in the *expected collaboration schema* (step 2.*a* of Figure 2). Due to the lack of observers during the enactment, face to face interactions are not monitored; nevertheless, the software used in the learning scenario registers when users access to the shared resources. Based on this, the evidences that are used to detect collaboration are shown in the *monitorable collaboration schema* (step 2.*b* of Figure 2).

[1] `http://docs.google.com/`

[2] `http://www.dabbleboard.com/`

The third level of Figure 2 presents four hypothetical situations that might happen during the enactment of the script. A number appears next to each arrow, representing how many times the user has accessed to each resource and, in that cases where there is no evidence of collaboration, it has been pointed out by a red cross over a dotted line. The first case (*3.a*), shows a scenario where both members of the dyad have accessed the shared resources, so it seems that they have co-worked as it was expected. The second one (*3.b*), exemplifies a situation where students have not worked with all the resources they have to. Despite there is evidence of collaboration, the second phase of the learning situation does not seem to have been carried out yet, so the teacher may consider to intervene. The last two examples (*3.c* and *3.d*) illustrate two potentially critical situations where there is no collaboration evidence: the former shows a learning scenario that seems to have been developed by just one participant; in the latter, the activities probably are not being carried out by any of them.

Using this information, and enriching it with their knowledge about face-to-face sessions, teachers should be aware of the participants' progress during the course of the activity, and be able to react on time to undesired situations.

3 Case Study: Monitoring an Authentic Learning Scenario

In this paper, the authors address one main research question: "does the combination of pattern based scripting and monitoring collaboration contribute to provide useful information about the evolution of the learning process?" In our attempt to understand in depth this research question, an instrumental case study [22] was carried out on an authentic CSCL scenario. The authors followed the presented approach in a learning experience in which a CLFP-based script was applied. First, the teacher was guided during the design process and, later on, participants' activity was monitored during the enactment of the experience.

This section is structured as follows: first, we present briefly the main characteristics of the case; then, following the steps presented in the approach, we explain how the design guided the monitoring process, as well as some examples of how some undesired collaboration situations were detected; and finally, the section ends with the discussion of the results obtained from the case study.

3.1 Context and Methodologies of the Study

The case study presented here was developed in November 2010, and took place within a third year course (out of five) on "Network traffic and Management" of Telecommunications Engineering degree, at the University of Valladolid (Spain), with 46 students attending the course. During this course, students had to develop a chat tool using data network protocols. In order to help them to plan and anticipate problems for a subsequent programming assignment, they were asked to elaborate a sequence diagram of their software design. To elaborate this diagram, students worked in a blended CSCL situation, interleaving *face-to-face* with *distance activities mediated by ICT tools*.

The collaboration script implemented a two-level *Pyramid* CLFP [10]. At level-1, *groups* of 2 participants attended to a face-to-face lab session to carry out the first activity. In this activity, students had to draw a preliminary version of the sequence diagram and write a report with a summary of the main decisions and open issues. At level-2, groups joined to conform *super-groups* (composed of 4 groups) that had to accomplish both a distance and a face-to-face activity. During the former, each *group* had to review and provide feedback on the reports produced by their *super-group* mates; in the latter, they had to discuss and produce a joint version of the diagram, as well as perform an oral presentation with a common version of the conclusions and open issues.

Regarding the technological support, teachers used the VLE *Moodle* [3] to centralize the access to all the resources and activities. To accomplish the drawing tasks, students were provided with a shared board (*Dabbleboard*), and in order to explain, review and discuss, they had at their disposal shared documents and presentations (*Google Documents* and *Google Presentations*). Since these tools can not be automatically integrated in Moodle, the *GLUE!* architecture [1] was used to integrate them into the VLE.

With this case study we aimed at illuminating three issues that would help us to gain insight into our research question. These issues were: "does the CLFP description help to identify the constraints of the collaboration script?", "does the comparison between desired an actual collaboration provide teachers with relevant information about the current state of the learning process?" and, in a learning scenario characterized by the integration of ICT tools, "which are the required conditions for collecting relevant information about the key collaborative aspects?".

Our results were triangulated with data coming from observations of the face-to-face activities; a questionnaire that was handed to the students at the end of the experience; a group interview with students that volunteered to participate at the end of the activity, and an interview to the teachers.

3.2 Collaboration Schemas of the Proposed Script

According to the process proposed in section 2.1, the teacher is guided during the design to define two schemas, namely: the expected collaboration schema, which defines how collaboration should occur during the enactment, and the monitorable collaboration schema, which describes the evidence of that collaboration that can be gathered from the technological environment. It is a long-term goal to support the designer in these tasks so, for this pilot study, the authors themselves defined these schemas.

First of all we will go through the pattern constraints. The *Pyramid* CLFP defines a collaborative learning flow for a context in which several participants face the same problem, whose resolution implies the achievement of gradual consensus [10]. Therefore, the pattern defines a sequence of at least two phases with some particular collaboration objectives, as well as a rule for structuring groups (see Figure 3a).

These pattern constraints are translated into the heuristics shown in Figure 3b, which constitute what we have called *abstract collaboration schema*. This information will allow us to establish connections between those students or groups that are expected to collaborate in each phase of the *Pyramid*.

Secondly, to particularize and instantiate the pattern to the especific scenario, several details were fixed: the levels of the *Pyramid*, the definition of learning objectives, the set of activities and the interactivity type (e.g. face-to-face, distance or resource mediated),

[3] http://www.moodle.org

1. Phase $i = 1$: individual (or initial group) study of the problem and solution proposal
2. Phase $i > 1$: groups compare and discuss their proposals, and propose a shared solution
3. Rule: groups join in larger groups in each phase

(a) *Pyramid* constraints

$\forall phase$ { if $level = 1 \Rightarrow$ collaboration between participants; if $level > 1 \Rightarrow$ collaboration between sub-groups }

$\forall group$ { if $level = 1 \Rightarrow$ conformed by one or more participants; if $level > 1 \Rightarrow$ conformed by two or more sub-groups of $level - 1$ }

(b) Abstract collaboration schema: heuristics

Fig. 3. Pattern analysis

the provision of necessary resources (contents and tools), and decisions about completion of activities (e.g. control of time), as well as the creation of actual groups and the binding of participants to the groups of each level of the *Pyramid*, etc. The values of the most relevant constraints are summarized in Figure 4a.

1. *Pyramid levels: 2*
   - Level 1: 24 groups of 2 participants
   - Level 2: 7 super-groups of 4 groups
2. *Activities description:*
   - A1: groups draw a sequence diagram and write a report
   - A2: groups review and provide feedback on their super-group mates' reports
   - A3: super-groups discuss, produce a diagram, and perform an oral presentation
3. *Activities interactivity type:*
   - A1: face-to-face, distance and resource mediated (Dabbleboard, GDocuments)
   - A2: distance and resource mediated (GDocuments)
   - A3: face-to-face and resource mediated (Dabbleboard, GPresentation)
4. *Resources (contents and tools):*
   - 1 shared board (Dabbleboard) and document (GDocuments) per group
   - 1 shared board (Dabbleboard) and presentation (GPresentation) per super-group

(a) Main particularization and instantiation constraints

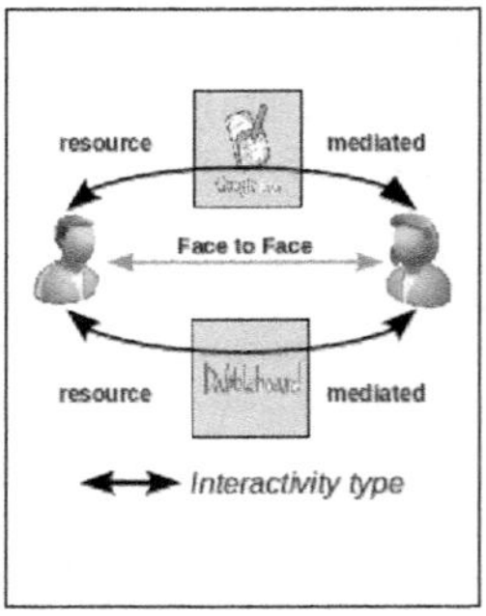

(b) Expected Collaboration Schema: 1st activity

Fig. 4. Particularization and instatiation analysis

While the pattern constraints define how to establish collaborative connections (abstract collaboration schema), the constraints derived from the particularization and instantiation let us draw the *expected collaboration schema*. This schema, as Figure 4b shows, specifies for each activity how the collaboration should occur (face-to-face, distance or both) and by means of which resources (shared documents and boards). However, not all kinds of collaboration are computationally monitorable. That is why, the next step deals with the analysis of the capabilities of the ICT tools involved experience to produce data that can be used for monitoring collaboration.

As mentioned in section 3.1, the software used to support the aforementioned learning scenario included a virtual learning environment (Moodle), a number of external Web 2.0 tools (Dabbleboard, Google Documents and Google Presentations), and an implementation of the GLUE! architecture for the integration of external tools into the VLE. Table 1 describes both the kind of monitoring data and the way in which these tools provide access to this information.

Table 1. Description of the software used in the case study in terms of (1) the information provided about the user activity and (2) the way this information is accessible

Software	User activity information	Retrievable from
Dabbleboard	None	Not possible
Google Tools	Document revision history: *user, date, time and document version*	User interface
Moodle	Event history: *date, time, IP address, user name, action, resource used*	User interface or database
GLUE!	History of accesses to the integrated resources *user, date, time, resource*	Event logs

Table 1 shows the high heterogeneity of both the data provided and the harvesting methods: for example, Dabbleboard does not provide any kind of user activity data and Google's tools do not offer a programmatic interface to obtain this information. Regarding Moodle, it would be necessary to query the database manually in order to complete the history information offered by the VLE. Taking into account that GLUE! offers largely the same information that would be obtained from Moodle, we decided to use GLUE! as the sole source of activity data. Since GLUE! registers every access to the instances of the integrated tools (Dabbleboard and Google suite in this case), the computational monitoring of the collaboration, in this case study, is based on the evidence of instance accesses provided by GLUE!.

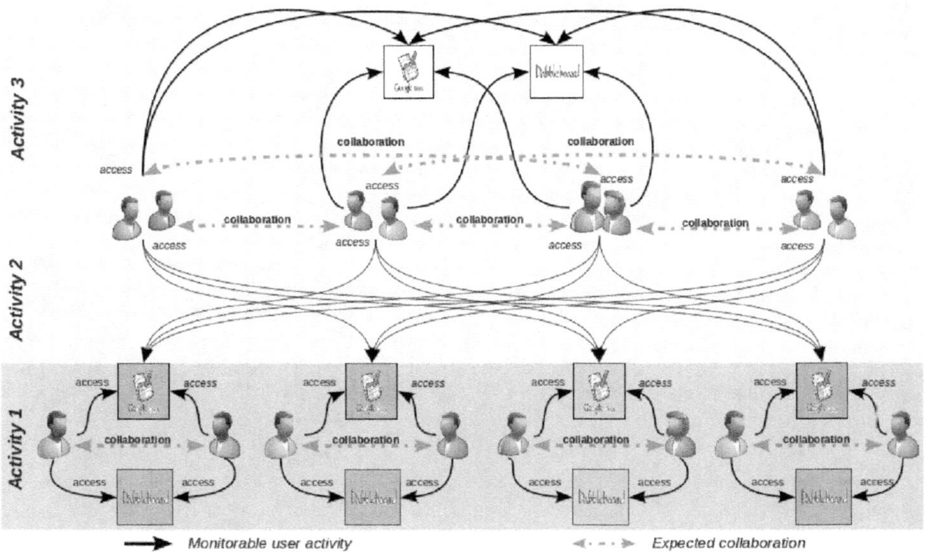

Fig. 5. Monitorable collaboration schema of the experience studied in this case

Figure 5 summarizes the outputs of all the previous mentioned steps. The collaboration heuristics that conform the abstract collaboration schema are represented by means of the dashed yellow arrows, highlighting those key collaborative aspects to be monitored. The especific sequence of activities, the resources required for each one of them, and the way the students are grouped are derived from the particularization and instantiation constraints (see Table 4a). Finally, among all the expected ways of collaboration (expected collaboration schema), continuous black arrows point to the information that may be collected in order to detect collaboration evidence. This view of the learning scenario depicts what we have called *monitorable collaboration schema*.

3.3 Comparing Monitorization Data against the Collaboration Schema

During the enactment phase of the learning experience, we retrieved and interpreted the content of GLUE!'s logs. The analysis of the logs focused on detecting evidence from the key collaborative aspects highlighted in the abstract collaboration schema (dashed yellow arrows in Figure 5) by means of the accesses to the shared resources (continuous black arrows in Figure 5). Figure 6 displays the actual collaboration of one super-group. The labels on the links specify how many times each participant (or group) accessed the resource. Red crosses over a dotted line represent those cases where no evidence of collaboration could be deduced from the collected data.

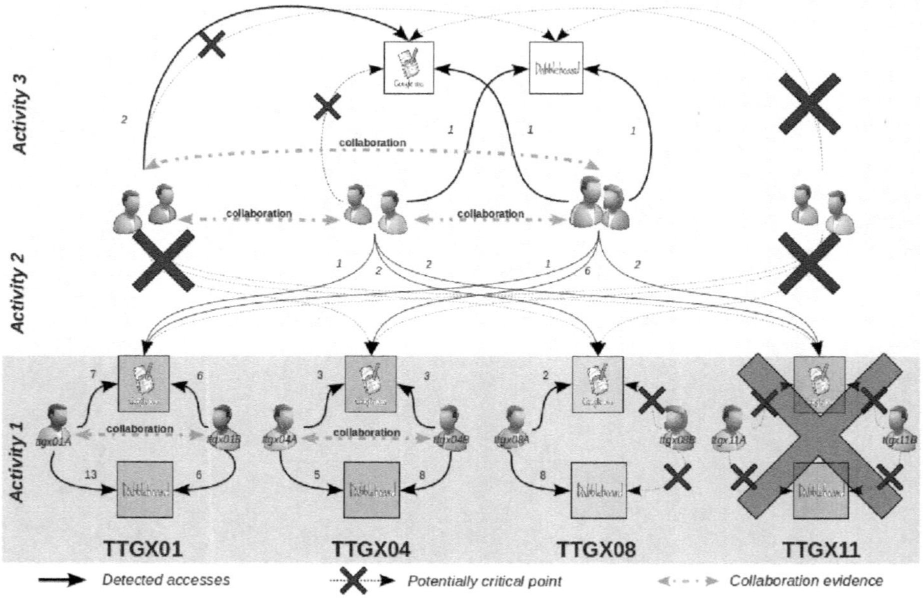

Fig. 6. Monitoring report of one super-groups that participated in the case study

By means of this information, we detected some potentially critical situations. For instance, during the first activity, two potentially undesired situations were identified.

As shown in Figure 6, in group TTGX08 just one of the group members interacted with the resources; and in TTGX11, none of the participants had accessed the resources, which was a pointer to a potential undesired situation.

Teachers where informed periodically about the potentially undesired situations. Being aware of them, they reviewed carefully the resources of those groups that showed irregularities to confirm (or not) our assumptions. In case of confirmation, teachers took preventive measures, for instance, by sending to the students a reminder of the deadlines or asking them whether they had found any problem during process to finish the activity.

From the researchers' perspective, in order to validate whether our assumptions based on the monitorable data were correct, we triangulated these results with the rest of the data sources. We found out that in those cases where monitorable data showed expected behaviour, participants had actually collaborated following the script (as extracted from the interviews with both teachers and students). Furthermore, all the cases of collaboration breakdown had been detected based on the monitorable data. However some cases that had been interpreted as potentially critical situations were indeed "false positives". These were cases where learners collaborated in unexpected ways. For example, from the observations and questionnaires we could see that students interacted face to face when they were expected to collaborate by means of the ICT resources. Despite this limitation, teachers stated in the interview that this was a minor problem, since it had been easy for them to confirm or disprove the initial assumptions, based on their own knowledge of the classroom dynamics from their informal observations.

3.4 Findings and Discussion

As mentioned in the Introduction, the main goal of the case study was to get evidence on whether the combination of structuring (by means of pattern based scripting) and monitoring collaboration contributes to provide an automated and higher level view about the evolution of the learning process. For achieving that goal we proposed three research issues (see section 3.1). We discuss the results according to them.

The first issue considers the usefulness of CLFPs for identifying the key collaborative aspects of the script. Results indicated that the analysis of the *Pyramid* constraints (abstract collaboration schema) has eased the identification of those constraints of the script that require accomplishment (expected collaboration schema). Indeed, with the information provided about the ICT tools involved in the learning scenario, it has been possible to identify which collaboration evidence could be obtained in order to inform about the key aspects of the script (monitorable collaboration schema). And finally, this schema guided the monitoring process to focus on those aspects of the expected interaction that could jeopardize the script collaborative purposes. It must be highlighted, that this analysis of the *Pyramid* pattern could be applied in different contexts. For this reason, we expect to carry out this analysis for each CLFP in order to guide designers automatically by means of authoring tools.

The second issue referred to whether the comparison between the desired and actual collaboration helps teachers to be aware of the progress of the learning scenario. The evidence points out that it is feasible to detect and inform teachers about potentially critical situations, facilitating the adaptation tasks. The proposed way of visualizing the learning process has been a fruitful first attempt to provide teachers with feedback. In this paper we have presented a prototypical solution that has served to test the proposal. The positive results obtained from the study lead us to plan future work, dealing with the automatic generation of the collaboration models for each CLFP.

The third and last issue is related to technological restrictions that appear in learning scenarios characterized by the use of diverse ICT tools, i.e., which are the required conditions for collecting relevant information about the key aspects of collaboration in these settings. We have shown that, just by monitoring the access to the tool instances, it has been possible to offer a general view about the collaborative activities in this specific scenario. With this simple information, we have given response to a very common problem: the lack of monitoring data in some ICT tools [18] (as it happened with Dabbleboard). As shown in the case, this information can still be enriched, including for example the data collected from each individual tool or even with interactions registered by means of face-to-face observations. Thus, one of the main issues in our upcoming work will delve into the problem of data integration.

Finally, we collected some impressions and feedback from both students and teachers. From the teachers point of view, they argued that, despite the limitations of the reduced variety of monitored events, results were pretty close to the real facts. By means of the monitored schemas developed for each super-group, teachers could realize at a glance which participants seemed not to be working, which ones were isolated and which was the best way to re-distribute them. Moreover, this information prevented teachers from going through all the 64 instances to check the activities progress, saving them a long time. On the other hand, students complained about the lack of information about their mates work, which forced them to connect and review their resources frequently to check whether there was any change. Therefore, this points to the potential usefulness of the monitoring results to facilitate students collaboration.

These findings provide evidences on the capabilities of the presented approach to generate relevant information about the evolution of the learning process. This information has had educational value, especially regarding the improvement of the regulation tasks of CSCL activities in real practice.

4 Conclusions and Future Work

This paper proposed an approach where CLFP-based scripting and monitoring methods and tools are combined to enhance collaboration. We have shown how the analysis of pattern constraints helps to identify those aspects of collaboration that should be looked after when monitoring the activity. This way, monitoring can focus on these aspects, which are easier to detect and to follow than fine-grained analysis of the activities. Finally, by comparing the actual and desired collaboration schemas, teachers can identify undesired situations and regulate the collaborative process.

This proposal has been put in practice through a case study, carried out in a higher education learning scenario, that has shown initial evidence on the suitability of the approach for the detection of potentially critical situations. The approach has been followed manually so next steps will deal with the process automation. The analysis of other CLFPs constrains to guide designers during the process, and the visualizations of the collaboration schemas constitute two main lines of future work.

These results could be helpful for facilitating the regulation tasks. For instance, the work done in [5] [2] and [15] analyzing the scripts flexibility and the ways to adapt them when eventualities appear, could be simplified by our proposal. Once teachers know which constraints are not being satisfied in real time, it will be easier for them to address the issues hampering collaboration (e.g. solving technical problems, extending the deadlines or modifying the student distribution). This benefit is even more remarkable using heterogeneous scenarios where several ITC tools are involved.

This work is connected to the effort towards the integration of VLE and external ICT tools in [1]. The use of an architecture that integrates all the software used to support the learning scenario, simplifies data gathering and provides relevant high level information about the progress of the learning activities. This trend underlines the need of developing architectures for the retrieval of monitoring data from decentralized learning environments. However, the integration of different data sources is a complex task due to the heterogeneity of their content [18]. These two pending issues, gathering data from different data sources and their integration, constitute other lines of future work.

Acknowledgements. This research has been partially funded by the Spanish Ministry of Science and Innovation Projects TIN2008-03-23 and IPT-430000-2010-054, and the Autonomous Government of Castilla and León, Spain, Project VA293A11-2. The authors would also like to thank the rest of GSIC/EMIC Group at the University of Valladolid for their support and ideas, and all the students who participated in the evaluation process.

References

1. Alario-Hoyos, C., Wilson, S.: Comparison of the main Alternatives to the Integration of External Tools in different Platforms. In: Proceedings of the International Conference of Education, Research and Innovation, ICERI 2010, Madrid, Spain, pp. 3466–3476 (2010)
2. Demetriadis, S., Karakostas, A.: Adaptive Collaboration Scripting: A Conceptual Framework and a Design Case Study. In: Proceedings of the 2nd International Conference on Complex, Intelligent and Software Intensive Systems, CISIS 2008, pp. 487–492. IEEE Computer Society, Washington, DC, USA (2008)
3. Dillenbourg, P.: Over-Scripting CSCL: the risks of blending collaborative learning with instructional design. In: Three Worlds of CSCL. Can We Support CSCL?, pp. 61–91. Open Universiteit Nederland, Heerlen (2002)
4. Dillenbourg, P., Jermann, P.: SWISH: A model for designing CSCL scripts. In: Scripting Computer-Supported Communication of Knowledge Cognitive, Computational, and Educational Perspectives, pp. 1–10. Springer, New York (2007)
5. Dillenbourg, P., Tchounikine, P.: Flexibility in macro-scripts for CSCL. Journal of Computer Assisted Learning 23(1), 1–13 (2007)
6. E-LEN: Design expertise for e-learning centers (2004), project website, `http://www2.tisip.no/E-LEN/` (last visit: April 2011)
7. Fischer, F., Kollar, I., Haake, J., Mandl, H.: Perspectives on collaboration scripts. In: Scripting Computer-Supported Communication of Knowledge Cognitive, Computational, and Educational Perspectives, pp. 1–10. Springer, New York (2007)
8. Goodyear, P.: Educational design and networked learning: patterns, pattern languages and design practice. Australasian Journal of Educational Technology 21(1), 82–101 (2005)

9. Gómez Sánchez, E., Bote Lorenzo, M., Jorrín Abellán, I., Vega Gorgojo, G., Asensio Pérez, J., Dimitriadis, Y.: Conceptual framework for design, technological support and evaluation of collaborative learning. International Journal of Engineering Education 25(3), 557–568 (2009)
10. Hernández-Leo, D., Villasclaras-Fernández, E., Asensio-Pérez, J., Dimitriadis, Y.: Diagrams of learning flow patterns' solutions as visual representations of refinable IMS Learning Design templates. In: Handbook of Visual Languages for Instructional Design: Theories and Practices. IGI Global,
11. Hernández-Leo, D., Villasclaras-Fernández, E., Asensio-Pérez, J., Dimitriadis, Y., Retalis, S.: CSCL Scripting Patterns: Hierarchical Relationships and Applicability. In: Proceedings of the 6th IEEE International Conference on Advanced Learning Technologies, pp. 388–392. ICALT, Kerkrade (2006)
12. IMS Global Learning Consortium: IMS Learning Design Specifications (2003), http://www.imsglobal.org/learningdesign/ (last visit: April 2011)
13. Jermann, P., Soller, A., Lesgold, A.: Computer Software Support for CSCL. In: Strijbos, J.W., Kirschner, P., Martens, R. (eds.) What We Know About CSCL.. and Implementing it in Higher Education, Computer-Supported Collaborative Learning, vol. 3, pp. 141–166. Kluwer Academic Publishers, Norwell (2004)
14. Kahrimanis, G., Avouris, N., Komis, V.: Interaction analysis as a tool for supporting collaboration. An overview. In: Technology-Enhanced Systems and Tools for Collaborative Learning Scaffolding. Springer, Heidelberg (2011)
15. Karakostas, A., Demetriadis, S.: Adaptation patterns in systems for scripted collaboration. In: Proceedings of the 8th International Conference on Computer Supported Collaborative Learning, CSCL 2009, vol. 1, pp. 477–481. International Society of the Learning Sciences, Rhodes (2009)
16. Kobbe, L.: Framework on multiple goal dimensions for computer-supported scripts (2005), Kaleidoscope Noe. Deliverable D29.2.1
17. Koschmann, T.: Paradigms Shift and Instructional Technology. In: CSCL: Theory and Practice of an Emerging Paradigm, pp. 1–23. Lawrence Erlbaum Associates, Mahwah (1996)
18. Martínez-Monés, A., Dimitriadis, Y., Harrer, A.: An interaction aware design process for the integration of interaction analysis in mainstream CSCL practices. In: Analyzing Collaborative Interactions in CSCL: Methods, Approaches and Issues, pp. 269–291. Springer, Heidelberg (2011)
19. PPP: The Pedagogical Patterns Project (2005), project website, http://www.pedagogicalpatterns.org/ (last visit: April 2011)
20. Soller, A., Martínez, A., Jermann, P., Muehlenbrock, M.: From Mirroring to Guiding: a review of the state of the art in interaction analysis. International Journal on Artificial Intelligence in Education 15, 261–290 (2005)
21. Stahl, G., Koschmann, T., Suthers, D.: Computer-supported collaborative learning: an historical perspective. In: Cambridge handbook of the learning sciences, pp. 409–426. Cambridge University Press, Cambridge (2006)
22. Stake, R.: The art of Case Study Research. Sage Publications, London (1995)
23. TELL: Towards Effective networked supported coLLaborative activities (2005), project website, http://cosy.ted.unipi.gr/tell/ (last visit: April 2011)
24. Villasclaras-Fernández, E., Hernández-Gonzalo, J., Hernández-Leo, D., Asensio-Pérez, J., Dimitriadis, Y., Martínez-Monés, A.: InstanceCollage: A Tool for the Particularization of Collaborative IMS-LD Scripts. New Directions in Advanced Learning Technologies. Educational Technology & Society 12(4), 56–70 (2009)

Automatic Discovery of Complementary Learning Resources

Vicente Arturo Romero Zaldivar[1,2], Raquel M. Crespo García[2], Daniel Burgos[1,3], Carlos Delgado Kloos[2], and Abelardo Pardo[2]

[1] Atos Origin SAE,
Albarracín 25, 28037 Madrid, Spain
{vicente.romero,daniel.burgos}@atosresearch.eu
[2] Department of Telematic Engineering,
University Carlos III of Madrid, Spain
{rcrespo,cdk,abel}@it.uc3m.es
[3] Universidad Internacional de La Rioja,
Gran Vía Rey Juan Carlos I 41, 26002 Logroño, Spain

Abstract. Students in a learning experience can be seen as a community working simultaneously (and in some cases collaboratively) in a set of activities. During these working sessions, students carry out numerous actions that affect their learning. But those actions happening outside a class or the Learning Management System cannot be easily observed. This paper presents a technique to widen the observability of these actions. The set of documents browsed by the students in a course was recorded during a period of eight weeks. These documents are then processed and the set with highest similarity with the course notes are selected and recommended back to all the students. The main problem is that this user community visits thousands of documents and only a small percent of them are suitable for recommendation. Using a combination of lexican analysis and information retrieval techniques, a fully automatic procedure to analyze these documents, classify them and select the most relevant ones is presented. The approach has been validated with an empirical study in an undergraduate engineering course with more than one hundred students. The recommended resources were rated as "relevant to the course" by the seven instructors with teaching duties in the course.

Keywords: Personalisation, recommendation, adaptive mentoring, learning analytics, information retrieval.

1 Introduction

This paper focuses on how students search for useful resources in Internet while participating in a learning experience. Searching for information is a task commonly required in course activities of any kind (distant, face-to-face or blended learning). Even if a set of exhaustive course notes is produced, students still search for additional information in Internet. Thus, a community of students emerges that is simultaneously visiting a potentially large number of resources,

C. Delgado Kloos et al. (Eds.): EC-TEL 2011, LNCS 6964, pp. 327–340, 2011.

some of which could be relevant to the course. These resources could be automatically detected by analyzing this process. But there are two major hurdles to analyze the activity of such community. First, the resources viewed by the students are not trivial to monitor without the student intervention (for example, with a rating system). And secondly, from the visited resources, only those truly relevant for the course should be detected and eventually reported back to the community.

Current Learning Management Systems (LMSs) typically store usage and event information in databases and log files that can later be accessed and analyzed using data mining techniques [12]. Students, however, tend to use certain tools that lie outside the scope of the LMS[16,19]. Skills commonly included in courses such as "managing various information sources", or "search for relevant information", are developed mainly in the user personal space. This type of activities are difficult to supervise both in remote as well as in face-to-face learning.

When a student starts working in a course activity, it is highly likely that she searches the Web for information that is not present in the course resources or perhaps not adequately explained. This behavior may be originated by the need of solving a very specific problem or in other cases by the need of obtaining a tutorial type of document explaining step by step how to solve a given problem. During this process, students may encounter resources related to the course that could be interesting for their peers.

The main question addressed in this document is whether the resources viewed by the students (apart from conventional course notes) can be automatically processed to obtain a set of resources related to the course.

Ideally, the procedure by which these additional resources are identified should be executed without student intervention. Although students clearly embraced the Web 2.0 philosophy [2], current state-of-the-art technologies such as information retrieval and data-mining show that useful information can be automatically derived by processing user observations.

These techniques can be used to automatically detect similarities between documents without manual classification or labeling. The idea in this work is to create a reference corpus with the course documents, and use it to determine how additional resources visited by the students in Internet are related to them. Those resources that are similar to the course material can be recommended to the students. The steps to obtain these recommendations are: track the student browsing activity, process the set of obtained URLs, characterize the course resources offered by the teaching staff, compare both document sets, and determine which additional resources are most related to the course.

The rest of the paper is organized as follows. Section 2 discusses related work and technologies used in the presented approach. The adopted solution is described in Section 3. An empirical study followed a statistical valuation is presented in Section 4. Finally, conclusions and future work are discussed in Section 5.

2 Related Work

The work described in this paper is related to the following areas: user monitorization, recommender systems, and information retrieval.

Monitorization of student activity is typically done by analyzing the logging facilities included in LMSs or other learning platforms (e.g. [18]). There are numerous applications in which this information is used to anticipate student behavior and take some corrective actions (see [12] for an example). But the trend in learning environments is toward using a variety of tools among which the learning activitites are distributed. The tendency is for users to design their own Personal Learning Environment by combining multiple applications [3]. This increase of the activities outside the LMS is particularly high when methodologies such as problem-based or project-based learning are applied. Complementary mechanisms for gathering information about such activities would thus be useful for improving the understanding of the student's learning process.

In the field of Technology Enhanced Learning, monitoring student activity has been proposed as a tool for promoting self-reflection [10,22], awareness [10,14], student assessment [14], course management [15], course evaluation [25], and, of course, learning adaptation and personalization [26], among other potential applications.

Attention.XML [24,6] was an early approach to capturing and storing attention metadata, that is, to represent attention. Due to its insufficient granularity and the lack of context information, the CAM (Contextualized Attention Metadata) Schema has been defined as an extension of Attention.XML [26], focusing the most important extensions on actions that occur on data objects [23]. The procedure described in this paper is based on monitoring student's web surfing activity in order to gather a collection of potentially interesting resources to complement course materials.

The number of learning resources available now to students has increased massively. As it has been the case in other fields, such increase prompts the need for improved methods to find and retrieve these resources [13]. Recommender systems were originally defined as those in which "people provide recommendations as inputs, which the system then aggregates and directs to appropriate recipients" [17]. The term now has a broader connotation, describing any system that produces individualized recommendations as output or has the effect of guiding the user in a personalized way to interesting or useful objects in a large space of possible options [5].

Recommendation techniques are classified into two main categories: model-based and memory-based [1]. As explained in [9], model-based techniques periodically cluster the data in estimated models, using techniques such as Bayesian models, neural networks or latent semantic analysis. Memory-based techniques continuously analyse all user or item data to calculate recommendations. The latter can be further classified into collaborative filtering, content-based and hybrid. Collaborative filtering techniques follow a social approach, recommending items based on the preferences of similar users. Content-based recommendation systems try to recommend items similar to those a given individual user has liked in the past [11].

In this paper, the content-based approach has been adopted. The course material provided by the teaching staff is used as the input information for characterizing student interests and needs. Resources visited by the students are then evaluated against this set of relevant items in order to identify their adequateness for course purposes.

This adequateness is obtained using Information Retrieval techniques, more concretely, the algebraic vector space model proposed by Salton in [21]. In this model each document is characterized by a vector of term weights. Each vector component represents the weight of the corresponding term in the document and is calculated based on local and global parameters. Such model is known as term frequency-inverse document frequency (TF-IDF), and computes the weight of a given term based on the frequency of the term in the document itself and in the complete collection. By weighting the local term frequency with the global inverse document frequency, the TF-IDF model avoid associating high values to common terms that do not characterize the document when compared to the rest of the collection, despite potentially high local frequencies in the document. Once modeled by vectors, the similarity between two documents (or between a document and a query) can be easily calculated using techniques such as the cosine of the angle between the vectors.

Information retrieval algorithms are complemented in this work with natural language processing, in order to achieve a more accurate set of descriptive terms. In this field they are several tools than can be used for example GATE [7] which is a General Architecture for Text Engineering. GATE, according to its authors, is intended to apply to whole families of problems within the language processing field, and to be like a toolbox in service of construction and experimentation. GATE provides not only the algorithms to analyze text but also tools for the visualization of the results which is of great help in the process of discovering similarities between documents.

In this paper, we use information retrieval techniques combined with natural languages algorithms to process a set of resources visited by the students while working on a course and select the most relevants so that they can be recommended to the entire course. The whole procedure is done automatically and requires no interaction with the students.

3 Detecting Resources Outside the LMS

Learning experiences are including an ever increasing number of resource types: documents, videos, audio clips, Web pages, etc. In principle, these resources are published by the instructors so that students use them to carry out the course activities. But in the information age, it is not uncommon for activities to require *explicitly* the search for additional resources. Even in courses where the resources made available cover exhaustively the considered topics, when faced with any course task, students resort to Internet to find additional information. This search for information is spreading both to individual and collaborative activities. This paper proposes a technique to automatically select the most relevant resources from those visited by the students while working on a course.

These *unofficial* resources visited by the students may complete or explain with other words any of the course concepts. Sometimes students find in these resources solutions to tasks related to those required in the course material. Additionally, students (specially digital natives) are used to look for tutorials and hands on documents explaining how to solve all types of problems. Independently of the pedagogical strategy used, students tend to solve problems using not only the course documentation but also, and in a significant percentage, external resources available in Internet. Ideally, all these searches and discoveries could be collected and processed to select a subset of "most relevant" resources and recommend them back to all the students in a course. Furthermore, the procedure could be deployed transparently to the students with no need for rating schemes.

The generic scenario considered for this paper is a learning experience where a set of resources previously selected (or produced) by the instructors are made available to the students. When working on the course activities, students search for auxiliary material in Internet. These additional visited resources are then collected and analyzed. It follows a description of the strategies adopted in a real case that fits into this generic scenario.

3.1 Data Collection Strategy

The main challenge when collecting the sites that a student visits while carrying out activities related to a course is precisely to restrict the observation only to that situation. In a typical scenario, students use a personal computer in which *at certain times*, they use the browser to check the course material and maybe search for additional material. Access to course material is typically monitored using the logs stored in the LMS hosting the course. But the focus of this study is to quantify and analyze those sites that are visited outside the scope of the LMS while working on a course.

The adopted strategy consisted on creating a virtual machine (a file containing an entire computer that can be run as another application in a personal computer) offered to the students at the beginning of the course. The virtualization platform used was VirtualBox[1]. The advantage for the students is that this machine contained a fully configured working environment including all the additional tools required in the course. This machine was instrumented so as to store and then rely to a remote server the URLs that were introduced in its internal browser as described in [16]. To comply with personal data privacy legislation, students are informed of this instrumentation as well as the procedure to deactivate it if they wish to do so. Additionally, students were advised by the instructors at the beginning of the semester to restrict the work in this machine only to activities directly related with the course. This strategy offers a

[1] www.virtualbox.org

reasonable trade-off between a monitoring scheme that would record every single action on their personal computers, and a system that can only analyze their interaction with the LMS.

When a student boots the virtual machine *and* its internal browser is used, every time a new URL is opened, a line containing the URL and the time stamp is stored and later relayed to a remote server. As a result, the set of URLs visited by all the students (that used the virtual machine and decided to maintain enabled the recording mechanism) is collected. Separating the URLs pointing to the course material is simple because their location is known in advance. The objective, then is to automatically analyze the remaining URLs and select those pointing to resources that are most similar to the course notes, thus maximizing the probability of detecting useful complementary material.

3.2 Calculating the Similarity

The proposed algorithm receives as input two sets of URLs: those obtained from the students, henceforth called the "student set", and those pointing to the course resources or "reference set". The steps in the algorithm to process these sets are depicted in Figure 1.

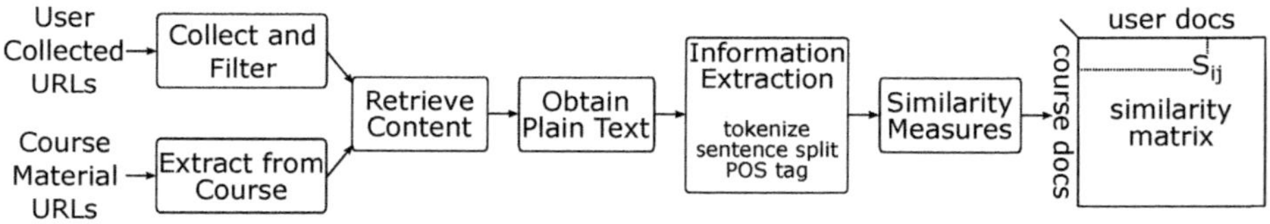

Fig. 1. Data processing steps to compute the similarity matrix

An initial filtering step is applied to the user collected data to remove unwanted URLs. For example, URLs pointing to course material are not considered because they are already part of the reference set. Additionally, a fixed set of domains corresponding to social networks, email services, etc. must be removed as they provide irrelevant information for the detection process[2]. The analogous process in the reference set amounts simply to extract the URLs pointing to the course resources from the proper source (the official course material repository, a list of all the handouts, etc.)

During the second step the content is automatically retrieved from Internet and classified attending to the type of resource (HTML page, image, PDF document, plain text, Microsoft Word, etc.) This is done to select only resources that can be translated to plain text to apply lexical analysis and information

[2] Although occasionally discussions related to the course can happen in these domains, they are password protected and thus private beyond the scope of the analysis.

extraction techniques. During the translation to plain text, the algorithm applies different tools depending on the type of the resource and both sets are then translated into plain text documents.

The information extraction step is carried out using Gate, a toolkit for natural language processing [7]. First, the documents of both the student and reference sets are used to create and populate a "collection" or "corpus". This new set is needed by the Gate Document Manager to perform the text analysis and obtain the proper annotations. Although Gate offers some support for processing non-plain documents, mainly in XML format, due to the presence of multiple formats and sometimes incorrect HTML documents in the student set, the translation to plain text was performed by an ad hoc step.

Once the corpus is created, an automatic sequence of so-called "processing resources" is applied to all the documents. Gate includes a set of algorithms for natural language processing called "ANNIE" (A Nearly New IE System) consisting of tools such a tokenizer, a sentence splitter or a Part-of-speech (POS) tagger, etc. (see [8] for a more detailed description). The application discussed in this paper uses these tools to obtain an annotated view of each document where nouns and proper nouns are identified.

The input data for the last stage shown in Figure 1 is a set of annotated documents in XML in which the nouns and proper nouns have been identified. The output is a matrix M of size $n \times m$ where n and m are the number of documents in the student and reference sets respectively and such that M_{ij} is how similar document D_s from the student set is to the reference document D_r.

The similarity coefficients are obtained using ranking techniques conventionally used in information retrieval [4]. First, each document is translated into an n-dimensional vector where n is the number of terms (nouns and proper nouns) identified in the previous step. Two documents can be compared by computing the cosine of the angle of their corresponding vectors. Although there are numerous well established and robust techniques to compute this coefficients, some specific features of the considered context need to be taken into account.

Typically, ranking systems solve the problem by which given a set of documents and "a query", the subset of most relevant documents for that query needs to be computed. The scenario discussed in this document is slightly different. The input data are two sets of documents, but one of them, the reference documents, limit the scope in which new relevant resources need to be discovered. In other words, given the set of reference documents, select those documents from the student set that are more similar.

Similarity measures and ranking techniques have been widely studied in the area of information retrieval. The selected approach uses term weighting techniques that combine the distribution of a term within a document, but also within a collection. However, in order to force the algorithm to detect documents relevant to the course topic, for computing both the vocabulary and the weight of a term within the collection, only the reference documents are considered. With this adjustments, given the vectors $D_s = (w_{s1}, \ldots, w_{st})$ and $D_r = (w_{r1}, \ldots, w_{rt})$

representing two documents in the student and reference sets respectively, the similarity coefficient is computed as follows [21]:

$$Similarity(D_s, D_r) = \frac{\sum_{i=1}^{t}(w_{si} \times w_{ri})}{\sqrt{\sum_{i=1}^{t}(w_{si})^2 \times \sum_{i=1}^{t}(w_{ri})^2}}$$

The set t contains all the terms identified by processing *only* the nouns in the reference documents. The weights of each term for each of the documents (w_{ri} and w_{si} standing for the weight of term i for a document in the reference or student set, respectively) have been calculated following the $tf*idf$ product [20]. The document vectors are obtained as follows:

$$\begin{aligned} w_{ri} &= tf_{ri} * idf_{ri} \\ w_{si} &= tf_{si} * idf_{ri} \\ tf_{ri} &= freq_r i/|terms_r| \\ idf_{ri} &= \log_2 \frac{N - n_i}{n_i} \\ tf_{si} &= 0.5 + \frac{0.5 freq_{si}}{maxfreq_s} \end{aligned}$$

where N is the number of documents in the reference set, n_i is the number of documents in the reference set where term i appears, $freq_{ri}$ and $freq_{si}$ are the appearances of term i in a reference or student document respectively, $|terms_r|$ is the number of terms in a reference document, and $maxfreq_s$ is the maximum number of appearances of a term in a student document.

The similarity values are numbers between zero and one where zero means maximum distance and one means minimum distance. It thus provides a metric for measuring the relevance of each resource related to the course topic(s), automatically filtering out the non-relevant ones and allowing to selecting the closest ones for recommendation.

4 Empirical Study

In order to validate the hypothesis enunciated in Section 1, a monitoring mechanism was deployed in a face-to-face, second year engineering course with 220 students over an 8 week period. The course topic is C programming but students are supposed to use additional tools such a memory profiler, debugger, integraged development environment, and a version control system. The cohort is divided into sections of 40 students that meet twice a week. An active learning methodology is used by which students need to work on activities *before* each class. The virtual machine described in Section 3 contains all the required tools to work in these activities.

The course lasted for 14 weeks in the Fall semester of 2010, but the URL recording mechanism was used only for the initial 8 weeks. The reason for this time window is to show that, if the approach is successful and a set of meaningful auxiliary resources is obtained automatically, they can be incorporated into the on-going course edition. Table 1 summarizes the information obtained during this period.

Table 1. Summary of the data collected in the study

Students participating	125
Total number of URLs collected	17,787
URLs pointing to course material	6,926 (38,94%)
URLs of institutional services (LMS)	8,901 (50,04%)
Unique URLs pointing outside the institution after filtering	1,018

On average, the system recorded slightly under 150 URLs per user. The initial filtering step significantly reduced this number as more than one third of them were pointing to course material, and the number increases to half of them when considering other sites within the university (student email account, virtual folders, etc.)

The content retrieval step where the URLs are used to effectively fetch the document from the net was performed over the 1,018 unique URLs produced by the filtering step. The course under consideration is part of a bilingual program (English/Spanish). As a consequence, the retrieved content is analyzed to detect the language and the documents are divided into two collections: English and Spanish. The lexical analysis phase was applied to a collection of 25 reference documents and 252 student documents in English, and 25 reference documents and 599 student documents in Spanish. The results reported in this paper are for the English documents.

The entire processing phase, from collecting the URLs until the similarity matrix is obtained took less than 15 minutes. Although a lexical analysis of a large set of documents could become a bottleneck, the previous stages of filtering and classification reduced the number of documents to a size manageable by current techniques within reasonable execution times.

As a result, the similarity matrix of size 25×252 is obtained. In order to select the most relevant documents from the student set, two different approaches were considered. The first was to calculate the accumulated similarity of a student document with respect to all the reference documents. This method will be referred as the ACC method. Alternatively, the documents were sorted in decreasing order of its maximum similarity to any reference document. This method will be referred as the MAX method. As a sample, Table 2 shows the ten URLs with the highest accumulated similarity (ACC).

When using the MAX method, the ten URLs with the maximum similarity to a single reference document had only two URLs in common with the set shown in Table 2. The low number of URLs in common between these two methods suggested a validation strategy to try to characterize this difference.

Table 2. Ten resources with the highest accumulated similarity

N	Title	Host	Comment
1	Linked list	en.wikipedia.org	Page explaining what is a linked list. The course uses this structure in numerous activities.
2	Preprocessor directives	www.cplusplus.com	This compiler functionality is included in the course material and used throughout the course
3	Debugging with Gdb	sourceware.org	This tool is essential for the course and must be used frequently
4	Unicode	en.wikipedia.org	The course does not explicitly mention Unicode
5	Maemo Development Environment	maemo.org	Students develop an application that must execute in the Maemo platform.
6	Index of stdlib in C	www.thinkage.ca	The functions in this library are used in most activities
7	Linux essential keyboard shortcuts	linux.about.com	The development platform runs on Linux and the command line is used frequently
8	What is a pointer?	pw1.netcom.com/t̃jensen	Pointers are one of the most studied topics in the course
9	Apache Subversion	subversion.apache.org	Subversion is used to exchange documents between team members and instructors.
10	Linux Command Line Reference	www.pixelbeat.org	Useful to work in the Linux environment.

4.1 Statistical Validation

Automatically obtaining a set of documents that are similar to the course notes is not enough to validate the proposed approach. It remains to be proved that the selected documents are indeed relevant to the course. The adopted validation approach consisted on a survey given to the seven instructors with teaching duties in the course. A questionnaire was given with the URL of the 20 resources obtained with the *ACC* and *MAX* methods respectively. For each URL, the instructor was asked to review its content and answer the following two questions:

1. How is the resource related to the course? The answer was given in a five levels scale (5 = Very related, 4 = Somewhat related, 3 = Neutral, 2 = Not related, 1 = Totally unrelated).
2. If the resource is related (or somewhat related), this relation applies to 1 = the entire course, 2 = a subset of activities or topics, 3 = a single activity.

The first question is oriented toward proving that the set of automatically obtained URLs are relevant to the course (the algorithm works). The second question was posed to test if the two methods would offer different resources in terms of their scope of application. Intuitively, if a resource has a high accumulated similarity (selected by the ACC method), it is probably more adequate for the general course topic. Analogously, if a resource is selected by its high similarity to one single reference document, it is probably more related to the specifics of the activity in that document.

Thus, for each instructor, at most 40 data points were obtained (only those resources ruled "related" or "somewhat related" were considered in the second question. The first step in the validation was to obtain a single mark per instructor for the relevance of the two sets of URLs, so the average score was chosen. The obtained samples had a mean of $43,57$ and a standard deviation of $3,64$. Next, a null hypothesis test that the average relevance of the URLs in the ACC group is not larger than 40 (somewhat related) was performed The t-test returned a significance p-value of 0.02, thus the hypothesis can be rejected and the sample assumed to have a mean above the "somewhat relevant" level. The normal distribution of the sample was verified with the Shapiro-Wilk test ($W = 0.8969$, p-value $= 0.313$).

The analogous test was carried out for the URLs obtained with the MAX method. In this case, the normal distribution of the sample could not be verified, thus a Wilkoxon signed rank test was performed. In this case, the null hypothesis of the average not larger than 40 (somewhat relevant) could not be rejected (p-value $= 0.13$) suggesting that the relevance of these URLs is lower than those in ACC. Additionally, the hypothesis of the average to be not larger than 30 (neutral) can be rejected, confirming that this second set of resources have some relation with the course.

The second question in the survey was included to investigate if the two methods ACC and MAX discovered resources with a different granularity in its relevance. For each of the 20 resources, the average of the valuations given by the 7 instructors was obtained. The two samples (each with $n = 10$) had means 0.84 and 1.35 respectively, suggesting that method ACC discovers more generic URLs whereas MAX contains resources related to a single task. The two samples were compared with the Wilcoxon Signed-Rank Test. A significant p-value $= 0.01782$ was obtained concluding that the granularity valuations indeed come from two different populations.

As a summary of this validation, both discovery methods, ACC and MAX discover somewhat relevant resources, but in the case of ACC, they are conclusively relevant. Furthermore, the URLs discovered with ACC seem to be more relevant to the entire course, and those discovered with MAX typically apply to a more reduced set of course activities.

Once the validity of the automatic discovery procedure has been shown, the resources can be used as recommendations to the students during the course. Two approaches have been envisioned. First an unsupervised variant by which the recommendations are automatically added to the course material. The second approach follows a "moderated recommendation scheme" by which instructors must approve a resource before being included as a recommendation for the students. Because the discovery procedure is implemented along the course, this second approach improves the confidence of teachers on the system, at the expense of increasing instructor load.

5 Conclusions and Future Work

Student's support on problem-solving processes is improved by real time recommendations. In order to increase the performance ratio and the accuracy of such recommendations, automation of document discovery and classification is required. In addition, the recommendation must follow an automated method, able to provide the best match out of the huge amount of retrieved data. In this paper, we provide a first-hand experience with students, which carry out a blended learning activity. During the problem solving activity, students surf the Internet and look for a number of resources. However, this activity provides too many visited documents, from which there is no clear evidence of usefulness to the activity as they can be both related as well as non-related to the course contents. Task monitoring and document retrieval become a challenge, then. The authors design, implement, and test a method to deliver an automatic selection process of resources, which will be suggested in the form of recommendations to the students. Specifically, the method is characterized by a step-by-step process, namely: user tracking, URLs retrieval, document comparison, document fitness to the course, and eventually, recommendation.

By using ANNIE (a set of algorithms for natural language processing), which is part of the Gate toolkit, the implemented method obtains an annotated view of each document, which feeds a matrix with coefficients that supports the ranking system for the documents, following an algorithm that provides specific vectors for every document, that show the similarity factor to the course reference material, and therefore, usefulness to the students.

Out of the empirical study carried out with 220 students over a 8-week period, and in order to select the most relevant resources, we use two discovery methods: The ACC method (calculate the accumulated similarity of a student document with respect to all the reference documents) and the MAX method (sort the document decreasingly out of its maximum similarity to any reference document). We conclude that both discover relevant resources to the course and are thus useful for document recommendation. However, ACC seems to increase the validity and fitness ratio to the entire course, while MAX is more efficient with a limited set of activities. Once a document is selected as relevant, it can be upgraded to a recommended one either by an automatic process, or a supervised process in which the teacher grants the selected document, at his workload expense.

In summary, this paper presents a supporting system for selecting the most relevant and useful resources among the ones browsed by the students beyond the course materials. In order to maximize transparency and minimize interference and workload for students, a content-based approach is proposed for determining the relevance of the potential resources. Evaluation results from empirical data confirms the validity of the approach. First, relevance of the selected resources has been confirmed by teaching staff's validation, providing a sound base for recomendations based on the proposed method. Additionally, appropriateness of selected resources is further analysed depending on the granularity required; alternative methods are thus suggested for selecting the most appropriate resources when considering either global course recommendations or more specific ones for a certain task. Experimental evaluation validates also the performance of the system for its real time application during a course. Finally, based on the successful rates for relevance and opinions provided by the teaching staff, we are confident that the system earned their trust for using it in real classes.

Future research will work on the implementation of a recommender system that will be integrated in the learning environment delivered to the students (a virtual machine in this case). This recommender will be based on the algorithms discussed in this paper, providing potentially useful resources in real time during the course. Further evaluation is also planned, in particular considering students' point of view. This study will allow us to validate aspects like the quality of real time recommendations, and if students and course staff actually consider the recommender useful for a better development of the course. Finally, alternative strategies like collaborative filtering or hybrid approaches combining both content-based and collaborative filtering are also being considered.

Acknowledgments. The authors are truly indebted to Luis Sánchez Fernández for his insightful comments about the lexical analysis, as well as the teaching staff that participated on the experience. Work partially funded by the Learn3 project, "Plan Nacional de I+D+I TIN2008-05163/TSI", the Acción Integrada Ref. DE2009-0051, the "Emadrid: Investigación y desarrollo de tecnologías para el e-learning en la Comunidad de Madrid" project (S2009/TIC-1650) and TELMA Project (Plan Avanza TSI-020110-2009-85).

References

1. Adomavicius, G., Tuzhilin, A.: Toward the next generation of recommender systems: a survey of the state-of-the-art and possible extensions. IEEE Transactions on Knowledge and Data Engineering 17(6), 734–749 (2005)
2. Ashraf, B.: Teaching the Google-eyed YouTube generation. Education+ Training 51(5/6), 343–352 (2009)
3. Auinger, A., Ebner, M., Nedbal, D., Holzinger, A.: Mixing content and endless collaboration–MashUps: Towards future personal learning environments. Universal Access in Human-Computer Interaction. Applications and Services (2009)
4. Baeza-Yates, R., Ribeiro-Neto, B.: Modern information retrieval, vol. 463. Addison Wesley/ACM Press (1999)
5. Burke, R.: Hybrid recommender systems: Survey and experiments. User Modeling and UserAdapted Interaction 12(4), 331–370 (2002)

6. Çelik, T.: Attention.xml Technology Overview (2005)
7. Cunningham, H.: GATE, a general architecture for text engineering. Computers and the Humanities 36(2), 223–254 (2002)
8. Cunningham, H., Maynard, D., Bontcheva, K., Tablan, V.: GATE: an architecture for development of robust HLT applications. In: Proceedings of the 40th Annual Meeting on Association for Computational Linguistics, vol. 54, pp. 168–175. Association for Computational Linguistics (2002)
9. Drachsler, H., Hummel, H.G.K., Koper, R.: Personal recommender systems for learners in lifelong learning networks: the requirements, techniques and model. International Journal of Learning Technology 3(4), 404–423 (2008)
10. Govaerts, S., Verbert, K., Klerkx, J., Duval, E.: Visualizing Activities for Self-reflection and Awareness. In: 9th Int. Conf. on Web-based Learning (2010)
11. Lops, P., Gemmis, M., Semeraro, G.: Content-based Recommender Systems: State of the Art and Trends. In: Recommender Systems Handbook, pp. 73–105. Springer, US (2011)
12. Macfadyen, L.P., Dawson, S.: Mining LMS data to develop an "early warning system" for educators: A proof of concept. Computers & Education 54(2) (2010)
13. Manouselis, N., Drachsler, H., Vuorikari, R., Hummel, H., Koper, R.: Recommender Systems in Technology Enhanced Learning, ch. 12, pp. 387–415. Springer, Heidelberg (2011)
14. Mazza, R., Dimitrova, V.: CourseVis: A graphical student monitoring tool for supporting instructors in web-based distance courses. International Journal of Human-Computer Studies 65(2), 125–139 (2007)
15. Mazza, R.: A graphical tool for monitoring the usage of modules in course management systems. In: Lévy, P.P., Le Grand, B., Poulet, F., Soto, M., Darago, L., Toubiana, L., Vibert, J.-F. (eds.) VIEW 2006. LNCS, vol. 4370, pp. 164–172. Springer, Heidelberg (2007)
16. Pardo, A., Delgado Kloos, C.: Stepping out of the box. Towards analytics outside the Learning Management System. In: Int. Conf. on Learning Analytics (2011)
17. Resnick, P., Varian, H.: Recommender systems. Communications of the ACM 40(3), 58 (1997)
18. Romero, C., Ventura, S., Garcia, E.: Data mining in course management systems: Moodle case study and tutorial. Computers & Education 51(1), 368–384 (2008)
19. Romero Zaldívar, V.A., Burgos, D., Pardo, A.: Meta-rule based Recommender Meta Systems for Educational Applications (2011)
20. Salton, G., Buckley, C.: Term-weighting approaches in automatic text retrieval. Information Processing & Management 24(5), 513–523 (1988)
21. Salton, G., McGill, M.: Introduction to modern information retrieval, vol, vol. 1. McGraw-Hill, New York (1983)
22. Schmitz, H.-C., Scheffel, M., Friedrich, M., Jahn, M., Niemann, K., Wolpers, M.: CAMera for PLE. In: Cress, U., Dimitrova, V., Specht, M. (eds.) EC-TEL 2009. LNCS, vol. 5794, pp. 507–520. Springer, Heidelberg (2009)
23. Schmitz, H.C., Wolpers, M., Kirschenmann, U., Niemann, K.: Contextualized Attention Metadata, ch. 8. Cambridge University Press, Cambridge (2009)
24. Sifry, D., Marks, K., Çelik, T.: Attention.XML Draft Specification (2004)
25. Valsamidis, S., Kazanidis, I., Kontogiannis, S., Karakos, A.: Course Ranking and Automated Suggestions through Web Mining. In: 2010 10th IEEE International Conference on Advanced Learning Technologies, pp. 197–199. IEEE, Los Alamitos (2010)
26. Wolpers, M., Najjar, J., Verbert, K., Duval, E.: Tracking actual usage: the attention metadata approach. Technology Education & Society 10(3), 106–121 (2007)

Usage Pattern Recognition in Student Activities

Maren Scheffel[1], Katja Niemann[1], Abelardo Pardo[2], Derick Leony[2], Martin Friedrich[1], Kerstin Schmidt[1], Martin Wolpers[1], and Carlos Delgado Kloos[2]

[1] Fraunhofer Institute for Applied Information Technology FIT, Schloss Birlinghoven, 53754 Sankt Augustin, Germany
[2] Universidad Carlos III de Madrid, Avenida Universidad 30, E-28911, Leganés (Madrid), Spain
{maren.scheffel,katja.niemann,martin.friedrich}@fit.fraunhofer.de, {kerstin.schmidt,martin.wolpers}@fit.fraunhofer.de, {abel,dleony,cdk}@it.uc3m.es

Abstract. This paper presents an approach of collecting contextualized attention metadata combined from inside as well as outside a LMS and analyzing them to create feedback about the student activities for the teaching staff. Two types of analyses were run on the collected data: first, key actions were extracted to identify usage patterns and tendencies throughout the whole course and then usage statistics and patterns were identified for some key actions in more detail. Results of both analyses were visualized and presented to the teaching staff for evaluation.

Keywords: attention metadata, usage monitoring, teacher feedback.

1 Introduction

Collecting data about how students interact with each other, the teaching staff and learning objects is becoming increasingly complex. Traditional approaches place a Learning Management System (LMS) at the center of these interactions [1]. Such systems are typically used as a container to store learning resources supplied by the teaching staff and offer a variety of services to capture the interaction between the users and other elements, e.g. discussion forums, quizzes, wikis, blogs, etc. This approach can be called a "LMS-centric" one. With the arrival of the Web 2.0, however, this approach is no longer functional [2]. Learning activities, sometimes even without the consent of the instructors, often take place within numerous applications that are hosted outside the LMS. Some of these activities may explicitly require the use of certain tools that must be installed outside the LMS, e.g. an editor, and others require students to search for auxiliary material that is not stored in the LMS.

The use of tools not hosted in the LMS is specially common in experimental sciences [3] where students are usually required to learn procedures with specific tools either installed on their personal computers or on equipment made available

C. Delgado Kloos et al. (Eds.): EC-TEL 2011, LNCS 6964, pp. 341–355, 2011.

by the educational institutions. This fact and the tendency of students to access an increasing number of resources outside an institution's LMS [4] gives rise to the need for more sophisticated data collection mechanisms.

This paper presents an approach of collecting usage data from inside a LMS (i.e. forum interactions) and outside a LMS (i.e. use of a virtual machine as a self-contained course environment) and analyzing them to create teacher feedback about the course they were taken from. In section 2 we motivate our approach and present previous works connected to this topic while section 3 describes what kind of usage data we collect, how we collect it and how we store it. After explaining the analysis methodologies we employed in section 4, analysis results are presented in section 5. Finally, section 6 concludes our paper and discusses further work.

2 Related Work

Modifying a learning scenario based on user observations is a research topic that has been attracting significant attention. Areas such as intelligent tutoring systems [5] propose the design of applications that observe and react to the user actions. But these applications have the same drawbacks as LMSs: If a learning experience occurs in a context not only involving the tutoring system, the truly valuable observations would most likely be those of the student interacting with the whole environment, and not only the ones with the tutor.

Zinn and Scheuer [6] conducted a survey among teachers trying to identify requirements for student tracking tools. Among the information deemed mostly important were the students' overall success rate, the mastery level of concepts, skills, methods and competencies as well as the most frequently diagnosed mistakes. As for the reasons why they would employ student tracking at all, most teachers said that they would like to be able to respond to individual students, adapt their teaching and identify problems of understanding the material. One such approach to provide teachers with feedback about their courses is described by Jovanovic et al [7]. They present a tool called LOCO-Analyst that analyses user tracking data based on an ontological framework. It was developed specifically for online learning courses and thus only collects data from within the LMS. Kosba et al [8] also present a framework that generates feedback for teachers from tracking data collected within a course management system. Here, however, the generated advice bases on student, group and class models which are computed with the help of certainty factors and fuzzy set theories. Again, the system was applied to an online (distance) learning course. TADA-Ed (Tool for Advanced Data Analysis in Education) is a data mining tool by Merceron and Yacef [9] with the aim to support teachers in discovering pedagogically relevant patterns in user activity data collected in online exercises using a web-based tutoring system. Therefore, it offers visualizations of simple statistics (e.g. mistake frequency) and results of data mining algorithms (e.g. clustering students based on the concepts of their mistakes). However, TADA-Ed presupposes that the users (i.e. teachers) are familiar with data mining techniques and competent enough to choose and apply them as well as to analyze their results which can be quite complex.

Extending the range of observable events in learning environments is not a trivial task. In principle, a student may engage in a course activity interspersed with other activities or at unexpected times. In the rest of the paper the concept of *observability* will be restricted to the realm of computer use. Given the wide variety of applications and systems that can potentially be used in a course, it seems obvious that an important improvement would be to move the recording capabilities outside of the LMS and onto the students' personal computer. There are already certain applications that record these events which can then be shared in a community of users (e.g. Wakoopa[1]). In a learning environment though, the need is to monitor only those events that are related to the course.

3 Data Collection

This section describes what usage data was collected and how. First, the schema in which the collected data is stored is explained. We then present the data collection processes from several sources and finally address data privacy issues.

3.1 Contextualized Attention Metadata

The Contextualized Attention Metadata (CAM) schema allows modelling a user's handling of digital content across system boundaries [10]. As CAM was developed to describe as many types of attention metadata as possible, CAM records of a user cannot merely describe the user's foci of attention but rather his entire computer usage behaviour. The latest version[2] is better suited for evaluating and analyzing user observations due to its event centredness. Figure 1 shows an image of the new schema.

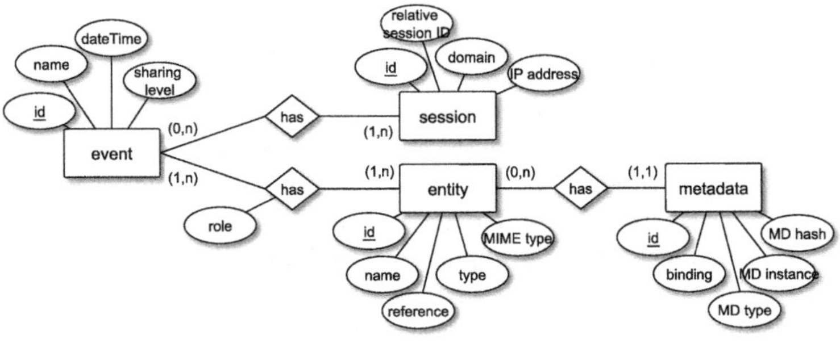

Fig. 1. Contextualized Attention Metadata (CAM) Schema

While the older versions of the CAM schema revolved around the user, the new schema has the `event` element as its entry point. Every user action is described by one `event`, e.g. *UserA sends MessageX to UserB in ApplicationY*

[1] http://wakoopa.com/

[2] More information and software about CAM can be found at https://sites.google.com/site/camschema/home

and several attributes define this event in more detail: the `name` attribute (name of the event, e.g. open - for opening a document) and the `dateTime` attribute (timestamp of the event, e.g. 20111504-123428). As opposed to these two attributes, the `sharing level` attribute is optional. Every `event` is connected to at least one `session` and every `session` can be defined by attributes for `IP address`, `domain`, and a `relative sessionID`.

An `event` can be connected to several `entities` that represent the participants of this event. The connected `entity` has to be further specified with a `role`, e.g. in the action mentioned above *UserA* has the role *sender*, *UserB* has the role *receiver*. `Entity` elements are defined by attributes for `name`, `MIMEtype`, `type` and `reference`. An `entity` in turn can have `metadata` attached to it. This `metadata` element is defined by a `binding`, `MDtype`, `MDinstance` and `MDhash`.

Due to the simplicity of this schema, a lot of knowledge has been moved to the `role` attributes. Also, the schema now requires rules to be enforced on the instances that go beyond simple cardinality restrictions, that is, the cardinality depends on the type of the value of `role`. For example: if there is a related `entity` with the `role` attribute *receiver*, there needs to be exactly one related `entity` with the `role` attribute *sender*.

CAM can be analyzed to provide an overview about where (i.e. with which application) and when an action takes place and what happens in the environment. CAM analyses enable the discovery of popularity, usage bursts and trends of tools. Information about when an action takes place can be useful in controlled environments such as formal learning environments where activities are usually scheduled. It is also quite useful in less controlled or more blended environments to understand when learners are actually active. The next sections therefore describe how CAM was collected during programming courses at the Universidad Carlos III de Madrid and then analyzed.

3.2 Collecting Data from the Virtual Machine

The concrete scenario where our approach has been deployed was a second year course of an engineering program. The course topic was C programming using mobile devices, and students were supposed to use a set of industry-type tools as well as work in teams for the course activities. The proposed approach consisted of offering the students a fully configured virtual machine containing all the tools required to carry out the course activities. A virtual machine is a self-contained computer captured in a single file that can be emulated by a client application in a different computer. This emulation is also called "virtualization" and has been considered in some learning scenarios to eliminate configuration problems [11].

The virtual machine was created by the teaching staff before the beginning of the course with a configuration specially tailored to the course activities. As part of that configuration, the main tools inside the virtual machines have been modified so as to record certain events in an internal database. Those recorded events were regularly relayed to a central server that collects them. The virtual machine made available to the students has been configured with the following tools: compiler, debugger, memory profiler, file editor, version control system, emulator, command line interface, browser and an integrated virtual environment (IDE).

A significant number of course activities require writing code fragments and testing their execution. The compiler, debugger and memory profiler are required to carry out these tasks. For the compiler the date and time when the compilation is invoked, the error and warning messages, and the final status of the compilation are recorded. The debugger is essential when developing C applications and students must be proficient in its use. The date and time the tool was started and closed, as well as all the commands used during the session are stored. The memory profiler executes and detects memory leaks. The invocation and the messages produced by the application are recorded together with a time stamp. The Command Line Interface is being used in the course to invoke most of the tools as well as carrying out regular operations such as creating folders, moving files, etc. Therefore, all commands are stored together with a time stamp. C programs can be edited either using a regular file editor or an integrated virtual environment. The programs are shared among team members and with the teaching staff using the version control system Subversion[3]. The date and time when the editor or IDE are invoked and closed are recorded. Additionally, the browser can be considered as yet another development tool that students will use frequently while working in the course. For that reason, it is also used to retrieve the visited URLs.

Each tool stores the recorded events at a fixed location in the user's home folder. Users should be able to easily enable/disable the recording mechanism. This feature has been implemented as a simple test for the existence of the folder where the events are recorded. If such folder is removed, the tools no longer record any event. The client application of the version control system Subversion was modified so that every `commit` operation that sends the current version of a file to the server would actually send two sets of changes: The normal set of changes in the files of the specified directory, and a second set of changes including all the files with the events recorded from the last transmission. As a consequence, every time a student submits a new version of a project, all recorded events are submitted to the server. The transmission has been deployed so that the system works incrementally and only those events produced from the last transmission are sent. Typically, a Subversion repository requires user authentication. As they are part of the same subversion session, the recorded events are submitted to the server with the same user authentication as the rest of files. Once the recorded events are submitted they are transformed into CAM instances.

3.3 Collecting Data from the LMS

Extending the observation of student tasks outside of the LMS does not mean that the data collected inside the LMS is no longer valid. On the contrary, the events taking place in the LMS are yet another ingredient to add to the overall observation mechanism.

[3] http://subversion.apache.org/

The analysis of usage behaviour within the LMS is made possible by reworking the Web server logs of the system. Every line in the log file represents one event and contains information such as date and time, the initiator's user id, type of event, etc. After collecting the logs a grammar consisting of regular expressions was constructed. With the help of this grammar the logs can be grouped into meaningful user actions. For example, the complex event of posting a new message in a forum consists of several single user-initiated actions in the logs (i.e. the user needs to click on a *new post* button which initiates the loading of a new page, where he can write the post to then click on the *send* button which initiates the loading of the forum page with the new message).

Given this file and the stored logs, an event was created whenever a pattern was identified and transformed into a CAM instance. Any static information needed for later analysis but not contained in the log files, e.g. name of a course, were pulled from the database the LMS is based on. Using this approach, events such as posting in a forum, replying to a post, looking at a message, logging in or logging out can easily be detected in the server logs.

Due to synchronized server time between the virtual machine and the LMS merging of the CAM instances from the two was unproblematic. The combined data were then stored in a relational database from which all further analyses were done.

3.4 Data Privacy Issues

Extending the observation of student activities with the described approach poses certain concerns on data privacy. The first observation is that by using a virtual machine, only those events that take place inside the machine (and not on the host) are being recorded. Students are instructed at the beginning of the course to use this machine *only* for course related tasks. By simply adhering to this policy (which cannot be strictly enforced), the recorded events are all relevant.

But current legislation about data privacy imposes additional constraints that have to be reflected in the deployed system. The first one is that users should be made explicitly aware of the type of data that is being collected. The virtual machine was configured so as to start with a user session with the browser showing a page explaining the details of the recording mechanism. Second, the recording system must be easily disableable by the user. As it was described in the previous section, the presence or absence of a concrete folder is used to switch the system on or off respectively. And third, and most importantly, users need to be made aware of how to contact the person/institution in charge of the data because they maintain the right to query and request the deletion of any of the data.

4 Analysis Methodologies

Two types of methodologies have been applied to the collected CAM. In a first step we extract key actions to identify usage patterns and tendencies throughout the whole programming course. Then, in a second step, some sequences of events are looked at in more detail and usage statistics and patterns for these sequences are identified.

Our first approach, extracting key actions, is to analyze CAM with techniques used in the research field of computational linguistics, that is, we transfer methodologies from text analysis to action analysis and try to find patterns within the recorded activities. Languages are rule-governed [12], i.e. they are based on patterns and structures. We exploit this fact and apply methodologies to detect and analyze such patterns to CAM by mapping linguistic concepts to the respective parts of CAM instances which is, however, not a simple one to one mapping. The concepts word and action can quite easily be seen as analogous but sentence and session cannot: While sentences are fixed linguistic categories, sessions describe a concept of time with a variable beginning and end (e.g. an hour, a few minutes, a month, a year, etc.). Mapping session to the concept of text, i.e. a collection of sentences, possibly makes more sense. As the word-to-action-mapping is very reasonable, we start our approach of detecting meaningful patterns from CAM by transferring methods from keyword extraction to key action extraction. The content of a document can be semantically represented by keywords [13]. We thus assume that key actions can semantically represent the session they are taken from. In order to find repeated string patterns we analyze the collected CAM with the so-called n-gram approach [14]. The following example illustrates the technique in a simplified way:

$$_0\,\mathtt{A}\,_1\,\mathtt{B}\,_2\,\mathtt{C}\,_3\,\mathtt{A}\,_4\,\mathtt{B}\,_5\,\mathtt{D}\,_6\,\mathtt{B}\,_7\,\mathtt{C}\,_8\,\mathtt{A}\,_9\,\mathtt{B}\,_{10}\,\mathtt{A}\,_{11}\,\mathtt{A}\,_{12}\,\mathtt{C}\,_{13}\,\mathtt{D}\,_{14}$$

The letters represent a sequence of actions while the numbers indicate the position of the actions within the sequence. In a first step, all possible monograms are extracted:

```
A [0,1],[3,4],[8,9],[10,11],[11,12] C [2,3],[7,8],[12,13]
B [1,2],[4,5],[6,7],[9,10]          D [5,6],[13,14]
```

The merging of n-grams is possible if the frequency of the new key action is above a set threshold. Lets assume the threshold in this example was set to 2. As no monograms are below that threshold, none get discarded from further calculations. In a next step, all possible bigrams are extracted by trying to combine the monograms with one another:

```
AA [10,12]               BC [1,3],[6,8] CA [2,4],[7,9]
AB [0,2],[3,5],[8,10]    BD [4,6]       CD [12,14]
AC [11,13]               BA [9,11]      DB [5,7]
```

The bigrams `AA`, `AC`, `BD`, `BA`, `CD` and `DB` only occur once. Hence, they are discarded from further calculations and can consequently neither be a key action nor part of one. Additionally, if shorter sequences are contained in longer ones, the shorter ones are no longer part of further calculations and thus no key action candidate. In our example the monograms `A`, `B` and `C` are discarded but the monogram `D` stays as it is not contained in any of the remaining bigrams `AB`, `BC` and `CA`. These now pose as the basis for trigram extraction, those are turned into tetragrams, then pentagrams and so forth. Calculations end when no further n-grams can be merged. Our example ends with two n-grams, namely the twice occurring tetragram `BCAB` and the monogram `D`, and thus two key actions.

Once the key actions are identified, our second approach is to look at some of them in more detail. Depending on the domain of interest or the focus of the analysis, some actions can be deemed more important than others and a detailed analysis can help the teacher to become more aware of what problems students might run into. Once the sequences to be looked at in more detail are identified, they can be analyzed further and displayed to the teachers, e.g. in the form of text files, tables or diagrams. Interesting analyses can, for example, be the distribution of certain events over time or students, common subsequent actions of relevant actions in the current domain, anomalies, tendencies or regularities in the usage behaviour.

5 Analysis Results

To analyze and evaluate our approach we collected data from the C programming course mentioned above. First we extracted key actions from the collected data on several granularity levels and presented our results to the teaching staff. Then we ran several pattern detection analyses on those actions identified as worth looking at by the teachers in more detail and presented the results to the teaching staff again.

5.1 Key Action Extraction

The C programming course that our analyses base on took place from September 6th until December 16th, 2010 and 244 students attended the course. In that period of time, around 10,000 sessions were recorded comprising a total of 119,652 events which in turn contained 19 different event types. There are some events that do not contain any additional information as the event name itself as well as the date and time of occurrence (e.g. `startDebuggerGDB` or `endTextEditorKate`). However, most events contain one additional attribute, e.g. the ID of the viewed message for the event `viewMessage`, the error message for an unsuccessful compile event or the URL for the event `visitURL`.

In order to find key actions, i.e. frequent patterns in the data, we distinguish between two granularity levels of the data: a short and a long version. In the short version, only the events and their timestamps are considered but no further attributes. In the long version, all attributes (e.g. forum IDs and warning messages) are completely used for the calculation of key actions. There is one exception for the event `visitURL` that can be combined with the short or with the long version, i.e. only the domain of the URL can be considered instead of using the whole URL (long version) or no URL at all (short version). For example, we would assume the pattern of several students getting the same error and conducting a google search subsequent as very meaningful, even if they do not use the exact same query terms. If we want to detect such patterns, we need to use the long version, but shorten the URLs of the event `visitURL` to their domain.

As exploratory investigation, all variations were calculated with a threshold of 5 as well as 10. We extracted key actions on the basis of the whole course, that is, all sessions from all users were taken into account at the same time. Thus, if an action was identified as a key action with 15 occurrences, it could be due to several different users executing the action but also to one user executing it 15 times.

A visualization for the results was implemented and given to the teaching staff for evaluation. The version deemed most useful to deduce meaningful things from was the long version with the threshold set to 10. By analyzing the sequences of event patterns, the teachers discovered several situations interesting to them as they denoted incorrect procedures. Figure 2 shows an example of such a visualization. The darker boxes mark the first event of a key action sequence. The key action sequence in the upper right corner for example starts with a `gotoForum` event, followed by `viewMessage`, another `gotoForum` event and two more `viewMessage` events.

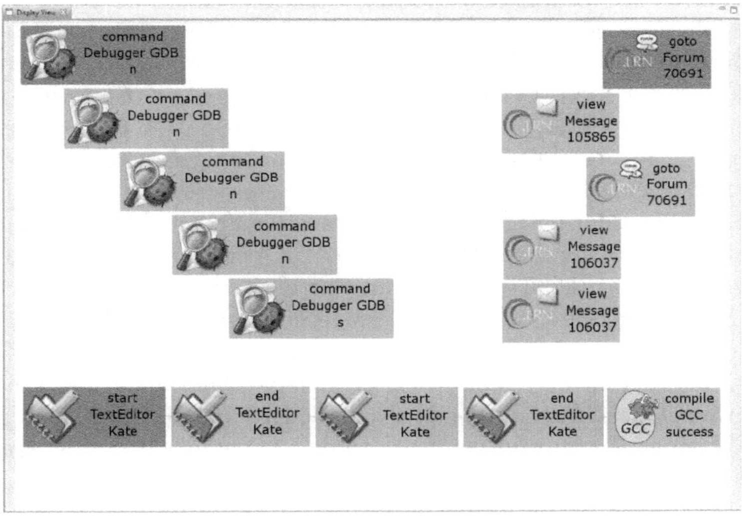

Fig. 2. Visualization of extracted key actions

Some sequences identified as interesting by the teaching staff contained events where the text editor was started and ended. This points to development flows in which for each compilation, students opened a file with the editor and closed it again before compiling. According to the teaching staff this translates into a significant increase in the development time and should be corrected. Other sequences of events to access the course material also pointed to some possible corrections. The course material was available through a direct link as well as through a redirection at a second official location. The second route was followed by a significant number of users, thus pointing to the need for a teacher to emphasize how to quickly access the course material. Another example identified as an important one were sequences where tools such as the debugger or the memory profiler were used several times which denotes the students' difficulty in understanding the anomalies detected by these tools. As the teaching staff considered error messages and the students' reaction to them as especially valuable, we analyzed these events in more detail.

5.2 Frequent Error Patterns

One of the most significant difficulties students face in introductory programming courses is how to understand the messages returned by the compiler when writing a program. There is an important gap between learning a programming language and quickly identifying an anomaly spotted by the compiler during development. This gap typically requires an unusually high amount of time from students to be closed. Tutor assistance in these situations is very important. However, even if the students are well aware of the fact that they are observed, we cannot assume them to be aware of the effects that the analysis of their data can have. We assume that showing a teacher some statistics of a user who for example compiled several times getting the same (trivial) error over and over again, might lead to the teacher getting negatively affected. Therefore, we only showed statistics of errors that occurred at least 50 times in total for a minimum of 10 users. Additionally, we made the students' usage patterns more abstract, e.g. `too few arguments to function usernameTest` was shortened to `too few arguments to function...`. This process helps combining activities when looking for patterns but also keeps the anonymity of the students.

Within the collected data there were 17,266 `compile` events. 6,443 of them were error messages, 3,190 were warnings and 7,633 events compiled successfully. Figure 3 shows those errors with the highest frequency and that at least 25 different user received ordered by the number of students that got it. From this diagram, instructors were able to easily identify those anomalies that require clarification. Teachers liked the fact that with the obtained data, they may even select a sample of the errors and use them as examples to guide students to detect and correct the most frequent anomalies.

We also calculated the distribution over time of the most frequent errors and asked the teaching staff to evaluate whether such a visualization can be useful. Figure 4 shows the distribution of the most frequent error, i.e. `'foo' undeclared (first use in this function)`, over time and the number of students that got this error. In total the error occurred 1048 times for a total of 92 users. As it can be seen, students encountered this error even in the final stages of the course (when there was an increase on student load due to a project deadline). According to the teaching staff corrective actions taken at the beginning of the course might have helped to lower the error's appearance.

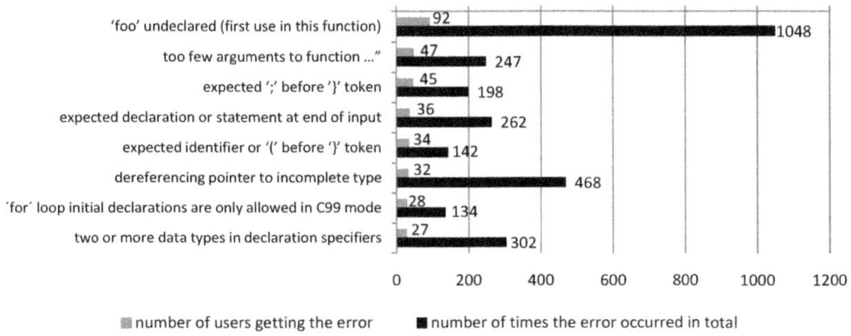

Fig. 3. Distribution of the most frequent errors

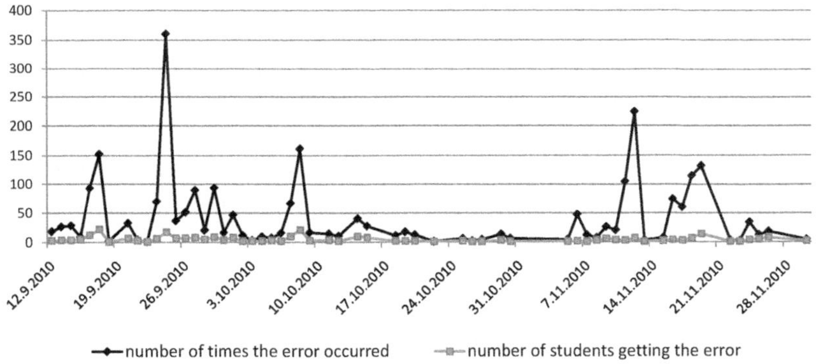

Fig. 4. Error `'foo' undeclared (first use in this function)` over time

Figure 5 shows the distribution over time of the error `two or more data types in declaration specifiers`. In total it occurred 302 times for a total of 27 users. The error disappears over time which, according to the teaching staff, shows that the main problem of understanding why the error occurred got solved by the students. However, as the error occurred very frequently at the beginning of the course, teacher support could have helped in improving the learning curve faster.

Another interesting aspect the teaching staff wanted to have more detailed information about was what the students did after getting error messages, i.e. what kind of events were `compile_ error` events followed by. We therefore created some diagrams to help identify more precisely the nature of the errors encountered by the students. Figure 6 displays the actions that students performed directly after receiving the error message `'foo' undeclared (first use in this function)`. It confirms that the error is persistent, that is, it tends to appear again in the next compilation. This information is essential for teachers to pinpoint those anomalies that students are not understanding, thus giving rise to important time delays.

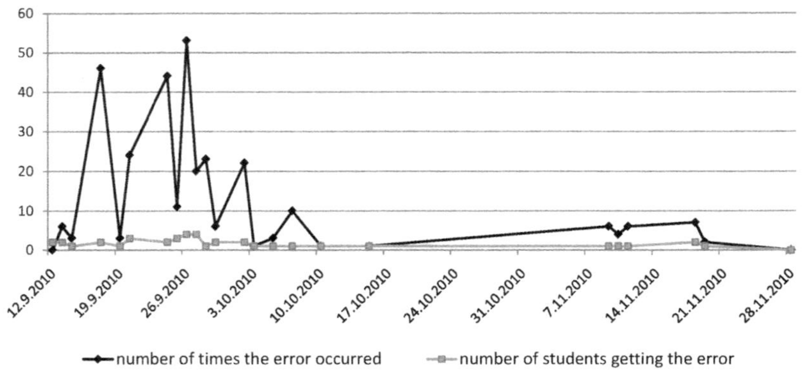

Fig. 5. Error `two or more data types in declaration specifiers` over time

Figure 7 shows the detail of a third compilation error, i.e. `expected ';' before '}' token`. It occurred 198 time in total for a total of 45 users. In this case we can see that the next events are sometimes a correct compilation, sometimes the appearance of some warnings, and in some other cases the error re-appears. This error clearly does not have the "road-block" nature, i.e. not knowing what to do next and trying random changes in the code, identified in the previous two.

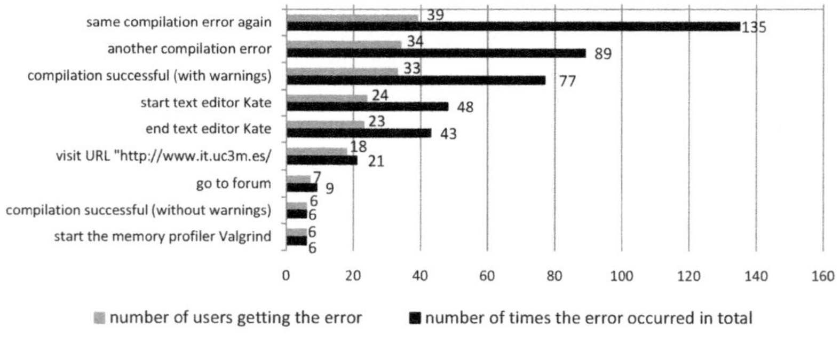

Fig. 6. First actions after error `'foo' undeclared (first use in this function)`

These diagrams show the power of offering visualizations of events occurring in every-day student work sessions to the teaching staff. If these events occurred in a face-to-face session, the instructor would quickly point the students to techniques to identify and solve the anomaly. With the proposed paradigm, these patterns can still be detected and brought to the attention of the instructor when they occur in any course working session and as a consequence of the analysis of these data, several corrective actions can be deployed by the instructors. They may range in complexity from discussing the most common mistakes in class and making sure students identify the errors correctly, to an automatic support system that detects when certain specific errors are produced by the compiler and some additional explanation is given live to the student (a pop-up message, an email, a URL pointing to a more detailed explanation, etc.).

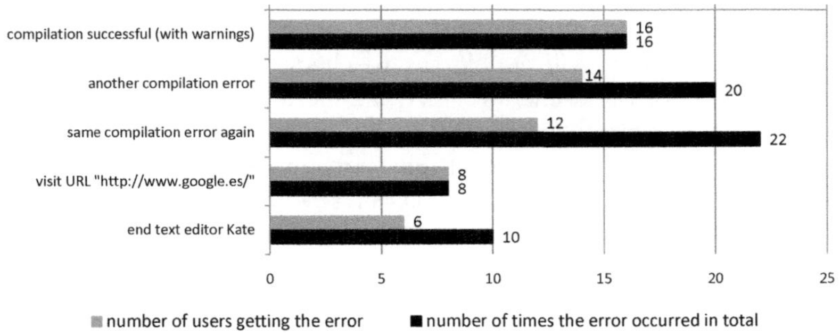

Fig. 7. First actions after error `expected ';' before '}' token`

6 Conclusion

In this paper an approach to analyze user activities and detect usage patterns in the context of a programming course has been presented. Contextualized Attention Metadata (CAM) was used to represent the multiple events occurring in a learning environment. The capturing mechanism employed extends the conventional methods that rely solely on those events recorded in the logs of a Learning Management System. A virtual machine specifically configured for the course was distributed among students for the purpose of being used in all the course activities. This machine deployed a high coverage recording mechanism for the events occurring in the previously installed tools. These observations were also complemented with the corresponding events derived from the LMS.

Once the events have been captured, techniques were applied to first extract key actions from the collected data, and then to identify usage patterns by exploring in detail sequences of events. With these techniques discovering patterns in large collections of events was successful where no manual technique is feasible.

The procedure was validated by applying it to an undergraduate engineering course on introductory programming in C and a large number of events was collected (almost 120,000 events of 244 students). First a visualization of the key actions (i.e. frequent patterns of events) was produced. The visualization showed certain sequences that clearly point to corrective actions to be deployed. Second an analysis of the most frequent error patterns was performed. Based on this analysis, the teaching staff identified some errors as most problematic that should be discussed in more detail in the course.

Two approaches are considered as future research lines. The first is to support reflection and awareness of students. Self-reflection allows students to metacognitively assess, analyze and evaluate their learning processes. It is also an essential part of self-regulated learning [15]. Furthermore there is the need for direct feedback to the students in their learning environment [16]. This can be accomplished by using the version control system to transfer information directly into the user desktop.

For the second approach, the reflection and awareness support of the teachers, we will embed the graphical analyses that the teaching staff graded as especially helpful in our tool. This will enable the teachers to directly react in class on the usage patterns of the students. Once the merit of the approaches has been established by the experimental results, techniques for the automation of the corrective actions can be explored. Taking advantage of the use of the virtual machine, an algorithm that reacts when certain patterns are detected can be designed and deployed. Furthermore, the detection algorithms can be refined as to tackle specific problems within a course. The teaching staff may describe activities that might require a closer analysis so that more targeted corrective actions can be discovered.

Acknowledgements. The research leading to these results has received funding from the European Community's Seventh Framework Programme (FP7/2007-2013) under grant agreement no 231396 (ROLE project). Work was also partially funded by the Learn3 project (TIN2008-05163/TSI), the eMadrid project (S2009/TIC-1650), and the Acción Integrada DE2009-0051.

References

1. Ertugrul, N.: Towards virtual laboratories: A survey of LabVIEW-based teaching/learning tools and future trends. International Journal of Engineering Education 16(3), 171–180 (2000)
2. Chatti, M.A., Jarke, M., Frosch-Wilke, D.: The future of e-learning: a shift to knowledge networking and social software. International Journal of Knowledge and Learning 3(4), 404–420 (2007)
3. Auinger, A., Ebner, M., Nedbal, D., Holzinger, A.: Mixing content and endless collaboration–MashUps: Towards future personal learning environments. Universal Access in Human-Computer Interaction. Applications and Services, pp. 14–23 (2009)
4. Waycott, J., Bennett, S., Kennedy, G., Dalgarno, B., Gray, K.: Digital Divides? Student and Staff Perceptions of Information and Communication Technologies. Computers & Education 54(4), 1202–1211 (2010)
5. Woolf, B.P.: Building intelligent interactive tutors. Morgan Kaufmann, San Francisco (2008)
6. Zinn, C., Scheuer, O.: Getting to Know Your Student in Distance Learning Contexts. In: Nejdl, W., Tochtermann, K. (eds.) EC-TEL 2006. LNCS, vol. 4227, pp. 437–451. Springer, Heidelberg (2006)
7. Jovanović, J., Gašević, D., Brooks, C.H., Devedžić, V., Hatala, M.: LOCO-Analyst: A Tool for Raising Teachers' Awareness in Online Learning Environments. In: Duval, E., Klamma, R., Wolpers, M. (eds.) EC-TEL 2007. LNCS, vol. 4753, pp. 112–126. Springer, Heidelberg (2007)

8. Kosba, E., Dimitrova, V., Boyle, R.D.: Using Student and Group Models to Support Teachers in Web-Based Distance Education. In: Ardissono, L., Brna, P., Mitrović, A. (eds.) UM 2005. LNCS (LNAI), vol. 3538, pp. 124–133. Springer, Heidelberg (2005)
9. Merceron, A., Yacef, K.: TADA-Ed for Educational Data Mining. Interactive Multimedia Electronic Journal of Computer-Enhanced Learning 7(1) (2005)
10. Schmitz, H.-C., Wolpers, M., Kirschenmann, U., Niemann, K.: Contextualized attention metadata. In: Roda, C. (ed.) Human Attention in Digital Environments, pp. 186–209. Cambridge University Press, Cambridge (2011)
11. Gaspar, A., Langevin, S., Armitage, W.D., Rideout, M.: March of the (Virtual) Machines: Past, Present, and Future Milestones in the Adoption of Virtualization in Computing Education. Journal of Computing Sciences in Colleges 23(5), 123–132 (2008)
12. Sag, I.A., Wasow, T.: Syntactic Theory – A Formal Introduction. Center for the Study of Language and Information, Stanford (1999)
13. Rose, S., Engel, D., Cramer, N., Cowley, W.: Automatic Keyword Extraction from Individual Documents. In: Berry, M.W., Kogan, J. (eds.) Text Mining, pp. 1–20. John Wiley & Sons, Ltd., Chichester (2010)
14. Hulth, A.: Improved Automatic Keyword Extraction Given More Linguistic Knowledge. In: Proceedings of the 2003 Conference on Empirical Methods in Natural Language Processing, EMNLP 2003, Stroudsburg, PA, USA, pp. 216–223. Association for Computational Linguistics (2003)
15. Zimmermann, B.J.: Self-regulated learning and academic achievement: An overview. Educational Psychologist 25(1), 3–17 (1990)
16. Butler, D.L., Winne, P.H.: Feedback and self-regulated learning: A theoretical synthesis. Review of Educational Research 65(3), 245–281 (1995)

Cross-Lingual Recommendations in a Resource-Based Learning Scenario

Sebastian Schmidt, Philipp Scholl, Christoph Rensing, and Ralf Steinmetz

Multimedia Communications Lab - Technische Universität Darmstadt
64283 Darmstadt - Germany
{sschmidt,scholl,rensing,ralf.steinmetz}@kom.tu-darmstadt.de
http://www.kom.tu-darmstadt.de

Abstract. CROKODIL is a platform supporting resource–based learning scenarios for self–directed, on–task learning with web resources. As CROKODIL enables the forming of possibly large learning communities, the stored data is growing in a large scale. Thus, an appropriate recommendation of tags and learning resources becomes increasingly important for supporting learners. We propose *semantic relatedness* between tags and resources as a basis of recommendation and identify Explicit Semantic Analysis (ESA) using Wikipedia as reference corpus as a viable option. However, data from CROKODIL shows that tags and resources are often composed in different languages. Thus, a monolingual approach to provide recommendations is not applicable in CROKODIL. Thus, we examine strategies for providing mappings between different languages, extending ESA to provide cross–lingual capabilities. Specifically, we present mapping strategies that utilize additional semantic information contained in Wikipedia. Based on CROKODIL's application scenario, we present an evaluation design and show results of cross–lingual ESA.

Keywords: Explicit Semantic Analysis, Cross–Language Semantic Relatedness, Wikipedia, Reference Corpus, Recommendation.

1 Introduction

With ever changing working environments and a decreased life-span of knowledge, learning becomes a lifelong process. The learning process which complements institutional education (including school, apprenticeship, university etc.) is characterized by the learner's self-responsibility and self-monitoring. Learning materials are available and accessible (e.g. on the Web), but not necessarily prepared for learning by a teacher like in traditional learning environments. Self-directed learning using learning materials is called *Resource–Based Learning* (RBL). In *RBL* settings, a major challenge for learners is finding relevant *Learning Resources* (LRs). One common strategy for this form of learning is to use a web search engine or specialized digital libraries. In learning settings, however, where a community like a learning group, a class of students or a group

C. Delgado Kloos et al. (Eds.): EC-TEL 2011, LNCS 6964, pp. 356–369, 2011.

of colleagues does already exist, the probability that other members of the community have already found relevant information is high. In order to discover this information, recommender systems that recommend information based on different features can be useful.

In this paper, we examine the data from a user study with the *RBL* platform *CROKODIL* with special regard to the language of the stored information (Section 2), analyze existing approaches for semantic relatedness with regard to cross-linguality (Section 3) and propose a feasible approach to enable cross-lingual recommendations (Section 4). Further, we evaluate this approach within our scenario (Section 5) and give a conclusion and prospect on further work (Section 6).

2 Multilinguality in the Usage Scenario

2.1 CROKODIL

The e–learning platform *CROKODIL* supports learners in finding, collecting and organizing learning materials from the Web in *RBL*. All data within the platform that is inserted by the users is stored in information items. Those can be *LR*s, learners, tags or various other types. Information items can be interconnected via relations. The overall resulting graph represents a kind of folksonomy. An *LR* in *CROKODIL* can be a whole document (web-page, pdf-file, textfile, etc.) or a short text fragment. Learners have a profile describing the represented person. The tags that are used by learners to describe the *LR*s are terms consisting of one or several words.

CROKODIL uses a recommendation engine that attempts to provide tags or *LR*s that are likely to be of interest to the learner. These recommendations bridge the gap between the information need of the learners and already existing possibly matching information items in the system. However, up to now *CROKODIL* only provides recommendations based on structural properties of the network built by information items and relations, e.g. whether there are explicit connections between two *LR*s over a defined set of tags. This means that if there is no explicit relation between two *LR*s, the recommendation engine is not able to infer this connection. Therefore, the formation of separate partitions of the networks is favoured, especially as different users commonly have a different terminology for denoting related information (eg. “Technology Enhanced Learning” and “e–learning”). Information items containing identical or almost identical semantic information to those the user has already stored are only of minor interest. Therefore, we aim to enable recommendation of semantically *related* information items.

Another challenge that *CROKODIL* has to meet with regard to recommending relevant items is that the overall knowledge base is expected to be sparse. In contrast to social bookmarking applications like Delicious, *CROKODIL* does not have millions of users and therefore collaborative filtering [1] might not be appropriate for recommending items. Therefore, a content–based recommendation paradigm is targeted in our work.

2.2 Language of Tags and Learning Resources

We examined the data from a user study [2] with ELWMS.KOM, a predecessor of *CROKODIL*, in order to determine the used language of *LR*s and tags. In the study, 21 knowledge workers at Technische Universität Darmstadt used ELWMS.KOM over a period of several weeks. Table 1 shows the language distribution of the stored web resources. A majority of the resources are in English, which is contradictory to the local language and the mother tongue of most of the participants (17 of the participants are German native speakers).

Table 1. Web resources contained in the knowledge base grouped by language and fraction of resource language chosen by German and non–German native speakers

Language	Web Resource Count	Web Resource Percentage	by German native speakers (4)	by Non–German native speakers (17)
English	333	75.33%	73.31%	79.45%
German	98	22.18%	23.99%	18.49%
French	2	0.46%	-	1.37%
Page forbidden (403)	1	0.22%	-	0.69%
Page unavailable (404)	8	1.81%	2.70%	-
Total	442	100.00%	100.00%	100.00%

Further, the language of the tags was evaluated. The results (cf. Table 2) let us infer that the language of a resource does not necessarily correspond to the language of the attached tags (e.g. French is not contained in the tag languages). Finally, in 70.2% of the cases the languages of tags and tagged resources correspond.

Table 2. Tags used for web resources in different languages in *ELWMS.KOM* sample. Note that German and non–German native speakers were involved and the number of participants does not match the numbers in the resource language experiment, as only 18 participants applied tags to resources.

Type	Tag Count	Tag Count in %	by German native speakers (15)	by Non–German native speakers (3)
English	300	30.70%	25.40%	42.91%
German	183	18.73%	22.17%	10.81%
Ambiguous Language	194	19.87%	20.41%	18.58%
Named entity (uni–lingual)	240	24.56%	28.63%	15.20%
Date or year	60	6.14%	3.39%	12.50%
Total	977	100.00%	100.00%	100.00%

This analysis shows that in real–world settings, the usage of *RBL* often crosses language borders. For content–based recommendations, this adds a dimension of complexity, as the language of *LR*s and tags has to be taken into account additionally. The learners' choice of tags is often influenced by the language a *LR* is composed in. This means that the same learner could use different tags for the same concept, eg. one user tagged related *LR*s with the English "visual" and the German "Visualisierung". This adds to the aforementioned challenges.

3 Related Work

Content-based recommendations can be performed by analyzing the semantic content of the available information items and by recommending those with the highest semantic closeness to other items which are known to be relevant for the user. According to Budanitsky and Hirst [3], there is a considerable difference between the two notions of semantic closeness *semantic similarity* and *semantic relatedness*. Semantic similarity denotes the degree of two different terms describing the same concept, e.g. the terms "cash" and "dough" have a high semantic similarity, because "dough" is a colloquial synonym for "cash". In contrast, semantic relatedness mimics the associative perception of humans. E.g. the terms "cash" and "bank" do not have a semantic similarity, but are semantically related because they often occur in a common context. Especially in our scenario, the concept of semantic relatedness is appropriate as recommendation of information items with a related content is targeted.

3.1 Semantic Relatedness

Milne and Witten [4] state that "any attempt to compute semantic relatedness automatically must also consult external sources of knowledge". Thus, all approaches to determine semantic relatedness utilize additional information by employing reference corpora in order to provide additional general knowledge. WordNet [5] is often used as an external source of knowledge to enable the calculation of semantic relatedness [3,6]. However, in recent work the focus has shifted to Wikipedia as a knowledge base because of its corpus size and its up-to-dateness.

Explicit Semantic Analysis (ESA). A promising approach for calculating semantic relatedness called *ESA* has been proposed by Gabrilovich and Markovitch [7]. Here, documents are not represented by means of terms but by their similarity to concepts derived from a reference collection of documents. *ESA* is based on the assumption that in the reference document collection, an article corresponds to a semantically distinct concept. Thus, by comparing documents based on their terminology to all articles in the document collection that have been pre–processed by tokenization, stemming, stop word removal and a term weight metric, a vector is obtained that contains a similarity value to each of the articles. This process step is called *semantic analysis*. The vector, called *semantic interpretation vector*, abstracts from the actual term occurrences and thus represents a semantic dimension of that document. A major advantage of *ESA* is that semantic relatedness can be calculated for terms and documents alike, providing good and stable results for both modes [7].

Formally, the document collection is represented as a matrix M with the dimensions $n \times m$ (called *semantic interpreter*), where n is the number of articles and m the number of occurring terms in the corpus. M contains (normalized) *TF–IDF* document vectors of the articles.

For calculating the similarity between the document and the corpus, the *cosine similarity measure* is employed. Analogously, two documents represented as semantic interpretation vectors can be easily compared by using cosine similarity again.

Despite its primary usage with Wikipedia, *ESA* is also applicable to different reference corpora. Gabrilovich and Markovitch have used the *Open Directory Project* (ODP) as well as Wikipedia, showing that *ESA* using Wikipedia outperforms the *ODP* reference corpus. They state that Wikipedia is especially practical for *ESA* as each of Wikipedia's articles ideally describes one concept.

Gabrilovich and Markovitch show that *ESA* using Wikipedia as a reference corpus outperforms other approaches like WikiRelate! [8], approaches based on WordNet [3] or Roget's Thesaurus [9] and *Latent Semantic Analysis* [7].

Extensions for ESA, which consider Wikipedia's internal link structure and its category system have been proposed and successfully evaluated [10]. Thus, further usage of structural information obtained from Wikipedia seems to be promising for the calculation of semantic relatedness.

3.2 Cross-Language Semantic Relatedness

As the Web contains documents composed in a variety of languages the need for cross-lingual approaches in semantic relatedness determination emerges. Semantic relatedness across language borders has therefore become a focus of research in recent years.

Schönhofen et al. [11] investigate the usage of Wikipedia for *Cross-Language Information Retrieval* (CL IR), aiming to query and retrieve English documents by German and Hungarian queries. For that purpose they first do a "word-by-word translation by dictionary", yielding in many cases a large set of word pairs for a single word in the source and the possible translations in the target language. In order to overcome this issue, they first aim to maximize the bigram similarity between the different translation combinations of adjacent words, consulting statistics obtained from the English Wikipedia as a reference corpus in the target language. Then, the links between pairs of articles containing the two translated terms in the article title are used to rank the translations. After having obtained the ranks for the translation pairs, Schönhofen et al. combine both measures to a final rank which results in an order describing the most probable terms. Although this approach benefits from the networked structure of Wikipedia which mirrors the semantic relatedness of concepts, it is still a term based approach which does not take the global term distribution, a measure of global term relevance, into account.

Potthast et al. [12] and Sorg et al. [13] both investigate a usage of *ESA* in cross-lingual contexts. Potthast et al. focus on the field of automatic cross-lingual plagiarism detection. They consider a language-independent concept space to which for each supported language a document-collection is aligned via a one-to-one mapping. This requires a reference corpus which contains articles describing the same set of concepts in different languages. Hence, only a subset of articles can be considered for the semantic relatedness computation in the case of Wikipedia usage. Because of their assumption of a bijective article mapping function and their restrictive usage of disjunction of all articles, the direction of their mapping does not matter to the results. Sorg et al. present a slightly more elaborated approach which does not assume a one-to-one mapping between articles in the corpus but a many-to-one mapping for articles in the source language to articles in the target language. So each target article might be targeted from different articles in the source language. In their approach, they first compute the ESA vector in the source language and map it to the target language afterwards by summing up the relatedness values from all concepts in the source language pointing to a single concept in the target language. They indicate a a good correlation to human rankings. But in case of topics being underrepresented in a specific Wikipedia language version, their approach is not capable of mapping the interpretation vectors without loosing semantic information.

Dumais et al. [14] present an approach using *Latent Semantic Indexing* (LSA) for *CL-IR* with a common space containing terms from different languages and show to perform well in CL-IR tasks. In contrast to *ESA*, *LSI* does not allow to give human-readable explanations about the reasoning for a particular relatedness rating.

4 Our Approach

Due to the promising results of ESA in mono- and multilingual settings we decided to apply this method in *CROKODIL* and extend it by an approach which can handle underrepresented topics in the Wikipedia. This challenge has not been addressed within the work presented in the previous section. Therefore, in the following, adaptations made to *ESA* in order to use it in multilingual environments are explained. Steps 1, 2, 3 and 5 are basically the steps executed in the original ESA procedure. The index l indicates the respective language.

1. The reference corpus (in our case the Wikipedia in language l) is preprocessed with stemming, stopword removal, article filtering[1], *TF–IDF* calculation and normalization. The result is a language-specific matrix M_l, the *semantic interpreter*, with the shape $n_l \times m_l$, where n_l is the number of articles and m_l the number of terms.
2. For each document d_l (in language l) that is to be compared, the same preprocessing steps have to be executed, so that the result is the document vector v_d^l with the form $m_l \times 1$, where m_l is the number of terms.
3. As all document vectors are normalized, the interpretation vector $i_{esa}^l = M_l \cdot v_d^l$ that represents the cosine similarity of v_d^l with all article vectors of M_l is simply computed by applying the inner product with the matrix M_l. The result is the interpretation vector i_{esa}^l with the dimensions $1 \times n_l$.

[1] In previous work [15] the systematic filtering of articles has shown to lessen the concept space without impairing *ESA*'s quality.

4. All interpretation vectors $i_{esa}^{l_s}$ now have to be transferred (mapped) from their respective source language space into one common target language space l_t and result in vectors $i_{esa}^{l_t}$. We will explain two different mapping strategies later on.
5. Finally, the resulting interpretation vectors $i_{esa}^{l_t}$ can be compared pairwise using the cosine similarity.

The first step has to be executed for all languages which are supported (in our case German and English). Step 4 is not contained in the original ESA implementation. It needs to be executed in our cross-lingual approach because step 5 requires the interpretation vectors to have the same dimensions. A mapping has to be applied to equalize the number of dimensions of the interpretation vector (corresponding to the number of Wikipedia articles in the respective language). This mapping is the crucial step in cross-lingual semantic relatedness calculation with ESA. The more concepts can be mapped from one language to the other, the closer the quality of interlingual semantic relatedness matches that of monolingual ESA. We now describe the two mapping techniques we apply.

The recommendation quality is expected to be constant over the time. Preliminary experiments have shown that the use of updated Wikipedia versions does not have a major influence on the computation's quality, only in case of topics not being represented in the Wikipedia up to now.

4.1 Direct CL Mapping

Wikipedia articles can be interlinked to corresponding articles in another language version of the Wikipedia by *interlanguage links* (called ILLs). Ideally, the interlinked articles describe exactly the same concept. If such a link exists for the article whose dimension needs to be mapped, it is used and the weight (determined in step 3 of the ESA algorithm) is transfered to the target article of the link. This matches the approach presented by Potthast et al. and Sorg et al. [12,13].

If for each article in the source language a linked article in the target language exists, the mapping can be expected to be ideal. Unfortunately, this is usually not the case. Considering the German and the English Wikipedia (cf. Table 3) the total numbers of articles differ strongly, thus, a one-to-one article mapping can not be achieved. For example, only around 18% of the English articles have a (linked) German correspondent, thus 82% of the concept dimensions would be discarded during the mapping step using only *ILL*s.

4.2 Meta CL Mapping

We present a novel approach which aims at overcoming CL mapping issues due to articles not being interlinked to other languages. First, all weights in $i_{esa}^{l_s}$ of articles a_{l_s} being interlinked via a *ILL* to an article a_{l_t} are directly transferred into the new vector $i_{esa}^{l_t}$. Afterwards, for each article a_{l_s} which does not have an outgoing *ILL*, the set of articles within the same language version of the

Table 3. Key figures of the size of the English and the German Wikipedias

	English Wikipedia	German Wikipedia
Date of Dump	2011-01-15	2010-06-03
Number of articles	3,601,228	1,095,678
CLLs to other language	657,874	606,160

Wikipedia that are interlinked from a_{l_s} are considered. For all those articles, the articles which have an *ILL* to an article in the target language are taken into account. The resulting set of articles in the target language represents a *meta concept* for the article a_{l_s}. The link between a_{l_s} and the meta concept is called in the following *meta interlanguage link* (MILL). The weight for a_{l_s} is equally distributed to all elements in the *meta concept*. Figure 1 shows exemplarily the *MILL* for the English article "Clark Kent" to its meta concept in the German Wikipedia. The original weight 0.8 is divided and each of the four elements in the meta concept get assigned a weight of 0.2.

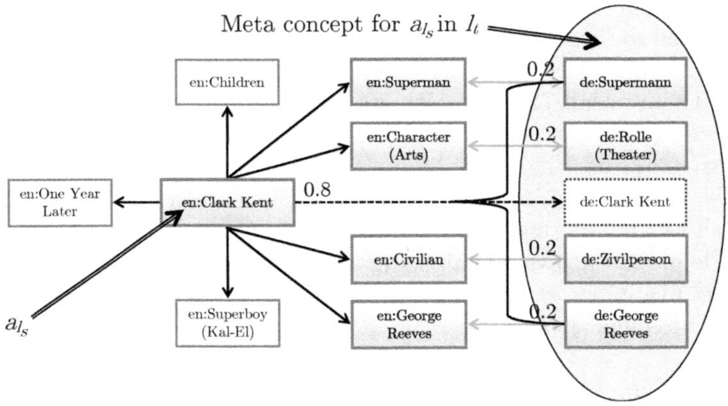

Fig. 1. Example of a *meta interlanguage link* from the English article *Clark Kent* that does not have a corresponding article in German

5 Evaluation

5.1 Corpora

In order to evaluate our approach we needed an appropriate evaluation corpus. Following the considered usage scenario for our approach, we identified the following requirements:

- The main languages used within the *CROKODIL* system in the user study were English and German. Thus, the evaluation corpora should consist of documents in those languages.

- The information items to recommend are tags and multi-term documents. Thus, we need corpora containing terms and multi-term documents.
- In most cases, nouns have been used as tags, thus the corpus containing terms should include primarily nouns.

We observed, that often *LR* and their assigned tags share the same language (cf. Section 2.2). Thus, cross-recommendation between tags and *LR*s are not focus of our work because this is mainly a monolingual task. The term-term relatedness and the document-document relatedness are evaluated separately due to the different conditions and strongly differing results for those two evaluations in other studies [10]. For the different scenarios, we use different corpora which are explained in the following.

The Document Corpus. To the best of our knowledge, there exists no appropriate multilingual document corpus that contains relatedness values for the documents. We decided to focus on a parallel corpus, containing exactly similar documents in different languages. The corpus *Europarl test2007*[2] is a testing subset of the *Europarl* [16] corpus containing 2000 parallel sentences in four languages. The content of the *Europarl* corpus has been extracted from proceedings of the European Parliament. For our evaluations, we use a subset of the English-German parallel corpus with 300 documents.

The Term Corpus. The multilingual dataset *Schm280* [15] is adapted from the English *WordSim353* dataset created by Finkelstein et al. [17]. It contains 280 English word pairs (mainly nouns) with their German equivalents translated by 12 participants, each value pair with a value of semantic relatedness (between 0.0 and 10.0) rated by at least 13 subjects. The words are very generic and not restricted to a single topic.

5.2 Evaluation Methodology and Results

In order to evaluate our approach we execute different experiments. We aim to reduce storage space for the created interpretation vectors and to minimize the computation time for the cosine similarity computation. Therefore, we define and apply the `selectBestN` function [10,13] which is aimed at choosing only the most relevant concept dimensions (in dependency of their weights). N should be chosen as small as possible without having a considerable negative impact on the results. Therefore, all evaluations are executed with a varying number of dimensions of the interpretation vectors. We parametrize the function with 16 values for N between 10 and 10000.

Document-Document Similarity. Using an Information Retrieval scenario we evaluate the applicability on cross-lingual document similarity. Our system is queried with a sentence from the *Europarl test2007* corpus in one language and the parallel document from the other language is expected as result. Because

[2] Available via http://www.statmt.org/wmt07/shared-task.html, retrieved 2011-04-04

one document has exactly one correspondent document in the other language, a *Top-k* evaluation is applicable in this scenario. The scenario is evaluated using both the *ILL* and *MILL* mapping to map the interpretation vectors of the documents to the query language l_t. Additionally for comparison, an evaluation with monolingual ESA is applied where all documents are in advance translated to the respective query language using Google Translator[3].

In the evaluation the mapping approaches do not seem to achieve a similar precision like the comparison approach using monolingual ESA with Google Translator (cf. Figure 2). The correct English document is returned as the best ranked one when passing a German query in at most 80% of the cases for the translational approach, in 60% for the *ILL* mapping approach and in 26% for the *MILL* approach. The correct German document is returned as the best ranked one when passing an English query in at most 84% of the cases for the translational approach, in 57% for the *ILL* approach and in 40% for the *MILL* approach. Table 4 shows a detailed analysis for different parameterizations of k. Shown is the maximal precision for each approach over all parameterizations of N. According to the previous analysis for $k = 1$, the monolingual setting is always superior to the *ILL* and the *MILL* mapping. Further, the *MILL* mapping can not achieve a precision similar to the results of the *ILL* mapping.

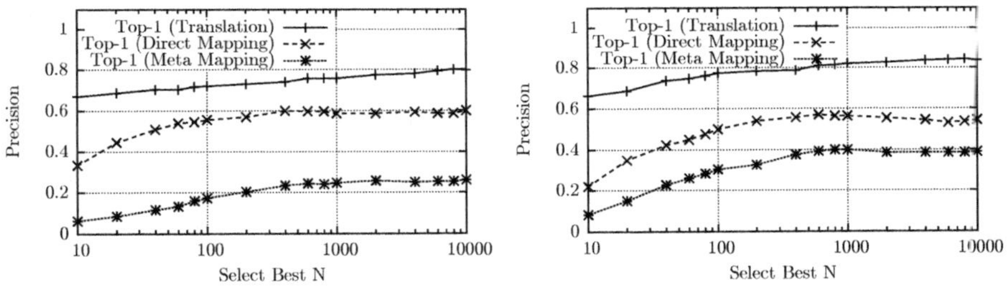

(a) German target language space (b) English target language space

Fig. 2. Precision for Top-1 evaluation

Term-Term Relatedness. In order to determine the statistical dependence of pairwise relatedness results, we use the Spearman's rank correlation, also known as Spearman's rho [18]. It considers ranks of the single values instead of the values itself as the Pearson correlation does. As for relatedness measures, the ranking of values is more important than the exact relatedness value, we prefer this measurement. Other authors also used this procedure to evaluate their semantic relatedness approaches (i.e. [12]). In our evaluation, we use this approach to determine how close to human relatedness rankings the rankings determined by our automatic relatedness determination techniques are.

[3] http://translate.google.de/, retrieved 2011-04-04

Table 4. Maximum precisions for different k in *Top-k* evaluation and considered number of dimensions of the interpretation vectors

		$k = 1$		$k = 5$		$k = 10$	
German to English	Translation	0.84	8000	0.94	2000	0.96	1000
	ILL mapping	0.57	600	0.82	1000	0.88	4000
	MILL mapping	0.4	800	0.72	4000	0.83	2000
English to German	Translation	0.8	8000	0.92	8000	0.92	4000
	ILL mapping	0.6	10000	0.8	2000	0.87	800
	MILL mapping	0.26	10000	0.54	6000	0.64	2000

The significance of the difference between two correlations can be determined by using t_{diff} [18]. It is used to check whether the correlation between the pairs of variables (x, y) and (z, y) is significantly different. It is defined as:

$$t_{\text{diff}} = (\rho_{xy} - \rho_{zy})\sqrt{\frac{(k-3)(1+\rho_{xz})}{2(1-\rho_{xy}^2-\rho_{xz}^2-\rho_{zy}^2+2\rho_{xy}\rho_{xz}\rho_{zy})}} \quad (1)$$

The resulting values for t_{diff} are compared with the critical values of the t–distribution.

The evaluation is executed using the *Schm280* corpus to determine the rank correlation between the relatedness values determined by humans and our approach. Within the evaluation of the cross-lingual experiments, the relatedness is determined for each pair with the first word being in language l_t and the second word being in language l_s and mapped to l_t. As a comparison, we determine the monolingual correlation with both words existing in l_t's language space.

Figure 3 shows Spearman's rank correlation ρ between relatedness determined by human raters and relatedness determined by the *ILL* and *MILL* mappings. Further, results of the monolingual setting are shown.

In the monolingual setting, the Spearman's rank correlation coefficient ρ is maximal for *selectBestN* with $N \approx 2000$ for both languages (with $\rho_{en}(i_{esa}^{2000}) = 0.71$ and $\rho_{de}(i_{esa}^{2000}) = 0.60$). In cross–lingual settings, the experiment in the

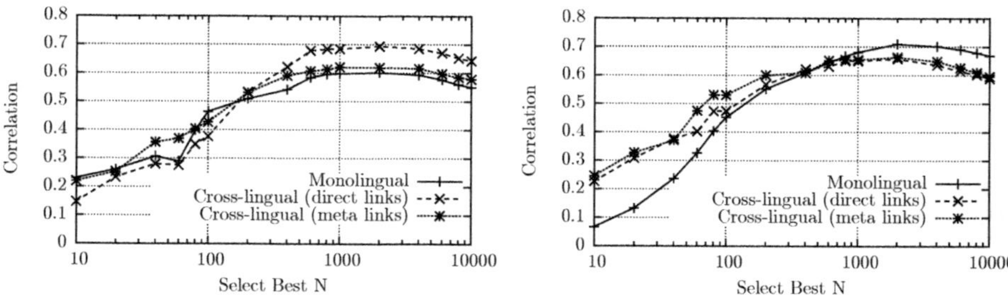

(a) German target language space (b) English target language space

Fig. 3. Correlation for Semantic Relatedness between human judgement and our approach

English language space l_{en} outperforms the experiment using the German language space by approximately 18% in terms of correlation. When mapping the second term of each term pair from German to English by *ILL*, the correlation is significantly lower than the monolingual experiment (for $\rho_{en}(i_{esa}^{2000}) = 0.66$, $t_{\text{diff}} = 2.01$, $p < .05$). In the German language space however, the *ILL* mapping results in a significantly higher correlation (with $\rho_{de}(i_{esa}^{2000}) = 0.69$, $t_{\text{diff}} = 3.37$, $p < .01$) compared to the monolingual approach. This shows that the quality of *ESA* is highly dependent on the used language space and a cross–lingual mapping transfers qualitative properties to the target language space. Remarkable is the slight improvement for *MILL* mapping compared to *ILL* mapping and monolingual evaluation for small N. We explain the better performance of *MILL* compared to *ILL* mapping with an accumulation of article links to relevant concepts (which usually are general and therefore have a statistically higher probability of being the target of article links), boosting the values for these relevant concepts due to spread. In fact, an analysis of the data shows that the relevant concepts benefit from the incomming *MILL*s. This effect can be used when N has to be set to a low value for performance and storage reasons. In the English language space, *ILL* and *MILL* mappings perform similarly for large N. However, especially in the German language space, the correlation of *MILL* mapping is lower for high values of N.

6 Conclusions and Further Work

In this paper, we presented a multilingual scenario of resource-based learning using web resources. *CROKODIL* enables users to store various types of learning resources and assign tags to them without being confined to a single language. In order to provided content-based recommendations across language borders we proposed an adaption to Explicit Semantic Analysis which does not only take interlanguage links into account but additionally Wikipedia's internal link structure: the *MILL* mapping. In the evaluation of term-term relatedness, it showed better correlation for low dimensional interpretation vectors, even compared to the monolingual evaluation. Thus, due to the reduction of computational complexity the *MILL* mapping can be helpful for real-time applications like the on-line recommendation of newly integrated information items in *CROKODIL*. Further evaluation did not show improvements when compared to the mapping using only *ILL*s. We however still see potential of this approach, especially regarding underrepresented topics in the Wikipedia. When topics are underrepresented for a certain language, documents covering those can not be mapped into this language space using only *ILL*s. This can lead to a loss of important information if strongly weighted concepts are discarded. For document-document similarity tasks we currently recommend the usage of *ILL* mapping due to the restrictive usage policy of the Google API which does not allow the translation-based approach. The goal of document recommendations in *CROKODIL* is not to recommend similar resources but rather semantically related resources, which is not directly addressed by our evaluation. We imagine the following scenario when including cross-lingual content based recommendation in *CROKODIL*:

1. The users insert learning resources and attach tags to them, both can be in any of the supported languages.
2. For each inserted learning resource and tag, a semantic interpretation vector is created, mapped to a common language, reduced via `SelectBestN` and stored in the system.
3. Users get those tags/learning resources recommended which have the highest semantic relatedness to the tags/learning resources they have used.

In future work we will focus on further adaptions and evaluations of our approach as we consider the achieved results as motivation for further research in this field. By distributing the weight of a mapped concept well-directed to the different elements of the *meta concept*, we hope to enable the amplification of important linked concepts. Further, if only strongly weighted concepts in the original interpretation vectors are mapped via *MILL*, noise might be attenuated. In order to satisfy the requirements for the described application area more precisely, an evaluation with further corpora is needed. In addition for the evaluation of the recommendation of documents, a corpus containing relatedness values is required. As we have seen [10], approaches relying on Wikipedia's link structure can outperform plain *ESA* in scenarios where documents are rather related than they are similar. Considering the recommendation of tags, a more specific corpus would represent the scenario better. The used corpus contains highly generic terms, which are strongly represented in Wikipedia and thus the approach relying on meta concepts cannot demonstrate its benefits. Finally, for adapting the approach to general cross-lingual recommendation, evaluations with further languages are needed. Especially in scenarios with languages of considerably different structure (like English and Arabic) interesting results can be expected as our mapping approach only relies on the quality of the Wikipedia reference corpus in the respective language and not the quality of a translation engine.

Acknowledgments. This work was supported by funds from the German Federal Ministry of Education and Research under the mark 01 PF 08015 A and from the European Social Fund of the European Union (ESF). The responsibility for the contents of this publication lies with the authors.

References

1. Jannach, D., Zanker, M., Felfernig, A., Friedrich, G.: Recommender Systems: An Introduction, 1st edn. Cambridge University Press, Cambridge (2010)
2. Böhnstedt, D., Scholl, P., Rensing, C., Steinmetz, R.: Enhancing an Environment for Knowledge Acquisition based on Web Resources by Automatic Tag Type Identification. In: Proceedings of International Conference on Computer-aided Learning 2010 (ICL 2010), pp. 380–389. Kassel University Press, Kassel (2010)
3. Budanitsky, A., Hirst, G.: Evaluating WordNet-based Measures of Lexical Semantic Relatedness. Computational Linguistics 32(1), 13–47 (2006)

4. Milne, D., Witten, I.H.: An Effective, Low-Cost Measure of Semantic Relatedness Obtained from Wikipedia Links. In: Proceedings of AAAI Workshop on Wikipedia and Artificial Intelligence: an Evolving Synergy, pp. 25–30. AAAI Press, Chicago (2008)
5. Fellbaum, C.: Wordnet: An Electronic Lexical Database. MIT Press, Cambridge (1998)
6. Jiang, J.J., Conrath, D.W.: Semantic similarity based on corpus statistics and lexical taxonomy. In: Proceedings of International Conference Research on Computational Linguistics, ROCLING X (1997)
7. Gabrilovich, E., Markovitch, S.: Computing Semantic Relatedness using Wikipedia-based Explicit Semantic Analysis. In: Proceedings of the 20th International Joint Conference on Artificial Intelligence, pp. 6–12 (2007)
8. Strube, M., Ponzetto, S.P.: WikiRelate! Computing Semantic Relatedness Using Wikipedia. In: Proceedings of the National Conference on Artificial Intelligence, vol. 21, p. 1419. AAAI Press; MIT Press, Menlo Park, CA; Cambridge, MA; London (2006)
9. Jarmasz, M., Szpakowicz, S.: Roget's Thesaurus and Semantic Similarity. In: Recent Advances in Natural Language Processing III: Selected Papers from RANLP, vol. 1, p. 111 (2004)
10. Scholl, P., Böhnstedt, D., Domínguez García, R., Rensing, C., Steinmetz, R.: Extended Explicit Semantic Analysis for Calculating Semantic Relatedness of Web Resources. In: Wolpers, M., Kirschner, P.A., Scheffel, M., Lindstaedt, S., Dimitrova, V. (eds.) EC-TEL 2010. LNCS, vol. 6383, pp. 324–339. Springer, Heidelberg (2010), `http://www.springerlink.com/content/978-3-642-16019-6#section=781081`
11. Schönhofen, P., Benczúr, A.A., Bíró, I., Csalogány, K.: Cross-Language Retrieval with Wikipedia. In: Peters, C., Jijkoun, V., Mandl, T., Müller, H., Oard, D.W., Peñas, A., Petras, V., Santos, D. (eds.) CLEF 2007. LNCS, vol. 5152, pp. 72–79. Springer, Heidelberg (2008)
12. Potthast, M., Stein, B., Anderka, M.: A Wikipedia-Based Multilingual Retrieval Model. In: Macdonald, C., Ounis, I., Plachouras, V., Ruthven, I., White, R.W. (eds.) ECIR 2008. LNCS, vol. 4956, pp. 522–530. Springer, Heidelberg (2008)
13. Sorg, P., Cimiano, P.: Cross–lingual Information Retrieval with Explicit Semantic Analysis. In: Working Notes for the CLEF 2008 Workshop (2008)
14. Dumais, S.T., Letsche, T.A., Littman, M.L., Landauer, T.K.: Automatic Cross-Language Retrieval Using Latent Semantic Indexing. In: AAAI Spring Symposium on Cross-Language Text and Speech Retrieval, pp. 15–21 (1997)
15. Schmidt, S.: Language-Independent Semantic Relatedness Using Wikipedia as Reference Corpus. Master's thesis, Technische Universität Darmstadt (2010)
16. Koehn, P.: Europarl: A Parallel Corpus for Statistical Machine Translation. In: Machine Translation Summit 2005. European Association for Machine Translation, vol. 5, pp. 79–86 (2005)
17. Finkelstein, L., Gabrilovich, E., Matias, Y., Rivlin, E., Solan, Z., Wolfman, G., Ruppin, E.: Placing Search in Context: The Concept Revisited. ACM Transactions on Information Systems (TOIS) 20(1), 116–131 (2002)
18. Field, A.P.: Discovering Statistics Using SPSS. SAGE publications, Thousand Oaks (2009)

Student Behavior in Error-Correction-Tasks and Its Relation to Perception of Competence

Lenka Schnaubert[1], Eric Andrès[2], Susanne Narciss[1], Anja Eichelmann[1], George Goguadze[2], and Erica Melis[2]

[1] Technische Universität Dresden, Dresden
[2] Deutsches Forschungszentrum für Künstliche Intelligenz, Saarbrücken

Abstract. This paper investigates students' behavioral patterns within web-based multi-trial error-correction-tasks. By analyzing logfiles and considering students' initial perception of competence, we contribute to micro- as well as macro-adaption. We describe and visualize task processing data of 159 students, considering performance as well as attempts to solve a task. Taking preceeding behavior into account, it was possible to identify (maladaptive) behavioral patterns. Furthermore we compare the behavior of students with low vs. high perceptions of competence subsequent to a failure. In line with research regarding the influence of self-concept on performance and motivation, our findings suggest that students with a low perception of competence perform poorer and tend to skip trials more often after a failure, indicating motivational losses. Further research should build upon this study to enhance technology-based learning by designing learning environments and adaption strategies that take student behavior and prerequisites into account.

Keywords: student modeling, web-based learning, micro-adaptation, perception of competence.

1 Introduction

Adaptability is considered to be one of the main characteristics of ubiquitous learning [1]. Adaptability allows learners to access the right information, at the right place, in the right way and form [2]. In doing so it supports students in learning effectively. A core issue for instructional research and development on ubiquitous learning is thus to examine in detail students' behavior in technology enhanced learning environments in order to investigate under which situational and individual conditions what information should be provided for various learners. Technology enhanced learning environments offer the opportunity to provide learners with a large variety of information, including various kinds of learning tasks helping students to apply, elaborate and deepen their (newly) acquired knowledge. Current approaches to instructional design consider learning tasks to be fundamental components of technology enhanced learning environments (e.g., [3,4]). The focus of the present work is thus on investigating students' behavior within tasks-with-typical-errors (TWTE). TWTE are specifically designed multi-trial-learning-tasks which contain one (or several) specific

C. Delgado Kloos et al. (Eds.): EC-TEL 2011, LNCS 6964, pp. 370–383, 2011.

task-requirement(s) *and* one (or several) typical error(s) related to these task requirements. The selection of these task-requirements and typical errors is based on cognitive task- and error-analyses. Within such a TWTE students have first to identify the error, then to correct it. To support them in error correction, students are provided with tutoring feedback. (Informative) tutoring feedback is feedback which provides strategically useful information and guides the student towards successful task completion [5,7,8,9]. Developing and evaluating tutoring feedback which is adaptable to learner- and task-characteristics is one of the major goals of the project "Adaptive Tutoring Feedback" (ATuF) funded by the German Research Foundation (DFG). To adapt feedback to learner- and task-characteristics, it is important to examine how students behave while completing a TWTE, how they react to feedback and how their reactions differ according to their (cognitive and motivational) prerequisites, and to situational conditions (e.g., task requirements; successful or unsuccessful problem solving steps).

2 Purposes and Research Questions

The goals of the present study are twofold: First, it aims at describing behavioral patterns of students working on multi-trial tasks-with-typical-errors. Second, it aims at comparing the behavioral patterns of students taking into account individual student characteristics which have been found to have an important impact on how students react to feedback in a learning situation. Students' perception of their own competences in a given domain has been found to be such an important individual factor. The importance of this factor is reflected in several lines of research using a variety of constructs which have been conceptualized either as global and generalized attitudes about capabilities (e.g., academic self-concept; [10]) or task-, domain- or subject-specific beliefs (e.g., self-efficacy; [11]). In the present study domain-specific competence perceptions are at the focus of interest. Prior research suggests that such domain-specific perceptions influence students reactions to negative feedback [12,13]. For example, Eckert, Schilling and Stiensmeier-Pelster found that students' performance on different cognitive tests was significantly lower after experiencing failure, if they perceived their competences as low than as high in the given domain [13].

Our study aims at extending this prior research by investigating not only students' performance but also students' behavior after failures depending on their competence perceptions. Self-efficacy researchers have found that task-specific competence perceptions influence effort expenditure on an activity, persistence and resilience in the face of obstacles or failure [5,11,14,15]. Thus, we not only assume that students with low perceptions of competence generally tend to perform poorer after failure, but also that they tend to quit trying after they have failed to complete a correction-step in a TWTE. Therefore our research questions are: 1. How do students behave within multi-step error-correction tasks? 2. Do students' behavioral patterns differ according to their initial perception of competence, especially after failure? We hypothesize (due to the research mentioned above), that students with low perception of competence perform poorer and also tend to skip trials more often, in particular subsequent to failure.

3 Experimental Design

3.1 Working Environment

ActiveMath [16] is an intelligent learning environment, that adapts the learning experience of a student at the macro-level, i.e., it can assemble courses (including learning material, tests as well as questionnaires) according to the knowledge and goals chosen by the learner. Furthermore, it can run multi-step exercises with scripted feedback and uses computer algebra systems to evaluate the student's input.

Typically, interactive exercises in ActiveMath contain multiple steps that combine various types of tasks such as choice (in multiple choice, single choice and mapping exercises) and formula input (fill-in-blank). However, conducting experimental studies on feedback using such exercises can lead to methodological problems. For instance, a single computation problem may combine different requirements, making it difficult to attribute an erroneous step to a specific failed requirement. This makes the assignment of corresponding feedback rather complicated. In order to bypass these problems, we decided to use single-step error-correction tasks for the provision of feedback. The user interface for this type of tasks is a slight modification of the standard exercise user interface used in ActiveMath and will be described below.

Exercises in ActiveMath are represented as finite state machines including edges representing correct and typical incorrect student input and nodes representing the system's (re)actions including feedback, hints, and tasks to solve. For the present study, ActiveMath has been revised to enable multiple strategies to run on a single exercise [17]. The tutorial strategies developed for the study specify what kind of feedback each learner receives and in which sequence [18].

3.2 Empirical Study

The behavior protocols of 159 students were examined in the course of the study. All students attended the 5th to 7th grade of a saxonian secondary school and were recruited via announcements in newspapers and local schools. 79 students were female and 80 were male, aged 10 to 15 years (with a mean of 12 years). All logs of the tasks these students worked on were included in the sample, which made a total of 1855 task completions. Students were payed 10 Euros for participation.

The study was conducted in the media-lab of the Learning and Instruction Research Group of the Department of Psychology at the TU Dresden. Up to 10 students worked simultaneously on the material. All tests and questionnaires were run within ActiveMath, only the introduction to the program was given by one of the researchers in person (supported by a slide-show presentation). The study consisted of three phases interrupted by short breaks of approximately 10 min: First was the 30 min pre-test-phase, in which the students got acquainted to the program, filled out two online-questionnaires regarding general information (e.g., age, sex) and their fraction-specific motivation (attainment value, intrinsic value, perception of competence and fear of failure), and took a pre-test

(15 min) in which their prior knowledge and skills regarding adding fractions were assessed. The second phase was the treatment-phase, in which the students worked 45 min on multi-trial tasks-with-typical-errors (cf. Fig. 1) and received feedback according to their performance and a pre-determined feedback-strategy. The post-test-phase (20 min) was quite equal to the pre-test-phase except there was no introduction and no general-information-questionnaire.

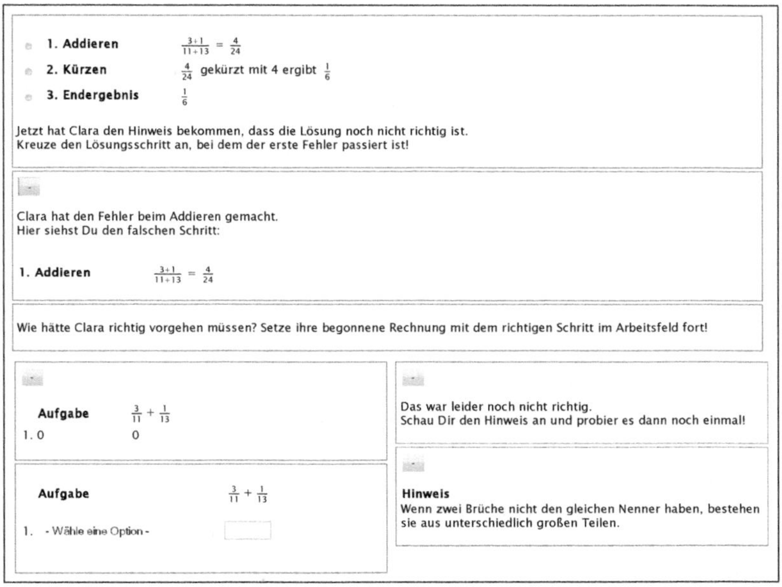

Fig. 1. Screenshot of a task-with-typical-error

During the treatment-phase the students worked self-paced on the tasks provided (cf. Fig. 1). The tasks were constructed to address some well-defined fraction-specific competencies (detect errors, apply rules regarding adding fractions, represent fractions, compare fractions), as described in the two- dimensional competency-model of Eichelmann and colleagues [19]. They consisted of two parts: an error-detection- and an error-correction-part. The students first were presented a worked-out-example of a fraction computation or representation problem, in which a typical error (as described in the literature concerned with fraction-learning, e.g., [20,21,22,23]) was included. For the error-detection part, the students were asked to mark the location of the error, for the error-correction part they were asked to replace the erroneous step by a correct solution (cf. Fig. 1, second box from bottom). In order to do so, the students had to choose the desired step from a drop-down menu (e.g., "expand" or "add") and execute it accordingly. If this correction failed, they received tutoring feedback and were asked to try again. After the third trial or a correct solution, they received a worked-out solution and moved on to the next task. The behavior protocols of the error-correction-part will be analyzed in this study.

Since error-detection is a prerequisite to error-correction, all students received feedback (knowledge-of-correct-result, KCR) after trying to detect the error (cf. Fig. 1, second box from top). The feedback-strategy for the second (error-correction) part consisted of a tutoring hint after the first failed error correction trial, and a tutoring explanation after the second trial (cf. Fig. 1, downright) and a worked-out solution after the third failed trial as well as after every correct solution (each containing knowledge-of-result, KR). The students were assigned randomly to four different feedback-strategies, which combined procedural and conceptual hints and explanations.

To answer the question if there are differences in the students' behavior (dependent variable) depending on their levels of perceived competence (independent variable), we measured students' perceptions of competence in the pre-test-phase with three items, which have been adapted to the domain of fractions from Narciss [5] (*I am very satisfied with my performance regarding fraction-tasks, I think my performance regarding fraction-tasks is very good, I think I am very skilled in solving fraction-tasks*). The construct validity of this scale has been confirmed in several studies, e.g. [5], [6]. The students had to mark their approval to the statements on a anchored 6-point-rating-scale from "not true at all" to "totally true". Since the internal consistency of the perceived-competence-items assessed in the pre-test-phase proved to be quite high (Cronbachs α = .86), we computed the arithmetic average for the three items per person to get a perceived-competence score. A median-split categorized the students as low-perceived-competence or high-perceived-competence. Students lying exactly on the median were excluded from the comparisons (but not from the general analyses of the behavioral patterns). This procedure left us with 69 students in the low-perceived-competence group (low-group) and 71 in the high-perceived-competence group (high-group).

3.3 Data Coding and Analyzing

To answer our research questions, we focused on the logfiles of the responses the students submitted during the treatment phase (error-correction-tasks). Since we were interested in the process of the students' task completion, we differentiated between a certain behavior, a certain behavior at a certain point within the task (behavior-trial-combination), a certain state (defined by actual and previous behavior), and within these states we distinguished endstates (states which mark the end of a task), all of which are illustrated in Fig. 2. It needs to be mentioned that in this paper the term "trial" refers to a sequence of system-provided events and the subsequent (observable) student reaction (cf. Fig. 2), which could be either entering a correct response (try/ succeed), an incorrect response (try/ fail), or not entering a response (skip/ fail) before submitting, whereas "to try" refers to a specific form of behavior itself.

The program automatically analyzed the entered answers as being correct (1) or incorrect (0). To get a closer look at the students' behavior, we scanned the log-files and recoded whether a student had entered anything at all before submitting his/ her answer for evaluation (try = 1 vs. skip a trial = 0). Since

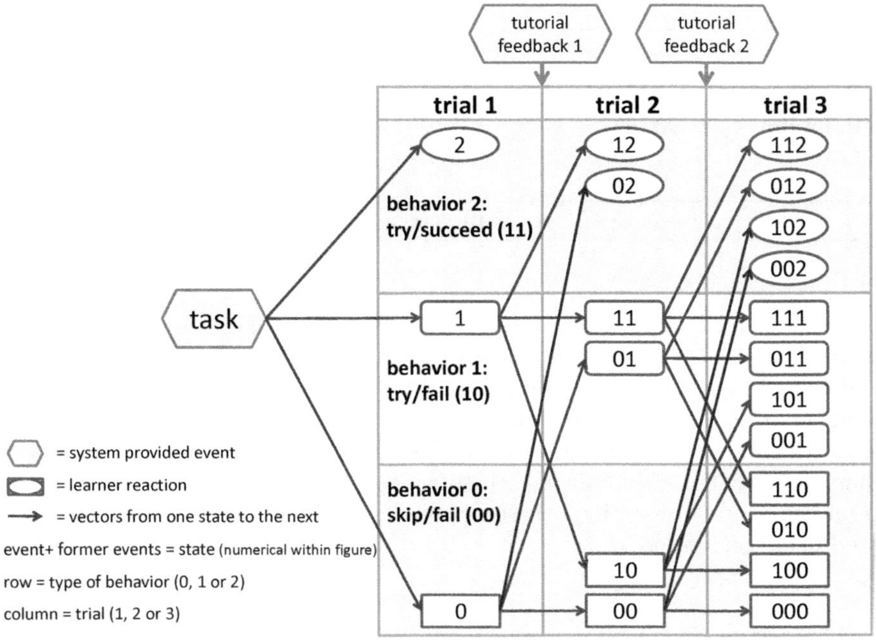

Fig. 2. Coding scheme of behavioral patterns within error-correction task

students could only solve a task by *entering* the correct solution, it was possible to assign a distinct (bijective) number to every possible behavior to represent the tuples on an ordinal scale (skip/ fail: 00 = 0, try/ fail: 10 = 1, try/ succeed: 11 = 2). This way, every behavior can be represented by a distinct value (0,1,2; cf. rows in Fig. 2). To add information, every behavior can be assigned to the point within the task (trial) it is performed (cf. Fig. 2, columns). Since there were up to three trials possible, each trial was assigned a number (1,2,3). We then matched behavior and trial (system-provided events and learners reactions, cells in Fig. 2).

All trial-behavior-combinations within each task-completions were combined to form a vector describing the state of a student within a task. For the (up to three) trials being discrete and at least ordinal, we can describe any state within a task as an ordered triple with every axis representing a trial (or likewise with up to three try-success-doublets in a defined order, which is easier to illustrate on paper, cf. Fig. 2, for an example cf. Tab. 1). Consequently, every completed task and every point within each task can be described with one distinct vector. Taking into account that within each task, every student received up to three trials to solve the task depending on the students' success, there are 21 possible states, 15 of which being endstates.

The empirically found behavior patterns were coded as described to resemble these endstates (three vectors in a defined order), which contained all behavioral information needed. We developed a tool to automatically generate the diagrams

Table 1. Illustration of 3-dimensional state-vector

Example:		trials (axis)			
try/fail - skip/fail - try/succeed		1	2	3	
behavior (value)	0 (00)		x		**vector** (10; 00; 11) = (102)
	1 (10)	x			
	2 (11)			x	
	vector	1:10	2:00	3:11	

as described in Fig. 2 from log data. It labels the nodes with the number of steps observed in the corresponding state in the underlying data set. The tool can be used to filter the original data according to certain criteria (e.g., high/ low perception-of-competence group). For commodity, it computes the conditional probabilities for transitions to other nodes and labels the edges accordingly. All behaviors and states were visualized seperately for each group.

Our (uni-directional) hypotheses about group differences in behavior were tested using one-tailed χ^2-tests.

4 Results

Two questions were covered in this study: We wanted to describe students' behavioral patterns within multi-trial error-correction tasks as well as compare these patterns between two groups of learners: low-perceived-competence learners and high-perceived-competence learners.

4.1 Description of Behavioral Patterns

To describe the behavior, we first computed how many percent of the conducted trials were classified as try/ succeed, try/ fail or skip/ fail, likewise we computed percentage-scores for every possible trial-behavior-combination (Tab. 2). We also examined how many times each endstate was reached (Fig. 3).

Table 2. Trial-behavior combinations

	trial							
	1		**2**		**3**		**∑**	
	N	(%)	N	(%)	N	(%)	N	(%)
try/ succeed	1145	(61.7%)	111	(15.6%)	41	(6.7%)	1297	(41.0%)
try/ fail	640	(34.5%)	324	(45.6%)	181	(30.2%)	1145	(36.2%)
skip/ fail	70	(3.8%)	275	(38.7%)	377	(62.9%)	722	(22.8%)
∑	1855	(100%)	710	(100%)	599	(100%)	3164	(100%)

All states and endstates (over all students and tasks) are shown in Fig. 4. It reveals that only 13 of the 15 possible endstates were observed. Considering the quantities given in Fig. 3, five to six endstates can be identified that were reached most frequently: The most frequent endstate was to try and succeed at the first trial (1145, that's 62% of all endstates). 108 tasks were solved correctly after one failed attempt (try/fail - try/succeed, 6%), 31 tasks after two failed attempts (2%). Overall, 70% (1297) of all tasks were solved correctly. 150 times the students failed to solve the tasks, but submitted an answer each time (8%), overall 77% (1434) of all tasks were completed without a single skipped trial. On the other hand, 63 tasks (3%) were completed without the student entering anything at all, 171 times the students tried and failed once and then skipped both following trials (9%), 141 times the students tried (and failed) twice and only skipped the third trial (8%).

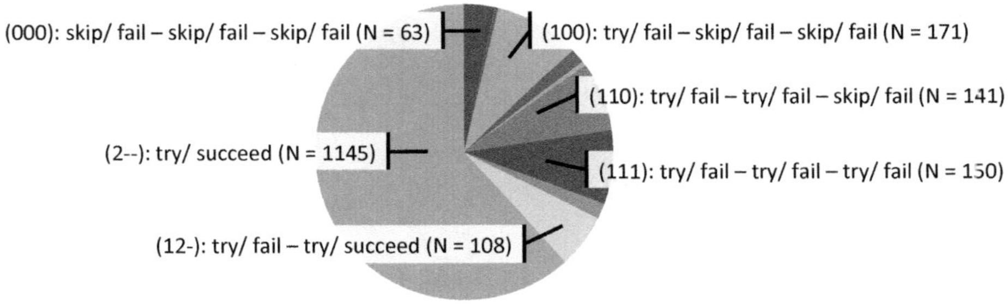

Fig. 3. Empirical frequencies of endstates

If the first attempt to solve a task failed (try/fail), there was a .67 probability that the student would try again in the second trial with 25% of these (tried) trials beeing successful. Consequently, there also was a .33 chance the second trial was skipped and a probability of .27 that *every* following trial would be skipped (cf. Fig. 4). After two failed attempts there was a probability of .56 that there also was a third attempt to solve the task (successfull in 17% of these cases). However, that also means, that in 44% of the cases there wasn't. If the second trial was skipped after one failed attempt, there was a .81 chance of the third trial beeing skipped likewise. But still in 19% of these cases, there was a further attempt to solve the task (4% being successfull, 14% not). If the first trial was skipped, there was a probability of .93 that the second trial would be skipped and a probability of .90 that after skipping the first trial, all following ones were skipped, too.

In order to take into account that not all students needed all the trials provided, we computed the probabilities to be in a certain state if students did not succeed previously. For the second trial (N = 710) there were three possible behaviors and six possible states to be in, all of which did come up, even though not equally frequently. The most probable state in the second trial can be described by the vector (11x) (try/fail - try/fail, with a probability of .45) and the

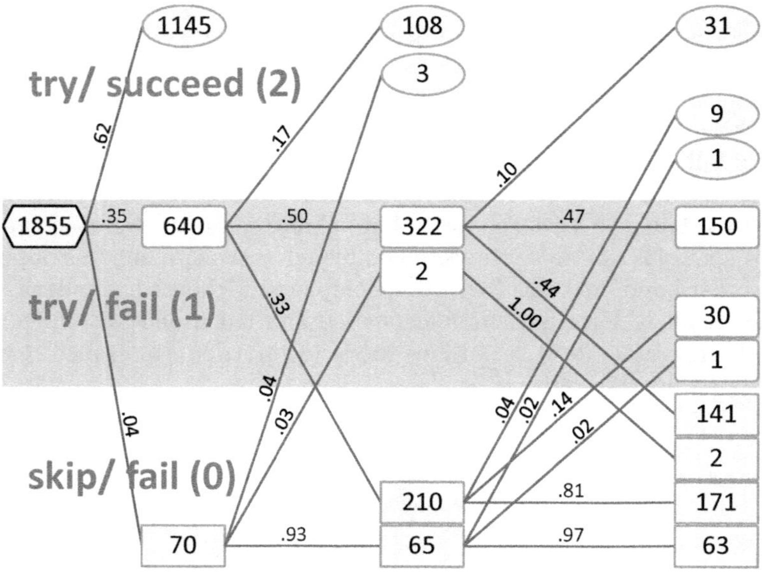

Fig. 4. Behavioral patterns all students all tasks

most probable way from there was to try and fail again (.47) or skip the next trial (.44). The overall probability of the second trial being successfully finished was .16, with the most probable state being a try/fail - try/succeed combination (.15). The probability that there was no attempt to solve the second trial was .39, the most probable state here being try/fail - skip (.30) with an 81% chance of skipping the following trial, too. If we look at the third trial ($N = 599$), more possible states exist. But not all of these 12 (end-)states did come up empirically. If one reached trial 3, the most probable behavior was skipping the trial (.63), and the most probable states to be in were try/fail - try/fail - try/fail (.25), try/fail - try/fail - skip (.25) and try/fail - skip - skip (.29). Also there was a 11% chance to be in the state skip - skip - skip, while there was only a 5% chance of being in try/fail - try/fail - try/succeed, and only an overall chance of 7% of succeeding in trial 3 at all.

4.2 Behavior According to Initial Perception of Competence

The behavioral patterns of the high- and low-perceived-competence groups are shown in the Fig. 5. The first thing that is noticeable is that both patterns look rather similar. Both groups reach basically the same endstates (except for (001), though this did only occur in one single task), but they do differ in their frequency.

First of all, the high-group completed more tasks ($N = 879 > 780$), but needed less trials ($N = 1356 < 1437$, cf. Tab. 3). Furthermore, they solved 78% of all tasks, whereas the low-group only solved 63%, the difference being highly

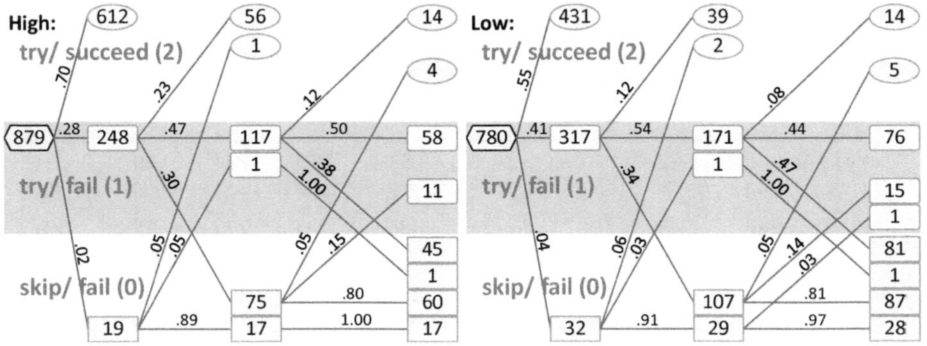

Fig. 5. Behavioral patterns: high (left) and low (right) perceived competence

significant ($\chi^2_{1659;0.05} = 46.43$, $p < .001$). In the high-group, 70% of the tasks were solved within the first trial. 21% of the tasks that were not solved within the first trial were solved in trial 2, and 9% of the third trials were solved correctly, whereas the percentages of the low-group are constantly lower (trial 1: 55%, trial 2: 12%, trial 3: 6%, cf. Tab. 3). While 7% of the tasks in the high-group were finished without success and without skipping a single trial (111), it is 10% in the low-group. 28% of the tasks in the low-group contain at least one skipped trial, whereas this is only true for 16% of the high-group tasks. 4% of the tasks in the low-group are finished without any attempt (000), compared to 2% in the high-group, but the probabilities to skip every following trial after skipping the first one are quite similar (low: .88, high: .89). If we look at the pattern try/fail - skip - skip, it is obvious that although the probability of skipping the last trial if trial 2 was skipped after a failed attempt in trial 1 is in both groups around .8 (low: .81, high: .80), overall 11% of the tasks in the low-group are handled that way and only 7% in the high-group. A comparable observation can be made within the try/fail - try/fail - skip pattern: It accounts for 10% of the tasks in the low-group and only for 5% of the high-group. Of all conducted steps, the low group skipped 25%, wheras the high-group skipped only 17%. This difference is statistically significant ($\chi^2_{2793;0.05} = 27.46$, $p < .001$).

The main question concerning group differences was whether the groups differ in their behavior after failure. Consequently, we observed the probabilities of a certain behavior after tried and failed trials as illustrated in Fig. 5. Subsequent to a tried but failed first trial, there was a probability of .54 for the low-group to try but fail the second trial and a probability of .34 to skip the second trial, while the probabilities of the high-group are lower for both behaviors (try/fail: $p = .47$, skip/fail: $p = .30$). Consequently, the probability of successfully finishing a task in the second trial subsequent to a tried but failed first trial was greater for the high-group. Concerning the state (11x) - trying but failing twice - the most probable behavior for the low-group was to skip the next trial ($p = .47$) or to try and fail again ($p = .44$), whereas for the high-group the most probable behavior was to try and fail ($p = .50$) and (to a smaller extent) to skip the

Table 3. Trial-behavior combinations per group

	high perceived competence				low perceived competence			
	trial				trial			
	1	2	3	$\sum$	1	2	3	$\sum$
try/ succeed	612	57	18	687	431	41	19	491
% of trial	69.6%	21.4%	8.6%	50.7%	55.3%	11.8%	6.2%	34.2%
try/ fail	248	118	69	435	317	172	92	581
% of trial	28.2%	44.2%	32.9%	32.1%	40.6%	49.3%	29.9%	40.4%
skip/ fail	19	92	123	234	32	136	197	365
% of trial	2.2%	34.5%	58.6%	17.3%	4.5%	39.0%	64.0%	25.4%
$\sum$	879	267	210	1356	780	349	308	1437

next trial (p = .38). Compared to students behavior after the first failure, the chances of trying and succeeding after failing twice were considerably lower in both groups (low: .08, high: .12), cf. Fig. 5.

To test our hypothesis that the low-group skips significantly more trials subsequent to failure than the high-group, we conducted two χ^2-tests (one concerning the probability of skipping trial 2 after trying but failing trial 1, and one concerning the probability of skipping trial 3 after trying and failing twice before). The difference after one failed attempt was not statistically significant ($\chi^2_{565;0.05}$ = 0.79, p = .19). After two (tried but failed) trials, this difference increased, yet it did not reach statistical significance either ($\chi^2_{288;0.05} = 2.24$, p = .07). We further tested whether the low-group failed (try/fail+skip/fail) more often than the high-group subsequent to failure. The χ^2-tests revealed that this was true for trial 2 ($\chi^2_{565;0.05} = 10.51$, p < .001), but not for trial 3 (failing after trying but failing twice before) ($\chi^2_{288;0.05} = 1.13$, p = .14).

5 Discussion

This study aimed at describing students' behavior in multi-trial error-correction tasks and at testing if students perceiving a low competence react differently to failure than students perceiving a high competence. Integrating performance measures (success vs. failure) and information about students attempts to solve the tasks (try vs. skip) to describe behavior and states (condsidering previous and actual behavior) offers a great opportunity to analyze behavioral patterns within the learning context. We found that even though most tasks were solved correctly, the chances of succeeding after failure lessened with every trial, while simultaniously the chances of skipping a trial increased. Within the last trial (trial 3), skipping even was the most probable behavior. But these probabilities dependend highly on the states the students were in in trial 2 (e.g. skip/fail-skip/fail: .97; try/fail-try/fail: .47), therefore, taken previous behavior into account proved to be essential for predicting behavior. These results yield a lot of potential if we consider micro-adaption. If we know what (maladaptive) behavior is most likely to follow a certain state, we are able to intervene accordingly.

To find out if these predictions depend on certain learner-characteristics, we considered the students' initial perception of competence, thus extending the research conducted by e.g. Eckert an collegues [13]. Confirming their findings concerning performance dropping after failure especially within groups perceiving their competence as being comparably low, we did not find clear evidence that these students skipped trials significantly more often after failure (although they did skip trials more often in general). However, even though the differences were not statistically significant, we did find a trend that is in line with research concerning self-efficacy and motivation [5,11,14,15], suggesting that perception of competence is an important factor regarding persistence after (multiple) failures. Interestingly we found that while the differences in performance (success vs. failure) were found after trying but failing once and seemed to ease out, the differences in attempting the subsequent trial (try vs. skip) became greater with each failed trial. The decrease of difference in performance might be partially due to actual differences in ability between the groups (after all the high-group did succeed in significantly more tasks) as well as to the fact that succeeding after failing twice was generally very unlikely in both groups. The increase of difference in skipping a trial suggests that if students do not manage to succeed, a higher perception of competence might be one factor to keep them trying even after repeated failures. It is crucial for further studies to take into account that some behavioral differences might need repeated failure to show.

Within technology enhanced learning environments, core instructional components are learning tasks [3,4] and the feedback provided [24]. To guide the learner to regulate his/ her learning process successfully (which is seen as one of the major functions of feedback, e.g., [7,25,26]), these components need to meet the needs of the learner, individualizing and therefore optimizing the learning process [27]. By analyzing behavioral logfiles and considering students' initial perception of competence, we contribute to micro- as well as macro-adaption. We presented a method to gain important information about learning behavior and detected behavioral patterns as well as differences between high and low perception of competence learners. Still, there is much research to be done to take task-characteristics, other learner-characteristics (initially assessed as well as other behavioral indicators like time-on-trial) and of course the feedback into consideration. Although we did use different tutoring feedback strategies in the empirical study conducted, investigating their impact on behavior was not the purpose of this paper, but clearly should be considered in future work.

Acknowledgements. This work was supported by the DFG - Deutsche Forschungsgemeinschaft under the ATuF project ME 1136/8-1 and NA 738/10-1.

References

1. Ogata, H., Yin, C., Yano, C.: Context-aware support for learning japanese polite expressions. In: Proceedings of the IEEE International Workshop on Wireless and Mobile Technologies in Education, WMTE 2004 (2004)

2. Fischer, G.: User modeling in human-computer-interaction. Journal of User Modelling and User-Adapted Interaction (UMUAI) 11(1/2), 65–86 (2001)
3. Merrill, M.: First principles of instruction. Educational Technology Research and Development 50(3), 43–59 (2002)
4. van Merriënboer, J.J.G., Dijkstra, S.: The Four-Component Instructional Design Model for training complex cognitive skills. In: Tennyson, R.D., Seel, N., Dijkstra, S., Schott, F. (eds.) Instructional Design: International Perspectives, pp. 427–445. Lawrence Erlbaum Association, Hillsdale (1997)
5. Narciss, S.: The impact of informative tutoring feedback and self-efficacy on motivation and achievement in concept learning. Experimental Psychology 51(3), 214–228 (2004)
6. Narciss, S.: Informatives tutorielles Feedback. Waxmann, Münster (2006)
7. Narciss, S.: Feedback strategies for interactive learning tasks. In: van Merriënboer, J.J.G., Spector, J.M., Merill, M.D., Driscoll, M.P. (eds.) Handbook of Research on Educational Communications and Technology, pp. 125–144. Lawrence Erlbaum Association, New York (2008)
8. Narciss, S.: Feedback in instructional contexts. In: Seel, N. (ed.) Encyclopedia of the Learning Sciences, Springer Science & Business Media, LLC, New York (in press)
9. Narciss, S.: Feedback strategies. In: Seel, N. (ed.) Encyclopedia of the Learning Sciences, Springer Science & Business Media, LLC, New York (in press)
10. Marsh, H.W.: A multidimensional, hierarchical self-concept: Theoretical and empirical justification. Educational Psychology Review 2, 77–172 (1990)
11. Bandura, A.: Self-efficacy: The exercise of control. Freeman, New York (1997)
12. Feather, N.T.: Attribution of responsibility and valence of success and failure in relation to initial confidence and task performance. Journal of Personality and Social Psychology 13(2), 129–144 (1969)
13. Eckert, C., Schilling, D., Stiensmeier-Pelster, J.: Einfluss des Fähigkeitsselbstkonzepts auf die Intelligenz- und Konzentrationsleistung. Zeitschrift für Pädagogische Psychologie 20(1/2), 41–48 (2006)
14. Pajares, F.: Self-efficacy beliefs in achievement settings. Review of Educational Research 66, 543–578 (1996)
15. Schunk, D.H.: Self-efficacy and education and instruction. In: Maddux, J.E. (ed.) Self-efficacy, Adaptation, and Adjustment: Theory, Research, and Application, pp. 281–303. Plenum Press, New York (1995)
16. Melis, E., Goguadze, G., Homik, M., Libbrecht, P., Ullrich, C., Winterstein, S.: Semantic-aware components and services in ActiveMath. British Journal of Educational Technology. Special Issue: Semantic Web for E-learning 37(3), 405–423 (2006)
17. Goguadze, G.: Representation for interactive exercises. In: Carette, J., Dixon, L., Coen, C.S., Watt, S.M. (eds.) MKM 2009, Held as Part of CICM 2009. LNCS, vol. 5625, pp. 294–309. Springer, Heidelberg (2009)
18. Goguadze, G., Melis, E.: One exercise – various tutorial strategies. In: Woolf, B.P., Aïmeur, E., Nkambou, R., Lajoie, S. (eds.) ITS 2008. LNCS, vol. 5091, pp. 755–757. Springer, Heidelberg (2008)
19. Eichelmann, A., Narciss, S., Faulhaber, A., Melis, E.: Analyzing computer-based fraction tasks on the basis of a two-dimensional view of mathematics competencies. In: Zumbach, J., Schwartz, N., Seufert, T., Kester, L. (eds.) Beyond Knowledge: The Legacy of Competence. Meaningful Computer-Based Learning Environments, pp. 125–134. Springer, Dordrecht (2008)

20. Padberg, F.: Didaktik der Bruchrechnung: gemeine Brüche - Dezimalbrüche. Spektrum, Heidelberg (1995)
21. Brown, G., Quinn, R.J.: Algebra students' difficulty with fractions: an error analysis. amt 64(2), 28–40 (2006)
22. Gerster, H.-D., Grevsmühl, U.: Diagnose individueller Schülerfehler beim Rechnen mit Brüchen. Pädagogische Welt 37, 654–660 (1983)
23. Padberg, F., Krüger, H.: Bruchzahlverständnis - systematische Fehlerstrategien und Fehlvorstellungen. Praxis der Mathematik 5, 193–195 (1997)
24. Clariana, R.B., Ross, S.M., Morrison, G.R.: The effects of different feedback strategies using computer-administered multiple-choice questions as instruction. Educational Technology Research and Development 39(2), 5–17 (1991)
25. Hattie, J., Timperley, H.: The power of feedback. Review of Educational Research 77, 81–112 (2007)
26. Butler, D.L., Winne, P.H.: Feedback and self-regulated learning: A theoretical synthesis. Review of Educational Research 65, 245–281 (1995)
27. Shute, V.J., Zapata-Rivera, D.: Adaptive technologies. In: Spector, J.M., Merril, M.D., Van Merrienboer, J.J., Driscoll, M.P. (eds.) Handbook of research on educational communications and technology, pp. 277–294. Lawrence Erlbaum Association, Mahwah (2008)

Adaptable and Reusable Query Patterns for Trace-Based Learner Modelling

Lemya Settouti[1,2], Nathalie Guin[1], Vanda Luengo[2], and Alain Mille[1]

[1] Université de Lyon, CNRS,
Université Lyon 1, LIRIS, UMR5205, F-69622, France
{Lemya.Settouti,Nathalie.Guin,Alain.Mille}@liris.cnrs.fr
[2] Université de Grenoble, LIG, France
Vanda.Luengo@imag.fr

Abstract. This paper defines a framework to describe Learner Modelling (LM) process based on interactions traces. This framework includes an RDF-Based representation of knowledge models that can be used by a LM designer. The first model enables the LM designer to describe observations about learner's interactions with a TEL-system. The second model enables the LM designer to describe the structure of learner's profile. This framework supports also the description of reusable and adaptable SPARQL-based query patterns. These patterns enable the LM designer to calculate and infer learner profile elements for different TEL systems. We define the notion of query pattern and illustrate its application in the context of two TEL systems.

1 Introduction

Learner modelling (LM) is a complex task to achieve and remains a serious problem for the implementation of TEL systems. This complexity is due to the Learner Modelling process which must take into account the implementation of several components: (1) Representation of observations about learner's behaviour (including learner-system interactions traces and their interpretations), (2) Representation of domain knowledge (including topics concepts, facts, procedures or methods that experts used to achieve task or solve a problem), (3) Diagnosis and reasoning function allowing to infer assumptions about a learner and (4) Representations of assumptions about individual learner's characteristics (e.g. performance, skills, misconceptions, strategies, interests, etc.).

The LM requires the development of data structures to represent such different components within the considered TEL system. Once LM data are modelled and collected, it must be converted into a format compatible with knowledge representation and reasoning function of TEL systems. Faced with these requirements, LM data is often stored in proprietary, hard-to-access formats that do not encourage reuse or distributed exploitation. In addition, the LM components are often tightly coupled to TEL systems for which they were developed, rendering them useless when another TEL system is considered.

C. Delgado Kloos et al. (Eds.): EC-TEL 2011, LNCS 6964, pp. 384–397, 2011.

Recently, some LM approaches have begun to adopt ontological technologies to solve the problems mentioned above. Learner modellers are developing their domain models and learner models using ontology language such as RDFS[1] or OWL[2] [1,13,10]. Learner models developed with such languages have the advantages of easy reuse, easy portability, availability of effective design tools, and automatic serialization into a format compatible with popular logical inference engines. Although it seems that LM systems which are developed using languages for ontologies have the potential for easing reuse, most LM frameworks do not pay much attention on the reusing issues. At present, the issue of LM components reusing has been implicitly investigated in terms of data structures and ontologies involved by LM systems, however, without addressing the issue of reusing the diagnosis part allowing to infer relevant information about learners. To the best of our knowledge, there is no LM framework supporting the diagnosis reusing specially in the context of their application within different TEL systems.

In this paper, we argue that the issue of reusing diagnosis is a complex problem mainly due to (at least) two reasons. The first aspect of the problem concerns the conceptual issue of LM which is mostly domain-dependent. In the practice, it is hard to reuse LM domain-dependent components specially diagnosis part. One solution is to identify diagnosis parts which are domain-independent from domain-dependent ones. We need a framework making a clear separation between such kind of diagnosis and defining a generic diagnosis rules domain-independent which can be exploited in several contexts.

The second aspect of the diagnosis reusing problem is more technical and concerns diagnosis language encoding. LM diagnosis are encoded in many forms making their reuse complex. We can identify two main classes for encoding diagnosis which are implemented in LM frameworks: (1) Diagnosis encoded with general and high level programming languages (i.e. the host programming language used for TEL as Java, Php, C++, etc.). (2) Diagnosis encoded with declarative languages equipped with reasoning capabilities as prolog [11], etc. Both forms of encoding are ill-adapted to diagnosis reuse. For the first class of languages, the access and comprehension of diagnosis are difficult and even impossible in most TEL (e.g. no open source TEL). For the second class, there exist many syntaxes (and particularly several semantics) relying on textual representation which make difficult the reusing issue. We need a framework dealing with diagnosis representation facilitating reusing.

To deal with these problems, this paper aims at defining a framework supporting the construction of learners' profiles based on interactions traces. Our contribution is twofold: firstly, we define an RDF-based framework defining knowledge models to construct Learners' Profiles. While our framework supports description of LM dependent and independent on specific domain [17], we will focus in this paper on LM domain-independent part. Actually, the separation between

[1] Resource Description Framework Schema.

[2] Web Ontology Language.

models that are domain dependent and independent is a key element in our framework, which make reusing diagnosis a tractable issue.

Our second contribution to achieve a pertinent reuse of LM components presents the notion of *query pattern*. Query patterns are reusable solutions describing diagnosis queries that can be adapted to a specific context. A query pattern is a generic solution which can be used in several TEL systems. To describe such generic query patterns, we use SPARQL query as encoding language. Instead of relying on the textual representation of SPARQL expressions, we define an RDF representation that can be easily converted to textual SPARQL queries. The main advantage of RDF-based SPARQL representation is that queries expressed by the LM designer can be stored as RDF together with the knowledge models proposed in our framework. This enables linking models elements with the associated SPARQL queries as well as easy sharing and reuse of rules and queries as part of learner modelling framework.

The remainder of this paper is organised as follows: Section 2 describes our Learner Modelling framework based on interaction traces. Section 3 describes the notion of query pattern that can be reused in several TEL Systems. Section 4 presents a generic query pattern for detecting when the learner is gaming with the TEL system and describes how to adapt this pattern in the context of two TEL systems. In section 5 we relate our approach to other work. Finally, we draw some conclusions and proposals for further work in section 6.

2 Trace-Based Learner Modelling Framework

Learner Modelling is the process of creating learner model from observed learner's behaviors [5]. Learner's behaviors refer to historic of actions, results of these actions including intermediates ones [19]. To acquire such learner's behaviors, a common approach is to observe and produce trace of learner's interactions with TEL systems. In this section, we propose a Learner Modelling framework based on interactions traces.

The Trace-Based Learner Modelling framework is an extension of Trace-Based System (TBS) framework [18]: an approach defined to describe, reason and exploit interaction traces in severals domains. We extend TBS framework by defining new knowledge models to support Learner profile building. Before define the process of building learner profile, we introduce the main notion for Trace-Based Systems: the notion *modelled trace*. A *Modelled trace* is defined as a sequence of observed elements tracked within systems, associated explicitly to a *Trace Model* [18]. The *Trace Model* is a general ontology for modeling traces. This Model defines and structures explicitly the content of interaction traces from various activities not specifically the learning ones.

TBS offers severals services. A *collecting service* elaborates the *primary trace* by capturing and collectings observed data from different input sources (log files, streamed actions, interface events, etc.) according to a *Trace Model*. The *primary trace* is the first *modelled trace* created into TBS. A *Querying service* enables the extraction of patterns from the traces. A *Transformation service* can perform rule-based interpretations on traces like rewriting and aggregating elements, computing elements attributes, etc. The result is called transformed modelled traces that can be more easily reusable and exploitable in a given context than primary trace. While TBS framework[18] has been successfully applied in severals domains (TEL observation personalisation [18], activity reflection and analysis [7], etc.), its exploitation to Learner Modeling is relatively new and needs an extension to support generic learner profile building. Specially, as showed by the figure 1, the building of learner profile requires carrying out three main processes not fully covered by TBS.

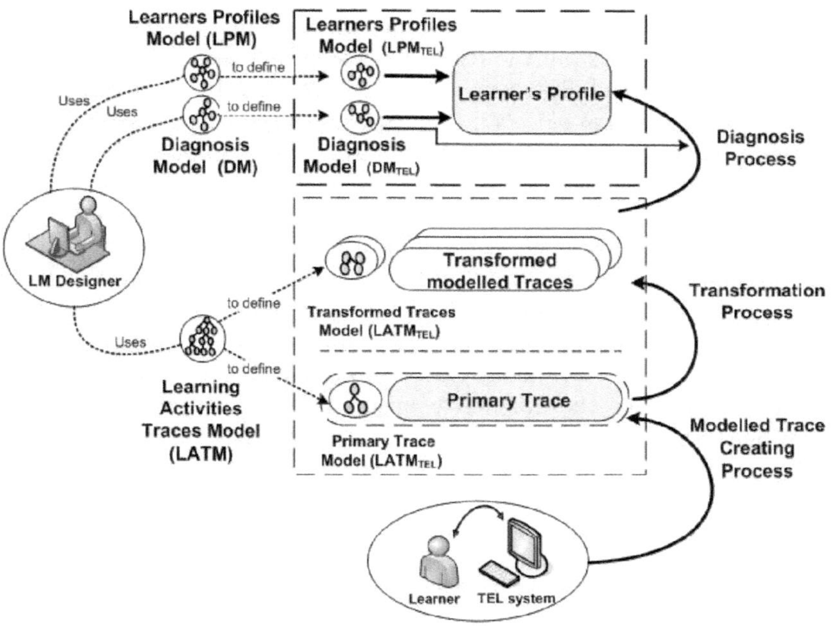

Fig. 1. The Trace-Based Learner Modelling Framework

Step 1: Modelled Traces Creating Process

The first process consists in creating a *primary trace* as a reification of observations obtained during the learning activities session. The primary trace is consistent with a specific trace model called Learning Activities Traces Model ($LATM_{TEL}$).

The LM designer defines the $LATM_{TEL}$ corresponding to the learning activities tracked by the TEL system by using and instantiating the Learning Activities Traces Model $LATM$. We defined in our framework the Learning Activities Traces Model ($LATM$) as an extension of the *Trace Model* of TBS. Indeed, the Trace model is used to describe traces collected from different systems and diverse activities, while $LATM$ enables to describe activities observed by TEL systems supporting individual learning. *LATM* is an RDF vocabulary used in order to models interaction traces observed by TEL systems by specifying how observed elements are typed and categorized, what relations may exist between them and what attributes further describe each of them.

In order to obtain the *LATM* vocabulary, we studied and analyzed traces obtained from several TEL systems (ANDES, AMBRE-add, Active-Math, TP-Elec, SQL-Tutor, TELEOS, copex-chimie, etc.). These TEL systems provide the learner with various activities such as reading courses, solve problems which require guidance and plan to solve them (resolution may be done in several steps and sub-steps). In most of these TEL systems, each problem or each problem step have a helping functions or assessments functions of learner's productions.

The modelling method used to construct the *LATM* ontology was a round-trip process that starts from the dataset representing the real traces to model the *LATM*. The studied TEL systems enabled to identify the learner's *actions* manipulating the *learning objects* during the learning session. Indeed, LATM assumes that observing a learning activity session consists in observing the several type of actions (e.g. consultation actions as *Consult Course* or *Consult Example*; request actions as *Ask For Help* or *Ask For Assessment*); micro-action as *mouse clicks, keystrokes*, etc.) performed by the learner, and in observing learning objects manipulated by actions (e.g. *Course*, *Problem*, *Problem Step*, *Problem Step Solution*,*Problem Step assessment*, *Problem Step Help*, etc.). Learning session observation may also include information about learner's identification (e.g. *id*, *login*, etc.) and information about the TEL system (ex. *OS*, *IP address*, etc.). In addition to describe the different concepts and relationships used to define information that can be observed about learner's activities performed within TEL system, *LATM* enables also to extract contextual information which can be used by the LM designer in order to infer the learner profile.

Step 2: Modelled Traces Transformation Process

The second process is optionally and consists in process $LATM_{TEL}$ according to TBS transformation service. The result is transformed modelled traces consistent with *LATM* which can be more reusable and exploitable than the primary trace in order to infer the learner profile elements.

Step 3: Diagnosis Process

The last process consists in performing diagnosis. The diagnosis process enables to construct the learner's profile by querying modelled traces, calculated data and inferring learner's strategies and/or learner's knowledge. The learner's profile is built from: the *modelled traces* elements (primary or transformed), the specific TEL system Diagnosis Model elements (DM_{TEL}) and from the Learner Profile Model of a specific TEL system (LPM_{TEL}).

The LM designer defines the LPM_{TEL} by using the Learner Profile Model (LPM). LPM is an RDF vocabulary which enables to describe the structure of learners profiles. LPM is composed of: (1) *Calculated Data* that describes information derived directly from $LATM_{TEL}$ or from other calculated data. These information are about learner's performances (e.g.number of learning objects which are consulted, durations of problem solutions, average time spends on solving a problem, etc.) or learner's observable behavioral patterns (e.g. consulting course before solving some problem, requesting help before answering, etc.). (2) the *Strategy* performed by the learner in order to solve problems and (3) qualitative or quantitative estimation about the learner *Knowledge*.

The LM designer defines the DM_{TEL} by using the Diagnosis Model (DM). DM enables to make separation between models that are dependent on a specific domain (i.e. using *queries and rules diagnosis* dependent on a defined *domain knowledge*), and domain independent (i.e. using only the *queries and rules diagnosis* without referring to a specific domain knowledge). The separation between models that are domain dependent and independent is key element in our framework [17], which make reusing diagnosis a tractable issue. The *domain knowledge* refers to the knowledge used in both overlay and buggy LM approaches.

The *queries diagnosis* refers to an RDF representation of SPARQL queries. The RDF-based description of SPARQL queries provides an alternative representation that goes beyond the classical and textual format that does not encourage reuse or sharing. One benefit of this choice is that it makes our framework possible to consistently store rules together with the models for LM. In addition, when ontologies and RDF have proven their usefulness to integrate and reuse several models for TEL systems, we use them also to support queries reuse and sharing for diagnosis by facilitating adapting and reasoning about the diagnosis queries themselves (in addition of reasoning about LM models).

3 Query Patterns for Learner Modelling

To deal with reusability, we propose the notion of *query pattern*. We define a query pattern as a general and reusable query that describes how to use, process and interpret learners' interactions in order to compute a relevant information. In our framework, the goal of query patterns is twofold: on the one hand query patterns specify how to obtain learners' individual observable behaviors during their learning activities and on the other hand they enable to infer learners' competences and problem-solving strategies. To be generic, query patterns must be able to be used in a wide range of TEL systems. In our approach, using query patterns requires that a TEL system should have a model of traces, a model of diagnosis and a model of learner profile structure consistent with the vocabularies and models defined in our framework (i.e. LATM, LPM and DM models). In order to apply query patterns in different contexts and domains, we propose an adaptation function δ enabling to adapt query patterns to specific TEL systems.

Formally, we define a query pattern Q_p as an ordered set of SPARQL queries $\{q_0, ..., q_n\}$ which use elements of knowledge models defined in our framework (i.e. elements of LATM, LPM and DM models).

The adaptation function δ is a function enabling to reuse SPARQL queries that compose the query pattern Q_p by constructing a new pattern Q'_p adapted to a specific TEL system. Concretely, the function δ adapts $\{q_0, ..., q_n\}$ by substituting each element from general models $LATM$, LPM and DM by elements from the specific models (respectively $LATM_{TEL}$, LPM_{TEL}, DM_{TEL}). The result is a set of new SPARQL queries $\{q'_0, ..., q'_n\}$ which are based on a specific TEL system models elements.

In this paper, for the sake of simplicity, we will give just a textual description of δ. The complete algorithm implementing the adaption function can be found in [16].

Informally, δ enables to substitute elements of $LATM$ by the corresponding elements of $LATM_{TEL}$ in pattern queries. The input data for δ are multiples:

1. an ordered set of general SPARQL queries $\{q_0, ..., q_n\}$,
2. an ordered set of couples (c_i, c'_i) meaning that each concept c_i of $LATM$ model should be substituted by the corresponding concept c'_i of $LATM_{TEL}$ (with c'_i is an instance of c_i),
3. an ordered set of couples (p_j, p'_j) meaning that each attribute or relation p_j of $LATM$ should be substituted by the corresponding attribute or relation p'_j of $LATM_{TEL}$ (with p'_j is an instance of p_j).

The general principle of δ can be described abstractly as follow: for each q_i composing the query pattern Q_p , the function substitutes every triple pattern $(subject_i, predicate_i, object_i) \in q_i$ that is linked to elements of $LATM$ by a triple pattern with elements related to a specific TEL system $LATM_{TEL}$. This substitution consists in either replacing the concept c_i of $LATM$ by the corresponding concept c'_i of $LATM_{TEL}$, or replace the attribute or relation p_j of $LATM$ by the corresponding attribute or relation p'_j of $LATM_{TEL}$. One aspect of the role of the LM designer consists in defining the parameters of the function by providing the specific elements c'_i and p'_j which substitute the general ones c_i and p_j. The output data for δ is an ordered set of adapted SPARQL queries $\{q'_0, ..., q'_n\}$.

We have defined in our framework several query patterns (about 50 patterns) enabling to obtain relevant information manipulated and stored in LM systems. These query patterns can be classified into three main classes :

1. Patterns for detecting learner's observable behaviors, e.g., detecting consulting course before solving a problem, requesting help before answering, asking for assessment after giving a solution, etc.
2. Patterns for detecting learner's performances, e.g., total learner attempts on problems, success rate on problems, average time spent on solving one/all problems, etc.
3. Patterns for detecting learner's strategies as gaming the system, etc.

In this paper, we will present as an example the query pattern enabling to detect when the learner is gaming with a TEL system. While the defined query pattern can be used in several TEL systems, we exemplify its usage only in the context of two TEL system. This exemplification demonstrates our proposal specially by showing how to reuse and adapt a query pattern in TEL systems which are domain different. The first TEL system is Copex-chimie [4] and the second is AMBRE-add [14].

4 Query Pattern to Design Gaming the System

According to Baker *and al.* [2], gaming the system is defined as *attempting to succeed in an educational environment by exploiting properties of the system rather than by learning the material and trying to use that knowledge to answer correctly.* Gaming the system is a strategy that is a sign of learner's lack of motivation, or a sign of difficulties encountered by the learner in the problem resolution steps. Generally, gaming the system can be detected from incorrect answers given quickly by the learner with ask for help, for example when the learner asks for help quickly and in a repetitive way until the system gives him the good answer or when he/she gives quickly and automatically answers to the system until the system identifies the good answer and thus allows the learner to advance in the exercise.

In this section, we begin by defining the *gaming query pattern* based on the elements of $LATM$, DM and LPM. Then, we describe the traces models of the two TEL systems: copex-chimie and AMBRE-add. Finally, we show the process to adapt the gaming query pattern for both TEL systems.

4.1 Query Pattern to Detect Gaming the System

To detect gaming the system, we defined the gaming query pattern denoted by P_g. The gaming pattern P_g is composed of an ordered set of SPARQL queries $\{q_1, q_2, q_3\}$.

The first query q_1 (figure 2) matches the $LATM$ elements (in the where part) in order to calculate durations of solutions which are followed directly by an asking for assessment. This query constructs an intermediate RDF graph derived from `CalculatedData`, the element of LPM that is composed of solutions and their durations. The second query q_2 (figure 2) matches the result of the first query q_1 in order to filter only solutions with durations which are less than a threshold duration predefined by the LM designer (`?threshold1`). This query constructs an other intermediate RDF graph derived from the element `CalculatedData` representing the short solutions and their durations. The last query q_3 (figure 2) matches the result of the second query q_2. This query q_3 enables to calculate the number of solutions which their durations are less than the threshold duration, and compare it with an other threshold defined by LM designer (`?threshold2`). If the count of solutions is higher than the specified threshold then the gaming is detected by constructing as `Strategy` a `gaming the system` element with the value *true* in the learner profile.

```
CONSTRUCT{ _:ag1 a lpm:CalculatedData.
           _:el1 a _:ag1.
           _:ap1 a lpm:AttributeProfile.
           _:a1 a _:ap1.
           _:rp1 a lpm:RelationProfile.
           _:r1 a _:rp1.
           _:el1 _:r1 ?solution.
           _:el1 _:a1 ?duration. }
WHERE {?o1 a latm:Problem .
      ?p1 a ?o1.
      ?a1 a latm:ProblemAttribute.
      ?p1 ?a1 ?id1.
   1: ?o2 a latm:ProblemStep .
   2: ?ps1 a ?o2.
      ?a2 a latm:ProblemStepAttribute.
      ?ps1 ?a2 ?id2.
      ?r1 a latm:hasProblemStep.
      ?p1 ?r1 ?ps1.
      ?o3 a latm:ProblemStepSolution.
      ?solution a ?o3;
      tbs:hasBegin ?begin1;
      tbs:hasEnd ?end1.
      ?r2 a latm:hasProblemStepSolution .
      ?ps1 ?r2 ?solution.
      ?o4 a latm:ProblemStepAssessment .
      ?ass1 a ?o4.
      ?ass1 tbs:hasBegin ?begin2 .
      ?r3 a latm:hasProblemStepAssessment .
      ?ps1 ?r3 ?ass1.
      FILTER (?end1 <= ?begin2) .
      NOT EXISTS {
          ?o a tbs:Obsel .
          tbs:Obsel tbs:hasBegin ?begin3.
          tbs:Obsel tbs:hasEnd ?end3.
          FILTER ((?end1<= ?begin3) && (?end3<=?begin2) )
      }
      LET (?duration:= ?end1 - ?begin1) }
```

q1

```
CONSTRUCT { _:ag2 a lpm :CalculatedData.
            _:el2 a _:ag2.
            _:ap2 a lpm :AttributeProfile.
            _:a2 a _:ap2.
            _:rp2 a lpm :RelationProfile.
            _:r2 a _:rp2.
            _:el2 _:r2 ?solution.
            _:el2 _:a2 ?duration. }
WHERE { ?ag1 a lpm:CalculatedData.
        ?el1 a ?ag1.
        ?ap1 a lpm:AttributeProfile.
        ?a1 a ?ap1.
        ?rp1 a lpm:RelationProfile.
        ?r1 a ?rp1.
        ?el1 ?r1 ?solution.
        ?el1 ?a1 ?duration.
        ?threshold1 a dm:Threshold.
        ?threshold1 rdf:value ?v1.
        FILTER ( ?duration <= ?v1) }
```

q2

```
CONSTRUCT { ?stl a lpm:Strategy.
            ?el3 a ?stl .
            ?el3 rdf:value ?gaming .}
WHERE {
  SELECT (COUNT (?solution) AS ?nb)
  WHERE { ?ag1 a lpm:CalculatedData.
          ?el1 a ?ag1.
          ?ap1 a lpm:AttributeProfile.
          ?a1 a ?ap1.
          ?rp1 a lpm:RelationProfile.
          ?r1 a ?rp1.
          ?el1 ?r1 ?solution.
          ?el1 ?a1 ?duration.
        }
  ?threshold2 a dm:Threshold.
  ?threshold2 rdf:value ?v2.
  LET (?gaming:= (?nb >= ?v2 )) }
```

q3

Fig. 2. General SPARQL Queries to detect Gaming the system

In order to detect the gaming within a TEL system, the LM designer reuse the general queries (q_1, q_2, q_3) composing the gaming pattern P_g. The reuse of SPARQL queries is possible by adapting (q_1, q_2, q_3) to the specific models elements according to the adaptation function δ . The adaptation function δ generates a new ordered set of SPARQL queries described with elements from trace model and learner profile model of the specific TEL system.

Before describing the adaptation, we begin by describing the trace models considered in AMBRE-add and copex-chimie TEL systems.

4.2 Traces Models for Copex-Chimie and AMBRE-Add TEL Systems

Copex-chimie is a Web-based TEL system offering learners the means to design experimental procedure in the context of chemical practices. The learner must for instance determine the concentration of the red dye in a grenadine syrup using spectrophotometry by writing an experimental procedure [4]. The experimental procedure is structured into three steps (preparation of a series of standard solutions; obtain the points of the standard curve; determine the concentration of the dye in the grenadine syrup). For each step, the learner selects adequate actions among a list of eight actions (rinse an equipment, make a dilution, etc.). For each action added in the experimental procedure step, the parameters describing the action have to be set by the learner. The learner can ask the system for assessment of the procedure. The system evaluates the procedure, step by step and points out to the learner the errors detected in the procedure.

To observe a copex-chimie learning session, several elements are tracked. The figure 3 (B) shows a part of copex-chimie trace model corresponding to the observation of the first step of an experimental procedure (preparation of series of standard solutions). This model is composed of : `Procedure`, `ProcedureStep`, `ProcedureAction` and `ProcedureStepEvaluation` as observed elements types, `idProcedureStep` as attribute and `hasProcedureStep`, `hasAction` and `hasEvaluation` as relationships.

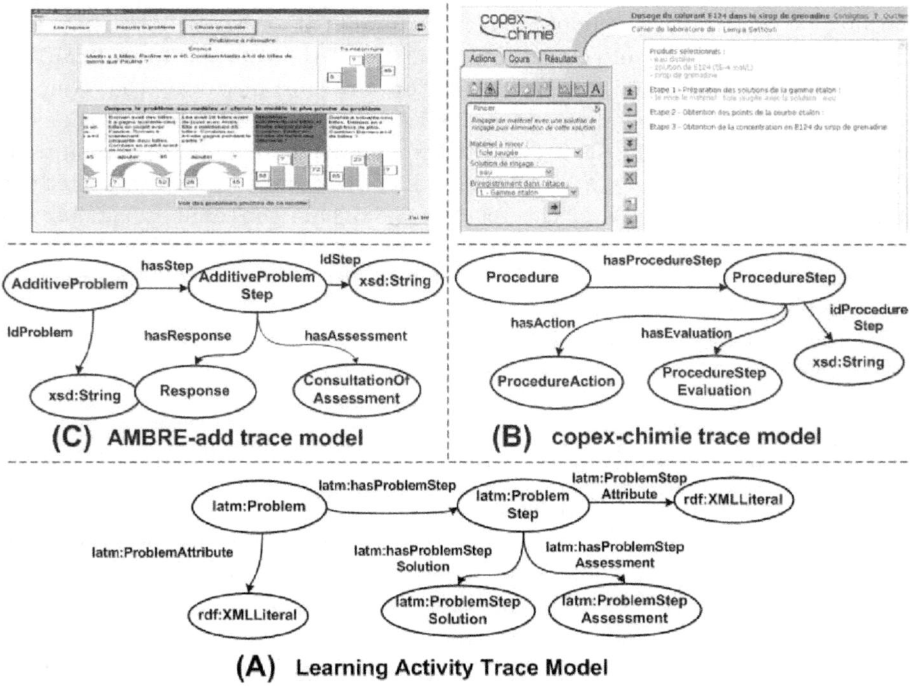

Fig. 3. Traces models for copex-chimie and AMBRE-add TEL systems

AMBRE-add is an ITS that enables learners to learn methods based on a classification of problems in order to solve additive problems [14]. During a AMBRE-add learning session, the learner begin by observing typical examples of solved problems, then he/she tries to solve problems in five steps. In the first step, the learner reads the problem statement and tries in the second step to reformulate the problem to be solved by identifying the relevant features of problem statement. The third and fourth steps consist in choosing among the examples that he/she has seen before, the problem that seems the nearest to the problem to be solved and adapt the solution. In the last step, the learner stores the new problem with a typical problem that represents a group of existing problems of the same class. For each step, the learner can ask the system for help or to evaluate his/her responses, then the system gives him/her appropriate feedbacks.

As copex-chimie, AMBRE-add is fully tracked system and several elements are observed. The figure 3 (C) shows a part of AMBRE-add trace model corresponding only to the step 3 that consists in choosing an example of a typical problem. This model is composed of: `AdditiveProblem`, `AdditiveProblemStep`, `Response` and `ConsultationOfAssessment` as observed elements type, `idProblem` and `idStep` as attributes and `hasStep`, `hasResponse` and `hasAssessment` as relationships.

The two defined trace models are both derived from the part of the learning activity trace model *LATM* showed in 3 (A). More precisly, `Procedure` and `AdditiveProblem` are instances of the concept `latm:Problem`; `ProcedureStep` and `AdditiveProblemStep` are instances of the concept `latm:ProblemStep`, `ProcedureAction` and `Response` are instances of the concept `latm:ProblemStep- Solution`. `idProcedureStep` and `idStep` are instances of the attribute `latm: ProblemStepAttribute`, `hasProcedureStep` and `hasStep` are instances of the relation `latm:hasProblemStep`, `hasAction` and `hasResponse` are instances of the relation `latm:hasProblemStepSolution` and `hasEvaluation` and `hasAsse- ssment` are instances of the relation `latm:hasProblemStepAssessment`.

In order to detect when learners are gaming with these two systems, the adaptation function specifies the generic gaming pattern by using these traces models.

4.3 Adapting Gaming the System Query Pattern

To demonstrate the concept of query pattern, we describe only the process of adaptation of q_1 (figure 2) for AMBRE-add and copex-chimie TEL systems. The other queries (q_2, q_3) are adapted with the same function δ. The result of the adaptation function δ (figure 4) represents two SPARQL queries noted q'_1 and q''_1 ready to be used with the two systems.

In order to show how the function δ adapts the gaming query pattern, we give as an example the adaptation of the general triple pattern `?o2 a latm:ProblemStep` from q_1 (see figure 2 line 1). The general triple pattern enables to match instances of the observed element type `ProblemStep` from *LATM*, and the LM designer is the one who decides which instance must be adapted. In this context, the LM designer defines the parameters of the function δ by providing the elements of a specific TEL system which substitute the general ones. In this example, the general element c_1 from *LATM* is `latm:ProblemStep` and the specifics elements for AMBRE-add and copex-chimie TEL systems which are instances of `latm:ProblemStep` are respectively: c'_1 = `ambre-tm:AdditiveProble- mStep` and c''_1 = `copex-tm:ProcedureStep`.The results of the function δ is the triple pattern `?ps1 a ambre-tm:AdditiveProblemStep` for AMBRE-add TEL system (see figure 4 line a for q'_1) and the triple pattern `?ps1 a copex-tm:ProcedureStep` for copex-chimie TEL system (see figure 4 line b for q''_1).

```
CONSTRUCT{ _:el1 a ambre-lpm:SolutionDuration.
           _:a1 a ambre-lpm:hasDuration.
           _:r1 a ambre-lpm:hasSolution.
           _:el1 _:r1 ?solution.
           _:el1 _:a1 ?duration. }
WHERE {                                              q'1
    ?p1 a ambre-tm:AdditiveProblem.
    ?p1 ambre-tm:idProblem ?id1.
  a: ?ps1 a ambre-tm:AdditiveProblemStep.
    ?ps1 ambre-tm:idStep ?id2.
    ?p1 ambre-tm:hasStep ?ps1.
    ?solution a ambre-tm:Response ;
    tbs:hasBegin ?begin1;
    tbs:hasEnd ?end1.
    ?ps1 ambre-tm:hasResponse ?solution.
    ?ass1 a ambre-tm:ConsultationOfAssessment.
    ?ass1 tbs:hasBegin ?begin2 .
    ?ps1 ambre-tm:hasAssessment?ass1.
    FILTER (?end1 <= ?begin2).
    NOT EXISTS {
        ?o a tbs:Obsel .
        tbs:Obsel tbs:hasBegin ?begin3.
        tbs:Obsel tbs:hasEnd ?end3.
        FILTER ((?end1<= ?begin3) && (?end3<=?begin2) )
}
LET (?duration:= ?end1 - ?begin1)
FILTER ((?id1="problem1") && (?id2="step3")) }
```

The first SPARQL Gaming query adapted for AMBRE-add TEL system

```
CONSTRUCT{ _:el1 a copex-lpm:SolutionDuration.
           _:a1 a copex-lpm:hasDuration.
           _:r1 a copex-lpm:hasSolution.
           _:el1 _:r1 ?solution.
           _:el1 _:a1 ?duration. }
WHERE {                                              q''1
    ?p1 a copex-tm:Procedure.
  b: ?ps1 a copex-tm:ProcedureStep.
    ?ps1 copex-tm:idProcedureStep ?id2.
    ?p1 copex-tm:hasProcedureStep ?ps1.
    ?solution a copex-tm:ProcedureAction ;
    tbs:hasBegin ?begin1;
    tbs:hasEnd ?end1.
    ?ps1 copex-tm:hasAction ?solution.
    ?ass1 a copex-tm:ProcedureStepEvaluation.
    ?ass1 tbs:hasBegin ?begin2 .
    ?ps1 copex-tm:hasEvaluation?ass1.
    FILTER (?end1 <= ?begin2).
    NOT EXISTS {
        ?o a tbs:Obsel .
        tbs:Obsel tbs:hasBegin ?begin3.
        tbs:Obsel tbs:hasEnd ?end3.
        FILTER ((?end1<= ?begin3) && (?end3<=?begin2) )
}
LET (?duration:= ?end1 - ?begin1)
FILTER ((?id2="step1")) }
```

The first SPARQL Gaming query adapted for copex-chimie TEL system

Fig. 4. Adaptation of the general SPARQL gaming query q_1 to two TEL systems

5 Related Work

The work presented in this article is closely related to research in several areas. We review them briefly. Several research works have focused on the observation of learner's behaviors and how to extract and calculate them from the learner's interactions with a TEL system, in order to infer their performances and skills [12] or to identify their learning styles [15,3]. These behavioral patterns are calculated for a specific TEL system and there are difficult to apply and reuse in other TEL systems.

In the area of traces representation, Mostow *and al.* [12] have given some methodological recommendations and guidelines on relevant information that can be observed and tracked by an intelligent tutoring system. However, they did not address the issue of modeling and operationalisation of such guidelines, which are covered by our proposal model *LATM*. In addition, the issues of reusabillity LM models and/or diagnosis are not addressed by the framework defined by Mostow.

In area of Learner/user modelling, many works have addressed the issue of reusability in the context of Ontology-Based LM (e.g. [1,20,13]) however without dealing with diagnosis support and reuse. Even if they use semantic web ontology languages as RDF and OWL which offer many advantages (formal semantics, easy reuse, easy portability, availability of effective design tools, etc.) such frameworks lack to support LM designer task specially to offer him the reusing services.

For the part of learner profile, there are different standards which cover an aspect of Learner description as IEEE standards as PAPI [8] and IMS-LIP[9]. In our framework, we can use the RDF-vocabulary Learner Profile Model *LPM* in order to integrate others learner's characteristics (e.g. preferences, goals, etc.) by importing RDF-based representation of standards as PAPI[6] and IMS-LIP. Our framework enables this integration thanks to RDF schema that supports integration of standards.

6 Conclusion and Future Work

In this paper we have presented a framework to describe Learner Modelling (LM) process based on interactions traces. This framework includes an RDF-Based representation of knowledge models that can be used by a LM designer. The first model enables the LM designer to describe observations about learner's interactions with a TEL-system. The second model enables the LM designer to describe the structure of learner's profile. This framework supports also the description of reusable and adaptable SPARQL-based query patterns. These patterns enables the LM designer to calculate and infer learner profile elements for different TEL systems. Learner profile elements are obtained from observations model elements for a specific TEL system and/or from learner profile elements. We defined formally the notion of query pattern and illustrate its application in the context of two TEL systems in order to detect gaming the system.

Our future works must deal with several aspects. Actually, our proposal have been applied only to LM with domain independent patterns as detecting learner's observable behaviors, learner's performances and learner's strategies as gaming the system, etc. However, we plan to define more diagnosis patterns in the context of LM dependent domain patterns as diagnosis of learner knowledge, detection of misconceptions, etc. We are also interested in combining deductive approach based on SPARQL with inductive diagnosis approaches (e.g. Bayesian inferences) by using the magic properties of SPARQL.

Finally, we are engaged in an implementation of our framework as Learner Model Server based on RESTful Web Services as means to manage, access and use the modelled traces and learners profiles. Actually, all LM models of our framework are defined with RDF, tested and experimented with several TEL systems. Some aspects of our framework are already developed, e.g., the web services[3] concerning management and transformation of modelled traces which are actually developed by the team SILEX[4].

References

1. Aroyo, L., Dimitrova, V., Kay, J. (eds.): PerSWeb 2005 Workshop. Personalisation on the Semantic Web at 10th International Conference User Modelling 2005 (2005)
2. Baker, R.S.J.d., Corbett, A.T., Koedinger, K.R., Evenson, S., Roll, I., Wagner, A.Z., Naim, M., Raspat, J., Baker, D.J., Beck, J.E.: Adapting to when students game an intelligent tutoring system. In: Ikeda, M., Ashley, K.D., Chan, T.-W. (eds.) ITS 2006. LNCS, vol. 4053, pp. 392–401. Springer, Heidelberg (2006)
3. Bousbia, N., Labat, J.M., Rebai, I., Balla, A.: Indicators for deducting the learners' learning styles: Case of the navigation typology indicator, pp. 385–389 (2009)

[3] http://liris.cnrs.fr/sbt-dev/ktbs/doc/
[4] http://liris.cnrs.fr/equipes?id=44&set_language=en

4. d'Ham, C., Girault, I.: Scaffolding the students' activity of experimental design with a dedicated software: copex-chimie. In: Proceeding of ESERA 2009 (2009)
5. Dillenbourg, P., Self, J.: A framework for learner modelling. Interactive Learning Environments 2(2), 111–137 (1992)
6. Dolog, P., Gavriloaie, R., Nejdl, W., Brase, J.: Integrating adaptive hypermedia techniques and open RDF-based environments. In: WWW (Alternate Paper Tracks) (2003)
7. Georgeon, O., Mille, A., Bellet, T.: Analyzing behavioral data for refining cognitive models of operator. In: Society, I.C. (ed.) Philosophies and Methodologies for Knowledge Discovery, Seventeenth International Workshop on Database and Expert Systems Applications, pp. 588–592 (2006)
8. IEEE: Ieee learning technology standardization committee(ltsc). draft standard for learning technology – public and private information (papi) for learners (papi learner) (2001), `http://www.cen-ltso.net/main.aspx?put=230&AspxAutoDetectCookieSupport=1`
9. IMS: Ims learner information package specification (2005), `http://www.imsglobal.org/profiles/`
10. Kay, J., Lum, A.: Ontologies for scrutable student modelling in adaptive elearning. In: Proceedings of the Adaptive Hypermedia and Adaptive Web-Based Systems Workshop on Semantic Web for E-Learning (2004)
11. Michelet, S., Luengo, V., Adam, J.M., Mandran, N.: Experimentation and results for calibrating automatic diagnosis belief linked to problem solving modalities: a case study in electricity. In: Proceedings of the 5th European Conference on Technology Enhanced Learning Conference, EC-TEL 2010, pp. 408–413 (2010)
12. Mostow, J., Beck, J.: Why, what, and how to log? lessons from listen. In: EDM, pp. 269–278 (2009)
13. Munoz, L.S., Palazzo, J., Oliveira, M.: Applying semantic web technologies to achieve personalization and reuse of content in educational adaptive hypermedia systems. In: Proceedings of the SWEL Workshop at Adaptive Hypermedia, pp. 348–353 (2004)
14. Nogry, S., Guin, N., Jean-Daubias, S.: AMBRE-add: An ITS to Teach Solving Arithmetic Word Problems. In: Technology, Instruction, Cognition and Learning, pp. 53–61 (2008)
15. Popescu, E.: Learning styles and behavioral differences in web-based learning settings. In: IEEE ICALT, pp. 446–450 (2009)
16. Settouti, L., Guin, N., Luengo, V., Mille, A.: Towards Adaptable and Reusable RDF-based Query Patterns for Trace-Based Learner Modelling. technical report, LIRIS, University of Lyon 1 (2011), `http://liris.cnrs.fr/publis/?id=5054`
17. Settouti, L., Guin, N., Mille, A., Luengo, V.: A trace-based learner modelling framework for technology-enhanced learning systems. In: IEEE International Conference on Advanced Learning Technologies, pp. 73–77 (2010)
18. Settouti, L.S., Prie, Y., Marty, J.C., Mille, A.: A trace-based system for technology-enhanced learning systems personalisation. In: IEEE International Conference on Advanced Learning Technologies, pp. 93–97 (2009)
19. Sison, R., Shimura, M.: Student modeling and machine learning. IJAIED International Journal of Artificial Intelligence in Education 9, 128–158 (1998)
20. Winter, M., Brooks, C., Greer, J.: Towards best practices for semantic web student modelling. In: Proceeding of the 2005 Conference on Artificial Intelligence in Education, pp. 694–701 (2005)

Learning Analytics at Large: The Lifelong Learning Network of 160,000 European Teachers

Ergang Song, Zinayida Petrushyna, Yiwei Cao, and Ralf Klamma

Information Systems & Databases, RWTH Aachen University

Abstract. Until now, eTwinning supports 160,000 European teachers is their continuous professional development (CPD). Only computational tools of learning analytics at large like social network analysis and performance measurements are offering deep insight into competence development in the anonymous data sets from the eTwinning database. To this end, we have developed CAfe, a learning analytics tool for eTwinning data. After getting an identification number, teachers can monitor their performance in eTwinning with CAfe. Teachers may also draw conclusions and gain insight into their competences and their continuous professional development. We evaluated CAfe with teachers in eTwinning workshops in Germany.

Keywords: Lifelong Learning, Social Network Analysis.

1 Introduction

The interest in learning analytics in increasing [17,8]. Especially in lifelong learning (LLL) and other informal learning settings, there will be a growing need for extracurricular assessment methods of learning progress. However, there are some issues to widen application of learning analytics in LLL. First, the rapidly developing information and communication technologies (ICT) reveal new possibilities for lifelong learning (LLL). New analytic tools have to be developed for the integration with existing competence assessment methods in LLL. Second, there is a lack of large data sets comparable to data sets in digital libraries [16] or bioinformatics [12]. This is limiting the development of reliable analytic tools and sound visualization.

eTwinning[1] is an online community for European teachers. The aim of eTwinning is to facilitate collaborations among schools and to support the LLL activities of teachers. Thus, we can call it a digital mediated learning network. eTwinning supports teachers in their continuous professional development (CPD) by providing possibilities to find teachers with same interests, create projects with other teachers and exchange experience and information. The continuous professional development consists of key activities like self-monitoring/self-evaluation, goal setting, help seeking, and time management [10,11].

[1] `http://www.etwinning.net/en/pub/index.htm`

C. Delgado Kloos et al. (Eds.): EC-TEL 2011, LNCS 6964, pp. 398–411, 2011.

During the learning process, self-monitoring is necessary for inspecting learning achievements and adjusting learning goals. How can we support the learners' self-monitoring in a digital learning network like eTwinning? Obviously, self-monitoring is one of teachers' competences which has to be developed in CPD like other competences. Moreover, it seems to be also the prerequisite to identify learning gaps and goals. So, this special competence will be inquired further. In the context of eTwinning competences can be considered as a set of characteristics of teachers, such as domain or subject knowledge, skills, sociability, activity, creativity, etc. [14], which result in superior performance in their teaching but also in their competences using eTwinning itself. To capture this double recognition of competences, we define and measure activities of eTwinners, i.e. teachers registered in eTwinning, and visualize the results over time. Based on visualization and intended insight into their own performance we support teachers and investigators of eTwinning in analyzing competences of eTwinners and their development over time [4,9]. Especially, the existence of visualization in forms of networks and charts needs new competences of understanding complex visualization of data. In former research it was obvious that these competences are not well developed yet, also among teachers [2].

For the professional development of such competences, we developed a tool for competence management, namely Competence Analyst for eTwinning (CAfe) for the eTwinning network. CAfe works on data dumps gathered in the eTwinning project, provided by the European Schoolnet[2]. The data dumps are growing quite rapidly and are a unique source for the development of learning analytic algorithms. All data in the dumps is anonymous. When teachers or national contact points in eTwinning install the tool, they can request an identification number from the eTwinning administration for their account information stored in the data dumb to get access to their own data. Based on the data dumps of the eTwinning network, teachers can monitor their performance in activities that help them to reflect their competences. This is possible by a set of newly developed learning analytic methods and according visualization. Moreover, we designed a reusable template for defining an activity to be monitored. Teachers can create new activity patterns or revise the existing ones according to their requirements. Results from the learning analytics algorithms then are presented in different graphical forms (graphs and charts) and on multiple levels (personal level, community level, and network level).

The paper is organized as follows. In Section 2 we discuss related work. Section 3 is about how we modeled our competence assessment approach. Section 4 describes the implementation and evaluation of our tool. Finally, Section 5 concludes the paper and gives an outlook on further research.

[2] http://www.eun.org

2 Related Works

In this section, we introduce the theoretical background of the paper, including the definitions of a competence and a meta-competence, and the assessment approaches for competences.

2.1 Competences and Meta-competences

The original term of competence was derived from the Latin word "Competere" or "Competentia", which means to be suitable[3]. In 1973, McClelland [14] has introduced the concept of competence into the area of Human Resources Management (HRM). In his paper, the term competence is defined as: "*The knowledge, skills, traits, attitudes, self-concepts, values, or motives directly related to job performance or important life outcomes and shown to differentiate between superior and average performers.*" The competence-based HRM helps find the most appropriate matches between available persons and a given task.

Brown and McCartney [3] have suggested meta-competence as a higher order competence. Meta-competences can be learned but cannot be taught according to them. They have defined: "*A meta-competence is the overarching ability under which competence shelters. It embraces the higher order abilities which have to do with being able to learn, adapt, anticipate and create. Meta-competences are a prerequisite for the development of capacities such as judgment, intuition and acumen upon which competences are based and without which competences cannot flourish.*" Cheetham and Chivers [4] have defined: "*Meta-competence is the competence that is beyond other competences, and which enables individuals to monitor and/or develop other competences.*"

Thinking over competence and meta-competence terms, we consider reading as a competence. Some people can read faster than others. By learning some techniques, people learn to read faster. The ability of learning the techniques together with the desire to advance in the reading speed is called meta-competence. Although one don't probably want to read faster, some information may motivate him/her. These are the comparisons between his/her reading speed and others' reading speeds or benefits one can achieve by advancing the reading speed. Thus, sometimes some information is needed to trigger meta-competence and to define future steps one is going to take to advance his/her competences.

Processes like goal setting, self-monitoring, task strategies, help seeking, and time management contribute to a process of LLL [10,11] and some of the processes can invoke activate meta-competences. The self-monitoring process is one of such processes that plays an important role in the reflection of learning according to the monitored data. Self-monitoring may provide information to learners that is not obvious to them without further investigation. The ability of self-monitoring is a meta-competence in the context of LLL [4,9].

[3] `http://www.onrec.com/news/competency_profiling_fits_the_bill`

But there are some issues with current approaches:

- in modeling: there are a lot of debates in the related work even about the definition of "competences"and "meta-competences". There are some competence management frameworks like the European Qualification Framework (EQF), but we still have to think about a competence model fitting the eTwinning network data sets reporting very new competences and meta-competences. The gap between what frameworks are modeling and what is available as the data set is still big.
- in analyzing: competence assessment is a complicated endeavor. However, this can be simplified with the set of standard definitions developed by the European Union for competence assessment of teachers. In the eTwinning data gathering process, no all required data are gathered. Thus, we examine activities of teachers based only on available eTwinning data. Teachers can analyze their activities and draw conclusions on their competences. In the moment, there are no standardized tests available for professional assessment.
- in visualizing: the visual representation of data is much better than a set of data. Anyway, we have to consider the easiest and appropriate solutions for teachers that represent teachers activities clear and easy for making conclusions. In previous research [2] we found out that network data visualization and time-based visualization are not easy to understand for teachers. This is still an open issue.

2.2 Competence Assessment

Competence assessment is the core component of a competence management system. Usual assessment methods include direct observations, simulations, video observations, interviews, examinations of related documents, etc. [4]. In most cases, a combination of several methods is needed for an accurate assessment [5].

In [15], two competence assessment strategies are introduced: explicit assessment and implicit assessment. Explicit assessment methods determine the competences by asking people directly or indirectly. Three explicit assessment methods were suggested: the self-evaluation (a questionnaire), a theoretical test and a practical experiment.

With regard to implicit assessment methods, behaviors of people are monitored, competence-related events are tracked, and needed information is mined. According to the extracted information, algorithms are designed to calculate competence value.

A manual assessment is usually not very efficient, if the amount of competence respondents and competences is bigger than 25. In this case, automated assessment is better. In a workshop of automated competence assessment at the Prolearn Summer School in 2007, four approaches were introduced: a multiple choice approach, a simulation-based approach, a graph-based approach and a natural language processing approach. The first two approaches belong to explicit assessment strategies; methods such as online experiments, visual lab were suggested. The other two approaches belong to implicit strategies. The natural language processing approach analyzes the syntax and semantics of communications between people and derives competence information with certain algorithms. The graph-based approach focuses on the network structure and uses technologies such as Social Network Analysis (SNA) to assess competences.

In our research, we focus on implicit assessment methods, since explicit assessment method require the direct participation of teachers. With more than 160,000 teachers this is not feasible. Both graph-based approach and natural language processing approach can be considered. However, the natural language analysis approach is tough to be followed in eTwinning, as teachers in the eTwinning network speak many different languages all over Europe.

3 Competence Modeling in eTwinning

In this section, we build a hierarchical competence model for teachers in the eTwinning network. Then, we define a general assessment algorithm for measuring performance and activities of teachers that infer what competences the teachers possess. Moreover, we design a reusable template for defining the performance and activities, and their assessment algorithms.

3.1 A Classification Schema for Competence Management

In eTwinning we divide teachers' competences into three categories: *professional competences*, *social competences*, and *meta-competences*. We discuss which roles the competence categories play and which competences belong to which categories. Moreover, we also consider how the competences can be measured. The corresponding assessment methods are described in Section 3.2. Figure 1 shows the hierarchical structure of competences in eTwinning.

Professional Competence. As shown in Figure 1, professional competences are at the bottom of the model. Professional competences involve all abilities or skills of teachers, which are necessary for performing professional tasks in eTwinning. For example, sufficient knowledge of a subject and a language are essential for organizing a successful consortium that focuses on the subject. Professional competences are fundamental for organizing projects or teaching.

Usually, the information stored in the database is not sufficient for the accurate assessment of professional competences. In the paper, we define two indicators for teachers' professional competences, namely *project performance* and *project efficiency*. *Project performance* describes how a teacher performs in projects in eTwinning. It depends on the number of projects that a teacher has participated in, the number of awards that a teacher has gained etc. *Project efficiency* is a normalized value of project performance according to the project number that teacher has participated in. It shows the efficiency of a teacher in achieving sufficient performance in projects. We need to calculate the achievement of teachers in eTwinning, as we want to enable the teachers to monitor their achievements and to make conclusions about their competences. We measure the project performance and the project efficiency implicitly.

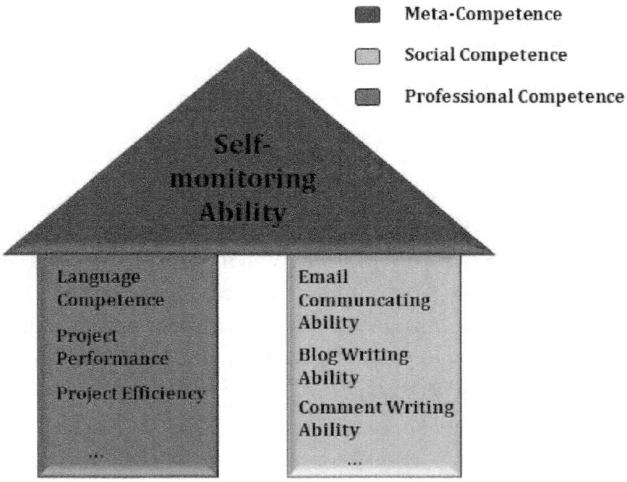

Fig. 1. Hierarchical Structure of Competences in eTwinning

Social Competence. We assume learning as a process of social interactions thus one of the best way to learn is by interacting in communities [18]. The social competence is defined as a key competence for LLL by European Commission [6]. Therefore we are looking at social competences within the eTwinning as well.

In this section we define several performance indicators of activities within eTwinning. The indicators suggest how competent teachers are in social interactions using eTwinning tools like emails and blogs.

- *Email communication ability*
  The email communicating ability describes the ability of teacher to communicate with others using an email tool embedded in eTwinning. Teachers with high scores for email communication ability have sent and received more Emails. When inspecting email communications from the SNA point of view, teachers sending and receiving more emails are well-connected in the email network, i.e. have a higher centrality [7] in the network. These teachers act as hubs in the eTwinning email network.
- *Blog writing ability*
  The blog writing ability defines an ability of a teacher in writing blog posts for a project. For the blog writing ability we consider not only the quantity of written blog posts, but also the feedback. Feedback can be measured by assessing popularity of posts and interest shown. Interest can be measured by the amount of received post comments. A well organized project blog can connect project members more densely and attract others.
- *Comment writing ability*
  The comment writing ability depends on the amount of comments a teacher has written.

– *Activity*
  Activity describes activities of a teacher in participating in all kinds of eTwinning interactions. The teacher with high scores of activity prefer to send emails to people, read blogs and comment on them, visit project and leave messages in guestbooks, and write comments if some teachers receive prizes.
– *Popularity*
  Popularity describes the ability to attract attention of other teachers. A teacher with high scores of popularity will more probably receive emails, and his/her blog posts are popular to be read and get more comments. Such teachers are quite famous in eTwinning, and a community having such teachers usually gets more attention and can attract more teachers to join creating scale-free networks [1].

Meta-Competences. Meta-competences encourage the improvement of other competences. As mentioned before, self-monitoring is a crucial meta-competence in the context of LLL. Our prototype, which provides various tools to help teachers to monitor their competence development, aims at enhancing *self-monitoring abilities* of teachers in eTwinning.

The self-monitoring ability describes teachers' ability to take advantage of monitoring their activities in the eTwinning network. It depends on various features, such as basic knowledge about SNA, understanding about graphs, charts, observation ability or abstract thinking ability etc. For accurate assessment of the self-monitoring ability, explicit methods such as a questionnaire have to be used additionally.

The self-monitoring ability of teachers was evaluated in teacher workshops by means of questionnaires and interviews. In the workshops, we asked teachers to work with CAfe, and then answer several questions about basic knowledge about SNA, graph theory etc.; average time cost to understand network and charts displayed in the CAfe; self-evaluation on impact of the CAfe on self-monitoring etc. Afterwards, teachers may track their activities and get insight into how the awareness about their activities has influenced their competences.

3.2 A First Modeling Attempt for Basic Competence Assessment

In Section 3.1 we have defined some indicators that may help get a clear idea about teacher competences. The assessment methods of them are different. Some of them have to be measured with explicit methods, such as examinations or questionnaires. Others are measurable implicitly i.e. we can monitor some factors that are included in the indicators and can infer competence development. In this section, we discuss how to measure indicators implicitly in three steps.

At first we divide indicators into several related factors. This factor design allows to extract it easily from the underlying relational database where the data dumps of the eTwinning network have been stored. Factors can be SNA results such as a centrality. Such measures are implemented as additional algorithms. Factors can be also a result of a SQL query such as number of written blog posts of a teacher. Such queries can be performed on top of the relational database directly with common user interfaces. Some predefined queries are also implemented in the CAfe prototype.

Factors need to be normalized due to their different value distributions, so that they can be combined together. In our approach, we choose the so-called z-score to normalize the factors (see [13,19] for more information on z-score normalization). After the normalization, the factor value varies between 0 and 10.

At last, according to the contribution to the indicator, each factor is assigned with a weight. The value of an indicator is then calculated as the linear combination of all related factors:

$$I = \sum_{f \in F} w_f \cdot Norm(f) \ ,$$

where F denotes the factor set related to the indicator I, f denotes single factor in the Set F, and $Norm(f)$ denotes the normalization of f.

We designed a reusable template for defining factors. Since an indicator consists of a set of related factors, we introduce the template for defining factors. A factor consists of the following attributes:

- *Identifier*: a globally unique label that identifies the factor definition.
- *Name*: a single mandatory text label for the factor. This is a short human-readable name for the factor.
- *Description*: an optional human readable detailed description of the factor.
- *Assessment*: the corresponding assessment method that must be realized in the program code.

For the specification of an indicator, the following attributes are needed:

- *Identifier*: a globally unique label.
- *Name*: a single mandatory text label.
- *Category*: this mandatory attribute describes the category of the indicator, i.e which competence the indicator may contribute to such as professional competence or social competence.
- *Description*: an optional human readable detailed description of the indicator.
- *Assessment*: the assessment involves a set of factor-weight pairs. A factor-weight pair consists of the ID of a related factor, and the corresponding weight of the factor.

Figure 2 shows the ER diagram of the reusable template for indicator modeling in the eTwinning. User can create new indicators, or modify the existent ones expediently. Indicators are stored in corresponding XML files which can be understood and analyzed by the CAfe prototype.

This is a very basic modeling approach leaving out any usage scenarics of indicators for assessment. It is a first attempt to build up a set of performance indicators which can be used for further professional competence assessment of teachers acting in digital networks.

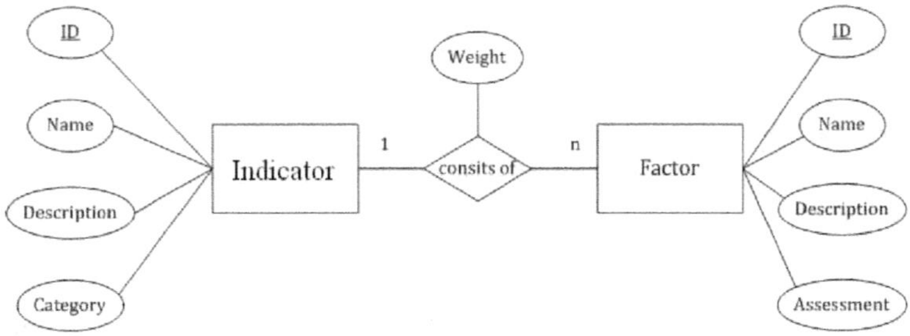

Fig. 2. ERD of the template for Defining Indicators

4 Implementation and Evaluation of the CAfe Prototype

Here we introduce briefly the implementation of the CAfe prototype, including the system architecture and some screenshots of the visualization. Then we present the evaluation result of the prototype on the TeLLNet meeting and the workshop for teachers registered in the eTwinning.

4.1 Implementation and Evaluation of the CAfe Prototype

CAfe serves two purposes:

- as a self-monitoring tool in the LLL of teachers. The prototype helps teachers monitor their positions in the eTwinning network, knowing their achievements, recognizing their weakness and shortages, and being motivated to minimize competence gaps.
- additionally, coordinators and researchers of the eTwinning project may use CAfe as a monitoring and analysis tool based on the anonymous data sets. As already mentioned no identification support is integrated in the data dumps or in the prototype. We are aware that with some effort the identity of teachers may be revealed. Therefore, the use of the prototype should be limited within the eTwinning network. Further research is needed to protect teacher identities. The prototype itself helps researchers monitor eTwinning networks over the years and examine development in networks, i.e. activities of teachers coming from different countries.

First of all, we created a relational database for the application in a secured environment. We update the database iteratively and automatically to keep aligned with the eTwinning database which is constantly updated by the activities of teachers. For the prototype evaluation, we performed two updates of the dump: in June and November 2010.

The Java client of CAfe consists of four modules. The database module is the interface connecting the database and the prototype. The module receives requests from the network and competence modules, transforms the requests to SQL instructions, and sends the instructions to the database. Furthermore, the module transforms a result set to the data format that the prototype can handle. The network module constructs the teacher network and handles all network related computation, e.g. SNA. The competence module handles all tasks for competence management. The visualization and GUI modules provide multiform presentations for network and competence analysis results, e.g. graphs, time series, bar charts, in order to support teacher's self-monitoring better. Figure 3 depicts the system architecture of the CAfe prototype. The Java client is depicted as a rectangle labeled "CAfe".

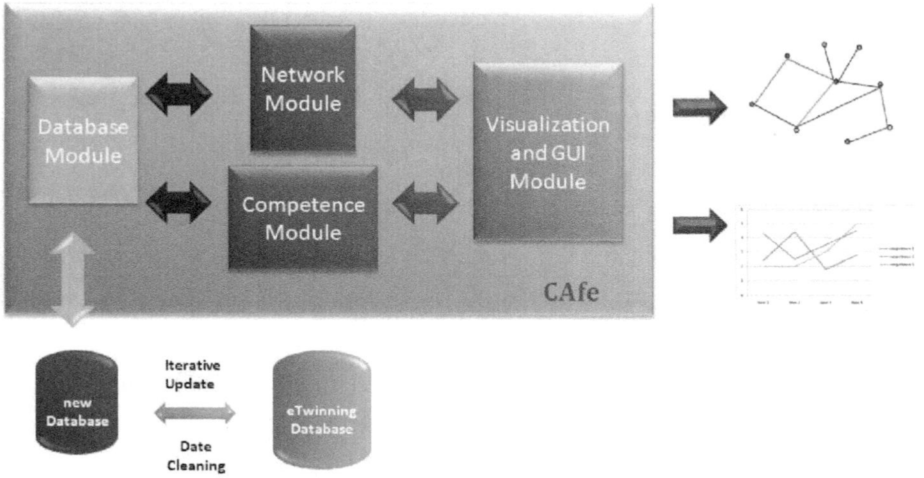

Fig. 3. System Architecture of the CAfe Prototype

The CAfe prototype supports the self-monitoring of teachers on three levels. On the first level, when a teacher launches the prototype, a graph representing the teachers' network is displayed, where nodes represent teachers and edges represent interactions between teachers. Teachers differentiate according to the date of their registration in the eTwinning portal. The interactions differentiate according to the date of the materialization of an activity: the date of project registration for projects; the date of sending an email for emails and so on. The example of the teacher network can be found on the Figure 4. Little blue arrows depict interface elements of the prototype like the network visualization (network), the network overview (overview), the tools for scrolling (scroll), and zooming (zoom), the toolbar (toolbar), some legends (node legend, edge legend), and finally the control panel (control panel) where the teacher can specify what he/she wants to explore.

When a teacher chooses a node, she can open the competence report of the chosen teacher. Moreover, teachers can compare competences among teachers with competence report on a community. A community is a concept from SNA depicting the nodes which are important for a teacher in the context of her interactions and projects in eTwinning. The notion of the community is considered here either as a school where a teacher works and her colleagues or as a project in which the teacher participate and her collaborators. For each level, we provide multiform visualization (e.g. graphs, time series charts, and bar charts) and user interfaces to customize the visualization.

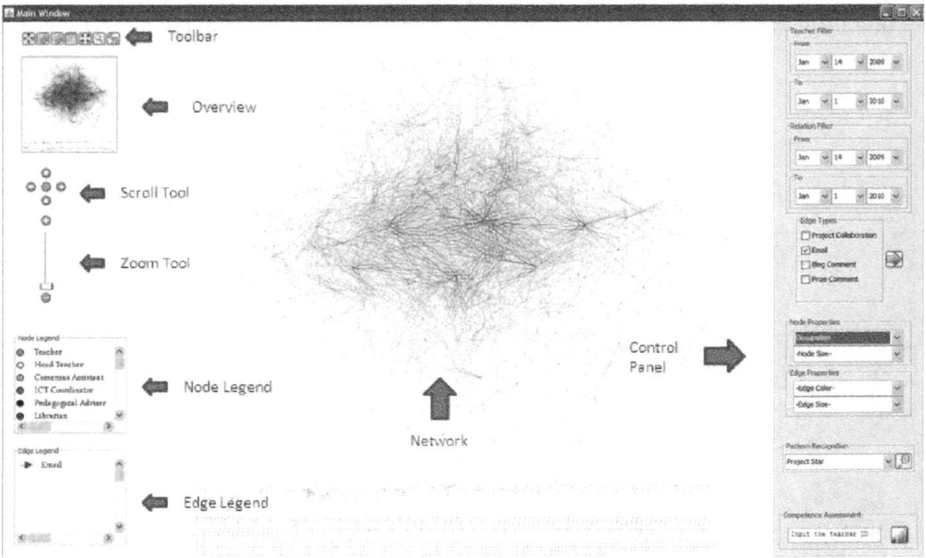

Fig. 4. Teacher Network in the eTwinning portal

4.2 Evaluation

An on-the-spot evaluation of CAfe beyond workshops with teachers was organized in a project meeting in early 2011. In the meeting, the CAfe prototype was demonstrated. In the course of the meeting the prototype was used by the participants of the project meeting with some tasks we defined. As usual, the participants were interviewed shortly and a questionnaire was filled by them. The questionnaire was designed to gather evaluation information from the project partners. In the questionnaire, we focused on the following features: basic knowledge of researchers; usability of the CAfe; evaluation of the CAfe; and further requirements.

- Basic knowledge, experience about SNA and competence of respondents are surveyed;
- Usability of the CAfe: Respondents are asked to perform some operations with the CAfe and to judge the time they need for completing the tasks;
- Further requirements and comments of the respondents are surveyed.

Besides the experienced researchers from the project which were all experienced with the eTwinning network, we were also interested in users without experience about the eTwinning. We chose some computer science students following the same procedures explained above.

The usability of the prototype is satisfactory for most of the users. Most users could perform our tasks successfully despite the very unequal distribution of knowledge about the eTwinning network and experience with tools. And most users considered the prototype informative and helpful for extending knowledge. The major drawback is the understandability of the result presentation in the prototype. Many users wanted to get more help functions to understand the competence report better. This was to be expected since the many graphical presentations created additional overload and our prototype aimed at a better understanding of these biases of presenting information in different formats. As already mentioned, experienced "eTwinners" have evaluated the CAfe at workshops, where ambassadors (active teachers) of the eTwinning network in a country gathered. The workshop has showed that the prototype should also be refined in its usability. Some functionalities were not clear for teachers and they asked for a better support and better interactivity of the system, i.e. more explanations of buttons. Anyway, 16 of 20 teachers showed their interest in observing their network of collaborations, the development of their competences and activities over the course of time and correlations between successful project, activities and networks. This is a very promising result, since teachers have no additional incentives in using the tool but their own personal interests in their professional development. Using ICT in general, and the eTwinning portal in particular is not a common part of teacher assessment in Europe and despite of the pedagogical and social success of the eTwinning network, teachers are not systematically supported in the career development if they participate in such activities. However, we are convinced that networking activities and the ability to monitor such networking activities are crucial competences (or meta-competences) of 21st century teachers and policy makers should have a closer look at professional development of teachers with respect to those competences.

5 Conclusions and Outlook

The ability for self-monitoring is a crucial competence in the LLL process of teachers. We have motivated that competences like the use of ICT tools for self-monitoring and meta-competences like striving for improvement by constant seeking of competence gaps can be supported by a deeper understanding of SNA results and insight into performance indicators in a collaboration data set. The goal of the CAfe prototype is to support the eTwinning teachers to monitor and enhance their activities and based on the information follow or change their strategies with respect to their professional development needs. We have just started to develop the means for handling large analytic data sets in technology enhanced learning. The eTwinning data set covering LLL activities 160,000 teachers is one of the largest available data sets in the moment.

We have designed assessment methods for the activities of eTwinners. The indicator of an activity in the eTwinning network consists of several weighted factors, which can be calculated based on available data in the eTwinning data set. The indicators are calculated with a linear combination of the related factors. Moreover, we have designed a standard template for describing the indicators and the corresponding assessment methods. An indicator can be described by the use of the XML language, so that indicators can be exchanged and enriched between different context. With regard to interoperability issues which is not addressed in this contribution but as usual in European contexts, the widespread use of measures depends heavily on standardization and interoperability of tools.

After calculating the indicators, we focused on implementing multiform charts to present indicators to teachers in different forms. They include bar charts, time series, comparison among community members etc. The aim of these visual representation is to support users in the self-monitoring process. For reducing the time of indicator computations and visualization construction we optimized our algorithm to reduce the response time and memory cost of the application. However, as a prototype we were not seeking perfection in all dimensions but feedback from real teachers and experts. Here, improvements are certainly to be applied.

In the future we are going to investigate competence models in more details and to find mappings from indicators to well-known but also new competences described in standards and frameworks. This will help utilizing the knowledge extracted from the eTwinning data sets. We want to gather as well the feedback from eTwinners to verify the accuracy of our indicators. New indicators are of interest for us as well, e.g. the eTwinners want to observe about teachers activities and performances.

Long-term tracking of competence development of eTwinners is necessary for evaluating the effect of the interventions we made. By making teachers aware, we also changed the use of the eTwinning portal and in consequence the data in the data dumps. To this end, we will develop a time series analysis model based on an extended data warehouse model of the eTwinning portal. We can then perform much more detailed and accurate analysis of dynamic processes in the portal. We have to distinguish effects of interventions from other developments and statistical noise in the data set.

Acknowledgment. This work was supported by the EU Lifelong Learning Programme - Agreement number- 2009-3574/001-001 - Project number- 505594-LLP-1-2009-1-BEKA4SRM - Teacher's lifelong learning network (TellNet). We would like to thank all project partners for their support and the fruitful discussion. In particular, we would like to express our gratefulness to the people running the eTwinning network, among others Riina Vuorikari.

References

1. Barabási, A.L.: Linked: The New Science of Networks. Perseus Publishing, Cambridge (2002)
2. Breuer, R., Klamma, R., Cao, Y., Vuorikari, R.: Social Network Analysis of 45,000 Schools: A Case Study of Technology Enhanced Learning in Europe. In: Cress, U., Dimitrova, V., Specht, M. (eds.) EC-TEL 2009. LNCS, vol. 5794, pp. 166–180. Springer, Heidelberg (2009)
3. Brown, R.B., McCartney, S.: Competence is not enough: Meta-competence and accounting education. Accounting Education 4(1), 43–53 (1995)
4. Cheetham, G., Chivers, G.: Professions, Competence and Informal Learning. Edward Elgar Publishing, Northampton (2005)
5. Dinham, S.M., Stritter, F.T.: Research on professional education. In: Wittrock, M.C. (ed.) American Educational Research Association: Handbook of Research on Teaching, pp. 952–970. McMillan Publishing Co, New York (1986)
6. Education Council: Recommendation of the European Parliament and the Council of 18 December 2006 on key competencies for lifelong learning (2006)
7. Freeman, L.C.: Centrality in social networks: Conceptual clarification. Social Networks 1(3), 215–239 (1979), http://dx.doi.org/10.1016/0378-8733(78)90021-7
8. Johnson, L., Smith, R., Willis, H., Levine, A., Haywood, K.: 2011 horizion report. Tech. rep., The New Media Consortium (2011)
9. Kalz, M., Klamma, R., Specht, M.: Kompetenzentwicklung in der Wissensgesellschaft mit Social Software (2008)
10. Kitsantas, A.: Test preparation and test performance: A self-regulatory analysis. Journal of Experimental Education 70(2), 101–113 (2002)
11. Kitsantas, A., Dabbagh, N.: Promoting self-regulation in distributed learning environments with web-based pedagogical tools: An exploratory study. Journal of Excellence in College Teaching 15, 119–142 (2004)
12. Klamma, R., Jarke, M., Hannemann, A., Renzel, D.: Der Bazar der Anforderungen - Open Innovation in emergenten Communities. Informatik-Spektrum 34(2), 178–191 (2011)
13. Larsen, R.J., Marx, M.L.: An Introduction to Mathematical Statistics and its Applications. Prentice-Hall, Englewood Cliffs (2000)
14. McClelland, D.C.: Testing for competence rather than for intelligence. American Psychologist 20, 321–333 (1973)
15. Nussbaumer, A., Fruhmann, K., Lachmann, P., Pachtchenko, S., Naeve, A., Kirschenmann, U.: Assessment procedures and guidance principles for supporting self-regulation. Tech. rep., ROLE Deliverable ID6.1 (T6.2)
16. Pham, M., Cao, Y., Klamma, R., Jarke, M.: A clustering approach for collaborative filtering recommendation using social network analysis. Journal of Universal Computer Science (j-jucs) 17(4), 583–604 (2011)
17. Verbert, K., Drachsler, H., Manouselis, N., Wolpers, M., Vuorikari, R., Duval, E.: Dataset-driven research for improving tel recommender systems. In: 1st International Conference on Learning Analytics and Knowledge (LAK 2011), Banff, Canada, February 27-March 1 (2011)
18. Wenger, E.: Communities of Practice: Learning, Meaning, and Identity. Cambridge University Press, Cambridge (1998)
19. Woolf, P., Keating, A., Burge, C., MichaelYaffe: Statistics and Probability Primer for Computational Biologists. Springer, Heidelberg (2004); Massachusetts Institute of Technology, BE 490/ Bio7.91

Simulating Learning Networks in a Higher Education Blogosphere – At Scale

Fridolin Wild[1] and Steinn E. Sigurdarson[2]

[1] Knowledge Media Institute, The Open University,
Walton Hall, Milton Keynes, MK7 6AA, UK
f.wild@open.ac.uk
[2] University of Iceland, Iceland
steinnes@gmail.com

Abstract. Blogging has become mainstream, even in Higher Education. How to successfully build and sustain dispersed cross-institutional learning networks, however, still remains a conundrum. Validation trials evaluating learning technology and practice lack possibilities to investigate this at scale with hundreds of individuals over yearlong periods of time. With this contribution, we therefore propose a model for simulating a Higher Education blogosphere in order to circumvent this cost and resources problem. The model is informed by actual, but smaller-scale trials conducted within the EC-funded iCamp project. This computational model is used to investigate the potential impact of a new educational intervention model and new blog network management facilities developed with the aim of facilitating social networking, self-direction, and collaboration among learners. The simulations conducted predict increased density and reciprocity of an open Higher Education blogosphere, when fitted with improved technology and when a subset of individuals is scaffolded along formal courses to form weak ties.

Keywords: blog networks, blogs, simulation.

1 Introduction

Blogs form a new way to communicate knowledge. Not astonishingly, they have met adoption in educational institutions worldwide: blogging has become mainstream (Nguyen-Ngoc & Law, 2009, p.14). Technologically, however, blog networks are not yet as mature as we would wish them to be. Furthermore, learning and teaching practices still fall short in methods for the efficient use of blog networks at scale and as an integrated part of routine knowledge work. For example, De Moor and Efimova (2004) call for "methods for gradually developing the required complex socio-technical systems around blogs, such as virtual communities" (i.e., collaborative management methods).

The EC-funded iCamp project has met these challenges with proposing and validating an educational intervention model to inform teaching and learning practice, supported by a technical blog network management component for increasing the density of the networks (Fiedler et al., 2009; Wild, 2008). Even though the final validation trial included 76 students from 11 countries supported by 10 facilitators

C. Delgado Kloos et al. (Eds.): EC-TEL 2011, LNCS 6964, pp. 412–423, 2011.

over a period of 14 weeks (Nguyen-Ngoc & Law, 2009), it remains unclear, what the effects of such approach would be at scale, i.e. in an open blogosphere of several hundred participants spread over a number of institutions and locations, with a small amount of courses scaffolding weak ties among a small subset of its participating individuals and in regular intervals.

To pull us over this obstacle by our own bootstraps, we have developed a simulation that models a higher education blogosphere consisting of students and facilitators in different universities and different places. This simulation enables us – given that its underlying assumptions are realistic – to investigate how changes in learning practice and technology impact on the structure of the networks formed in such an open higher education arena.

In this contribution, we use this simulation model to test the scalability of the service and course design that simplify the management and building of blog networks for learning by reducing both social and technical barriers in creating reciprocal relationships.

The conglomerate of all blogs available online, the so-called 'blogosphere', is somewhat elusive with actual statistics and demographics being very difficult to obtain. Early studies, such as Kumar et al. (2003) ascertain a bursty evolution at least since 2001 and identify an eruptive rise not only regarding metrics of scale, but also with respect to deepening community structures and higher degrees of connectedness. For the more recent years, reports seem to indicate stagnation in growth or even a decline, attributed to the rise of social media alternatives (Arnold, 2009; Economist, 2010). Still, the German ARD/ZDF online study for 2010, for example, reports 3% of the population of Germany to maintain or have maintained a blog, compared to 1% who posted on twitter (Busemann & Gscheidle, 2010).

It is even more difficult to estimate how many bloggers are learners. Studies report various differing shares over time. The professional blog-indexing service Technorati reports two to seven percent of the 7,200 surveyed bloggers to be students (Sobel, 2010). Other, less professionals-focused surveys identify a higher share of learners, although these are found not to use blogging for learning purposes. For example, Herring et al. (2004) identify in their study on blog genres that 57.5% of the blogs' authors investigated in a random sample of 203 blogs are students on a secondary and tertiary level. However, at the same time, the major share of 70.4% of the blogs are personal journals reporting on the lives of their authors, clearly less are deployed for filtering, i.e. commenting on external content, and knowledge sharing. Similar results are reported by Schmidt & Mayer (2007) in their end-2005 study among German speaking bloggers: users in education (pupils, students) are underrepresented among the blog writers with the (primary) aim for knowledge sharing (a.k.a. knowledge-bloggers = k-loggers), the major share of k-loggers stems from a work context.

The rest of this paper is organised as follows. First, we describe current problems of blog-based learning networks. We argue that these problems can be overcome with the educational intervention model and technological support components elaborated by the iCamp project. Based on literature research and informed by data from the trials performed in iCamp, we subsequently develop a generic simulation model for a higher education blogosphere and use this to investigate the potential implications of such new management technology and educational practice – at scale. An outlook discusses open questions and possible further applications of the simulation model.

2 Blog Networks for Learning

The reasons why people create and maintain blogs vary to a large extent, however, always also including community building and social networking among the key motivations substantiated through empirical studies (Nardi et al., 2005; Schiano et al., 2004). Besides the obvious – group blogs –, community structures and social networks were shown to exist between individual, but networked blogs (Chin and Chignell, 2006). However, when deploying blogs in collaboration, many obstacles can be found that have not been overcome so far.

To facilitate "productive blog conversations", which are necessary in knowledge management and learning, "more carefully tailored socio-technical systems are needed", as De Moor and Efimova (2004) claim. De Moor and Efimova identify the problem of notorious fragmentation of conversations to be responsible for (even the authors') difficulties in re-constructing conversations. Furthermore, they see (initial) response times as a problem that may slow down conversations, especially when comparing to push technologies such as mailing lists. Another obstacle to productive conversations in collaboration processes is identified to be the low number of links to blog posts: 30.5% of all blogs do not link to anything at all (Herring et al., 2004). The average entry receives 0.3 comments; the majority of entries received none. Multimodality poses yet another problem: replies and comments are often distributed across comment fields, but can also be found in blogs of the repliers. Krause (2004) identifies the fuzziness of the audience as a problem that may be responsible for failing discussion in his course experiment (his article is titled 'When Blogging Goes Bad'), as it is unclear whether the desired audience (the course participants) will be reached in time and at all.

The notorious fragmentation of conversations, slow response times, the fuzziness of the addressed audience – all these reasons indicate that both practice and technology support in using blogs as learning instruments still fall short. Although there are a large number of bloggers today and although there is increasing teaching and learning practice using blogs, these shortcomings in building and sustaining learning conversation in blog networks present major obstacles to growth and characteristic of a higher education blogosphere.

The iCamp project has tackled this shortcoming and produced an educational intervention model and a course design as a best practice example (Nguyen-Ngoc & Law, 2009; Law & Nguyen-Ngoc, 2008; Fiedler et al., 2009). The underlying intervention approach can be implemented in the process depicted in Figure 1: in the actual implementation phase, course-related ad hoc networks are lifted from the blogosphere where already existent and created where missing: following resolution of individual technical incapabilities ('estalishing PLEs') and an introductory phase ('introducing course structure & objectives' and 'getting to know each other'), groups are formed and established in form of a distributed network ('establishing distributed environment'). Based on this regulatory activity, the subsequent collaborative and individual activities take place – reflecting not only in blog based communication artefacts. The process culminates in the reflection for assessment.

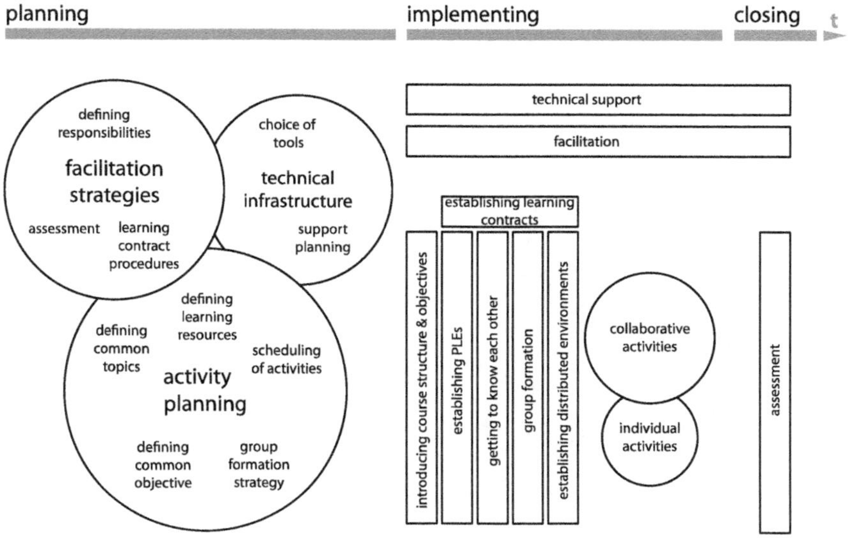

Fig. 1. Process view of iCamp approach (redrawn from Fiedler et al., 2009)

Technologically, the set-up and maintenance of this distributed learning environment can be supported with a management component – called 'FeedBack' (Wild, 2008). FeedBack was implemented (amongst others) as a WordPress plug-in. It assisted learners of the iCamp trials in offering their blogs to others, subscribing to the blog-feeds of others with an integrated feed reader, and providing active notifications about changes such as the availability of new postings. As such, the plug-in helps increasing awareness about followers and provides more efficient ways to offer blogs to others. This way, reciprocity of network connections is fostered: learners and facilitators become more aware of who is following the postings; and it is easier to propose to others to connect. This is particularly important in the implementation phase of a course, where groups are formed and shared, but distributed environments are created (see Fiedler et al., 2009, p.23; Law & Nguyen-Ngoc, 2008, p.15).

Figure 2 depicts such a resulting blog network as developed within the second iCamp trials: four universities were involved, bringing together 24 students and five facilitators (see Law & Nguyen-Ngoc, 2008). Within the three-month trial period, 68 offers were made, of which 49 were accepted. 94 requests for subscriptions had been made (see Law & Nguyen-Ngoc, 2008). The figure depicts the students and facilitators as the nodes. Offers, requests for subscriptions, each notification, and replies as directed links. Where two nodes are connected through multiple links, these are conflated and the link colour is darkened. The node size reflects the logarithmised calculated prestige score (Butts, 2009). R code is available from the first author.

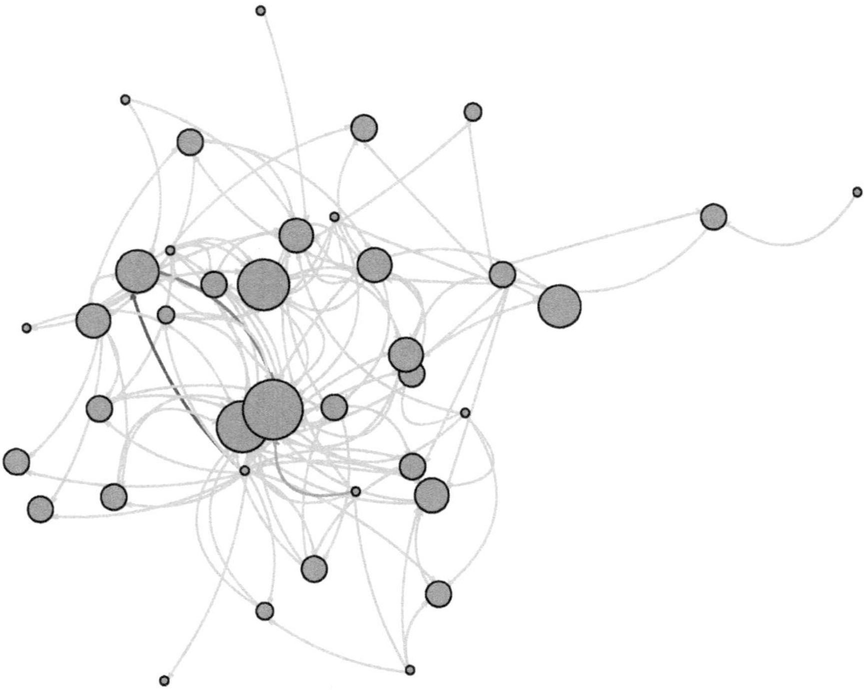

Fig. 2. Anonymised blog network in the second iCamp validation trial

3 The Simulation Model of an Open Higher Education Blogosphere

Based on the results found in the iCamp trials and combined with background data about the blogosphere in general and existing models thereof, a simulation of a learner blogosphere has been created in NetLogo[1] (Vidal, 2009).

Similar simulations have been conducted by Karandikar (2007) and Koper (2005). Karandikar has proposed and implemented a generative model to create blog graphs and post graphs from the interactions among bloggers using preferential attachment and uniform random attachment. The model uses local interactions among bloggers. It has been found to closely resemble large reference data sets studied. The model, however, considers only the generic blogosphere and not learning networks. Koper (2005) studied learning networks with the help of a similar simulation in NetLogo: he investigates whether indirect social interaction can be used to support learners with navigational support in sequencing units of learning by providing recommendations of units of learning that comparable learners have successfully mastered. The simulation found that indirect social interaction increases the share of learners acquiring the desired competence.

[1] http://ccl.northwestern.edu/netlogo/

For the simulation of a higher education blogosphere, the following scenario has been assumed. The number of individuals investigated was limited to 300 over the simulation period of one year. The scenario was assumed to be valid for a higher education landscape with ten universities and ten to twenty thousand students spread across ten to twenty disciplines. Within one subject area, this number of learners was assumed to be sufficiently large to show how a random number of 300 learners spread in this educational landscape would interact.

There were several assumptions about how and to whom learners create subscriptions, when they unsubscribe, and how often they read and post. Learners were treated as the vertices of a NetLogo network, their (directed) subscriptions as (directed) edges. Each learner vertex would note down for each step (tick) of the simulation its total number of subscriptions and subscribers, its total number of postings, a list of posting dates (in ticks), a posting quotient, and a read quotient. The posting quotient would be random exponential distributions converging around 1/t-postinterval (per default 1/5) to model how many posts this vertex makes per tick. Similarly, the reading quotient would be a random exponential distribution converging to the average number of postings read (n-readings = 10 per default) by the vertices. A maximum number of subscriptions was assumed to be 30, which is clearly below Dunbar's number of 150, which indicates the theoretical cognitive limit of people with whom one can maintain stable social relationships (Dunbar, 1992).

In each iteration (1 day = 1 tick), the number of postings assumed for each vertex would be created, following his average post quotient. The postings then would be immediately transferred to the subscribing vertices. Old postings would be removed from the feed, when they would be older than t-postingdeath (per default set to 105 days=ticks).

Additionally, new subscriptions would be created and old ones deleted based on the following procedures. Random subscriptions to anyone in the network would be created with a very low probability of p-subscription, per default set to 0.012. This means, that 1 out of 83 vertices would randomly subscribe to another vertex. This is used to model subscriptions motivated by other events, such as a user finding a blog post in a search engine based on which she decides to subscribe.

Since studies of the blogosphere have shown power law distributions of in-degrees (e.g. Karandikar, 2007), it is, however, much more likely that those vertices with already high in-degrees attract more subscriptions. Therefore, another subscription technique was modelled: authority subscriptions would subscribe a node with the same frequency p-subscription to nodes with high in-degrees. Subscriptions randomly die out with the frequency p-subscriptiondeath. Additionally, subscriptions age with a random exponential distribution around the average t-linkdeath: each aging act will reduce the link strength that increases with each posting by one. When a subscription is not very active, i.e. not many postings come through this channel, its probability to die away is higher. Subscriptions that expose a link-strength below one are removed in this step.

4 Results of the Simulation

Within this section, four different simulations and their results are presented, each of them slightly varied. The first simulation represents a typical Higher Education blogosphere, as constrained by the assumptions outlined above. The second simulation focuses on the effects of increased awareness and disburdened mutual subscription, as made possible with the FeedBack network management component. Subsection three adds modelling of courses (and their implied establishment of weak ties among a smaller number of individuals in the sphere) to the simulation. The final simulation investigates whether the effects on network density and reciprocity of realised connections would be also achieved without the technical support of FeedBack. In a sense, this last simulation is not fully realistic, as the result of established subscription structures can also be achieved with additional efforts of the facilitators of the courses. The iCamp trials, however, had shown, that this usually is not the case, particularly because monitoring actual subscriptions in a feed reader of choice in such an open scenario are beyond the control of a facilitator.

4.1 The Organic Higher Education Blogosphere

With this basic model, a blogosphere simulation is created that resembles typical link distributions found in generic blogosphere samples (cf. e.g. Karandikar, 2007). Figure 3 depicts the result of such a simulation run: in the centre, a force-directed lay-out of the network is depicted. White links indicate a unidirectional subscription, whereas red links mark up bidirectional ('reciprocal') ones.

The line width of the edges reflects the link strength. The numbers indicate the number of subscribers each vertex has. The vertex size grows with the number of subscribers they have. Vertices that have actively posted in this iteration change their colour from blue to green.

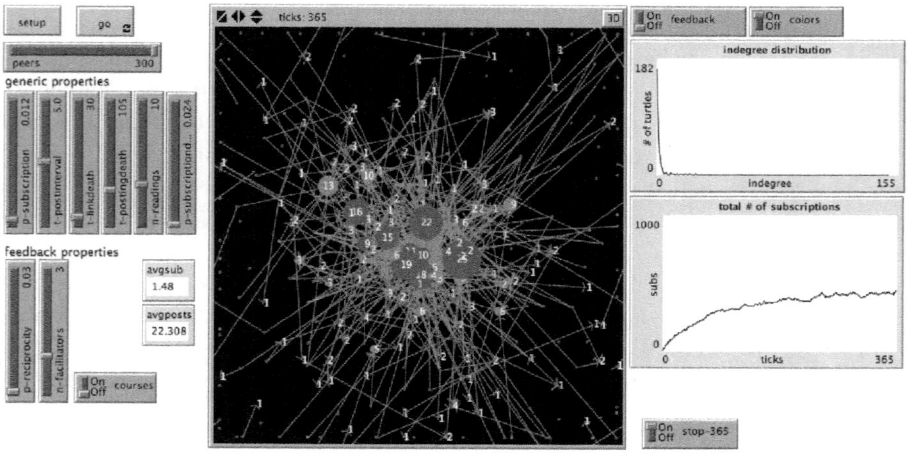

Fig. 3. Feed network simulation (without FeedBack)

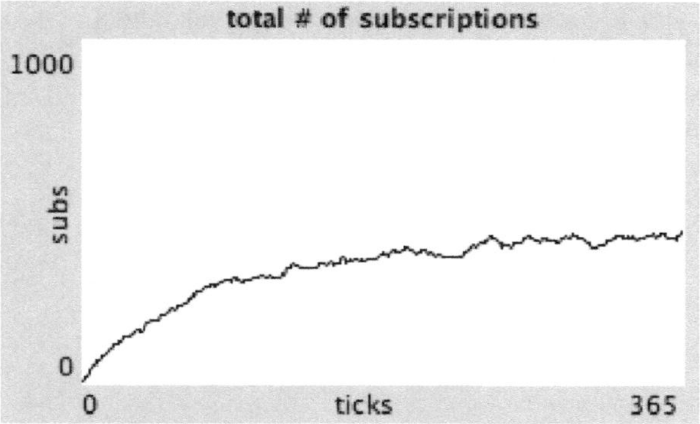

Fig. 4. Total number of subscriptions (w/o FeedBack)

4.2 The Reciprocity-Enabled Blogosphere

This simulation of a blog network among 300 random learners is now enriched in a second step with the FeedBack component, for which the assumption is made (based on the evaluation presented in Law & Nguyen-Ngoc, 2008), that it can increase reciprocity in subscribing. A very modest degree of p-reciprocity of 0.03 was assumed. This means that in average every 33rd subscription triggers the notified subscribed blogger to set up a reciprocal subscription. This is justified with the fact that informal contacts are easily possible with FeedBack, while articulation work and management efforts are reduced. Similarly, since bloggers are actively informed about their subscribers and the blog of their subscribers, chances to attract interest are higher. Previously, this knowledge about the identities of the subscribers had not been available, thus inflicting the probability to mutually subscribe. It is clear that mutual subscription is not identical with strong ties, as subscriptions age and randomly die. Still, with FeedBack every subscription becomes an implicit offer.

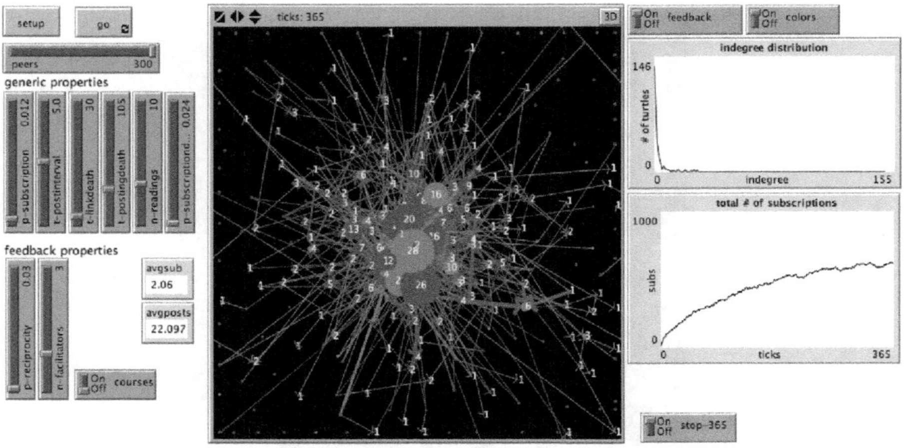

Fig. 5. Feed network simulation (with FeedBack)

As Figure 5 and Figure 6 show, the resulting total number of subscriptions also grows compared to the situation without FeedBack. In average, 2.06 subscriptions are generated, whereas the former simulation of the 'organic' blogosphere exposed a lower average number of subscriptions of 1.48 (see Figure 3).

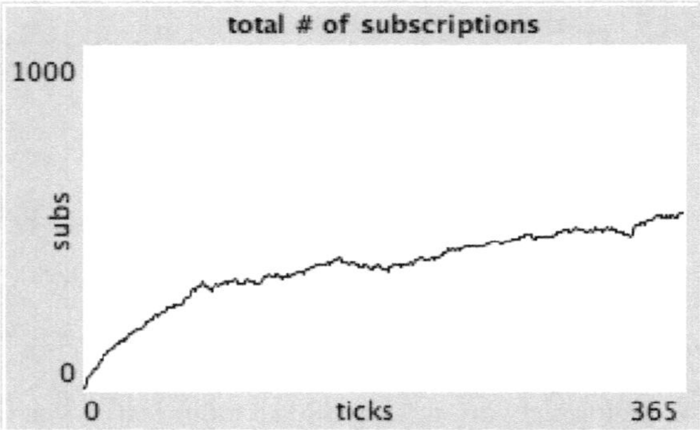

Fig. 6. Total number of subscriptions (with FeedBack)

4.3 Reciprocity-Enabled Blogosphere with Course Scaffolding

In a third step, this scenario is now additionally enriched with a formal education setting. It is assumed that three facilitators offer courses in this higher education landscape. With these courses, they generate weak structural ties and inject subscription structures, as they apply a learning design as described by Nguyen-Ngoc & Law (2009): they require their students to subscribe to each other's work blogs. When setting up their courses, there is one week in which students have an increased probability for reciprocity (0.03), as they are required to work with each other. Then

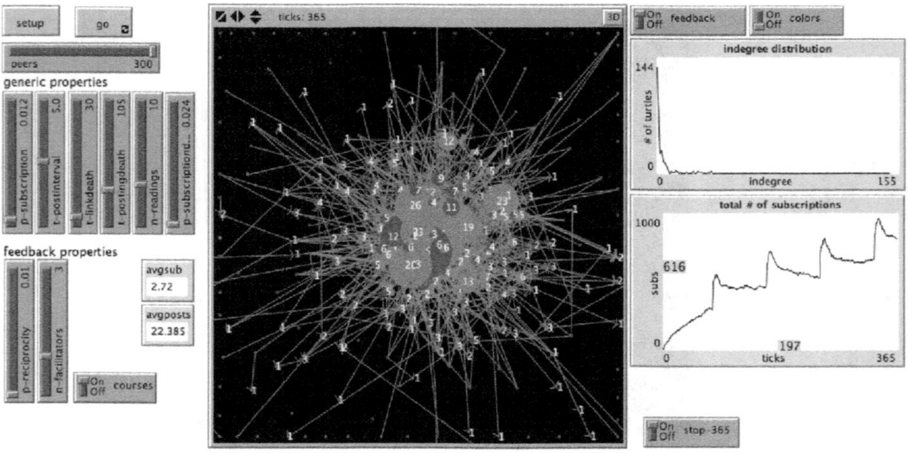

Fig. 7. Feed network simulation with FeedBack & 'course injections'

the probability for reciprocal connections is assumed to fall down to 0.01, as the learners are busy elaborating their projects. Each course randomly bundles together 16 students, a useful size for project-oriented seminars. Each student is assumed to subscribe to the facilitator blog, whereas students get injected mutual subscriptions only in small working groups of four.

In the simulation depicted in Figure 7, the subscriptions injected in the courses are shown in different link colours, reprinted in this paper in shades of grey. Over time, many of them die out again, but the structural injection given with the courses remains visible and reflects in a higher average number of subscriptions of 2.72.

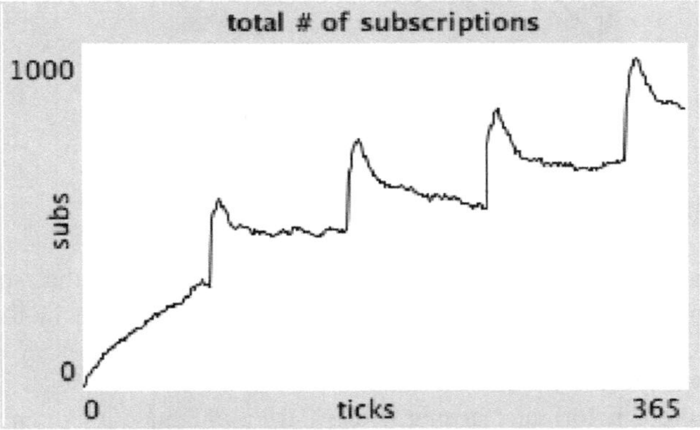

Fig. 8. Total # of subscriptions (+ FeedBack & courses)

4.4 Course Scaffolding without Reciprocity Support

Testing with the simulation whether this effect could be achieved also without the FeedBack component, shows that the spikes visible in Figure 8 in the total number of subscriptions are of course still salient (see Figure 10). The average number of subscriptions, however, is not rising the same way: it remains at a lower level of 1.93.

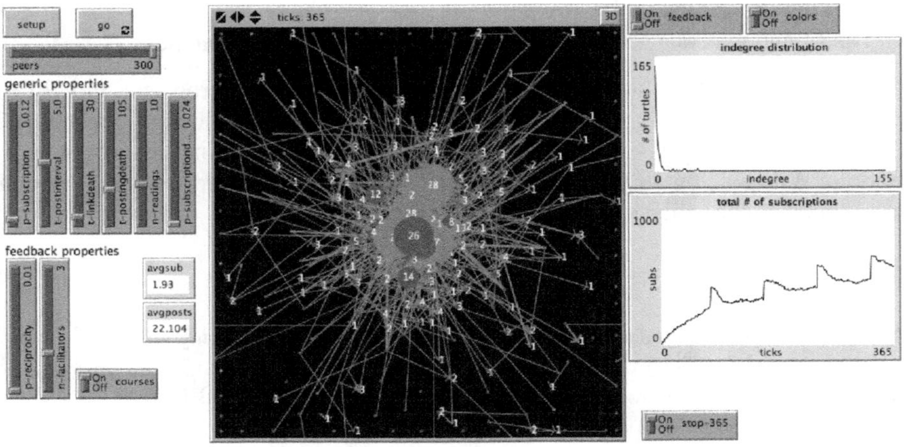

Fig. 9. Feed network simulation (courses only)

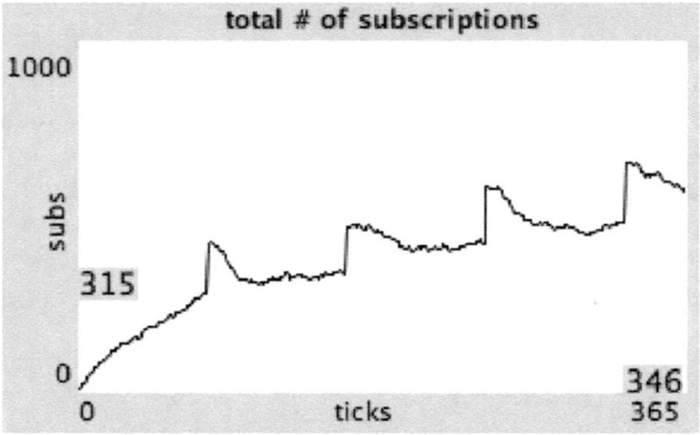

Fig. 10. Total number of subscriptions (courses only)

5 Summary and Outlook

Given that the assumptions are realistic, the simulation shows that supporting social networking via blogs along courses with improved management facilities could have a positive impact on the density and connectedness of learning networks in a Higher Education blogosphere.

When featured in formal learning offers, structural injections can support further integration of large learning network of hundreds of persons. Building professional networks in Higher Education will not only support rich professional competence development of the individual or learners studying at the same institution, it also creates effects of connectedness beyond the single organization.

The simulation model as such seems realistic as it is informed and in line with actual (but smaller) trials conducted. Within the model, effects of technology and practice can be studied – before developing technology and before conducting costly studies with real subjects. It would be interesting to see, if the proposed model can be extended to include or cover additional social media phenomena – such as social networking sites or microblogging.

The simulation code is available as open source from the first author.

Acknowledgements. This work has been co-funded by the European Union in the project iCamp and Stellar under the Information and Communication Technologies (ICT) theme of the 6th and 7th Framework Programme. Steinn Sigurdarson and Fridolin Wild collaboratively developed the code for the simulation, whereas Fridolin has conducted the analysis and written up this paper.

References

Arnold, B.: Blog statistics and demographics, Caslon Analytics on blogging (2009), http://www.caslon.com.au/weblogprofile1.htm

Busemann, K., Gscheidle, C.: Ergebnisse der ARD/ZDF-Onlinestudie 2010: Web 2.0: Nutzung steigt – Interesse an aktiver Teilhabe sinkt (2010), http://www.ard-zdf-onlinestudie.de/fileadmin/Online10/07-08-2010_Busemann.pdf

Butts, C.: sna: Tools for Social Network Analysis. R pck. version 2.0-1 (2009)

Chin, A., Chignell, M.: A Social Hypertext Model for Finding Communities in Blogs. In: Proceedings of the Hypertext 2006 (HT 2006), Odense, Denmark, pp. 11–22 (2006)

De Moor, A., Efimova, L.: An Argumentation Analysis of Weblog Conversations. In: Aakhus, L. (ed.) Proceedings of the 9th International Working Conference on the Language-Action Perspective on Communication Modelling (LAP 2004), New Brunswick, USA, pp. 197–211 (2004)

Dunbar, R.: Neocortex size as a constraint on group size in primates. Journal of Human Evolution 22(6), 469–493 (1992)

Fieder, S., Kieslinger, B., Pata, K., Ehms, K.: iCamp Educational Intervention Model, deliverable d1.3, iCamp consortium (2009)

Herring, S., Scheidt, L., Bonus, S., Wright, E.: Bridging the Gap: A Genre Analysis of Weblogs. In: Proceedings of the 37th Hawaii International Conference on System Sciences (HICSS-37). IEEE Computer Society Press, Los Alamitos (2004)

Karandikar, A.: Generative Model To Construct Blog and Post Network. In: Blogosphere, Master Thesis, UniversityofMaryland (2007)

Koper, R.: Increasing Learner Retention in a Simulated Learning Network Using Indirect Social Interaction. Journal of Artificial Societies and Social Simulation 8(2) (2005), http://jasss.soc.surrey.ac.uk/8/2/5.html

Krause, S.: When Blogging Goes Bad: A Cautionary Tale About Blogs, Email Lists, Discussion, and Interaction (2004), http://english.ttu.edu/KAIROS/9.1/praxis/krause/index.html

Kumar, R., Raghavan, P., Novak, J., Tomkins, A.: On the Bursty Evolution of Blospace. In: Proceedings of the WWW 2003, Budapest, Hungary, pp. 568–576. ACM Press, New York (2003)

Law, E., Nguyen-Ngoc, A.V.: Second Trial Evaluation Report, Seliverable d4.3, iCamp Consortium (2008)

Nardi, B., Schiano, D., Gumbrecht, M., Swartz, L.: Why We Blog. Communications of the ACM 47(12), 41–46 (2004)

Nguyen-Ngoc, A.V., Law, E.: Third Trial Evaluation Report, Deliverable d4.4, iCamp Consortium (2009)

Schmidt, J., Mayer, F.: Wernutzt Weblogs für collaborative Lern- und Wissensprozesse? In: Dittler, Kindt, Schwarz (eds.) Online-Communities als Soziale Systeme. Waxmann, New York (2007)

Sobel, J.: WHO: Bloggers, Brands and Consumers, Day 1, State of the Blogosphere 2010 (2010), http://technorati.com/blogging/article/who-bloggers-brands-and-consumers-day/

The Economist: An empire gives way. Blogs are growing a lot more slowly. But specialists still thrive (2010), http://www.economist.com/node/16432794

Vidal, J.: FundamentalsofMultiagent Systemswith NetLogo Examples (2009), http://jmvidal.cse.sc.edu/papers/mas.pdf

Wild, F. (ed.): An Interoperability Infrastructure for Distributed Feed Networks, Facultas, Vienna (2008) ISBN 9-783708-903354

Orchestration and Feedback in Lab Sessions: Improvements in Quick Feedback Provision

Israel Gutiérrez Rojas, Raquel M. Crespo García, and Carlos Delgado Kloos

Universidad Carlos III de Madrid, Avda. Universidad 30, E-28911 Leganés, Spain
{igrojas,rcrespo,cdk}@it.uc3m.es

Abstract. The benefits of formative assessment and the provision of quick feedback to the learners are unquestionable. A lot of effort needs however to be devoted by the teacher in order to assess the learner's status and providing quality and valuable feedback when actually implementing these techniques in real-world situations during active learning activities. The aim of this work is to offer teachers a tool that facilitates the quick provision of relevant feedback to the students, implemented as backchannel notifications in a lab session for ensuring scalability. Although being a work-in-progress, the aforementioned tool has been partially evaluated in a real course on programming multimedia applications.

Keywords: formative feedback, backchannel notifications, orchestration.

1 Introduction

Active learning is nowadays perceived as an engaging and effective learning methodology. When applied into actual class it can however be hindered by both intrinsic as well as circumstantial drawbacks. This work is inspired by the experience in an actual active learning-oriented course on Multimedia Applications. Although being addressed to 3rd year students, one particularly critical handicap in this course is students' reticence –active learning demands a bigger effort than just attending lecturers–, asking for step-to-step guidance by the teachers rather than willingly make an effort to solve the problems themselves and asking just for punctual help. This leads to scalability problems that jeopardize the feasibility of the methodology. Thus there is a need of better orchestration for the teachers to maximize the quality and efficiency of the interactions with the students in lab sessions.

Two ways are used for providing feedback to the students in the lab: personalized feedback directly delivered to individuals or teams; or explanations to the whole audience, intended to be used for common issues, general remarks and comments. The first type of interaction is highly effective (mainly due to the undivided attention provided by the students and its personalized nature), but not so efficient depending on the size of the audience. The second one is much more efficient as all the students are addressed simultaneously; its effectiveness will however depend on several aspects, among other, the attention devoted by the students and the relation of the explanation to the actual student work until that moment.

C. Delgado Kloos et al. (Eds.): EC-TEL 2011, LNCS 6964, pp. 424–429, 2011.

The objective of this work is to provide a supporting system for helping the teacher to scaffold the students during active learning hands-on sessions. The underlying idea is to provide a feedback mechanism that, while remaining efficient (addressing all the students) is more effective than generic explanations and is delivered to the students at the correct moment in time.

This work is founded on three main research lines:

- Use of communication backchannel (alternative communication channel supported by telematics) in class
- Benefits of the provision of quick formative feedback to the students
- Learning orchestration (a metaphor used to address the challenges of coordinating the complexity of conducting learning activities)

2 Related Work

Using backchannels for generating instant feedback within lectures is reported as a factor for potential success according to [2]. This is consistent with [1] who notes that: *"...the specific ways in which they can influence teaching pedagogy and learning opportunities are less well understood."* Although backchannels are typically used for bidirectional communication, the potential and learning opportunities are expected to be applicable to one-way notifications too.

Regarding pedagogical trends, the main goal of formative assessment is to provide feedback to the learner in order to improve his/her learning; thus the growing importance it has been receiving in the last decades [4]. Additionally, it promotes the continuous tracking of the learner's achievements. As explained in [5], formative feedback *"helps learners become aware of any gaps that exist between their desired goal and their current knowledge, understanding, or skill and guides them through actions necessary to obtain the goal"*.

Finally, different works report the benefits of appropriate learning orchestration. In particular, a distributed awareness system based on lamps (Lantern) that improves the orchestration of the class is presented in [3]. In our work, although the system is not distributed in the classroom it provides contextual information about the physical distribution of the room; hence, it preserves the fast decision making with a glance but adding more information obtained from the students' events monitoring.

3 Proposed Solution

The proposed solution is based on two artifacts: backchannel notifications (for quick feedback provision) together with an application for supporting orchestration.

Backchannel notifications are implemented using HTML5 technologies: web sockets, which permit real-time interaction between clients (browsers) and web notifications, which allow sending simple messages to the foreground of the clients' desktop. Combining both technologies a backchannel notification tool has been implemented, so that:

- When the students access to the wording of a lab session, they have the possibility to subscribe to the notifications channel using a "Connect" button in the same page of the wording.
- The teachers of the lab sessions are provided with a web interface used to send notifications to the subscribed students. This allows the teacher to provide feedback to the students *as demanded* according to the evolution observed in the class, as usually there are common doubts shared by the majority of students.
- Notifications are immediately shown in the foreground student screen and remain visible until the student chooses to close them.

The notifications application is being integrated in a higher level, more global, orchestration and feedback system. This system aims at providing relevant and efficient information to the teachers that allow them to improve the orchestration of the session. This awareness information is related to the performance of the students in the session as well as the required and provided teacher's support, and can be used by teachers to decide the most appropriate way of providing efficient and accurate feedback to the students.

The student interface is designed to be useful and easy to use in order to remain non-intrusive. It is integrated in the problem statement (wording) of the lab session (provided as a web page) and consists on a progress indicator including:

- a button to indicate student(s) have started working on this section
- a button to ask for teacher help
- a button to indicate the finalization of the section
- the expected time for the student to complete the exercise (on the right)

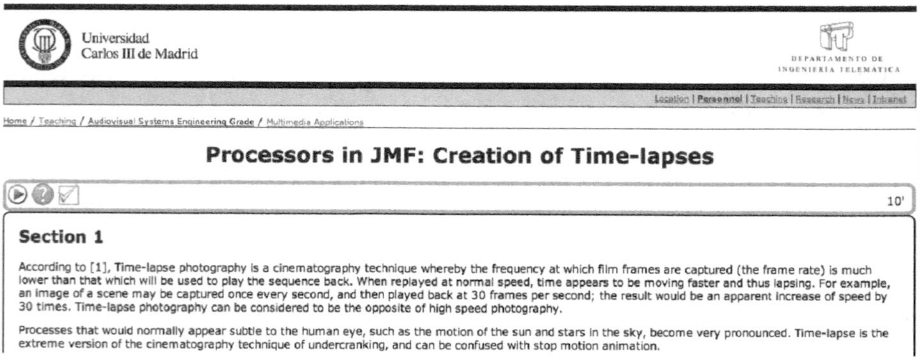

Fig. 1. Student interface

Teachers interface (Figure 2) has been designed for mobile devices that can be easily operated at the lab, mainly tablet PCs though smartphones could also be used. The tool summarizes awareness information over the map of the lab workstations:

- *Students' progress:* indicated as the number of the exercise they are currently working on, in the upper part of each computer icon.
- *Students asking for help:* background color of the upper part of the computer icon, ranging from orange to red as time passes.
- *Students with slow rate progress:* computer icon stroke turns orange for pointing out students delayed with respect to the main group.
- *Tutoring time already consumed:* shown in the lower part of the computer icon.
- *Consumed tutoring time relative to the most tutored student: the* bottom part of the computer icon is a progress bar colored in blue.
- Surrounded computers represent the *students being assisted* (blue) and the *next ones to be tutored* (the ones with highest waiting time, in red)

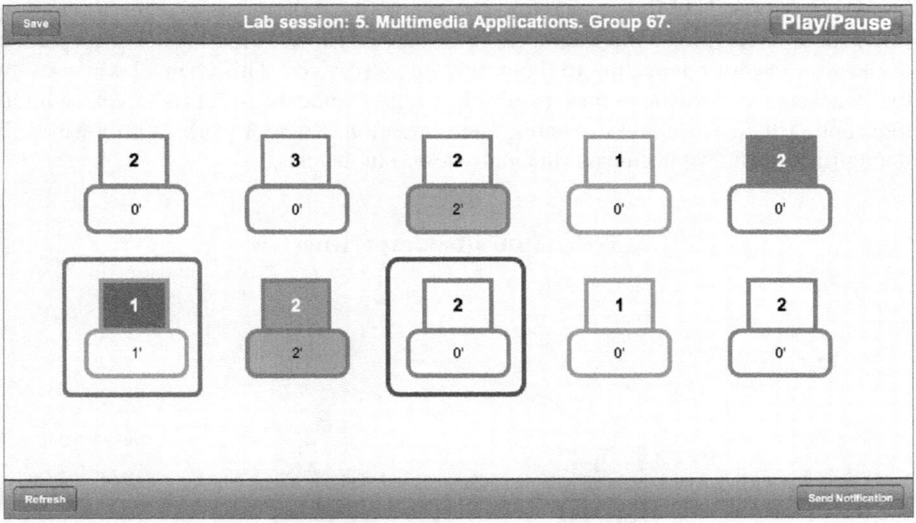

Fig. 2. Orchestration and feedback application interface

Any time, the teacher can catch a glimpse of the current progress of the class and the groups asking for help during the session. At a glance, (s)he can easily decide the next group of students to tutor, applying his/her own criteria. For example, higher priority can be given to either the students that have been waiting more time or to the students stuck at one of the first exercises (that is, delayed from the rest of the class), and also take into account the time already spent with each group. Once decided the group to assist, just touching the group will allow the system to track information about the time devoted to each group. If one problem is relevant for more students, and the progress state shows that it is the appropriate time, the notifications system can be used for making all the students aware of this relevant information. Having finished the tutoring action, the teacher decides the next group to attend or, if nobody is waiting for help, stop the timer that was measuring the time spent with the tutored group using the Play/Pause button in Fig. 2.

Following this workflow, several types of information will be stored in the system that will allow the teacher to reflect about the effectiveness of the session:

- Tutoring time, meaning time devoted in the session to solve doubts of the students personally, both for each student as the aggregate measure for the class.
- Performance of the students in the session, both individual and aggregated.
- Typical problems posed in this session (from submitted notifications).

4 Classroom Experience

The feedback and orchestration tool was used in the context of a Multimedia Applications course during two lab sessions. The students had to acquire some knowledge about HTML5, concretely the use of the 2D canvas and the drag&drop interaction, developing some examples in Google Chrome. A button was incorporated in the wording for connecting to the notifications services. This channel was used by the teachers to provide some feedback to the students at class level, without interrupting their work and disrupting their attention. Being a work in progress, only the notifications have been experimentally tested in the class.

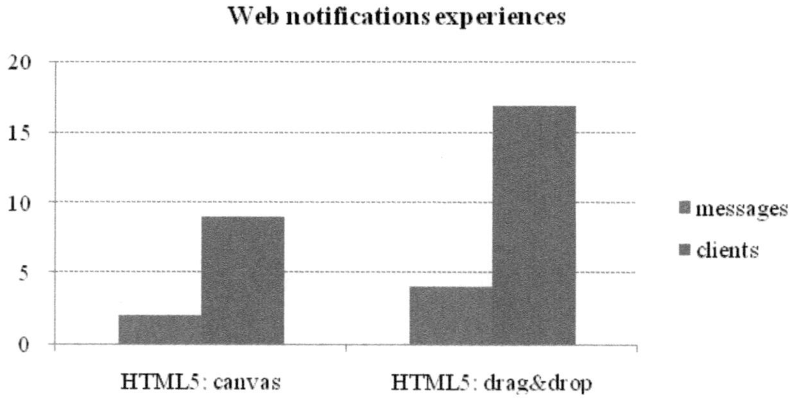

Fig. 3. Number of clients and messages sent in two experiences of websocket notifications

Figure 3 summarizes the use of notifications along the two sessions. In the first one, two notifications were sent and only 9 teams of 2 students each (18 students out of 40 in the class) connected to the notifications channel and received the feedback. In the second session, more feedback (broadcasted messages) was necessary (4 messages) as the drag&drop interaction in HTML5 is an advanced aspect. In this case, more students (17 groups, i.e. 34 students out of 40) connected to the server. There is an increasing trend in the addressed students (9 to 17) and the notifications sent (2 to 4), as users –both teachers and students– got familiar with the system and the needs for feedback increased.

5 Conclusions and Future Work

This paper presents a work-in-progress tool for lab sessions orchestration and provision of effective quick feedback to the participant students. The system usefulness is grounded by the state-of-the-art review and the fact that meets actual requirements by the course teachers. Apart from taking into account pedagogical aspects like orchestration and formative feedback, it also allows the collection of data useful for students assessment and quality improvement. It allows the teacher to be more aware of the problem-solving process followed by the students, and students can be assessed based on this process besides the product they deliver.

As future works in the orchestration and notification tool we have spotted two main lines of work. Regarding experimental validation, further evaluation of the expected benefits and utility of the system is to be done. Regarding application development, some improvements are being considered:

- Defer the delivery of notifications to the students depending on their progress in the activity, according to the information provided through the orchestration system; the underlying intention being to avoid diverting the students' attention.
- Create a visualization of the notifications flow; it will be much related to the difficulties that the students are encountering.
- Long term, create an intelligent tutoring system that sends feedback automatically to the students based on their progress in the lab session, their performance in previous exercises and common mistakes of the students.

Acknowledgments. This research has been partially supported by the project "Learn3: Towards Learning of the Third Kind" (TIN2008-05163/TSI) of the Spanish "Plan Nacional de I+D+i" and the Madrid regional project "eMadrid: Investigación y Desarrollo de tecnologías para el e-learning en la Comunidad de Madrid" (S2009/TIC-1650).

References

1. Yardi, S.: Whispers in the Classroom. In: McPherson, T. (ed.) Digital Youth, Innovation, and the Unexpected. The John D. and Catherine T. MacArthur Foundation Series on Digital Media and Learning, pp. 143–164. The MIT Press, Cambridge (2008)
2. Conole, G., Alevizou, P.: A literature review of the use of Web 2.0 tools in Higher Education. A report commissioned by the Higher Education Academy (2010)
3. Alavi, H.S., Dillenbourg, P., Kaplan, F.: Distributed awareness for class orchestration. In: Cress, U., Dimitrova, V., Specht, M. (eds.) EC-TEL 2009. LNCS, vol. 5794, pp. 211–225. Springer, Heidelberg (2009)
4. Sadler, D.R.: Formative assessment and the design of instructional systems. Instructional Science 18(2), 119–144 (1989), doi:10.1007/BF00117714.
5. Boston, C.: The concept of formative assessment. Practical Assessment, Research & Evaluation 8(9) (2002), `http://PAREonline.net/getvn.asp?v=8&n=9`

Using Online Digital Technologies to Build Knowledge: Lessons Learned from Three Initiatives

Marie Joubert

University of Bristol, UK

Abstract. This study draws on three initiatives in which groups of people collaborated, using online tools, to build knowledge. There are important differences between the initiatives; for example a) one involved school students and the others involved researchers b) the tools used were different and c) the aims and intentions were different. Despite these differences, there are interesting similarities between the outcomes of the initiatives; whereas all could be said to have reached their aims, two were less successful than the third in engaging the participants in collaborative knowledge building. It seems that the reasons for these differences are complex but are primarily related to two factors; the nature of the knowledge that was being built and the way the activity was set up. In particular it seems that ownership of the goals can account for the relative success of the third initiative.

Keywords: collaboration, knowledge building, wiki, discussion board.

1 Introduction

I believe that online collaboration, using technologies such as discussion boards and wikis, has much to offer in terms of sharing and building knowledge and learning. I see the social processes of sharing and building knowledge as closely related to learning, [1, 2], whether or not the activity is set up as a learning activity or as a work activity (e.g. research) where learning is a by-product of the activity. [3]

This paper introduces three different projects in which these technologies were used for collaborative knowledge building and varied in the degree to which they were successful. It explores the factors, and combination of factors, and that worked well and those that worked less well, comparing how knowledge-building activities were set up and how they developed.

2 Background and Related Work

The three initiatives were all set up carefully, taking into account research literature, which identifies potential pitfalls when using online technologies to collaboratively build knowledge. This literature falls into two key areas: the first is concerned with the processes of collaborating to produce knowledge, which crucially includes the

C. Delgado Kloos et al. (Eds.): EC-TEL 2011, LNCS 6964, pp. 430–435, 2011.

extent (number) of contributions, and the second with the nature of knowledge produced. The brief literature review below is framed within these two key areas.

The concern of much literature related to producing knowledge collaboratively using online technologies is about 'what works' in terms of the **extent of contributions.** The key points are that those concerned should be willing to share knowledge and are motivated to share knowledge [4, 5] and that trust is established between the collaborators. This seems to be particularly important in online collaboration where frequently the collaborators may not be known to one another [6, 7]. A further point is that in successful collaboration the goal is shared [8, 9].

Although online tools can be effective for aggregating knowledge, there is some debate about the **quality** of the knowledge aggregated. Some concerns relate to the accuracy of knowledge. Using Wikipeda as an example of knowledge aggregated using online tools, Don Fallis [10] suggests that 'serious concerns have been raised about the quality (e.g., accuracy, completeness, comprehensibility) of the information (p 1663). Fallis' article suggests that Wikipedia has been dismissed by much of the library and information science communities because it is seen as unreliable and argues that although Wikipedia fails rigorous tests of accuracy, it is 'quite reliable' and 'quite verifiable'(p 1669).

Other concerns relate to the superficiality of knowledge. For example, Anderson [11] argues that the 'Web of Content' (WoC) discourages 'a deep level of critical thinking' and the computer scientist, Jaron Lanier, in an essay about the dangers of elevating collectivism above merit suggests that he has seen 'a loss of insight and subtlety, a disregard for the nuances of considered opinions, and an increased tendency to enshrine the official or normative beliefs of an organization'[12, p 7].

3 Setting Up the Initiatives: A Comparison

This section describes the three projects discussed in this paper pointing out the similarities and differences between the three projects in the course of the discussion.

DAIS [13] was concerned with school students discussing ethical issues in science, assuming that discussion would help students develop both their argumentation skills and their understanding of different perspectives concerning ethical issues [14, 15]. Six experienced teachers initiated and encouraged online discussion (using a discussion board, http://beep.ac.uk) of these issues amongst their 16 – 17 year-old students, particularly encouraging the students to use high levels of argumentation [16]. The online discussions were analysed for this paper.

The second initiative aimed to develop a jointly held research vision and strategy for the STELLAR European Network of Excellence (http://www.stellarnet.eu/). In contrast to the DAIS example above, the participants were (mostly) researchers, and participation in the project was self-motivated, whereas to some extent the participation of the school students was required by their teachers. Further, although names used on the discussion boards were anonymous, in most cases the students knew one another, so it could be argued that there was little need to establish trust between them before the discussions took place. In STELLAR, however, the network was only in its fourth month, and hence many of the members did not know one another, so the work began with a day-long face-to-face meeting. The project, after considering alternatives, set up a wiki after the meeting for remote discussion related to the three main themes which underpin the scientific work of STELLAR. The contributions to the wiki provide the data for this paper.

In DAIS, the over-riding aim was to develop the argumentation skills of the students within the context of bio-ethics, and can be seen as being primarily about teaching and learning, whereas in STELLAR the aim was to produce a statement. It could be said that the students were aware of the goal of the activity, but whether or not this can be seen as a 'shared goal' is questionable. In STELLAR, however, the assumption was that all researchers shared the goal of the activity.

In DAIS, it was the teacher's role to start the discussion activity, to monitor contributions and encourage discussion. STELLAR, however, did not have an equivalent to the teacher but it was considered important to have one or more people to take on this role in order to mitigate against the risk of non-contribution and the 'D1.1 team' was set up to provoke discussion. In this way, these two initiatives addressed the risks of non-participation as described in Section 2 above. However, in terms of the nature of the knowledge built, the risk was less clear. The projects were not essentially about collecting facts and hence inaccurate contributions were not a risk. The projects were more about developing arguments, debate, insight and vision and did, perhaps, run the risks described by Anderson and Lanier in section 2.

The third example of a collaborating group was Trinity. In this group, two researchers specialising in education used online video analysis software to analyse video footage of two children engaged in a mathematical learning activity. They used a range of other online tools (e.g. online chat, wiki, blog) to support their research, and using an auto-ethnographic approach, investigated their own use of the various tools, particularly focusing on which tools (and combinations of tools) they chose to use and what for. In common with the STELLAR example, the participants were researchers and could be seen as self-motivated. However, whereas in STELLAR all members of the network (at the time 92) were potential contributors, in the case of Trinity there were only two researchers. As with STELLAR, it was assumed that both researchers shared the goal of the project. Unlike STELLAR, and more like DAIS, the researchers knew each other well. However, they did not have previous experience of using wikis and blogs and they recognised the potential risks associated with building knowledge together. They discussed in detail how they should handle the difficulties they foresaw in sharing thoughts and ideas as they were emerging (rather than well formed ideas, carefully argued and backed up). They developed a set of protocols or 'housekeeping rules', against which all their research activity took place.

Another variation between the groups was whether tools were chosen before the project began. In contrast to DIAS and STELLAR, the researchers in Trinity were free to use whichever tools they chose. They used mainly a messaging tool (MSN), a blog, a wiki and the video analysis tool. Over the period of about nine months, they worked together to analyse the video footage, meeting both face-to-face and online, as appropriate. All online discussions, emails, annotations on the video and blog and wiki posts, together with interview data, provide the data for this paper. In relation to the risks described in Section 2, non-participation was not considered to be a risk, but with respect to the nature of the knowledge produced, the risks were less clear.

4 What Happened: Results

The results are explored in terms of the extent of the participants' engagement and the nature of their engagement.

I turn first to DIAS to describe the **extent of contribution**. For five of the seven discussions in DAIS, the average number of posting per participant was fewer than 2. In general these postings were also very short and were often made within seconds of the previous posting. Two other discussions had a slightly higher average number of postings (almost 3). The factors that underlie this differences may relate to conditions for which the collaboration was established: one was set as homework and the comments were almost all made out of school hours, while the other, set up by the teacher in class, was distinct in that most contributions were made within about an hour of its having been started and therefore were likely to have been made in class.

The pattern of contributions to the wiki in the STELLAR initiative is comparable to that of DIAS, in the sense that few contributions were made and they tended to be over a very short time period. Analysis of the editing histories of the wiki pages indicates that most contributions took place within a short space of time and were usually very short. Overall about 20 people from STELLAR contributed to the wiki in the period of development from 22nd June to 6th July 2009. The majority of the contributions were made by a small number of people, usually within a short time frame. For example, the contributions to the three main pages (relating to the three research themes) had a total of six contributors, with an average of between three and four contributions each. In both these initiatives, the pattern of contributions can be described as 'scattergun' and although there were some exceptions, generally discussion not sustained and contributions were not very substantive.

In Trinity initiative, the pattern was very different. The data was analysed in terms of 'learning conversations', an idea borrowed from Gudeman & Rivera, [17] where each conversation was related to an aspect of the research process (e.g. analysis, housekeeping). The data suggests that the researchers used the tools they chose regularly and frequently, building up long conversations which would often begin in the morning and continue throughout the day.

The second key point was the quality or **nature of contributions**, in DAIS, the majority of comments were between one and two sentences. Generally they did not build on the contributions of others they did not draw on the scientific knowledge of the students and they did not display high levels of argumentation. Similarly, in STELLAR, contributions tended to take the form of paragraphs setting out the perspective of an individual. Contributions were seldom expanded with arguments, examples or references, by the original authors or other colleagues. There was little evidence of individuals challenging other people's contributions or questioning what they had said, but typically were more concerned with phrasing and style.

Once again, the Trinity example contrasts with the other two initiatives. Here the contributions built on previous contributions, answering questions and extending the discussion. In their online conversations, the researchers developed mutual understandings, both of the ways in which they were working and the content of the video footage they were analysing. The researchers took into account the contributions of the other and directly responding, hence collaboratively building their shared knowledge and understanding.

5 Discussion and Conclusions

The results section above suggests that, although the first two initiatives were successful in that they achieved what they set out to achieve (discussions on bio-ethical issues and a vision statement respectively), they were less successful than was hoped. Both the extent and the quality of contributions were disappointing. Trinity, was more successful. The collaboration between the two researchers involved was sustained and their contributions were substantive; although they were often short they demonstrate that the researchers were fully engaged. The difference is that in the first two, the contributions tended to be 'one-off' posting, whereas in Trinity the researchers were engaged in a series of conversations.

I suggest that the reasons for these differences are complex but are primarily related to two factors; the nature of the knowledge that was being built and the way the activity was set up. The knowledge in the school example tends to be belief-driven, and it can be difficult to construct an argument, for example, against a statement of belief. In the STELLAR example, the knowledge was a vision and we suggest that it is unlikely that one academic (researcher) would challenge the vision of another. In the Trinity example, however, the knowledge was more concrete and was directly related to the everyday practices of the researchers and hence it was easier to engage with the contributions of the other.

In terms of the way the activity was set up, in the school setting the teachers tended to begin a discussion with a question to elicit beliefs and it is not perhaps surprising that the students contributed beliefs. In the STELLAR example, the wiki was populated before the beginning of the initiative with relatively long, well crafted paragraphs which were perhaps difficult to pick apart and question. However, in the Trinity example, building the knowledge was initiated by the researchers, and was developed within a carefully developed 'housekeeping' system, which explicitly acknowledged the potential difficulties of building completely new knowledge, which may have involved some risk taking and exposure.

Further, and also related to the way the activity was set up, it seems that in the first two examples, the participants did not fully share the overarching goals of the activity. This is perhaps not unusual in a school setting where students are frequently given a task and the success criteria are made clear, but the overarching goals are not discussed. In the STELLAR example, although the goal was clear, it seems that many of the researchers were not committed to this shared goal and saw the creation of the vision statement as the responsibility of others. In Trinity, on the other hand, both researchers were committed to the shared goal, which they had helped to shape when the project was first launched. It could be said that they had some 'ownership' of the shared goal, whereas in the other two initiatives there was less sense of ownership. Perhaps this is the most important lesson we can learn from this comparison of the three initiatives.

References

1. Sorensen, E.K., Takle, E.S.: Collaborative knowledge building in web-based learning: Assessing the quality of dialogue. International Journal on E-learning 1, 28–32 (2002)
2. Scardamalia, M., Bereiter, C.: Schools as Knowledge-Building Organizations. In: Keating, D., Hertzman, C. (eds.) Today's Children, Tomorrow's Society: The Developmental Health and Wealth of Nations, pp. 274–289. Guilford, New York (1999)
3. Eraut, M.: Informal learning in the workplace. Studies in Continuing Education 26, 247–273 (2004)
4. Nov, O.: What motivates Wikipedians. Communications of the ACM 50(11), 60–54 (2007)
5. Miller, P.: Web 2.0: Building the New Library. Ariadane 45, 1–6 (2005)
6. Kwan, M., Ramachandran, D.: Trust and Online Reputation Systems. In: Goldbeck, J. (ed.) Computing with Social Trust - Human Computer Interaction Series, pp. 287–311. Springer, London (2009)
7. Coleman, D., Levine, S.: Collaboration 2.0: Technology and Best Practices for Successful .... . Happy About, Cupertino (2008)
8. Wagner, C., Majchrzak, A.: Enabling Customer-Centricity Using Wikis and the Wiki Way. Journal of Management Information Systems 23, 17–43 (2007)
9. Wagner, C.: Breaking the Knowledge Acquisition Bottleneck Through Conversational Knowledge Management. Information Resources Management Journal 19, 70–83 (2006)
10. Fallis, D.: Toward an Epistemology of Wikipedia. Journal of the American Society for Information Science 59, 1662–1674 (2008)
11. Anderson, P., Hepworth, M., Kelly, B., Metcalfe, R.: What is Web 2.0? Ideas, technologies and implications for education. JISC Technology and Standards Watch (2007)
12. Lanier, J.: Digital Maoism: The hazards of the new online collectivism. The Edge 183 (2006)
13. Wishart, J., Green, D., Joubert, M., Triggs, P.: Discussing Ethical Issues in School Science: An investigation into the opportunities to practise and develop arguments offered by online and face-to-face discussions. International Journal of Science Education, Part B 1, 47–69 (2011)
14. Levinson, R.: Key Issues in Bioethics A Guide for Teachers. Routledge Falmer, London (2003)
15. Newton, P., Driver, R., Osborne, J.: The place of argumentation in the pedagogy of school science. International Journal of Science Education 21, 553–576 (1999)
16. Erduran, S., Simon, S., Osborne, J.: TAPping into argumentation: Developments in the application of Toulmin's argument pattern for studying science discourse. Science Education 88, 915–933 (2004)
17. Gudeman, S., Rivera, A.: From Car to House (Del coche a la casa). American Anthropologist 97, 242–250 (1995)

On Psychological Aspects of Learning Environments Design

Miloš Kravčík and Ralf Klamma

RWTH Aachen University, Informatik 5, Ahornstr. 55,
52056 Aachen, Germany
{kravcik,klamma}@dbis.rwth-aachen.de

Abstract. Psycho-pedagogical theories provide a valuable input for learning. Our goal is to identify requirements for learning environments, conceptually evaluate them, and integrate them into a psycho-pedagogically sound framework as a basis for the development of a highly responsive open learning platform. In this paper we first describe how personal learning processes are conceptualized in our Psycho-Pedagogical Integrated Model. To put our research into a context we review relevant literature on human decision making. Finally, we summarize the impact of our findings on the design of learning environments, considering principles of Self-Regulated Learning.

Keywords: Behavioral and Cognitive Psychology, Self-Regulated Learning, Design of Learning Environments.

1 Introduction

The research in behavioral and cognitive psychology [1, 2, 3] found out that humans err predictably and this knowledge can be harnessed to support them. The interplay between the different systems driving our thinking processes generates various fallacies. But choice architecture can influence options in a way that will support choosers to act in their own interest, preserving freedom of choice [3].

We see a challenge in investigation of possible consequences for requirements of learning environments. Our aim is to investigate relevant psychology outcomes for the design of future responsive learning solutions. Designers and developers of e-learning environments should take into account that humans have cognitive biases, which can make certain decisions very difficult, and therefore a suitable choice architecture is crucial. The current context provides clues for learning and an open issue is how technology should support adaptive learning in an appropriate way.

In the next paragraphs we first introduce our conceptualization of personal learning processes. A literature review shows principles of the human mind that generate heuristics and biases, but also outlines how choice architecture can help us to deal with uneasy choices. Afterwards we summarize the consequences for design of learning environments.

C. Delgado Kloos et al. (Eds.): EC-TEL 2011, LNCS 6964, pp. 436–441, 2011.

2 Learning Process Model

Current demands of lifelong and personalized learning made *Self-Regulated Learning* (SRL) [8] increasingly important in order to give the learner greater responsibility and control over all aspects of the learning process. SRL is guided by meta-cognition, strategic action (planning, monitoring, and evaluating progress), and motivation to learn. Self-regulated learners are aware of their academic strengths and weaknesses, and they have a repertoire of strategies they appropriately apply to tackle the day-to-day challenges of academic tasks.

One of the key objectives in the ROLE project [4] is to support the individual assembly of accessible learning services, tools and resources in responsive open learning environments, which permit personalization of the entire learning environment and its functionalities, i.e. individualization of its components and their adjustment or replacement by alternative solutions.

A psycho-pedagogically sound framework for supporting the individual composition of learning services [7] is being developed in order to support SRL. Five key interconnected aspects form the main principles of the ROLE *Psycho-Pedagogical Integration Model* (PPIM): Personalization and adaptability, Guidance and freedom, Motivation, Meta-cognition and awareness, and Collaboration and good practice sharing. The ROLE SRL process model is learner-centric, made up of three meta-cognitive learning phases [8]: Forethought, Performance, and Reflection. Including also the idea of self-profile for personalization, our SRL Process Model has four phases: Learner profile update, Selection of learning resources, Learning with selected resources, and Reflection on learning achievements.

From the implementation point of view the concept *Personal Learning Environment* (PLE) [10] is in line with the SRL requirements. PLE describes the tools, communities, and services that constitute the individual educational platforms that learners use to direct their own learning and pursue educational goals. Compared to course-centric solutions (like Learning Management Systems) PLE is learner-centric, i.e. students are in charge of their learning process, emphasizing meta-cognition in learning. From the technological point of view PLE is based on Web 2.0, usually implemented in the form of customizable portals or dashboards.

3 Behavioral and Cognitive Psychology

In this paragraph we review literature from several authorities in the field of behavioral and cognitive psychology. From our perspective these popularizing books provide a nice overview of the key outcomes in this field. They show principles of the human mind that generate heuristics and biases [1, 6], but also outline how choice architecture can help us to deal with uneasy choices.

3.1 Human Mind

Human cognitive biases usually have a pattern – they are lawful, regular, and systematic. We can utilize these findings for our benefit, because when humans err predictably, this knowledge can be harnessed to help them [3]. Psychologists and neuroscientists [5] distinguish two kinds of thinking:

1. *Automatic system (AS):* gut reaction, i.e. intuitive, instinctive (associated with the oldest parts of the brain) – uncontrolled, effortless, associative, fast, unconscious, skilled
2. *Reflective system (RS):* conscious thought, i.e. rational, deliberate – controlled, effortful, deductive, slow, self-aware, rule-following

These two systems compete and complement each other, depending on the context, including available resources (e.g. time, energy). Research results in neuroscience shows that human decision making is based on the cognitive struggle between the brain's emotional and rational systems [5]. They can correct each other and we should learn how to control our thinking on a meta-level.

Apparently, the immediate context of decision making matters very much. A subtle influence can radically shift how people act. But the influences are quite predictable. Usually cheap metaphors and stories capture our minds. This has an enormous meaning for learning, but there is also a danger of oversimplification and misleading. Another issue is that school contexts for learning are very often highly artificial.

3.2 Choice Architecture

Now when we are aware of the limitations of the human mind, we may ask what practical problems we encounter as a consequence and how we can address them. The main candidates for alerts are uneasy choices [3]: *Delayed effects* (choices and their consequences are separated in time), *Difficulty* (some problems in life are complex), *Infrequency* (some decisions are rare), *Poor feedback*, and *Unclear impact* (ambiguous relation between a choice and its consequence).

Choice architecture has a huge importance for our decisions [3]. As a solution *libertarian paternalism* has been proposed, which preserves liberty and tries to influence choices in a way that will make choosers better off, as judged by themselves. This influence can be realized via suitable alerts or nudges. Such a nudge should alert people's behavior in a predictable way and at the same time it should be easy and cheap to avoid. The golden rule of libertarian paternalism says: offer nudges that are most likely to help and least likely to inflict harm. These are the basic principles of choice architecture: *Default options* (usually a lot of people end up with it), *Expect error* (a system should be as forgiving as possible), *Give feedback* (the best way how to improve our performance), *Understand mappings* from choice to welfare (options should be comprehensible), *Structure complex choices* (use the judgments of similar people to filter options), and *Incentives* (put the right incentives on the right people). So choice architecture is pervasive, unavoidable, and can have massive effects on people's behavior. The rules of libertarian paternalism can help designers and developers to preserve freedom of choice by providing gentle nudges.

4 Design of Learning Environments

A conceptual evaluation of the collected results is a necessary prerequisite for further implementation steps in the ROLE project. In this section we aim to summarize the impact of the above mentioned findings on the design of learning environments, considering the ROLE learning process model. In each of its four phases the learner should be supported by suitable services mainly through appropriate recommendations [9] in order to follow the principles of self-regulation. As for the presentation of the recommendations choice architecture can play a crucial role, personalization and adaptability based on the learner's profile cannot be omitted.

Learners can encounter uneasy choices, like selection of a larger module, too many alternatives, lack of feedback and assessment, or unclear learning effect. In such cases it is required to simplify complex choices taking into account current constraints and learner's preferences. Suitable default options should be selected by experts and adapted accordingly to the learner and context. Social recommendations can narrow down the selection space automatically. Other suitable interventions include tutoring and peer assisted learning. The main principles include freedom of choice and a gentle nudge, when the freedom becomes overwhelming. A good solution requires identification of uneasy choices and application of a suitable choice architecture.

4.1 Learner Profile Update

In the first phase the learner profile information is defined or revised, considering input from various sources. This is crucial for personalization and adaptivity of the provided services. The profile data includes information on the learning goals, learner's competences, knowledge, skills, preferences, learning history and progress, etc. This data can be updated by the learner, her tutors and peers, as well as relevant services and tools. From the SRL perspective it is crucial that the learner herself decides who can modify her learner model and how. To get enough information on the learner, especially from the peers, it is required to put the right incentives on the right people. This feedback is a clear indication of the learner's performance. What is really critical here is appropriate visualization of the relevant changes, which has to be comprehensible for the particular learner, but should not cause unnecessary distractions.

4.2 Selection of Learning Resources

Reasonable *planning* of the learning process is needed, taking into account the learning objectives as well as existing constraints, like time and money. The main danger here is that people tend to make too optimistic plans, overlooking possible hurdles and unpredictable issues. In any case, the learner should be informed on the planned outcomes, consequences, and side-effects. As humans are emotional beings, they can be attracted and motivated by suitable design of all the dimensions of the learning experience – learning approach, environment, and all kinds of learning resources. It is crucial to choose an authentic context for learning and adjust it accordingly to the current constraints (available resources) as well as learner's abilities and preferences. This means that personalization, adaptation, and recommendation must be user and context dependent. In this phase of the learning process the learner searches for relevant learning resources to create her PLE. This includes communities, learning plans, activities, services, tools, content, etc. Based on her learner profile she obtains recommendations and from them selects the preferred ones. But too much freedom may become overwhelming and contra productive, therefore the learner can delegate some of these tasks to tutors, peers, and services.

Because of human fallacies the choice of the learner can be influenced by default options, understanding of mappings from choice to welfare, and structure of complex choices, where collaborative filtering based on experience of similar people in similar contexts can be extremely helpful [1].

4.3 Learning with Selected Resources

If the learning is to be deep, it is necessary to provide various perspectives on the same topic. The key knowledge and skills have to be repeated to be remembered. Therefore, redundancy is required on various levels – learning communities, plans, activities, resources, as well as their media representations. People like stories, but they may be misleading and oversimplified. Thus it is crucial to choose suitable analogies and metaphors. In addition to confirming examples, also counter-examples are important to demonstrate the validity of hypotheses and to assess risks. Both content and form of the learning experience are important and need to be chosen according to the learner's traits and the current context. SRL prefers recommendations to automatic adaptation, but structuring of complex choices should be personalized and adapted. Personalization is based on the learner's knowledge, experience, and preferences. Adaptation deals mainly with the context taking into account the current constraints and available resources. On one side it can mean selection of the best context to facilitate learning and make it authentic. On the other hand it includes also an appropriate adjustment of the current context according to the changing circumstances. In this phase the learner works on the selected learning resources. This is the actual *learning* phase, including also assessment activities that can change the learner profile. Good explicit assessment should provide neutral choices (e.g. multiple choice tests, free text answers). The learner has always the control over her profile and can adjust it accordingly. During the learning process the progress has to be monitored and the plan adjusted if needed. Default options and understanding mappings from choice to welfare play a key role here. Appropriate feedback to performed actions is the best way how to improve the performance of humans.

4.4 Reflection of Learning Achievements

Regular checking of learning outcomes is necessary to provide feedback by means of notifications on the progress, success, or failure. To find out the causes of learning success, one has to consider both successful and unsuccessful learners. The information retrieval principles and techniques can be used to find the distinguishing factors between those two. In this phase the learner is *reflecting* on the chosen learning strategies and her achievements, by trying to understand mappings from the former choices to welfare. She gets various kinds of feedback on her progress (e.g. learning history, assessment performance), processes it, and provides her own feedback, which may include updates in the learning plan or in the learner profile. This stimulates meta-cognitive activities of the learner.

5 Conclusion

Our aim was to investigate a possible impact of behavioral and cognitive psychology outcomes on design of learning environments. To overcome cognitive biases of humans a suitable choice architecture is crucial and libertarian paternalism has been suggested as a solution. It should preserve freedom of choice by providing suitable recommendations that are easy to avoid. We attempted to relate these findings to design requirements for learning environments, especially when self-regulated learning is considered. We think that also in this field decision making is essential for the quality of the learning outcome and should be supported properly to optimize learning for benefit of the learner. The conceptual evaluation of the achieved results is a necessary prerequisite for further implementation steps in our development of responsive open learning environments.

Acknowledgments. The research described in this paper is funded through the ROLE Integrated Project (grant agreement no 231396) of the European Union. We like to thank all project partners for the fruitful cooperation.

References

1. Gilbert, D.: Stumbling on Happiness. Knopf (2006)
2. Ariely, D.: Predictably Irrational. Harper Collins, New York (2008)
3. Thaler, R.H., Sunstein, C.R.: Nudge: Improving Decisions about Health, Wealth, and Happiness. Yale University Press, New Haven (2008)
4. ROLE Project web page, `http://www.role-project.eu`
5. Lehrer, J.: How we decide. Houghton Mifflin, Boston (2009)
6. Taleb, N.N.: The Black Swan: The Impact of the Highly Improbable. Random House (2007)
7. Fruhmann, K., Nussbaumer, A., Albert, D.: A Psycho-Pedagogical Framework for Self-Regulated Learning in a Responsive Open Learning Environment. In: Proc. International Conference eLearning Baltics Science, Rostock, Germany (2010)
8. Zimmermann, B.J.: Becoming a self-regulated learner: An Overview. Theory into Practice 41(2), 64–70 (2002)
9. Adomavicius, G., Tuzhilin, A.: Toward the Next Generation of Recommender Systems. IEEE Transactions on Knowledge and Data Engineering 17(6), 734–749 (2005)
10. van Harmelen, M.: Personal Learning Environments. In: Proceedings of the 6th International Conference on Advanced Learning Technologies (ICALT 2006). IEEE, Los Alamitos (2006)

Evaluating Adaptive Work-Integrated Learning Systems: From the Lab to the Field

Barbara Kump[1,2], Tobias Ley[3], and Stefanie Lindstaedt[1,2]

[1] Know-Center GmbH, Inffeldgasse 21a, 8010 Graz, Austria
bkump@know-center.at, slind@know-center.at
[2] Knowledge Management Institute, TU Graz, Inffeldgasse 21a, 8010 Graz, Austria
bkump@tugraz.at, slind@tugraz.at
[3] Center for Educational Technologies,
Tallinn Univ., Narva Mnt 25, 10120 Tallinn, Estonia
tley@tlu.ee

Abstract. Evaluation frameworks have been presented that suggest layered evaluation of adaptive systems along two dimensions: (i) the software development cycle, and (ii) the component of the adaptive system that shall be looked at. We argue that a third dimension is crucial: the question whether an evaluation should take place in the lab or in the field. We present a refined systematization of evaluation approaches using the evaluation of the adaptive WIL system APOSDLE as an example.

Keywords: Layered Evaluation, Adaptive Systems, Work-integrated Learning.

1 Evaluating Adaptive Systems in Naturalistic Settings

New possibilities of learning support are being explored in traditional educational settings and various other areas such as *work-integrated learning* (WIL) [1]. WIL refers to informal learning at the workplace that is truly integrated in work processes and makes use of existing resources (e.g., project results) as well as co-workers. What most of the suggested functionalities for WIL systems have in common is that they provide adaptive learning support, i.e. support that is adapted to the worker's needs.

It is now generally agreed that adaptive systems must be evaluated in order to check whether they provide the intended results [2]. Most existing evaluation approaches recommend experimental methods in lab settings as a way to cope with the complexity of adaptive systems and to identify these aspects of the adaptive system that need to be improved. But in case the system is to be deployed in a naturalistic setting like WIL, this strategy has some restrictions such as the generalizability of the findings from the lab to the 'real world'. We argue that a consideration between field and lab settings is needed to trade-off issues of external and internal validity in an evaluation study. None of the existing evaluation strategies takes into account the lab versus field dimension. In this paper we present a systematization of evaluation strategies that takes into account: (i) the development phase, (ii) the component that shall be evaluated, and (iii) the setting (lab vs. field).

C. Delgado Kloos et al. (Eds.): EC-TEL 2011, LNCS 6964, pp. 442–448, 2011.

We illustrate our systematization with the example of the empirical research program that we carried out with the APOSDLE system, an adaptive system to support WIL.

2 A Systematization of Approaches for Evaluating Adaptive Work-Integrated Learning Systems, and an Example

Since the development of adaptive systems, various scholars have recognized that in order to evaluate an adaptive system and to draw conclusions on how to improve it, it is necessary to reduce complexity. Two main approaches exist to reduce complexity: (a) component-wise evaluation of adaptive systems [2] [3] [4], and (b) evaluation along the development cycles of (adaptive) software systems [4] [3].

Table 1. A Systematization of Possible Evaluation Studies for Adaptive Systems

Phases		Software Engineering	Knowledge Engineering	Setting
Requirements and Design	Development	Requirements engineering and user-centered design methodologies	Scoping of domain and models Knowledge Acquisition and Formalization	Field
	Evaluation	Requirements Validation	Domain model evaluation (Validity and Reliability Checks)	
Low Fidelity Prototypes	Development	Paper-based mockups	Design of inference and adaptation algorithms	Lab
	Evaluation	Validation of Paper-based mockups	Paper-based experimental studies of domain, user and adaptation model	
High Fidelity Prototypes and Components	Development	Running System Prototypes or components	Implementation of models and algorithms	Lab
	Evaluation	Usability Studies with single components	Simulation studies (e.g. simulation of user model algorithms) Experimental user studies (e.g. control group designs of different adaptation model)	
System Introduction and Operation	Development	Introduction of full scale system	Tailoring of algorithms to application setting	Field
	Evaluation	Usability studies and field studies with running system	Large scale field evaluations, continuous evaluation with log files	

Influential scholars in this field have argued that lab evaluations are needed to evaluate adaptive systems (e.g., [2]). In our view, for systems that shall be used in the 'real world', a systematic evaluation strategy combining both lab and field studies is necessary as eventually the effectiveness of these systems can only be established in a naturalistic field setting. That is, experiments in the lab are needed in order to complement the information obtained from other empirical methods. It is important to acknowledge that different empirical methods and data sources (field data, survey data and experiments, both lab and field) are complements, not substitutes, and that there is no hierarchy among these methods [6].

We suggest a systematization of evaluation methods that are suitable at different stages of the development lifecycle, and that guide evaluators of adaptive work-integrated learning systems in designing rigorous evaluation strategies combining lab and field studies. This systematization is shown in Table 1. Following van Velsen et al. [2], one can distinguish between the following four phases in the development lifecycle: (1) Requirements and Design, (2) Low Fidelity Prototyping, (3) High Fidelity Prototyping, and (4) System Introduction and Operation. Apart from distinguishing software development phases, in our view, it is necessary to separate software development from model development. Adaptive systems make use of models that encode domain knowledge as well as adaptation rules in an explicit form. Concretely, adaptivity in a learning system results from a complex interaction between a *domain model*, a *user model* and an *instructional model*. As Table 1 suggests, we argue that evaluation in each of the development phases should focus on both, the software engineering and the knowledge engineering strand.

2.1 The Adaptive Work-Integrated Learning System APOSDLE

APOSDLE is an adaptive WIL system to support self-directed learning at knowledge intensive workplaces. APOSDLE has been instantiated in five different domains (like aircraft simulation and innovation consulting) and it displays domain specific learning hints according to the task the user is currently performing. The task is either automatically detected through analysis of keystrokes and opened desktop applications, or selected manually from a list. For a task at hand, learning goals are suggested to a user in an adaptive manner, taking into account the user's knowledge state as stored in the user model. Given a learning goal, APOSDLE suggests learning resources or people that shall support the user in acquiring the learning goal.

2.2 Evaluating Adaptivity throughout the APOSDLE Development Lifecycle

2.2.1 Phase 1: Requirements and Design

For Phase 1 of the software development lifecycle, the software design methodology included a number of user-centric methods, such as creativity workshops, use cases, scenarios, personas, and interface mockups. Requirements elicitation was carried out following a user-centric, innovation driven requirements engineering methodology.

For the aspect of model development in Phase 1, we have suggested a knowledge engineering methodology for adaptive learning systems [7]. Evaluation forms an inherent part of this methodology that has been further elaborated [8]; specifically it includes validity and reliability checks that can be integrated into the modeling process. The research questions in focus at this stage are whether the domain model has construct and application validity. For construct validity, it is possible to check the consistency of several experts providing independent versions of the domain model (or parts thereof), or to check whether one expert has been consistent in providing the model. In Phase 1 of APOSDLE's development process, each of the evaluation methods described in [8] was applied in one or several out of 4 case studies in 4 different domains such as statistical data analysis, and aircraft simulation.

Findings from the model evaluation led to improved domain models: models were more complete, more consistent, and because most model validation techniques involve multiple experts, the improved models were agreed among several experts.

2.2.2 Phase 2: Low-Fidelity Prototyping

In Phase 2 of the software development cycle, (Low-fidelity Prototyping), it may be the case that specific interactions have been designed (e.g. how items will be recommended and displayed), but not necessarily implemented yet. This also means no running system is available at this stage, but instead the interaction may have been sketched as a paper mockup or an interaction diagram.

For this phase, we have suggested paper-based experimental studies in which the domain, user and adaptation model can already be evaluated. Here, the research question is: Is the adaptation provided by the models appropriate? In [9], we asked a group of requirements engineering students to perform a number of typical exercises that were related to the tasks modeled in the APODLE domain model. Before and after completing the exercises, they were asked to complete task and skill self-assessments. They were also asked to indicate which additional knowledge they would have required to perform better. All this was done using a paper-based test procedure. The exercises were later graded by an expert on requirements engineering.

We had to conclude from this study that students were not able to realistically self-assess their performance before conducting the exercises and that there was only a moderate correlation of self-assessment and real performance after completion. From the learning goals they had specified, we could derive general support for the domain model (summative focus), i.e. the mapping between tasks and skills, and were able to identify additional skills that had been missing in the domain model (formative focus). Finally, we used the self-assessment scores and the task performance to test some algorithms for task-based skill assessment. Data from the study were analyzed using a low-fidelity prototype in order to identify the best candidates of a set of inference and adaptation algorithms for APOSDLE.

2.2.3 Phase 3: High-Fidelity Prototyping

At this stage it is possible to perform evaluations of single components with or without the input of the user. For APOSDLE, we have conducted simulation studies of user model algorithms which were useful to and fine-tune the algorithms.

Furthermore, experimental user studies in controlled settings can be conducted at this stage. The research question is whether the adaptation together with the interaction provided by the software is appropriate and whether the adaptation improves performance. In [10] we present a laboratory based experimental study of the adaptation mechanism conducted as a control condition design with APOSLDE's high-fidelity prototype. Three adaptation conditions differed in the type of learning goals that were recommended to the learners after they had selected a task: a *personalized condition* where learning goals were ranked according to domain model and user model, a *fixed condition* where learning goals were chosen according to the domain model, but randomly ranked, and a *control condition* where learning goals were randomly chosen. It turned out that both perceived support and task performance were higher in the personalized and fixed conditions than in the control condition. Users selected more learning goals in the fixed as compared to the personalized condition. That way, the study did lend support to the effectiveness of the adaptation model. The question remains if this also holds in the natural use of the system.

2.2.4 Phase 4: System Introduction

In Phase 4, the system is ready to be put to usage in a real setting, and long-term field evaluations can be carried out.

We have carried out a field evaluation of APOSDLE's third prototype consisting of three individual studies [11]. We evaluated the validity of the user model and the accuracy of the adaptation decision of APOSDLE's third (and final) prototype in a small innovation management company in Austria. In order to gather information about the level of accuracy of the user model of six participants who had used APOSDLE for three months, the inferred user model was compared with external values (self- and peer-assessments). The outcomes showed a moderate agreement of automatically detected working areas in the user model with the self- and peer-assessed working areas. A rather low agreement was found for the inferred expertise levels and the self- and peer-assessed expertise levels in these working areas.

With the same system, a frequency log data analysis was applied for investigating whether recommended items, i.e. 'learning goals' in a higher ranking position, were chosen more often than 'learning goals' with a lower ranking position. All involved participants were using APOSDLE for three months. The outcome confirmed the assumption that the participants more often chose learning goals which were preferentially suggested by the system and therefore had a higher ranking position.

Additionally, a semi-structured interview combined with a qualitative approach of a replay-of-use method was chosen where participants from one of the companies evaluated the usefulness of suggested learning goals. The aim was to verify the results of the log data analysis and gather specific appraisal of (in-) appropriate suggested learning goals relating to their ranking position as well as their expertise. The participants had to rate each learning goal as appropriate or inappropriate help for performing the selected task. In all, assessments were available for ten different tasks (each with a corresponding list of one or several 'learning goals'). Generally, it appears, that the lower the learning goal is ranked (bottom of the list), the less likely the learning goal is assessed as appropriate. The results indicated that the adaptive decision at this stage was satisfying for the users.

3 Conclusions and Outlook

Our work illustrates that the research questions were shaped over the lifecycle of development, and how with each phase additional complexity was introduced in the evaluation strategy. Initially, the focus has been on the domain model. Similar to an early usability design, the focus was on involving experts in checking basic assumptions of the model. In Phase 2, it was then possible to validate some of the basic mechanisms with a paper-based lab study to find out whether assessment and learning need analysis would perform as planned. Also, it was possible at this stage to test several alternative inference algorithms for task-based skill assessment. Once a running prototype is available, one can start to ask whether the use of the system actually improves performance or learning. Hence, interaction with the software enters into the equation. Experimental lab studies are useful at this stage because sufficient control can be exerted over the conditions of use. This controlled evaluation setting also fits well into the stage of system development (high fidelity prototyping) as the software may not be mature enough at this stage to allow more large scale introduction. Finally, in Phase 4, it is necessary to ask whether the results obtained in a controlled lab setting transfer to the field. In a naturalistic environment, a number of additional influencing factors and constraints complicate evaluation matters. For example, whether there is sufficient time for users to engage in the software or whether recommendations are actually useful in a work setting are questions that can only be answered in a field study.

Acknowledgements. The Know-Center is funded within the Austrian COMET Program - Competence Centers for Excellent Technologies - under the auspices of the Austrian Federal Ministry of Transport, Innovation and Technology, the Austrian Federal Ministry of Economy, Family and Youth and by the State of Styria. COMET is managed by the Austrian Research Promotion Agency FFG.

References

1. Lindstaedt, S.N., Aehnelt, M., de Hoog, R.: Supporting the Learning Dimension of Knowledge Work. In: Cress, U., Dimitrova, V., Specht, M. (eds.) EC-TEL 2009. LNCS, vol. 5794, pp. 639–644. Springer, Heidelberg (2009)
2. Brusilovsky, P., Karagiannidis, C., Sampson, D.: The Benefits of Layered Evaluation of Adaptive Applications and Services. In: Workshop Proc. of UM 2001, pp. 1–8 (2001)
3. Paramythis, A., Weibelzahl, S., Masthoff, J.: Layered Evaluation of interactive adaptive systems: framework and formative methods. UMUAI 20, 383–453 (2010)
4. Weibelzahl, S., Lauer, C.U.: Framework for the evaluation of adaptive CBR-systems. In: Proc. of GWCBR 2001, Baden-Baden, Germany, pp. 254–263 (2001)
5. Van Velsen, L., Van Der Geest, T., Klaassen, R., Steehouder, M.: User-centered evaluation of adaptive and adaptable systems: a literature review. The Knowledge Engineering Review 23(3), 261–281 (2008)
6. Falk, A., Heckman, J.: Lab experiments are a major source of knowledge in the social sciences. Science 326, 535–538 (2009)

7. Ley, T., Kump, B., Albert, D.: A methodology for eliciting, modelling, and evaluating expert knowledge for an adaptive work-integrated learning system. Int. J. of Human-Computer Studies 68(4), 185–208 (2010)
8. Kump, B.: Evaluating the Domain Model of Adaptive Work-integrated Learning Systems. Unpublished Dissertation. University of Graz, Graz (2010)
9. Ley, T., Kump, B., Maas, A., Maiden, N., Albert, D.: Evaluating the Adaptation of a Learning System before the Prototype Is Ready: A Paper-Based Lab Study. In: Houben, G.-J., McCalla, G., Pianesi, F., Zancanaro, M. (eds.) UMAP 2009. LNCS, vol. 5535, pp. 331–336. Springer, Heidelberg (2009)
10. Ley, T., Kump, B., Gerdenitsch, C.: Scaffolding self-directed learning with personalized learning goal recommendations. In: De Bra, P., Kobsa, A., Chin, D. (eds.) UMAP 2010. LNCS, vol. 6075, pp. 75–86. Springer, Heidelberg (2010)
11. Pabst, C.: Validating the Adaptation Model of a Work-integrated Learning Environment: A Field Study. Unpublished Thesis, Dep't of Psychology, Univ of Graz (2010)

Implicit and Explicit Memory in Learning from Social Software: A Dual-Process Account

Tobias Ley[1,2], Stefan Schweiger[2], and Paul Seitlinger[3]

[1] Center for Educational Technology, Tallinn University, Narva Mnt 25, 10120 Tallinn, Estonia
tley@tlu.ee

[2] Cognitive Science Section, University of Graz, Universitätsplatz 2, 8010 Graz, Austria
stefan.schweiger@edu.uni-graz.at, tobias.ley@uni-graz.at

[3] Knowledge Management Institute, Graz University of Technology, Inffeldgasse 21a,8010 Graz, Austria
paul.seitlinger@tugraz.at

Abstract. Inspired by a recent surge to understand social cognitive processes in collaborative knowledge building, we have devised an experiment in which students learned from contents of a wiki. One of the informative results we observed was a dissociation between implicit and explicit memory measures that we used to track student's learning: an association test, and the drawing of concept maps, respectively. We put these initial results in the context of experimental research in cognitive psychology and show how the co-evolution model (Cress and Kimmerle, 2008) could account for them. With several network measures, we also suggest some ways of how to measure assimilation and accommodation, both in internal and external knowledge representations.

Keywords: Co-evolution model, implicit knowledge, network analysis.

1 Collaborative Knowledge Building in the Use of Social Software

With the recent rise of social software and its use for educational purposes in schools, universities and companies, there has been a renewed interest in collaborative knowledge building. Cress and Kimmerle [3] have recently suggested a co-evolution model of collaborative knowledge building in wikis and other social software systems [6]. Knowledge building is understood as a co-evolution of internally and externally represented knowledge by means of internalization and externalization processes.

Drawing on Piaget's theory of cognitive development, it is assumed that in the face of making sense out of encountered information with prior knowledge happens through internal processes of assimilation and accommodation. In assimilation, an existing schema is activated to encode information encountered. In accommodating, on the other hand, an existing schema is restructured or a new one is generated because the encountered information does not fit any existing schema. Similarly, Cress and colleagues propose external mechanisms of collaborative knowledge building in the wiki which they also differentiate into (external) assimilation and accommodation where the former can be characterized by mainly quantitative changes to the wiki, and the latter by qualitative restructuring.

C. Delgado Kloos et al. (Eds.): EC-TEL 2011, LNCS 6964, pp. 449–454, 2011.

Inspired by this model, we devised an experimental study in which students learned from a wiki on the subject of narratology. Our particular focus in this study was on measuring internal assimilation and accommodation: we were interested in what students would learn from the content, how their internal knowledge representation would change as a result of interacting with the wiki, and how this could be measured. Encouraged by our previous research in memory processes in the use of social software [8], we used two tests from two distinct classes of memory tests to measure knowledge of our subjects. One test was assumed to be more sensitive towards explicit knowledge that students had gained (drawings of concept maps), and the other assumed to be more sensitive towards implicit knowledge. For the latter, we employed an association tests in which subjects are asked to freely associate words given a certain stimulus word [8]. The number of associations, then, is a measure of the strength of implicit representation in memory for the stimulus word.

One of the somewhat surprising results of this study was a dissociation between the implicit and the explicit memory measure. In this paper, we will discuss these dissociations because they can advance theoretical understanding of collaborative knowledge building and the underlying cognitive mechanisms. This has also practical implications. Particularly, if assimilation and accommodation could be (automatically) measured, it would be possible to devise feedback mechanisms that could guide group work or formative assessment by a teacher. In the next section, we will present the results of the experiment as they pertain to the dissociation. We then relate these results to research into implicit and explicit memory processes in cognitive psychology. From this, we suggest how the co-evolution model could account for these results. We close with discussing practical implications.

2 An Experiment on Learning from Wikis

In our experimental study, students learned from a wiki about Narratology, the theory and study of narratives and narrative structure. They had to learn about a model which introduced several different narrative modes based on the role of the narrator and the style of narrative. The model can be depicted graphically as a circle in which different narrative modes can be placed at different sections around the circle depending on where they fall on dimensions such as narrator vs. reflector, outside vs. inside perspective. For example, one narrative mode is that of the first-person narrator which is characterized by a narrator, who has an identity in the story and who takes an inside perspective, i.e. having access to inner feelings of that person.

After learning a part of the model, students had to classify short passages of prose according to its narrative mode. While one group of students received texts that were in accordance with the model they had learned, others received passages that included narrative modes not yet learned. We assumed the first group could assimilate the examples into the existing schema they had constructed, while the second group would have to accommodate their internal schema. Hence, assimilation vs. accommodation instilled in our subjects was our main experimental manipulation which we will call *learning condition (ACC vs. ASS)*. 16 subjects were tested in each condition. As dependent measures of their learning, students *drew concept maps* and were asked to freely associate terms given a certain stimulus term (*association test*).

The stimulus terms were taken from the study material. Both tests were applied prior (pretest), as well as after the manipulation at the end of doing all exercises (posttest).

Just looking at the number of concepts introduced in the concept maps and the number of associations reveals an interesting pattern: The number of concepts introduced in the concept maps increased from pretest to posttest for both groups, a clear sign of explicit leaning (ACC group: M_{pre}=12.3 vs. M_{post}=14.8; t_{15}=-2.36, p<.05; ASS group, M_{pre}=10.8 vs. M_{post}=14.6; t_{14}=-6.76, p<.001). When looking at the number of associations, on the other hand, these only increased for the ASS group (M_{pre}=3.8 vs. M_{post}=4.4; t_{15}=-2.92, p<.01), but not for the ACC group (M_{pre}=4.3 vs. M_{post}=4.4; t_{14}=-0.08, ns). This points to a dissociation of the two memory measures: given the experimental manipulation (i.e. inducing assimilative vs. accommodative processes), an effect on explicit memory was observed, but not uniformly on associative strength.

Because we were especially interested in the structural changes of students' knowledge, we employed several network measures as they have previously been found to detect changes in structural aspects of knowledge [2] [4]. These measures can be employed both to concept maps, as well as to association networks which are formed by the stimulus terms from the association test and a co-occurrence relation. We measured the density of the network (proportion of edges) and its overall betweenness centrality (which measures the average shortest path between all nodes).

Fig. 1 shows the results for density of the two networks. For the Concept Maps (left), the main effects were not significant, but the interaction effect time x learning condition was significant ($F_{1,29}$ = 4.84, p < .05, ω^2 = .15): while there was no change in density in the ACC group over time (t_{15} = -.44, ns, r = .17), there was a substantial drop of density in the ASS group over time (t_{14} = 4.56, p < $.001$, r = .50). For the association network (right), there was also no main effect, but the interaction approached statistical significance and the effect size indicated a moderate effect ($F_{1,29}$ = 2.32, ns, ω^2 = .10), suggesting that the sample size might have been too small to show statistical significance.

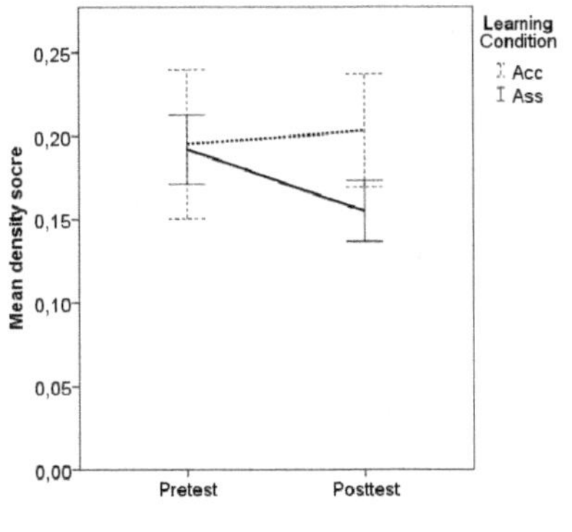

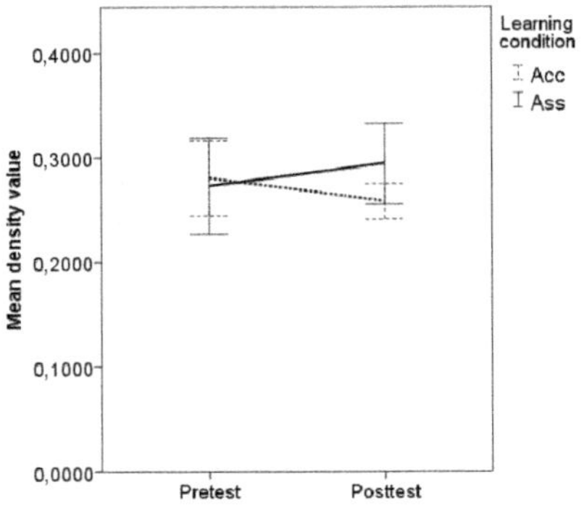

Fig. 1. Density of Concept Maps (left) and Association networks (right)

What is noteworthy about these results is that the two interaction effects for the two measures (one significant, one approaching significance) are in opposing directions. While in the assimilation condition, the drawn Concept Maps were sparser in the posttest than before, there was a tendency that density increased in the association networks. For the accommodation condition, these effects were reversed. A higher density in the concept maps points to the fact that more relations were drawn between concepts, while a higher density in the association networks may indicate less distinct internal categories (more similar associations given two stimuli).

Fig. 2 shows the same results for the measure of betweenness centrality. Hierarchical networks have higher degrees of betweenness centrality, while for networks that are more evenly connected, betweenness centrality is lower. For the Concept Maps (left), the time x learning condition interaction showed that the change in the ASS group was different to the change in the ACC group ($F_{1,29} = 14.69$, $p < 0.001$, $\omega^2 = .18$). While the drop of centrality in the ACC group was strong ($t_{15} = 9.44$, $p < .001$, $r = .62$), the ASS group did not show a significant change over time ($t_{14} = 1{,}24$, *ns*, $r = .29$). For the Association Networks (right), no effects were observed.

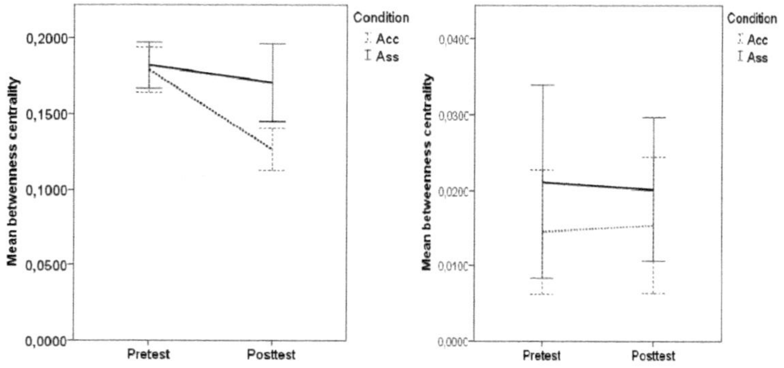

Fig. 2. Betweenness Centrality of Concept Maps (left) and Association networks (right)

Taken together, the results support the notion that structural changes in students' knowledge were different for the two groups. While for the assimilation group, concept maps became sparser and more hierarchical, the concept maps of the accommodation group were more evenly connected while having a comparable degree of density. This is generally in line with the co-evolution model. Accommodation (and hence restructuring or newly established schemata) would lead to more network-like structures in the concept map with a *lower* degree of betweenness centrality (see also [4]). In contrast, assimilation would lead to a simple adding of nodes to an existing hierarchical structure, and as a consequence betweenness centrality would be *higher*.

The study gives other evidence of dissociation of the two measures, explicit and implicit: we came up with several measures to determine whether students had learned more (e.g. number of errors in the concept map and number of correct associations in the association test). Correlational evidence suggest that for better students betweenness of the concept map *was higher* while betweenness of the network formed from co-occurring associations *was lower*. Again, distinctness of internal concepts would likely increase centrality of the network which was observed for the implicit measure. The explicit measure, on the other hand, showed a decrease in centrality which would point to the fact that better students had discovered more explicit connections between concepts.

3 Implicit and Explicit Memory in Knowledge Building

The results pertaining to the dissociation are informative form a theoretical perspective because they shed light on the underlying (non-observable) memory processes. While our results certainly are only initial findings, they may point to the important and differential role that implicit and explicit memory processes play in collaborative knowledge building. In experimental psychology, the dissociation between implicit and explicit processing in memory tasks has long been used to discover underlying memory processes of encoding and retrieval [1] [9]. While explicit memory preserves the context from the study episode, implicit memory does not support the conscious retrieval of the study context. Retrieving from explicit memory involves an experience of conscious and deliberate recollection, while retrieving from implicit memory is relatively automatic and effortless and is associated with an experience of familiarity or "just knowing" [5]. While explicit memory is mainly responsible for the conscious recollection in typical memory tests (e.g. recognition or recall tests of studied material), implicit memory facilitates performance on certain tasks (e.g. quickly associating words as in our association test) without conscious intention of recalling the study episode.

Why then would implicit and explicit memory processes play a role in collaborative knowledge building? If assimilation and accommodation can be understood as a process of schema activation and restructuring [3], we may deduce that implicit and explicit memory processes are differentially impacted: when making sense of encountered information by applying an existing internalized schema, the information is processed in a schema congruent manner. We assume that this is predominantly based on implicit processing. Activation of the schema in semantic memory should happen to a large degree unintentionally, especially in a situation of experienced performance. The application of the schema, then, should mainly be based on semantic or procedural knowledge – both again to a larger degree implicit. This is in line with research showing that under certain conditions the activation of a schema remains unconscious and schema-congruent information is only shallowly processed. In our experiment, this may have been the reason why in the assimilation group, the implicit memory measure (associative strength of the concepts learned) showed a stronger improvement than in the accommodation group.

When, on the other hand, one is confronted with schema-incongruent information and the activated schema fails to provide a suitable way to represent the external information, the information is processed more elaborately, both semantically and perceptually. This is in line with [7], who found higher levels of explicit recollection for incongruent items, while implicit familiarity was greater for congruent items. Information processing in this situation should therefore predominantly impact explicit memory. The co-evolution model assumes that as a result of this conflict, an existing schema is changed, or a new schema is generated. This should be largely dependent on explicit processing, as high levels of semantic elaboration are involved to perform this constructive task. Again, in our experimental results, there is clear indication of this restructuring in the accommodation group for the explicit measure (increased network density and decreased centrality of the concept maps). The implicit measure, on the other hand, does not show these same effects.

4 Conclusions and Outlook

We have presented some initial results of an experiment designed to test the co-evolution model of collaborative knowledge building. We have pointed out results that show a possible dissociation between implicit and explicit memory processes which are theoretically instructive. Two limitations of this research have to be mentioned. First and foremost, the results are exploratory in so far as we did not have any a-priori hypotheses about the dissociations. Hence, the results have provided the basis for hypothesis generation in section 3 that will inform future studies. Secondly, it has to be said that tests for explicit and implicit memory have shown different levels of reliability. This has to be kept in mind and accounted for in future research.

As a next step, we will be transferring the research to a situation where students collaboratively generate artefacts. The graph-based measures we have suggested in this paper are appropriate also for this purpose. For example, the link structure of generated wiki articles could be analyzed as a measure of explicit knowledge. Similar to the association networks, a term network could be constructed using term co-occurrence on a sentence level obtained from the text students produce in the wiki. This could be used as a measure for implicit knowledge in external representations. These measures would constitute theory guided and validated measures, and they should mirror the internal measures introduced in the present study.

References

1. Brainerd, C.J., Wright, R., Reyna, V.F.: Dual-Retrieval Processes in Free and Associative Recall. J. of Memory and Language 46, 120–152 (2002)
2. Coronges, K.A., Valente, T.W., Stacy, A.W., Coronges, K.: Structural comparison of cognitive associative networks in two Populations. J. of Appl. Soc. Psych. 37, 2097–2129 (2007)
3. Cress, U., Kimmerle, J.: A systemic and cognitive view on collaborative knowledge building with wikis. J. of Computer-Supported Collab Learning 3(2), 105–122 (2008)
4. Hay, D.B., Kehoe, C., Miquel, M.E., Hatzipanagos, S., Kinchin, I.M., Keevil, S.F.: Measuring the quality of e-learning. British J. of Educ. Technol., 1037–1056 (2008)
5. Kelley, C.M., Jacoby, L.L.: Recollection and Familiarity: Process-Dissociation. In: Tulving, E., Craik, F. (eds.) The Oxford Handbook of Memory, pp. 215–228. Oxford University Press, Oxford (2000)
6. Kimmerle, J., Cress, U., Held, C.: The interplay between individual and collective knowledge: technologies for organisational learning and knowledge building. Knowledge Management Research & Practice 8(1), 33–44 (2010)
7. Lampinen, J.M., Copeland, S.M., Neuschatz, J.S.: Recollections of things schematic: Room schemas revisited. J. Exp. Psych.: Learning, Memory and Cognition 27, 1211–1222 (2001)
8. Ley, T., Seitlinger, P.: A Cognitive Perspective on Emergent Semantics in Collaborative Tagging: The Basic Level Effect. In: International Workshop on Adaptation in Social and Semantic Web. CEUR Workshop Proceedings, pp. 13–18 (2010)
9. Nelson, D.L., et al.: How implicitly activated and explicitly acquired knowledge contribute to the effectiveness of retrieval cues. Memory & Cognition 35, 1892–1904 (2007)

Orchestrating Learning Activities in 3D Virtual Worlds: IMS-LD in Open Wonderland

David Maroto, Derick Leony, Carlos Delgado Kloos,
María Blanca Ibáñez, and José Jesús García Rueda

Telematics Engineering Department, Universidad Carlos III de Madrid,
28911 Leganés, Madrid, Spain
{dmaroto,dleony,cdk,mbibanez,rueda}@it.uc3m.es
http://www.it.uc3m.es

Abstract. Immersive environments such as virtual worlds and virtual reality platforms are being increasingly used for educational purposes. Possibilities of these environments are huge, but also are the technical challenges that have to be overcome in order for these platforms to become more usable. One of the main problems that educators have to face when designing an educational experience for virtual worlds has to do with the orchestration of the learning sequence. In this paper, we present a novel approach to support the deployment and execution of Units of Learning (UoLs), described with an Educational Modeling Language such as IMS-LD [1], into a virtual world, with the aim to facilitate the creation of well defined learning sequences for these platforms.

Keywords: 3D virtual worlds, Education, Learning design, Orchestration, Open Wonderland, CopperCore, IMS-LD.

1 Introduction

Immersive environments such as virtual worlds and virtual reality platforms are being increasingly used for educational purposes [2], [3]. These platforms present a very rich interface, where users can find many objects and resources that can be accessed at any time during the learning experience. But this behavior is often not desired, since the creator of the learning experience usually wants learners to perform the tasks and to use the resources in a particular order.

Educational Modeling Languages (EMLs) [4] have emerged as a very useful tool in order for the teachers to carry out this orchestration of the learning activities. In our case, we have chosen IMS-LD because it is the most wide spread EML, becoming the *de facto* standard adopted by the educational community.

This orchestration of the learning sequence inside a virtual world offers two main benefits:

1. Teachers and course creators can gain control over the content that is accessible inside the virtual world at every moment.

C. Delgado Kloos et al. (Eds.): EC-TEL 2011, LNCS 6964, pp. 455–460, 2011.

2. In turn, learners can see only the resources needed to perform the activity in which they are involved each time, which helps to reduce their sense of confusion when entering an environment in which they can move anywhere and access too many resources.

In this paper, we present the work we have carried out in order to integrate an IMS-LD engine into a virtual world based on Open Wonderland [5].

The paper is organized as follows: in Section 2 we introduce the main characteristics of IMS-LD. Open Wonderland and its main features are described in Section 3. Section 4 explains the three core elements that support our work. We present the integration of an IMS-LD engine with Open Wonderland in Section 5. Finally, we discuss the conclusions and future work in Section 6.

2 IMS Learning Design

IMS-LD is a specification that enables to describe an educational experience, whatever the pedagogy used, based on the metaphor of a play. IMS-LD provides a generic and flexible XML language that introduces the concepts of Acts, Activities, Environments, Roles and Resources to describe the learning sequence in a way that can be later understood by a runtime system.

Due to its complexity, IMS-LD specifies three levels of implementation and compliance (A, B, C), where each level adds new functionaities over the previous one. In our work, we offer support for UoLs designed using only level A.

All the elements used by IMS-LD (i.e. activities, environments, etc.) are written down in a manifest file called *imsmanifest.xml*. Later, this manifest file is packaged together with the resources. This package, known as a Unit of Learning (UoL), follows the IMS Content Packaging specification [6] on delivering reusable learning content. This specification establishes a life cycle for a UoL consisting on three phases: authoring, deployment and execution.

Fortunately, we do not need to write the manifest file and create the package on our own during the authoring phase. There are tools that help us in the process of creating a UoL compliant with the IMS Content Package specification, such as ReLoad[1] editor.

Once we have the package representing the UoL, it can be delivered and deployed into a runtime system such as CopperCore[2].

3 Open Wonderland

Open Wonderland is, in the words of its creators, "a 100% Java open source toolkit for creating collaborative 3D virtual worlds". This toolkit is completely extensible by means of modules that extend the functionality to create entire new worlds, new features in existing worlds, or new behaviors for objects and avatars.

[1] `http://www.reload.ac.uk/editor.html`
[2] `http://coppercore.sourceforge.net`

There have been many learning experiences developed over Open Wonderland, but all of them lack the same characteristic: there is no way to orchestrate a sequence of activities. The virtual world remains the same no matter the evolution of the learners. This is the main behavior we want to change with our developments. We will explain the tools and technologies used to carry out our work in the next section.

4 Elements Supporting our Developments

We have developed a module for Open Wonderland which is responsible for the second and third phases of the life cycle of a UoL: deployment and execution. The module we have developed introduces modifications both in the clients and server of Open Wonderland. We have also designed a new web interface in the web administration page of Open Wonderland, which allows managing all the tasks related to the deployment, deletion and configuration of UoLs.

But, before we explain the details of the implementation, it is important to know the three elements that support the rest of our developments. These elements are: the snapshots system of Open Wonderland, the security module of Open Wonderland and the CopperCore runtime engine.

Open Wonderland Snapshots System. The snapshot of an Open Wonderland virtual world is a collection of files, representing every single object that was present inside the virtual world just when the snapshot was created. These objects are represented as XML files (one file for each object). These files contain all the information needed for the platform to render each of the 3D virtual objects. We will use these files as the resources for the learning activities in which the course creator will divide the learning sequence.

Open Wonderland Security Module. 3D objects of the Open Wonderland platform can have some capabilities associated to them. Among these capabilities, there is one that can make the object visible or invisible for the users, depending on the group the user belongs to. This capability is expressed through the security module.

CopperCore Runtime Engine for IMS-LD. IMS Learning Design is a complex, semantically rich specification, so it is not trivial to provide full support for it. All the checking, synchronizing and personalizing of the learning process is called the business logic of Learning Design, and this is exactly what CopperCore handles for us. CopperCore is a J2EE runtime engine for IMS-LD which exposes a SOAP API [7] to communicate with it and to make use of its functionality. This engine is the basis for our integration of IMS-LD into an Open Wonderland virtual world.

5 Integration of Coppercore with Open Wonderland

The architecture of the system that we have developed is shown in Fig 1.

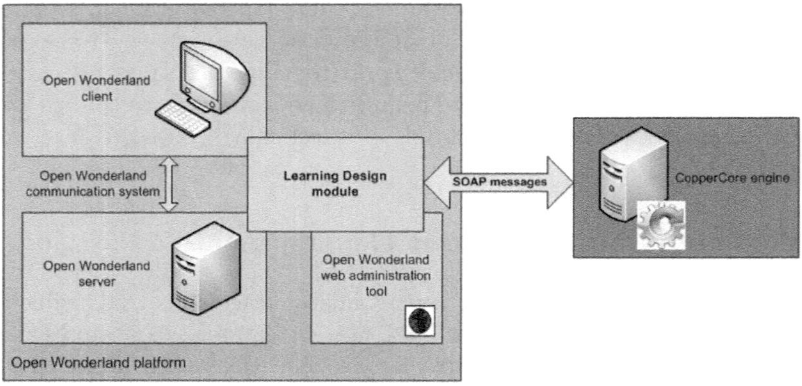

Fig. 1. System arquitecture

The best way to explain our work is to follow the sequence of steps that a teacher should take in order to have a learning experience, based on IMS-LD specification, deployed and running into Open Wonderland. We can divide these steps into *authoring*, *deployment* and *execution*.

5.1 Creating the UoL: Authoring Phase

This phase starts with an empty world in Open Wonderland. In this stage, the course creator must populate the virtual world with all the 3D objects needed to perform each of the activities, no matter the sequence, and later create the UoL. So in fact, this phase can be divided into two subphases: Creation of a snapshot with all the 3D objects needed to perform the whole learning experience, and creation of the learning sequence using the objects from the snapshot and an IMS-LD editor (e.g. ReLoad).

The result of this phase is a UoL containing the *imsmanifest.xml* which describes the sequence of learning activities for each role, and the resources associated to each activity.

5.2 Deployment of the UoL

Once we have the learning experience defined in the UoL, we need to deploy it into a runtime system to later execute it. For this purpose, the teacher can use our new Open Wonderland web interface to upload the UoL into CopperCore. When the teacher clicks the upload button, it triggers a series of functions in order for the platform to be configured for the next phase. These functions are:

- Validation and publication of the UoL into CopperCore.
- Creation of a new Open Wonderland snapshot. This snapshot contains exactly the same objects that the snapshot created during the authoring phase, but modified through the inclusion of the security module.

- Creation of new groups into Open Wonderland. Our module creates a new group for each of the activities, environments and roles defined in the UoL. We refer to groups for activities and environments as "control groups", since they will be used internally by our module to orchestrate the learning sequence.

5.3 Execution of the UoL

The first thing that the teacher has to do before the experience starts, is to place each user in his corresponding group (role) inside Open Wonderland. Our module is responsible for passing all the users registered in Open Wonderland to CopperCore, with the aim to also register them into the IMS-LD engine. This way CopperCore can track the itinerary of every single participant and react in consequence.

Once CopperCore has registered every user, our module asks it for the activity tree of each one of them. These activity trees are parsed in order to obtain the activity that every user has to perform first, with the aim to introduce them into the control group corresponding to that activity. This way, when a user accesses to the virtual world, he is able to see the objects needed to perform the first activity planned in the UoL for him, and only that objects.

We have also designed an enhanced client interface for Open Wonderland clients that allows users to manage their activities. Through this interface, users can obtain the description of the current activity or terminate it and move to the next. Every time a user pushes the button to terminate an activity, a message is sent to CopperCore in order to calculate the new activity tree for every user and introduce them into the corresponding control group for the activity that they have to perform next.

As we can note, all the tasks related to the execution of the UoL are handled by CopperCore. Our module is responsible, on the one hand, for communicating to CopperCore the events that take place into the virtual world which can produce a change in the state of the learning sequence (i.e. the termination of an activity). On the other hand, our module is also responsible for obtaining the new status of the UoL from CopperCore and move the users to their new corresponding control group into Open Wonderland.

6 Conclusions and Future Work

We have presented a novel work to integrate an IMS-LD engine into a virtual world, with the aim to orchestrate learning sequences in these kinds of platforms. We have only found a single reference about a similar work [8], but the approach followed in our developments is somehow different. We have designed a functionality that allows the complete reconfiguration of the virtual world during the runtime, as participants in the learning experience evolve, rather than designing some new virtual world objects that communicate with an IMS-LD engine to obtain the resources of the activities, as done in [8]. In our approach, 3D virtual objects are in fact the resources of the activities.

In the field of future developments, we must point out that although CopperCore offers full support for IMS Learning Design including levels A, B and C, our module currently supports only level A. Development of new functionalities to support levels B and C is in the roadmap. We would also like to test this work with real teachers and course creators, as well as students, with the aim to evaluate its usability and global performance.

Acknowledgments. This research has been partially supported by the following projects: The Spanish CDTI project "España Virtual" funded by the Ingenio 2010 programme, subcontracted by Deimos Space; The Spanish project "Learn3: Towards Learning of the Third Kind" (TIN2008-05163/TSI) funded by the Spanish "Plan Nacional de I+D+i" of the Ministry of Research and Innovation; The project "eMadrid: Investigación y desarrollo de tecnologías para el e-learning en la Comunidad de Madrid" (S2009/TIC-1650) funded by the Government of the Region of Madrid.

The authors want to thank Francisco José Sánchez for his valuable help during the implementation phase.

References

1. IMS Global Learning Consortium, Learning Design Specification (2003), `http://www.imsglobal.org/learningdesign` (accessed: June 2011)
2. European Community, EU project for language education in 2nd Life, `http://www.avalonlearning.eu` (accessed: June 2011)
3. Ibáñez, M.B., García, J.J., Galán, S., Maroto, D., Morillo, D., Delgado Kloos, C.: Multi-User 3D Virtual Environment for Spanish Learning: A Wonderland Experience. In: Proc. 10th IEEE International Conference on Advanced Learning Technologies (ICALT 2010), Sousse, Tunnisia, pp. 455–457. IEEE Computer Society Conference Publishing Services, Los Alamitos (2010), doi:10.1109/ICALT.2010.132
4. Koper, R., Manderveld, J.: Educational Modelling Language: Modelling Reusable, Interoperable, Rich and Personalised Units of Learning. British Journal of Educational Technology 35(5), 537–551 (2004), doi:10.1111/j.0007-1013.2004.00412.x
5. Open Wonderland, Open Wonderland 3D Virtual Worlds (2010), `http://openwonderland.org` (accessed: June 2011)
6. IMS Global Learning Consortium, Content Packaging Specification (2009), `http://www.imsglobal.org/content/packaging` (accessed: June 2011)
7. CopperCore, SOAP API Javadoc (2008), `http://coppercore.sourceforge.net/documentation/javadoc/soap/index.html` (accessed: June 2011)
8. Fernández-Gallego, B., Lama, M., Vidal, J.C., Sánchez, E., Bugarín, A.: OPENET4VE: A Platform for the Execution of IMS LD Units of Learning in Virtual Environments. In: Proc. 10th IEEE International Conference on Advanced Learning Technologies (ICALT 2010), Sousse, Tunnisia, pp. 472–474. IEEE Computer Society Conference Publishing Services, Los Alamitos (2010), doi:10.1109/ICALT.2010.137

The Effect of Dynamic Computerized Scaffolding on Collaborative Discourse

Inge Molenaar[1], Carla Van Boxtel[1], and Peter Sleegers[2]

[1] University of Amsterdam, Research Institute of Child Development and Education, Nieuwe Prinsengracht 130, 1018 VZ, Amsterdam, The Netherlands
i.molenaar@uva.nl
[2] University of Twente, Department of Education Organization and Management

Abstract. This paper explores the effect of computerized scaffolding and different forms of scaffolds on small groups' collaborative discourse. We developed a computerized scaffolding system that uses an attention management system to support metacognitive activities in small groups. We previously found that the scaffolding stimulates the group' metacognitive activities and enhances individual metacognitive knowledge. Moreover different forms of scaffolds have differential effects on learning that cannot be explained by quantitative differences in the groups' metacognitive activities. Therefore, we investigate to qualitative differences in the groups' collaborative discourse in this study. We found that groups receiving scaffolding had significantly less ignored metacognitive episodes. Groups receiving problematizing scaffolds had significantly less ignored metacognitive episodes and more co-constructed metacognitive episodes compared to groups receiving structuring scaffolds. These findings indicate that scaffolding indeed positively influenced collaborative discourse and intensive collaborative discourse seems to explain the differential learning effects of different forms of scaffolds.

Keywords: Metacognition, Attention management system, Collaborative learning, Elementary Education.

1 Introduction

Cognitive and metacognitive activities are key to self-regulating one's learning in computer-based learning environments (CLBE's)[1]. Students who orientate, plan, monitor and evaluate learn more and show higher motivation than students who engage in less metacognitive activities [2]. Additionally, metacognitive skills are an important predictor of students' learning abilities, possibly even more important than intelligence [3]. Even though the importance of metacognition is recognized, students generally reframe from sufficiently regulating their learning in CLBE's. Therefore, metacognitive scaffolding is used to support student's metacognitive activities and develop students' metacognitive knowledge and skills. Metacognitive scaffolding has also been applied in small group settings showing similar results as in individual settings [4].

C. Delgado Kloos et al. (Eds.): EC-TEL 2011, LNCS 6964, pp. 461–466, 2011.

Previously, we found that metacognitive scaffolding stimulates the group' metacognitive activities which in turn support the development of student' metacognitive knowledge. Moreover, different forms of scaffolds had a differentiating effect on group performance and student' domain and metacognitive knowledge [7]. These differential effects could not be explained by quantitative differences in the group' metacognitive activities. Therefore, we propose that qualitative differences in collaborative discourse could explain differential results. As under the right circumstances collaboration stimulates student's learning especially when students elaborate on each other's contributions [5,6]. Consequently, when scaffolding fosters more intensive collaborative discourse this can influence the effect of scaffolding on learning. Yet we know very little about the effect of scaffolding on the group' collaboration. The question raised in this paper is: How does computerized scaffolding and different forms of scaffolds influence collaborative discourse among collaborating students? In this study, we add to the literature by assessing the effects of scaffolding on student's collaborative discourse. First, we will shortly describe our computer-based scaffolding system and different forms of scaffolds. Second, we elaborate on how students learn from collaborative discourse.

1.1 Atgentschool: Dynamic Scffolding with an Attention Management System

Scaffolding can support students in tasks they cannot accomplish by themselves providing assistance when needed. Human tutors perform scaffolding analyzing the student' behavior in relation to the demands of the learning task (diagnosis). Based on their diagnosis, they select the right scaffold for the situation (calibrating). As diagnosis and calibration continuously happen, the scaffolds are reduced when the students becomes more comfortable with the activity (fading). Computer software has to perform the same processes to dynamically scaffold the learner. Systems' awareness and interpretation of the student' behavior in relation to the learning task is still restricted [8]. Attention management systems can capture the user's attentional focus and determine the costs associated with presenting certain information to a learner [8]. The utility of attentive systems for scaffolding is therefore to detect the students' attentional focus and interpret this information. This enables the continuous assessment of the students' attentional focus, its interpretation (diagnosis) and the selection of the appropriate scaffold (calibration) when needed (fading).

In the literature two different forms of scaffolds are distinguished namely structuring and problematizing [9]. Structuring scaffolds give examples to the group. An example of a structuring scaffold is to show students an exemplary plan for their mind mapping tasks "*What would you like to learn; let's make a mind map with important topics to learn, for instance the climate*", see figure 1. Problematizing scaffolds, on the other hand, pose questions that elicit students' activities. An example of a problematizing scaffolds is asking students to plan their mind mapping task "*How are you going to make the mind map?*".

We found that problematizing scaffolds yield better group performance, better transferable individual domain knowledge and more individual metacognitive knowledge. These differences in learning could not be explained by the quantity of metacognitive activities performed by the group and thus we believe differences in collaborative discourse could possibly explain this.

Fig. 1. Example of structuring and problematizing scaffolds

1.2 Collaborative Discourse and Learning

Collaboration is advantageous for learning when students modify their knowledge as a result of social interaction with their group members [5,6]. This is stimulated when students provide and receive explanations, co-construct ideas, reprove conflicts or negotiate meaning, thus engage in intensive collaborative discourse. Although, different perspectives on collaborative learning, namely the cognitive elaboration perspective, the socio-cognitive conflict perspective and the co-construction perspective, stress different mechanism that cause learning during collaboration (receiving explanations, co-construction or reproving conflicts) , they all emphasize that students' elaborations on one another's contributions support learning. Consequently, intensive collaborative discourse in which students relate to and engage in each others' contributions is expected to improve the group' activities and learning. Different social modes are distinguished based on the intensity of the group' collaborative discourse. We previously found the following social modes in social metacognitive episodes; group members that ignore attempts to regulate the group process; group members that accept these attempts applying them directly in their learning activities; group members that share and exchange metacognitive idea's and finally group members that co-construct new metacognitive activities together. These social modes specify the intensity of collaborative discourse among the group members. Ranging from no discourse in ignored episodes to the most intensive discourse in co-constructed metacognitive episodes.

2 This Study

We know that successful collaboration does not always materialize without help of instructional support. Instructional designs such as scripting, Jigsaw designs and role play support students to engage in successful collaboration [10]. Scaffolds are a form of instructional design supporting different learning activities, but possibly also influencing the students' collaborative discourse. Until now, few studies looked at the effects of scaffolding on students' collaboration. The question we adress is could the influence of different forms of scaffolds on collaborative discourse explain differentiating effects on learning. Structuring scaffolds are likely foster the groups' responds elaborating on given examples, thus sharing their metacognitive activities. Problematizing scaffolds, on the other hand, are likely to elicit the group' co-constructive activities as they have to create new activities together. We anticipate that metacognitive scaffolds stimulate more intensive collaborative discourse in metacognitive episodes and that problematizing scaffold elicit more co-construction whereas structuring scaffold foster more shared metacognitive activities.

We report an experiment in which students in elementary school collaboratively learned in triads. There was a control condition and two scaffolding condition, namely a structuring and a problematizing condition. The collaborative discourse will be measured during learning analyzing the dialogue of the triads. In the study, the sample consists of 54 students (23 boys and 31 girls) in 6 control triads, 6 structuring scaffold triads and 6 problematizing scaffold triads. These 54 students were in Grade 4 (9), Grade 5 (27) or Grade 6 (18) in 3 elementary schools.

The e-learning environment used in this study is called *Ontdeknet*. It focuses on supporting students in their virtual collaboration with experts. The experts provide students with information about their expertise, namely knowledge about their country for this study. The study consisted of 8 lessons, each lasting 1 hour. All students received the same instructions, and all triads spent the same time working on the assignment (6 hours). During these 6 lessons, the triads worked on an assignment called "Would you like to live abroad?" The goal of the assignment was to explore a country of choice (New Zealand or Iceland), write a paper on their findings and decide if they would like to live in this country. The triads worked on one computer and had access to an expert, namely an inhabitant of the country. All experimental groups received scaffolds as described in the introduction

We approached the discourse analysis in 5 steps. First, we detected the metacognitive activities in the groups discourse. Second, we formulated metacognitive episodes combining those metacognitive activities that share the same focus of regulation. Finally, we analyzed the group's collaborative discourse in the episodes to determine the social mode. The reliability was measured by two raters who independently coded two randomly selected protocols (2500 turns) all Cohen's Kappa for intercoder reliability were acceptable to good.

2.1 Results

As mentioned earlier, the purpose of our study was to determine the effect of scaffolding and the different forms of scaffolds on the groups' collaboration. As a consequence all analyses were done at the triad level. We assessed the collaboration of triads during learning using discourse analysis. The total dataset entails 108 hours of lessons and 51.339 utterances with 3702 metacognitive episodes. We used a Mann-Whitney test to evaluate the hypothesis.

We found that triads in the control group (mdn=0.20) showed significantly more ignored metacognitive episodes compared to the scaffolding group (mdn=0.18), (U=18, p=0.05(one sided), R=0.49). The other differences are not significant, but the trends are in the expected direction: triads in the control group (mdn=0.30) showed more metacognitive episodes that are accepted compared to triads in the scaffolding condition (mdn=0.25), (U=26, p=0.19(one sided), R=0.27); they (mdn=0.36) showed less shared metacognitive episodes compared to the scaffolding group (mdn=0.37), (U=30, p=0.31(one sided), R=0.16) and less co-constructed metacognitive episodes (mdn=0.04) compared to the scaffolding group (mdn=0.05) (U=27, p=0.22(one sided), R=0.24).

We found that triads in the structuring group (mdn=0.19) showed close to significantly more ignored metacognitive episodes compared to the problematizing group (mdn=0.17) (U=8, p=0.07(one sided), R=0.46) and close to significantly less co-constructed metacognitive episodes (structuring mdn=0.04 and problematizing mdn= 0.13), (U=8, p=0.07(one sided), R=0.46). The other differences are not significant: triads in the structuring group (mdn=0.40) show slightly more accepted metacognitive episodes compared to triads in the scaffolding condition (mdn=0.26), (U=17, p=0.46(one sided), R=0.05) and the structuring condition (mdn=0.40) showed more shared metacognitive episodes compared to the problematzing group(0.35)(mdn=0.37) (U=10, p=0.12(one sided), R=0.37).

3 Conclusion

This study we aimed to investigate the effects of scaffolding on students' collaborative discourse to explain differential effects on learning between problematizing and structuring scaffolds. We found that triads in the control condition showed more ignored metacognitive episodes than triads in the scaffolding condition. With respect to the differentiating effect of the form of scaffolds, we found that problematizng scaffolds have a positive effect on collaborative discourse. Triads in the structuring condition show more ignored and less co-constructed episodes than triads in the problematizing condition. Thus qualitative difference in metacognitive activities seems to explain the differential effects of problematizing scaffolds on learning.

These findings concur with earlier findings of Barron [11] concerning difference between successful and unsuccessful small groups; the distinction did not lay in the number of problem solving attempts of the group, but in the number of attempts that were *taken up* by the group in collaborative discourse. Thus all groups show attempts to regulate their learning, but groups in the scaffolding conditions also engage in collaborative discourse around these efforts. These findings indicate that under the right circumstances collaborative discourse can amplify the effect of scaffolding and that possibly scaffolding in small group settings is more beneficial then scaffolding in individual settings. Consequently, designing scaffolds that support successful collaboration could possible enhance this interaction. Thus we encourage further research to explore the relationship between scaffolding and collaborative discourse to design forms of scaffolds that optimize the collaboration. This opens up the line of thought about scaffolding as an instructional design to support successful collaboration among students.

References

1. Azevedo, R., Moos, D.C., Greene, J.A., Winters, F.I., Cromley, J.G.: Why is externally-facilitated regulated learning more effective than self-regulated learning with hypermedia? Educational Technology Research and Development 56(1), 45–72 (2008)
2. Azevedo, R., Hadwin, A.F.: Scaffolding self-regulated learning and metacognition - implications for the design of computer-based scaffolds. Instructional Science 33(5-6), 367–379 (2005)

3. VanDerStel, M., Veenman, M.V.J.: Development of Metacognitive Skillfulness; a Longitudional Study. Learning and Individual Differences 20, 220–224 (2010)
4. Azevedo, R., Cromley, J.G.: Does Training on Self-Regulated Learning Facilitate Students' Learning With Hypermedia? Journal of Educational Psychology 96(3), 523–535 (2004)
5. Chi, M.: Active-Constructive-Interactive: A Conceptual Framework for Differentiating Learning Activities. Topics in Cognitive Science 1(1), 73–105 (2009)
6. Webb, M.: The teacher's role in promoting collaborative dialogue in the classroom. British Journal of Educational Psychology 79, 1–28 (2009)
7. Molenaar, I., van Boxtel, C., Sleegers, P.J.C., Roda, C.: Attention management for Learning: Atgentschool. In: Roda, C. (ed.) Human Attention in Digital Environments, pp. 259–280. Cambridge University Press, Cambridge (2011)
8. Roda, C., Nabeth, T.: Supporting attention in learning environments: Attention support services, and information management. In: Duval, E., Klamma, R., Wolpers, M. (eds.) EC-TEL 2007. LNCS, vol. 4753, pp. 277–291. Springer, Heidelberg (2007)
9. Reiser, B.J.: Scaffolding Complex Learning: The Mechanisms of Structuring and Problematizing Student Work. Journal of the Learning Sciences 13(3), 273–304 (2004)
10. Weinberger, A., Stegmann, K., Fischer, F.: Knowledge convergence in collaborative learning: Concepts and assessment. Learning and Instruction 17(4), 416–426 (2007)
11. Barron, B.: When Smart Groups Fail. The Journal of the Learning Sciences 12(3), 307–359 (2003)

Verifying a Quantitative Model of Communities of Practice in a Computer Users' Community

Nicolae Nistor and Monika Schustek

Ludwig-Maximilians-Universität, Faculty of Psychology and Educational Sciences, Leopoldstr. 13, 80802 München, Germany
nic.nistor@uni-muenchen.de, moni_schustek@gmx.de

Abstract. The participation in communities of practice (CoPs) is an important way to construct and share knowledge and skills, including those related to the use of technologies. However, the relationships between the fundamental CoP notions are not yet sufficiently researched. While there are numerous qualitative CoP studies in the educational research, quantitative studies are still lacking. A recent study [1] proposes a quantitative model focusing on the causal chain domain knowledge – participation – expert status – artifact development. However, this study has limited validity due to the relatively small sample, i.e. missing replications in various settings. The present study replicates the first, involving a different knowledge domain, i.e. computer usage, and slightly different instruments. Based on a sample of N = 60 members of an academic CoP, we find results similar to the previous research [1], and thus confirm the hypothesized causal model, at the same time enlarging its empirical basis.

Keywords: communities of practice, informal learning, cultural artifacts.

1 Rationale

Communities of practice (CoPs) are groups of people sharing goals, activities and experience in the frame of a given practice. Participation in CoPs leads to the accumulation of experience, it stimulates the social construction of knowledge and the development of expertise [2, 3]. Examining the CoP literature, we observe that it is sustained mainly on qualitative research; there are hardly quantitative studies about learning and development in this context, particularly about the relationship between the fundamental notions of CoPs. Recent work by Nistor and August [1] proposes a causal model that includes the individual knowledge, the intensity of participation, the expert status, and the community member's contribution to artifact development. Due to the relatively small sample, i.e. missing replications, the study has limited validity. Therefore, it is the aim of this follow-up to replicate the first study and thus enlarge the empirical basis of the model. We begin with an overview of the central CoP variables and a presentation of the causal model. After this, we present the study we conducted to provide empirical evidence of the proposed CoP model.

C. Delgado Kloos et al. (Eds.): EC-TEL 2011, LNCS 6964, pp. 467–470, 2011.

2 Research Model

The proposed causal model of CoPs integrates the following variables (fig. 1):

Domain knowledge. Generally, expertise has the meaning of advanced and reproducible knowledge and skills in a specific domain. The longer the time-on-practice becomes, the more experience is acquired and the more individual expertise develops [4]. For our model, we regard domain knowledge together with the *time in the community* as independent variables, and predictors of participation.

Intensity of participation. Community member perform different degrees of participation, depending on the members' individual expertise. Lave and Wenger [2] emphasize the close relationship between participation and identity in the CoP context. Hence, we regard participation as determinant of expert status, and determined by expertise, i.e. as a mediator between the two variables.

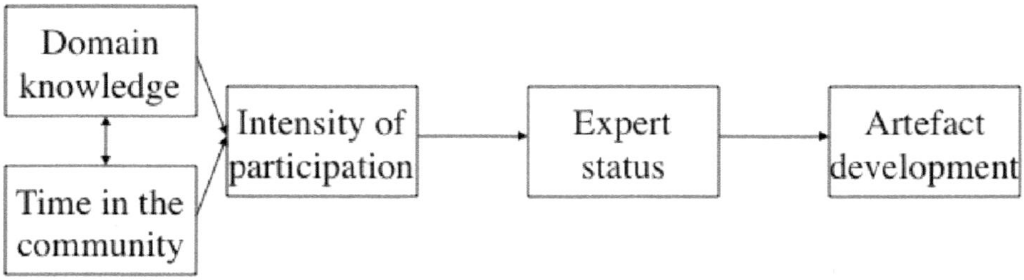

Fig. 1. Causal quantitative model of expertise, intensity of participation, identity and contribution to cultural artifact development in CoPs

Expert status. The theory of legitimate peripheral participation emphasizes the individual identity within CoPs, and distinguishes between full and peripheral community members. Expert status is the result of negotiation with and recognition of other CoP members, thus directly influenced by the intensity of participation in the community practice.

Cultural artifact development. Generally, cultural artifacts are defined as material or immaterial products of human activity with a specific meaning in the context of a given community practice. Wenger [3] describes artifacts as the product of knowledge reification. Nistor [5] proposes that knowledge construction and reification are more accessible to central than to peripheral community members. Therefore, members' contribution to the production of cultural artifacts will be in our CoP model determined by the expert status.

3 Methodology

In order to provide quantitative empirical evidence of the CoP model proposed above, we conducted a correlation study in a computer users' community located at a German university, in the field of Psychology and Educational Sciences. The sample consisted of N = 60 participants of different expertise levels. All variables were measured by means of a questionnaire. The subscales domain knowledge and intensity of participation referred to the use of several software packages, such as office software, statistics programs, electronic communication tools, literature databases etc. Expert status was measured by the number of persons who usually ask the responder for help with computer problems. The artifacts questioned were written descriptions of operations performed with computers.

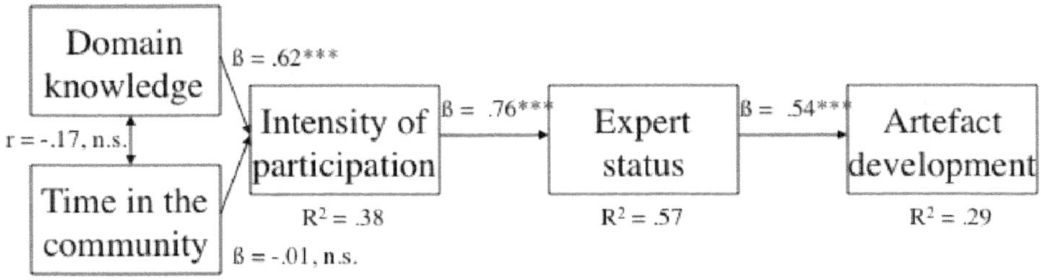

Fig. 2. Results of the regression analysis

4 Results

All the correlations hypothesized by the research model were found to be significant, with regression factors ranging from medium to high. The influence of domain knowledge on intensity of participation was strong (ß = .71, $p < .001$); the influence of time in the community on intensity of participation was medium (ß = .49, $p < .001$); the influence of intensity of participation on expert status was also strong (ß = .66, $p < .001$); the influence of expert status on artifact development was medium (ß = .45, $p < .001$). Intensity of participation mediated significantly the relationship between domain knowledge (i.e. both domain knowledge and time in the community) and expert status; expert status mediated significantly the relationship between intensity of participation and artifact development. The variance of intensity of participation could be explained to 50%, the variance of expert status to 43%, and the variance of artifact development to 20%. The research results are displayed in fig. 2.

5 Conclusions

The results of the regression analysis are in line with the theory, at the same time confirming the quantitative model hypothesized by Nistor and August [1], and enlarging its empirical basis. Future research may also use a mix of methods, including participant observation and artifact analysis.

References

1. Nistor, N., August, A.: Toward a quantitative model of communities of practice: Intrinsic expertise, participation, expert identity and artefact development in an academic community. Paper Presented at the EARLI SIG 2+3 Conference Moving Through Cultures of Learning, Utrecht, September 2-3 (2010)

2. Lave, J., Wenger, E.: Situated learning. Legitimate peripheral participation. University Press, Cambridge (1991)
3. Wenger, E.: Communities of practice. Learning, meaning, and identity. University Press, Cambridge (1999)
4. Ericsson, K.A.: The influence of experience and deliberate practice on the development of superior expert performance. In: Ericsson, K.A., Charness, N., Feltovich, P., Hoffman, R.R. (eds.) Cambridge Handbook of Expertise and Expert Performance, pp. 685–706. Cambridge University Press, Cambridge (2006)
5. Nistor, N.: Knowledge communities in the classroom of the future. In: Mäkitalo-Siegl, K., Kaplan, F., Zottmann, J., Fischer, F. (eds.) Classroom of the Future. Orchestrating Collaborative Spaces, pp. 163–180. Sense, Rotterdam (2010)

Help-Seeking in Communities of Practice: Design and Evaluation of a Help System Supporting Knowledge Sharing

Nicolae Nistor[1], Silke Schworm[2], and Matthias Werner[1]

[1] Ludwig-Maximilians-Universität, Faculty of Psychology and Educational Sciences, Leopoldstr. 13, 80802 München, Germany
{nic.nistor,matthias.werner}@psy.lmu.de
[2] Universität Regensburg, Institute of Educational Sciences, Universitätsstr. 31, 93040 Regensburg, Germany
silke.schworm@paedagogik.uni-regensburg.de

Abstract. Due to the widespread use of information technologies, call centers for technical support are increasingly demanded and often overloaded. Interactive online help systems are considered to be a fruitful supplement to traditional support systems. The paper at hand proposes a conceptual frame for this alternative, based on knowledge sharing in communities of practice (CoPs). A literature review is done bringing together theoretical issues and recent research from the fields of CoPs and academic help seeking. Both approaches are well represented in the empirical research literature. While they are complementary, hence compatible to each other, their combination is not yet sufficiently explored from the psychological and educational point of view. To explore and foster help-seeking in CoPs, we developed a help system based on active participation and knowledge sharing between CoP members. The empirical part of this paper presents the evaluation of the help system, which is based on acceptance theories. The evaluation results are encouraging, and suggest possibilities of further development of the help system in the CoP context, as well as means to foster CoPs and support knowledge sharing.

Keywords: communities of practice, informal learning, knowledge sharing, help-seeking.

1 Introduction

Since technology has become ubiquitous in the most domains of activity, it is a common fact that users encounter problems in handling it at their workplace and they ask for help. As a response, helpdesk services aim at sharing their knowledge with technology users, to enable them to manage their problems on their own. However, helpdesk services are increasingly in demand and often overloaded [1]. As a consequence users may turn to informal sources of support such as colleagues and friends or they may look for an online help system. Nevertheless, both alternatives may cause further problems [2]. Analyzing help seeking activities in different help

C. Delgado Kloos et al. (Eds.): EC-TEL 2011, LNCS 6964, pp. 471–476, 2011.

environments may help optimizing help systems and improving help interactions. However, studies on situational aspects of help-seeking behavior concentrated so far on the context of school and teaching [3]. Little research has been done regarding help-seeking activities within the context of workplace learning and communities of practice (CoPs). The paper at hand aims at exploring the common areas of help-seeking and CoPs. To explore and foster help-seeking in CoPs, we developed a help system based on active participation and knowledge sharing between CoP members. The empirical part of this paper presents the evaluation of the help system, which is based on a theoretical model inspired by acceptance theories. The evaluation results are encouraging, and suggest possibilities of further development of the help system in the CoP context and, more generally, means to foster CoPs and support knowledge sharing within these.

2 Theoretical Background

Communities of practice (CoPs) are large groups of people with common interests and activities, often emerging spontaneously from problem-solving activities [4]. Numerous communities are involved in developing tools or materials, which support important and challenging activities. Regarding CoP members' identity, Wenger [4] distinguishes between experts and novices, full and peripheral community members. The presence of different levels of expertise among the CoP members is a prerequisite of knowledge sharing. While accomplishing tasks within the community practice, novices learn from experts. Within this process, Wenger observes artifact production and describes it as reification of knowledge, claiming that the duality of participation and reification is the key to learning processes in the context of CoPs. Against this background, providing appropriate artifacts, to which the community may add constructed and created knowledge, appears to be an appropriate means of fostering CoPs [5].

Academic help-seeking. Knowledge sharing in CoPs can take various forms; one of these is initiated by help seeking, a resource-based learning strategy requiring interaction with persons expected to be more knowledgeable, or, in some cases, with computer-based learning environments such as online help systems. Mercier and Frederiksen [6] propose a model of the help-seeking process, which consists of five steps. (1) The recognition of an impasse indicates that a relevant task cannot be successfully completed, which leads to the awareness of need for help. (2) The diagnosis of the origin of the impasse leads to a specification of a need for help. (3) Consequently a help goal is set. (4) The learner looks for appropriate help. (5) Evaluating the received help completes the process.

In computer-based environments, help seeking can be either directed to human experts and therefore rely on computer-mediated communication, or it can comprise the interaction with a computer-based help system. For the purpose of this paper, we discuss in the following only the latter case, which can be placed on a continuum according to two dimensions: (1) static vs. dynamic, (2) written by experts vs. written by users [7]. Help-seeking in computer-based environments may be associated with several problems, such as requiring precise formulation of the requests and being intolerant to synonyms; help output may be not adequately adapted to the users' prior knowledge [8]; lengthy step-by-step instructions that overwhelm novices; "split-source" format of the help output may cause users' cognitive overload; or finally users can get "lost in hyperspace". In order to enable effective help-seeking, knowledge sharing and a positive learning outcome, it is necessary to understand these problems and to prevent them by appropriate instructional design.

3 Design of a Help System Supporting Knowledge Sharing

The literature review briefly presented above suggests in the first place the compatibility of the situated learning and the help-seeking approach. However, both approaches developed independently from each other, while scarcely taking into consideration the reciprocal influences between the social practice context and the individual help-seeking behavior. There is still little research built on this double basis.

As a setting for research on help-seeking within CoPs, we propose in the following a help system aimed at supporting knowledge sharing on technology use among the participants of academic CoPs. At universities, academic practice consists mainly of teaching and research. Both of these require however the existence of suitable tools such as computers and software, and the participants' ability to make use of these tools. More experienced users will be more likely to find efficient ways to use technology, and to communicate this newly constructed knowledge to less experienced colleagues. Thus, a technology users' community will emerge within and mostly overlapping with the original academic CoP.

The central component of the help system will be a FAQ collection, i.e. frequent problems and questions, together with corresponding explanations and descriptions. The help system involves technology users with different expertise, and collects user-generated e-content. The ground idea of the presented help system is the "propeller" model [9], based on the CoP concept with a focus on artifact production. It suits to CoPs that aim at developing e-content on technology use and include participants with different computer related knowledge and skills (cf. [4]). Nistor and colleagues [9] regard the FAQ collection as an artifact containing reified knowledge; therefore it is positioned in the centre of the activity. The participants are divided in three subgroups, i.e. users, experts and authors/trainers. Three activity zones arise in the interaction between them: (1) Helpdesk and content generation, where the users ask experts questions about their problems, and receive and record explanations; (2) development of e-learning content, where the authors/trainers put the expert explanation into a pedagogically meaningful form; (3) users training, where the authors/trainers teach users how to handle with technology. The use of the help system is expected to be initiated by a problem that a CoP member encounters at workplace and cannot solve by itself, which is described in the "propeller" model as helpdesk activity. As shown in fig. 1 from the user's perspective, several ways lead from the problem to the solution, and to e-content generation. In any case, the solution will be first evaluated, and then applied. If the solution was not available in the help system, it may be recorded either by the user or by the helpdesk expert. The "propeller" e-content development activity requires a simple way of recording solutions to the FAQ base. Once recorded, a solution will be reviewed and edited by content authors, so that the FAQ item is correct and self-explanatory. To prevent the problems described in the section 3.3, the entire FAQ environment should be designed according to the guidelines mentioned there. Finally, the developed FAQ base gives suggestions about frequent problems, and thus about users' need for technology training. After a needs analysis, the e-content authors can edit the FAQs into training material and conduct "propeller" training sessions in face-to-face meetings.

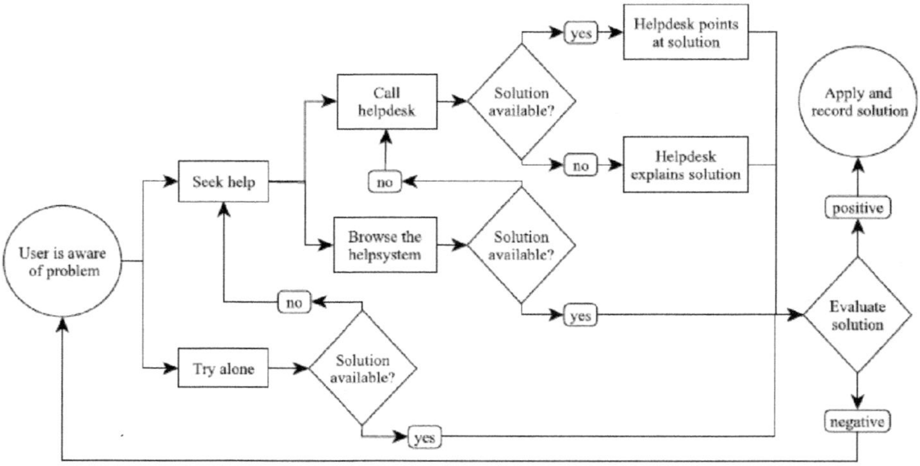

Fig. 1. Use scenario of the help system from the user's perspective

4 Evaluation of the Help System

In the first phase of the help system implementation, it is purposeful to analyze the systems's acceptance and use. We do so relying on the Unified Theory of Acceptance and Use of Technology (UTAUT)[10], in which Venkatesh and his colleagues describe technology use under the influence of the use intention, further determined by performance expectancy, effort expectancy and social influence. Applying the UTAUT for the evaluation of the help system requires an adaptation of the original model, which have to take into account the CoP context (i.e. the influence of the expert core membership), the particularities, i.e. problems of the academic help-seeking and those of the required activities (i.e. help use vs. help development).

The evaluation was first performed in an academic CoP at a German university. The CoP's shared enterprise was the technology use (electronic devices and special software) for professional purposes. The user group sample (N = 47) included faculty and helpdesk staff as IT experts. The evaluation model variables were measured by means of questionnaire survey (on a Likert scale from 1 = high acceptance to 5 = low acceptance), excepting the development behavior, which resulted from the content analysis of the developed materials for the help system. The preliminary results revealed moderate use intention (M = 2.77, SD = 1.20) and development intention

(M = 2.32, SD = 1.14), as well as the path coefficients of the structural equation model depicted in fig. 2. We observe that the intention to use the help system was determined to a high amount by the performance expectancy, which was further influenced by the effort expectancy. The intention to develop materials for the help system was determined in most strongly by the expert core membership, meaning that the helpdesk staff had the strongest development intention. The second predictor of the development intention was the encountered help-seeking problems, which probably points at users' searching for help information and not finding it, thus deciding to share their knowledge and develop corresponding help material. The help-seeking problems had also a strong influence on the effort expectation, which further influenced the performance expectation and the use intention on the one hand, and (however without reaching statistical significance) the development intention on the other.

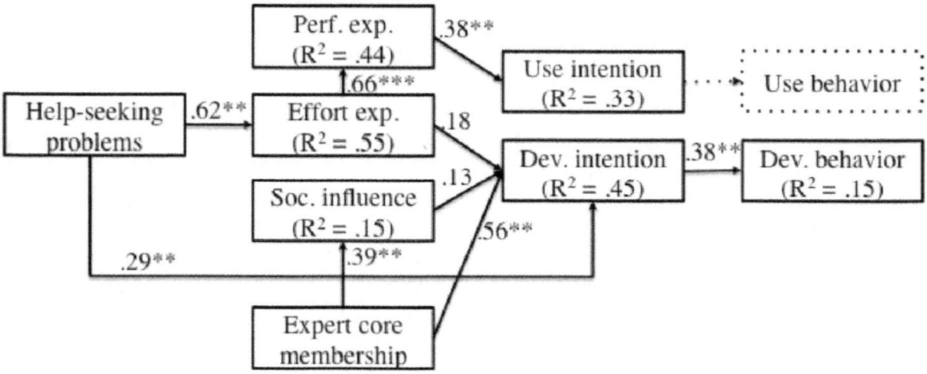

Fig. 2. Evaluation results of the help system

5 Conclusions

The paper at hand combines two approaches of the educational psychology, i.e. the CoP and the help-seeking approach. The CoP model provides a macroscopic view of learning, completed by the microscopic view from the help-seeking approach. Exploring the common zone of the two approaches allows us to draw some conclusions on fostering CoPs by focusing on academic help-seeking: (1) *Organization*. The initiation of a CoP appears to be successful, in this case as well, if a core of experts is provided from the beginning. If this core should not emerge spontaneously, it may be built by employing specialized staff. While the CoP grows, other members (in this case: faculty) may join the expert core and get involved in the community practice. (2) *Technology design*. Technology is the basic ground of a virtual CoP, so it has to be appropriately designed. This includes well understandable contents and self-explaining structures. While such attributes are almost natural at the beginning of the system implementation, when little content is available, they are likely to become hard-to-get features when users' active participation increases and the system grows. Therefore, an important task of the core experts – helpdesk staff or others – will be to review and if necessary revise the newly created materials, and keep them in structure, so that help-seeking problems can be avoided. (3) *Communication*. Obviously, a system needs to be popular in order to be intensively used. In our case, there was no significant social influence on the use intention of the help system, because the faculty did not know about it. This can be changed directly by advertising in the CoP, or indirectly by explicitly using the FAQs as mediating artifacts between helpdesk experts and technology users calling the helpdesk.

These findings are to be further studied in the near future with the help system outlined above in the frame of the academic community practice.

References

1. Leung, N.K.Y., Lau, S.K.: Information technology help desk survey: To identify the classification of simple and routine enquiries. Journal of Computer Information Systems 47(4), 70–81 (2007)
2. Aleven, V., Stahl, E., Schworm, S., Fischer, F., Wallace, R.: Help seeking and help design in interactive learning environments. Review of Educational Research 73(3), 277–320 (2003)
3. Karabenick, S.A., Newman, R.S.: Help seeking in academic settings: Goals, groups, and contexts. Erlbaum, Mahwah (2006)
4. Wenger, E.: Communities of practice. Learning, meaning, and identity. University Press, Cambridge (1999)
5. Nistor, N.: Knowledge communities in the classroom of the future. In: Mäkitalo-Siegl, K., Kaplan, F., Zottmann, J., Fischer, F. (eds.) Classroom of the Future. Orchestrating Collaborative Spaces, pp. 163–180. Sense, Rotterdam (2010)
6. Mercier, J., Frederiksen, C.H.: Individual differences in graduate students' help-seeking process in using a computer coach in problem-based learning. Learning and Instruction 17(2), 184–203 (2007)
7. Schworm, S., Heckner, M.: E-collaborative help-seeking using social web features. In: Ertl, B. (ed.) E-Collaborative Knowledge Construction: Learning from Computer-Supported and Virtual Environments, pp. 109–123. IGI-Global, Hershey (2010)
8. Wittwer, J., Renkl, A.: Why instructional explanations often do not work: A framework for understanding the effectiveness of instructional explanations. Educational Psychologist 43(1), 49–64 (2008)
9. Nistor, N., Rubner, A., Mahr, T.: Effiziente Entwicklung von E-Content mit hohem Individualisierungsgrad: Ein Community-basiertes Modell. In: Merkt, M., Mayrberger, K., Schulmeister, R., Sommer, A., van den Berk, I. (eds.) Studieren neu erfinden – Hochschule neu entdecken, pp. S.54–S.64. Waxmann, Münster (2007)
10. Venkatesh, V., Morris, M.G., Davis, G.B., Davis, F.D.: User acceptance of information technology: Toward a unified view. MIS Quarterly 27, 425–478 (2003)

Educational Technology and Culture: The Influence of Ethnic and Professional Culture on Learners' Technology Acceptance

Nicolae Nistor[1], Armin Weinberger[2], Ciprian Ceobanu[3], and Jan Oliver Heymann[1]

[1] Ludwig-Maximilians-Universität, Faculty of Psychology and Educational Sciences, Leopoldstr. 13, D-80802 München, Germany
{nic.nistor,oliver.heymann}@psy.lmu.de
[2] Saarland University, Saarbruecken, Educational Technology & Knowledge Management, Campus C5.4, D-66123 Saarbrücken, Germany
a.weinberger@mx.uni-saarland.de
[3] Universitatea "Alexandru Ioan Cuza" Iasi, Faculty of Psychology and Educational Sciences, str. Toma Cozma 3, RO-700554 Iaşi, Romania
ciprian@uaic.ro

Abstract. Technology-enhanced learning is increasingly used in international and multicultural contexts. However, little attention has been paid to learners' attitudes towards technology within different ethnical and professional cultures. This study attempts to integrate cultural dimensions (sensu Hofstede, [1]) into an established technology acceptance model (i.e. the Unified Theory of Acceptance and Use of Technology, UTAUT, [2]). Building on a relatively large sample (N = 2834) of educational technology users from Germany and Romania, we examine the differences of culture and technology acceptance between sample subgroups. The collected data reveals cultural differences both between countries and between professions. These differences are associated with different acceptance profiles. The theoretical significance of our contribution lies in the first steps towards integrating Hofstede's model of cultural dimensions into the UTAUT. For educational practice, it suggests a starting point of dealing with cultural differences in the context of technology-enhanced learning.

Keywords: educational technology acceptance, cultural values, ethnical culture, professional culture.

1 Introduction

Technology-enhanced learning is increasingly used in international and multicultural contexts. However, little attention has been paid to learners' attitudes towards technology within different ethnical and professional cultures. A usual, implicit assumption seems to be that learning technology use implies the same acceptance mechanisms in every culture; there is however not yet sufficient empirical evidence to what extent this is true. Therefore, this study examines the influence of learners'

C. Delgado Kloos et al. (Eds.): EC-TEL 2011, LNCS 6964, pp. 477–482, 2011.

culture on their learning technology acceptance (LTA) with the goal to arrive at an integrated acceptance model accounting for cultural dimensions.

The paper at hand reports on work in progress, aimed at verifying the Unified Theory of Acceptance and Use of Technology (UTAUT; [2]) on a larger basis, with a focus on the moderating influence of the culture values induced by ethnicity and profession. Firstly, we give a brief overview of the theoretical background with respect to technology acceptance and culture. Secondly, we report the actual stage and results of an ongoing LTA study conducted in Germany and Romania. Thirdly, we discuss the consequences of our findings for the theory and practice of technology-enhanced learning.

2 Educational Technology Acceptance and Culture

Educational technology acceptance. Technology acceptance models are mostly based on the view of acceptance as an attitude, applying the theory of reasoned action and its expanded version, the theory of planed behavior [3]. The Unified Theory of Acceptance and Use of Technology (UTAUT; [2]) is a powerful acceptance model describing technology use as a consequence of the use intention, further determined by performance expectancy, effort expectancy and social influence. Additionally, the facilitating conditions directly determine technology usage. Nistor and colleagues [4] report findings that are consistent with Venkatesh's [2]; they argue however that computer anxiety plays a decisive role for LTA. Therefore, computer anxiety should be explicitly included into the UTAUT model as an additional predictor of the use behavior.

Culture. Hofstede [1] defines cultural dimensions as patterns of thinking, feeling and potential acting, which have been learned throughout lifetime, therefore they are likely to be used repeatedly and unlikely (or difficult) to be changed by the individual. Cultural patterns are shared within a social environment. Hofstede identified five dimensions of culture: power distance, collectivism vs. individualism, femininity vs. masculinity, uncertainty avoidance, and long-term vs. short-term orientation.

Similarly to Hofstede's view of cultural dimensions, we regard attitudes towards educational technology as socially shared patterns of thinking, feeling and behavior towards technology. Thus, attitudes towards educational technology may co-vary with Hofstede's cultural dimensions [1]. Recent research attempts to identify connections between these dimensions and the LTA variables. The effect of performance expectancy is expected to be amplified by culture's masculinity [5], and high power distance [6], and reduced by uncertainty avoidance [6, 7]. The effect of effort expectancy is expected to be reduced in more masculine cultures [8]. Social influence on LTA should be stronger in collectivistic cultures [5]. Per definition, computer anxiety should have a stronger effect on LTA in uncertainty avoidant cultures.

3 Research Questions

Against the background of hypothesized influences of ethnic and professional culture on the UTAUT acceptance variables, we examine the following aspects:

The influence of ethnic culture. To what extent do Romanians and Germans differ with respect to (a) Hofstede's cultural dimensions, (b) attitudes towards educational technology, and (c) the correlations of the UTAUT model?

The influence of professional culture. To what extent do persons with technical and persons with non-technical profession differ with respect to (a) Hofstede's cultural dimensions, (b) attitudes towards educational technology, and (c) the correlations of the UTAUT model?

4 Methods

A correlation study was conducted, recording longitudinal data in a one-shot survey, from undergraduate and graduate students and faculty/teachers, from universities, technical colleges, and adult education centers. The collected sample (N = 2834) included Romanian (n = 1016) and German (n = 1818) students and professionals of technological (n = 861) and non-technological disciplines (n = 1972). The sample was balanced in terms of sex, age and profession. The research instrument consisted of a German and a Romanian translation of the questionnaire proposed by Venkatesh et al. [2] with variable values from 1 = very low to 5 = very high acceptance. Aimed at surveying general attitudes and intentions towards technology, the questions were generally related to "the computer as a learning tool", not to a more specific technology. The survey was partially conducted online, partially using pen-and-paper forms; the participation was voluntary.

5 Results

Overall, MANOVA results show that the influence of culture on Hofstede's cultural dimensions and attitudes towards educational technology is large, $F\ (4{,}2822) = 118.42$, $p < .001$, $\eta^2 = .25$, whereas profession influence is small, $F\ (4{,}5644) = 6.26$, $p < .001$, $\eta^2 = .02$, and interaction effects can be disregarded, $F\ (4{,}2822) = 1.92$, *n. s.*

The influence of ethnic culture. Descriptive results show that the Romanian participants are less power distant, more collectivistic, more masculine (i.e. they perceive greater differences between sexes), more uncertainty avoidant and more long-time oriented than the German participants (tab. 1).

As for participants' attitudes towards educational technology, Romanians have a more positive attitude, higher anxiety and a stronger intention to use educational technology than Germans (tab. 1). Inferential statistics show that all these single differences are significant ($p < .001$) and small in effect size – except for a medium-sized effect on masculinity.

In respect to the correlations of the UTAUT model, Romanians' use behavior is stronger influenced by computer anxiety and by the use intention than Germans. Romanians' use intention is also more socially influenced. Regarding performance and use expectancy, the former is a stronger predictor for Germans, the latter for Romanians (fig. 1).

Table 1. Hofstede's cultural dimensions indices and attitudes towards educational technology (mean values and standard deviations) of Germans and Romanians

	Romanians (n = 1016)	Germans (n = 1818)
Power distance index (PDI)	20.1	36.1
Collectivism vs. individualism (IDV)	67.4	92.5
Masculinity vs. femininity (MAS)	38.5	-29.9
Uncertainty avoidance (UAI)	65.2	76.9
Long-time orientation (LTO)	53.7	45.6
Attitude towards technology use	4.20 (.70)	3.85 (.89)
Technology anxiety	2.21 (.96)	1.80 (.88)
Use intention	4.09 (.90)	3.60 (1.21)

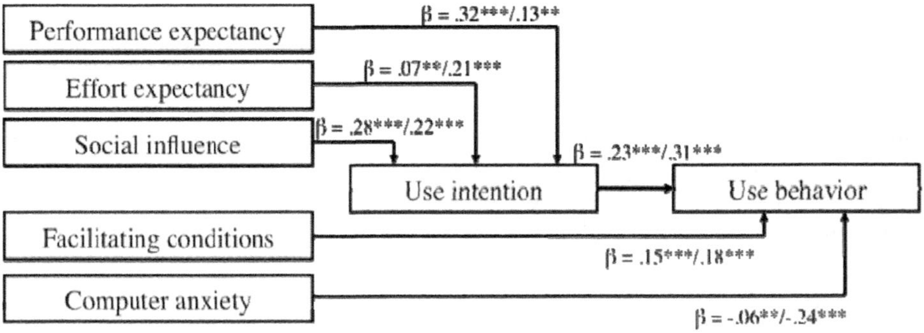

Fig. 1. The UTAUT model for German vs. Romanian learners

The influence of professional culture. Observing the differences between professional groups, the participants with technical professions are less power distant, more collectivistic, more masculine (i.e. they perceive greater differences between sexes), less uncertainty avoidant and more long-time oriented than the participants with non-technical professions (tab. 2).

As for participants' attitudes towards educational technology, the participants with technical professions have a more positive attitude, lower technology anxiety and a stronger intention to use technology than the participants with non-technical professions (tab. 2). Inferential statistics show only the differences of attitude towards educational technology, and anxiety between technical and non-technical professionals are significant on the level of $p < .001$ (and small in effect size).

In respect to the correlations of the UTAUT model, the use behavior of the learners having a technical profession is somewhat stronger influenced by the use intention. Regarding performance and use expectancy, the former is a stronger predictor for learners with technical, the latter for learners with non-technical profession (fig. 2).

Table 2. Hofstede's cultural dimensions indices and attitudes towards educational technology (mean values and standard deviations) of technical and non-technical professionals

	Technical professions (*n* = 1016)	Non-technical professions (*n* = 1818)
Power distance index (PDI)	27.3	32.1
Collectivism vs. individualism (IDV)	78.2	85.9
Masculinity vs. femininity (MAS)	12.7	-13.3
Uncertainity avoidance (UAI)	67.0	75.2
Long-time orientation (LTO)	50.1	47.8
Attitude towards technology use	4.05 (.86)	3.94 (.84)
Technology anxiety	1.82 (.91)	2.00 (.93)
Use intention	3.90 (1.08)	3.72 (1.15)

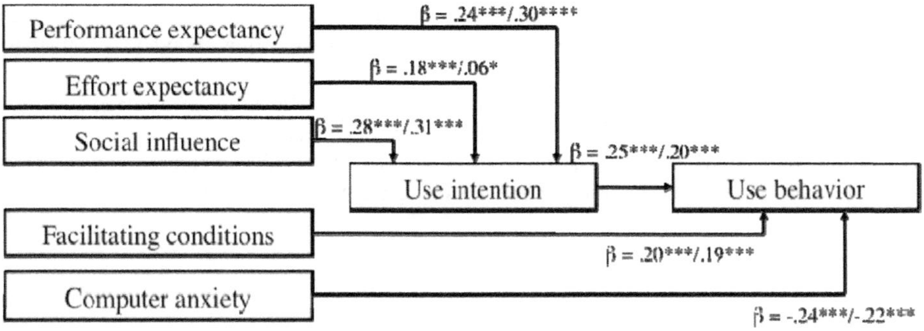

Fig. 2. The UTAUT model for technical vs. non-technical professionals

6 Discussion

The paper at hand examines the influences of culture on LTA, while making a distinction between ethnical and professional culture. A large sample including different ethnicities (i.e. German and Romanian) and different professions (i.e. technical and non-technical) offers evidence for different attitudes towards educational technology, and for different relationships between the variables of the UTAUT model. Ethnicity displays the strongest influence on LTA, both affecting learners' attitudes towards technology and moderating the effect of these attitudes on the use intention and use behavior. The influence of the professional culture is weaker, nevertheless statistically significant.

This study places main emphasis on theory, making the first steps towards integrating Hofstede's culture model [1] into the UTAUT [2]. In this respect, our findings are in line with the previous research [5], while enlarging the view of the culture influence. With respect to the educational practice, the study suggests acceptance and cultural dimensions as a starting point of dealing with cultural differences in the context of technology-enhanced learning.

References

1. Hofstede, G.: Culture's consequences: Comparing values, behaviors, institutions and organizations across nations. Sage, Thousand Oaks (2001)
2. Venkatesh, V., Morris, M.G., Davis, G.B., Davis, F.D.: User acceptance of information technology: Toward a unified view. MIS Quarterly 27, 425–478 (2003)
3. Ajzen, I., Fishbein, M.: Attitudes and the attitude-behaviour relation: Reasoned and automatic processes. European Review of Social Psychology 11, 1–33 (2000)
4. Nistor, N., Wagner, M., Istvanffy, E., Dragotă, M.: The Unified Theory of Acceptance and Use of Technology: Verifying the model from a European perspective. International Journal of Knowledge and Learning 6(2-3), 185–199 (2010)
5. Venkatesh, V., Zhang, X.: The Unified Theory of Acceptance and Use of Technology: U.S. vs. China. Journal of Global Information Technology Management 13(1), 5–27 (2010)
6. Zakour, A.B.: Information technology acceptance across cultures. In: Law, W.K. (ed.) Information Resources Management: Global Challenges, pp. 25–53. Idea, Hershey (2007)
7. Veltri, N., Elgarah, W.: The role of national culture in social networking adoption. In: Rogers, C.F., Chen, Y.C. (eds.) Proceedings of the 12th Annual Conference of the Southern Association for Information Systems, pp. 169–174. Georgia Southern University, Statesboro (2009), `http://sais.aisnet.org/2009/proceedings/SAIS%202009.pdf/Veltri,Elgarah.pdf` (last download: January 10, 2011)
8. Im, I., Kim, Y., Han, H.J.: The effects of perceived risk and technology type on users' acceptance of technologies. Information & Management 45, 1–9 (2008)

Transferring an Outcome-Oriented Learning Architecture to an IT Learning Game

Birgit Schmitz[1,2], Roland Klemke[2,1], Michael Totschnig[3], André Czauderna[4], and Marcus Specht[2]

[1] Humance AG, Goebenstr. 10-12, 50672 Köln, Germany
bschmitz@humance.de
[2] CELSTEC, Valkenburgerweg 177, Heerlen, The Netherlands
{roland.klemke,marcus.specht}@ou.nl
[3] Vienna University of Economics and Business, Augasse 2-6, 1090 Vienna, Austria
michael.totschnig@wu.ac.at
[4] Cologne University of Applied Sciences, Mainzer Str. 5, 50678 Köln, Germany
andre.czauderna@fh-koeln.de

Abstract. Today's technology enhanced learning scenarios focus on outcome-oriented delivery of learning processes, contents, and services. Also, learners increasingly demand for innovative and motivating learning scenarios that match their habits of using media. The European project ICOPER researches outcome-oriented learning infrastructures for higher education contexts. The German BMBF-project SpITKom aims at transferring such approaches to basic qualification. Based on a Browser Game, it uses ICOPER's technical infrastructure which combines learning object metadata repositories, learning outcome repositories, learning design repositories and learner profile repositories. This paper initially depicts the technical infrastructure of an outcome-oriented learning scenario that was developed in the course of ICOPER and then outlines its transformation to the game-based learning approach as realized in the course of SpITKom.

Keywords: game based learning, outcome-oriented learning, interoperability, IT-knowledge.

1 Outcome-Oriented Learning and Interoperability

One of the fundamental ideas of outcome-oriented education is to prepare students for the requirements of their professional life [13]. Rather than defining the resources to be used during the learning process, outcome-oriented learning scenarios focus on the results of the educational process, e.g. the skills and content students should be able to demonstrate.

Learning outcome orientation can be seen within a wider trend in educational technology recognizing the potential benefits of adaptivity. Paramythis and Loidl-Reisinger [9] argue that this "has been mainly driven by the realization that the ideal of individualized learning (i.e. learning tailored to the specific requirements and preferences of the individual) cannot be achieved, especially at a 'massive' scale, using traditional approaches" [p. 181]. Furthermore, the shift towards adaptivity has

C. Delgado Kloos et al. (Eds.): EC-TEL 2011, LNCS 6964, pp. 483–488, 2011.

been influenced by an increased awareness of learners' diversity which is a key consideration in both projects, ICOPER and SpITKom.

However, to be able to support such highly individualized learning scenarios, interoperability issues become important on the level of technical infrastructure: standards to describe learning outcomes, learner profiles, assessment items and learning materials are needed as well as tools to assess learner progress, maintain learner profiles, measure gaps and propose subsequent learning steps. Technical interoperability between different components of an overall learning delivery toolset is required in order to ensure seamless learning processes. Web-service based approaches help to simplify technical interoperability [14]. In the following, an exemplary application for outcome-oriented Learning will be described that was realized in the course of the European project ICOPER.

2 Learning Outcome-Orientation in ICOPER

The European ICOPER project [8] analyses and discusses state-of-the-art implementations of current standards as a base for the development of a comprehensive set of prototypes that support individual learning, teaching and authoring. In the course of ICOPER, the Open ICOPER Content Space (OICS) was developed which combines learning object metadata repositories, learning outcome repositories, learning design repositories and learner profile repositories [12].

Some of the standards the OICS works with comprise: *Sharable Content Object Reference Model (SCORM)* [1], *Learning Object Metadata (LOM)* [4] and *Open Archive Initiative's Protocol for Metadata Harvesting (OAI-PMH)* [6]. The *ICOPER Learning Outcome Definition (LOD)* is an application profile based on *IEEE Reusable Competency Definitions (RCD)* that can be used to create *Personal Achieved Learning Outcome (PALO)* profiles [7]. The OICS offers services that integrate concepts and data for the management of sharable educational resources as part of outcome-oriented learning processes (fig. 1).

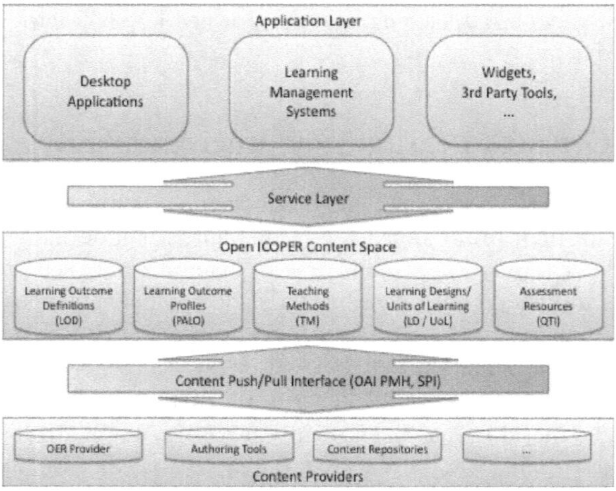

Fig. 1. OICS Architecture

The service-based architecture of the OICS enables its use for innovative learning scenarios such as game-based learning for instance: it defines web-services for all relevant learning events. The OICS does not require its own user interface to be in place for the learner though. Especially in game-based learning, the user interface of a learning management system would reduce the game-experience. Also, the OICS services are neutral in terms of the didactic processes they support. They serve as backend to fine-grained learning activities such as selection of learning outcomes, delivery of learning content, assessment execution, or monitoring and recording of learning progress. Thus, they allow efficient integration of heterogeneous applications, such as learning games, authoring environments or assessment engines.

In many learning games, learning processes are implicitly encoded into the game which makes the use of learning standards hard [5]. The use of a component such as the OICS enables to clearly separate the game logic from the learning activities and to control learning activities according to the game flow. In the following we will describe how the OICS architecture was applied to a learning game in the context of SpITKom.

3 Applying the OICS to SpITKom

SpITKom (**Sp**ielerische Vermittlung von **IT-Kom**petenz) builds on the motivational potential (cf. [6], [2]) and the learning potential of collaboration and reciprocal apprenticeship [11] computer games provide. Its main focus is to support the acquisition of IT-knowledge thus preparing and enabling educationally disadvantaged learners to find an apprenticeship. Furthermore, SpITKom aims at supporting the acquisition of practical knowledge related to the building industry. Therewith, SpITKom targets at bringing forward the participants' professional and social competence. Also, by directing the game to the building industry, it is intended to support the development of a professional identity.

SpITKom has chosen to integrate the European Computer Driving License (ECDL)[1] as a commonly accepted standard that reflects and certifies up-to-date skills and knowledge in the use of a computer and common applications. In its standard version 5.0, the ECDL syllabus comprises 478 learning outcomes, organized in seven modules.

Graphically, SpITKom was aligned to Browser Games such as FarmVille[2] as a reference for accessible, easy to use game environments that offer high level stimuli as well as ongoing feedback [10]. The learning game (see fig. 3) guides the learner through building- and construction-projects. Its main intention is to bring the target group (learners difficult to reach) "in touch" with the integrated IT-knowledge. A more elaborate engagement, i.e. the actual learning, takes place within the IT-Café (see fig. 4). It provides access to (a) the ECDL questions, (b) comprehensive knowledge tests that indicate individual knowledge gaps and (c) the recommended ECDL learning materials that help to close these gaps.

---

[1] http://www.ecdl.org
[2] http://www.farmville.com

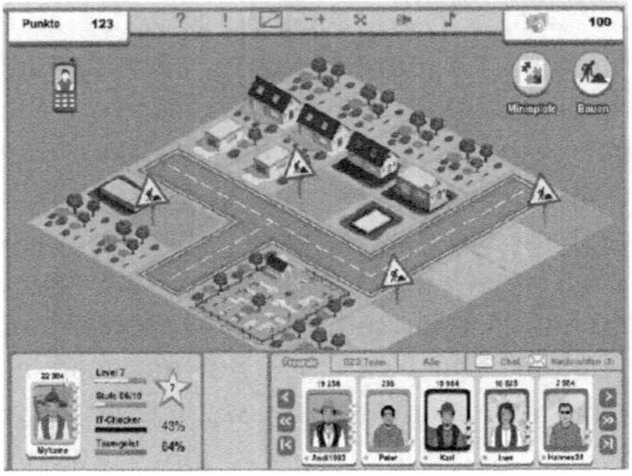

Fig. 2. SpITKom Browser-Game (Draft Version)

Fig. 3. IT-Café (Draft Version)

Technically, the SpITKom system is partly based on the OICS platform. The SpITKom architecture comprises two main components: the front-end community platform and the OICS learning service component. In addition, the open source QTIEngine[3] is integrated to visualize and evaluate assessments. The community platform is based on the LifeRay open source community server. It contains the flash-based game front-end (fig. 3) and the CCT (Competency Checker and Trainer) which is realized through the IT-Café (fig. 4).

During the game flow, certain situations trigger learning processes. These are delegated from the game to the CCT. The CCT is the communication bridge between the game process and the learning infrastructure. It is responsible for translating the learning process requests triggered from the game into the web-service calls of the OICS infrastructure.

The CCT also associates the community user with the learning profiles (PALO) stored in the OICS. Thus, it can individualize learning processes by selecting learning outcomes, learning contents and assessment items that are useful for the current learner. Additionally, the CCT controls the QTIEngine: According to the user's profile it selects the assessments to be played. The QTIEngine's assessment rendering components are integrated into the front-end.

SpITKom's learning requirements can be matched to the repositories and services offered by the OICS infrastructure (see fig. 2 for an overview of the complete SpITKom architecture including the OICS components). In this architecture, the OICS serves as a backend repository that encapsulates all contents and results relevant for the learning. Items of the ECDL syllabus can be represented as LODs stored in the learning outcome repository. ECDL learning contents are realized as SCORM units stored in the OICS content repository. ECDL tests are stored in the QTI format within the OICS assessment repository. The individual tests are played using the QTIEngine. Assessment results delivered from the QTI-engine are stored back into the profile repository in the PALO format.

---

[3] http://www.qtitools.org/landingPages/QTIEngine

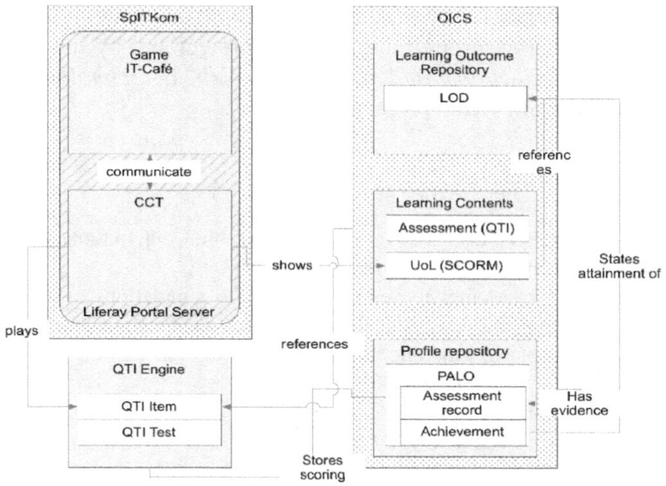

Fig. 4. OICS infrastructure as applied to SpITKom

In the course of the game, test items related to IT-Knowledge (ECDL) are displayed to the user. The questions are rated 1 (easy) to 3 (difficult). Depending on the learner's performance (right/wrong), the system's core backend component (CCT) chooses the follow-up question in an adaptive manner in order to rate the learner's knowledge. As technically provided by the OICS, SpITKom matches the ECDL learning outcomes against the learner's concrete abilities. This way, it analyses the learner's needs and offers questions and Units of Learning in the sense of concrete, contextualized units of education or training [3] that can each be traced back to a single learning outcome. The answers of learners are collected and stored back in the learner's profile.

4 Conclusion and Outlook

In this paper we have described SpITKom, a Browser Game designed to support the acquisition of IT knowledge. In order to support individualized learning and to enable technical interoperability between different components of an overall learning delivery toolset, SpITKom has implemented outcome-oriented learning and assessment components based on the ICOPER infrastructure (OICS). Thus, the solution offers all technical advantages of re-usability and integration inherent to standardized components.

An overall evaluation of the prototype with particular focus on the educational outcomes of the game is planned for early 2012. This evaluation will scrutinize (a) the effects of outcome-oriented delivery of content and services on the individual learning processes and (b) the learning outcomes. Besides, it will be of particular interest to evaluate the general capacity and motivation of learners (i.e. learners difficult to reach) to take over responsibility for their individual learning and respectively the corresponding learning outcome which is enabled by the transformation of the OICS to game based qualification. The results of this study will be published in due course.

Acknowledgments. Parts of this work are funded by the eContentPlus Programme of the European Commission through the ICOPER project and by the German BMBF-programme *Further Education and the use of Web2.0 technology in Vocational Education* through the SpITKom project.

References

1. ADL SCORM, Advanced Distributed Learning Sharable Content Object Reference Model, 2nd edn., Overview, Alexandria, VA (July 22, 2004)
2. Clark, A.: Learning by doing: the essential guide to simulations, computer games, and pedagogy in e-learning and other educational experiences. Pfeiffer, San Francisco (2005)
3. Derntl, M., Neumann, S., Oberhuemer, P.: Report on the Standardized Description of Instructional Models. In: ICOPER 2009 (2009)
4. IEEE LTSC, Final Draft Standard for Learning Object Metadata (2002), http://ltsc.ieee.org/wg12/files/LOM_1484_12_1_v1_Final_Draft.pdf (retrieved 01/15/10)
5. Kelle, S., Klemke, R., Gruber, M., Specht, M.: Standardization of Game Based Learning Design. Accepted for Proceedings of the International Conference on Computational Science and its Applications. Springer, Heidelberg (2011)
6. Lagoze, C., Van de Sompel, H.: The Open Archives Initiative: Building a Low-Barrier Interoperability Framework. In: Proc. 1st ACM/IEEE-CS Joint Conf. Digital Libraries, pp. 54–62. ACM Press, New York (2001)
7. Najjar, J., Derntl, M., Klobucar, T., Simon, B., Totschnig, M., Grant, S., Pawlowski, J.: A data model for describing and exchanging Personal Achieved Learning Outcomes (PALO). International Journal of IT Standards and Standardization Research (2010) (to appear)
8. Najjar, J., Simon, B.: Learning Outcome Based Higher Education: iCoper Use Cases. In: IEEE International Conference on Advanced Learning Technologies (ICALT), Workshop on European Workshop On E-Learning Standards Best Practice, Riga - Latvia, July 14-18 (2009)
9. Paramythis, A., Loidl-Reisinger, S.: Adaptive Learning Environments and e-Learning Standards. Electronic Journal of eLearning, EJEL 2(1), 181–194 (2004)
10. Schmitz, B., Czauderna, A.: Spielbasierte Kompetenzentwicklung für Bildungsbenachteiligte Jugendliche. In: Diener, H., Maciuszek, D., Malo, S., Martens, A., Urban, B. (eds.) Tagungsband der Abschlusskonferenz des Verbundprojektes Alphabit in Zusammenarbeit mit dem 4. Workshop Game-based Learning, Seiten 59-75, Fraunhofer (2010)
11. Steinkuehler, C.A.: Learning in massively multiplayer online games. In: Kafai, Y.B., Sandoval, W.A., Enyedy, N., Nixon, A.S., Herrera, F. (eds.) Proceedings of the Sixth International Conference of the Learning Sciences, pp. 521–528. Erlbaum, Mahwah (2004)
12. Totschnig, M., Klerkx, J. Klobučar, T., Law, E., Simon, B., Ternier, S.: D1.1 Open ICOPER Content Space Implementation of 1st Generation of Open ICOPER Content Space including Integration Mini Case Studies. ICOPER project deliverable (2009), http://www.educanext.org/dotlrn/clubs/icoper/new-lors/Deliverables/Deliverables_-_Submitted/D1.1/D1.1-final.pdf (retrieved online at 28.06.2010)
13. Vander Ark, T.: Toward Success at Scale. Phi Delta Kappan 84(4), 322–326 (2002), http://www.jstor.org/stable/20440344 (retrieved 12/13/2010)
14. Vossen, G., Westerkamp, P.: Why service-orientation could make e-learning standards obsolete. Int. J. Technology Enhanced Learning 1(1/2), 85–97 (2008)

Integrating Social Media and Semantic Structure in the Learning Process

Eline Westerhout[1], Thomas Markus[1], Vlad Posea, and Paola Monachesi[1]

[1] Utrecht University, Utrecht
[2] "Politehnica" University of Bucharest, Romania

Abstract. This paper presents the validation of an inFormal Learning Support System (iFLSS) that integrates social media and semantic structure within the learning process. The validation has shown that the system supports the learning process and results in an improved domain understanding.

Keywords: education, social media, ontology, learning.

1 Introduction

In our global world, learning is increasingly becoming a social process that goes beyond the physical boundaries of classroom activities: learners exchange information through social media, belong to various social networks and use the Internet to work together. The inFormal Learning Support System (iFLSS), that has been developed within the Language Technology for LifeLong Learning (LT-fLL) project[1], has been designed to meet the needs of these learners. The system makes full use of social media since it supports learners in finding content that has been uploaded within applications such as YouTube, Delicious, SlideShare and Bibsonomy and identifies it as appropriate learning material. The iFLSS has been implemented as a fully functional prototype and has been tested in classroom settings in order to verify whether learners are ready to exploit the power of social media not only in their free time but also to acquire new knowledge and ultimately to learn. In this paper, we focus on the results of a summative and formative evaluation of the iFLSS with learners and tutors after providing a short overview of its technical features.

Following Stahl [6], the iFLSS views the knowledge building process as a mutual construction of individual and social knowledge, seeking a balance between the Acquisition (individual) and the Participation (social) Metaphor. The system is also based on Vygotsky's idea that individuals develop personal representations and beliefs from their own perspectives, which are influenced and adapted by socio-cultural knowledge building, shared language and external representations [7]. As a result of social and individual learning, learners become

[1] http://www.ltfll-project.org

C. Delgado Kloos et al. (Eds.): EC-TEL 2011, LNCS 6964, pp. 489–494, 2011.

skilled members of a Community of Practice (CoP) [2]. The social learning process is in the iFLSS heavily supported by social networking sites, according to Redecker[2] one of the most common Web 2.0 applications in learning. Besides a social learning component, the iFLSS supports the knowledge discovery process by integrating semantic web technologies. Jovanovic [1] provides a comprehensive study on the application of semantic web technologies, folksonomies and the social semantic web for intelligent tutoring.

2 The inFormal Learning Support System

Learners searching for relevant and trusted information may get lost in the huge amount of learning materials and information when they use the Internet. The iFLSS addresses the challenges that learners face in their learning process using two components – a Knowledge Discovery component and a Social Learning component. The Knowledge Discovery component supports the learning process by offering a visual overview of the structure of a domain in a so-called ontology. These ontologies are linked to relevant learning materials from social media sites. Secondly, the Social Learning component employs the power of social media to support learners in selecting relevant and high-quality online learning materials within the vast amount of online resources.

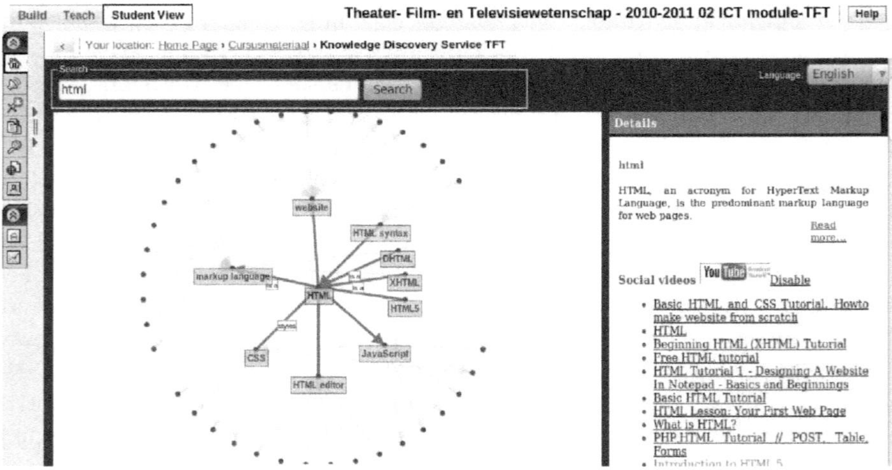

Fig. 1. The iFLSS Knowledge Discovery component: (1) "*Search box*", (2) "*Ontology browsing*", graph visualisation of a domain model, (3) "*Definition*" of the centralised concept, (4) "*Learning Materials*", dynamically retrieved from social networks

The central element of the Knowledge Discovery component of the iFLSS is the visualisation of an ontology (Figure 1). The ontology accommodates learners

[2] ftp://ftp.jrc.es/pub/EURdoc/EURdoc/JRC55629.pdf

with insufficient vocabulary by providing a visualisation based on the vocabulary of the CoP extracted from social media in addition to the vocabulary and structure provided by domain experts. To ensure up-to-dateness, completeness, flexibility, and an extended vocabulary, a method has been developed that enriches ontologies semi-automatically on the basis of tags [3]. The visualisation distinguishes between concepts coming from the expert-validated domain ontology (blue) and concepts that have been automatically added (green, in the figure the concept *CSS*). The search for learning materials is triggered either by entering a query or by clicking a concept in the ontology. The system searches dynamically on YouTube (videos), Delicious (bookmarks), SlideShare (slides), and Bibsonomy (scientific papers). A disambiguation component improves the results for ambiguous and poorly chosen keywords when searching in social media applications (e.g. python on YouTube).

After acquiring the basic domain conceptualisations, the learner needs to obtain specialised knowledge in order to become an expert. Learners usually have difficulties deciding which learning materials are appropriate and a common approach is to ask a tutor or a more knowledgeable friend for advice and references. On the social web, this process can be simulated by adding such persons to their networks. The Social Learning component of the iFLSS takes these connections as input and guides learners by providing trustworthy search results and recommendations of learning materials from learners' and tutors' social networks [5,4]. In addition, the system recommends knowledgeable peers for a given query. The Social Learning component can be used in two different settings. In the former setting, a learner uses his own network to search. The latter setting can be more appropriate in a higher education environment where learners are more limited in the use of social networks. In this situation, the recommendations of peers and materials are based on the social network accounts from the tutor instead of the ones from the learner.

3 Quantitative Validation with Learners

A small-scale validation of a preliminary version of the iFLSS yielded encouraging results suggesting that the iFLSS improves the learning process [8]. The feedback received from those learners has been taken into account in the final version, which has been validated with 80 undergraduate students within two existing courses at different universities. The first group consisted of 35 first year Humanities students from Utrecht University doing the course "*Introduction IT*" and a control group with 18 learners using Google instead of the iFLSS. The second group (45 Computer Science (CS) students) followed the "*Human Computer Interaction*" course at the University of Bucharest. The students have been using the iFLSS within their course for three to four weeks. The iFLSS has been embedded in the Learning Management System (LMS) employed within each course: Blackboard WebCT for the Humanities students and Moodle for the CS students. Because of the restricted use of social media for learning purposes by both the learners and tutors involved in the validation, we used the

Table 1. The learners' opinions on the iFLSS for learning purposes

		Humanities			Computer Science		
Aspect	**Service**	**Mean**	**Stdev**	**% Pos**	**Mean**	**Stdev**	**% Pos**
Visualised ontology	KD	3.12	0.88	38	4.10	0.94	70
Relevance learning materials	KD	3.20	0.93	43	3.25	0.91	35
	SL	3.32	0.91	47	2.88	0.83	28
	Google	3.40	0.71	29			
Relevance peers	SL	3.29	0.68	38	3.56	1.12	56

second setting of the social learning component and employed a tutor network containing experts and learning materials on the topics discussed in the course.

The potential of the iFLSS to enhance the learning process has been measured objectively by means of pre and post tests. Questionnaires and interviews have been used in addition to the pre and post tests to evaluate the iFLSS. The questionnaires consisted of statements on which the learners had to express their level of agreement on a five point Likert scale (1=strongly disagree–5=strongly agree). The outcomes of the questionnaires have been discussed in more detail in the interviews.

The results on the pre and post tests have shown that the iFLSS offers learners a broader view on the course domain. The progress of the learners in the test group was significantly higher than in the control group ($t\ (88) = 2.55$, $p < .05$). Table 1 summarizes the results on the questionnaires. For each of the aspects, the related component of the iFLSS is indicated (KD or SL). The CS students appreciated that the **visualised ontology** allowed them to learn by themselves, while the Humanities students often indicated that they needed more support (e.g. indicating (non-)obligatory course topics). The **relevance of the learning materials** returned by the iFLSS and Google has been judged in the context of their respective courses. In general, results of both the iFLSS and Google are not tailored to the learning task. There is neither a statistical difference between the two iFLSS components, nor between the test and control groups. The experiment set-up made it difficult for some students to assess the **relevance of the peers**, since they didn't know the recommended people. The positive responses were based on the trust the students had in their tutor and the fact that the students considered this an interesting feature that is currently missing in search engines.

The learners involved in the validation can be divided into two groups. There is a group that shows curiosity towards the iFLSS, whereas at the same time other learners are not yet convinced of the added value. The CS students are significantly more positive compared to the Humanities students for both the Knowledge Discovery ($t(52) = 3.61$, $p < 0.05$) and the Social Learning component ($t(57) = 2.67$, $p < 0.05$). Logging results have been collected as an objective measure of the likelihood of adoption. The iFLSS has been used by 49 students during the validation period. 26 out of these 49 students (53.1%) have accessed the system again after the pilot ended. In addition, 13 students who did not participate in the experiment have accessed the software.

4 Qualitative Validation with Tutors

The iFLSS has been validated with 11 tutors from different backgrounds at Utrecht University in a workshop setting to investigate whether they see it as an opportunity for enhancing the learning process. The workshop started with a short presentation about the iFLSS and the overall objectives of the LTfLL project. The tutors then experimented with the system individually for about 30 minutes, after which they filled in a questionnaire consisting of multiple choice and open ended questions. The workshop ended with a discussion about the potential of the iFLSS for learning.

The questionnaires and discussions have shown that many tutors consider the iFLSS useful for education, but that in general they were not ready yet to employ social media for teaching purposes. As a consequence, they were more positive about the knowledge discovery component. They agreed in the discussions that the use of ontologies for learning could positively influence the level of domain understanding of the learners. The most important drawbacks of the Knowledge Discovery service were the lack of tutor support and the fact that the design of ontologies for new domains requires a lot of expertise and time. Although the tutors did see advantages in social ontology enrichment, it became clear that they wanted complete control over their learning resources and domain model. As for the Social Learning service, they were afraid that it would require them to disclose personal information to students and feared that they would have to be available all day for students. They proposed alternatives, such as allowing them to choose which resources and messages to share with students. The discussions also revealed that students are very bad at determining the quality of resources. Further embedding social media in their teaching would increase the risk that students would collect information from sites like Wikipedia, which are generally considered to be unsuitable for scientific education. Another concern was that students need proper guidance during their studies and that the recommendation of student-provided resources might lower the quality of education. Some of the tutors, who generally were using social media already, did not consider it problematic to disclose their network to students. These tutors actually considered it to be an advantage, since students might come across interesting people. The majority of tutors (73%) saw market potential for a system like the iFLSS, while the remaining 27% indicated that they would like to use the system for a longer period before judging this.

5 Conclusions and Future Work

The integration of ontologies and social media in education is promising. The iFLSS has made an important contribution towards integrating semantic domain structures with social search, social media content and social networks in a learning environment. The learners appreciated the support offered by the iFLSS and the tutors were positive about the market potential of a system like the iFLSS. Especially the integration of semantic domain representations in the learning process is considered relevant for learning. In general, learners seem more ready to embed social media in the learning process than tutors. Tutors are worried that this would have consequences for the quality of education and their privacy. However, even though they were not ready to embrace social media themselves yet, the tutors expect a shift towards the integration of social media in education in the future. The learner validation showed that the background of learners is also an important factor that influences the willingness to adopt new tools and techniques: the learners with a Computer Science background were more eager to try new methods and tools than the Humanities learners.

The iFLSS is a first step in the direction of personalised learning. Each learner follows its own learning path, which changes dynamically during the learning process. Instead of offering the same ontology to each learner, a future version of the iFLSS could be used to create personalised ontologies for each individual learner on the basis of his past behaviour (e.g. bookmarks, videos, courses). The social search can then be used to identify learning materials within the learner's network that match his level of understanding and needs. The tutors indicated that they wanted to be able to adapt and modify the ontology and resources, which is not possible in the current version yet. We should include a tutor support feature in a next version of the iFLSS.

References

1. Jovanović, J., Torniai, C., Gašević, D., Bateman, S., Hatala, M.: Leveraging the social semantic web in intelligent tutoring systems. In: Woolf, B.P., Aïmeur, E., Nkambou, R., Lajoie, S. (eds.) ITS 2008. LNCS, vol. 5091, pp. 563–572. Springer, Heidelberg (2008)
2. Lave, J., Wenger, E.: Situated learning: Legitimate peripheral participation. Cambridge University Press, Cambridge (1991)
3. Monachesi, P., Markus, T.: Using social media for ontology enrichment. In: Aroyo, L., Antoniou, G., Hyvönen, E., ten Teije, A., Stuckenschmidt, H., Cabral, L., Tudorache, T. (eds.) ESWC 2010. LNCS, vol. 6089, pp. 166–180. Springer, Heidelberg (2010)
4. Posea, V., Trausan-Matu, S.: Bringing the social semantic web to the personal learning environment. In: Proceedings of ICALT 2010, IEEE Computer Society, Washington, DC, USA (2010)
5. Posea, V., Trausan-Matu, S., Mossel, E., Monachesi, P.: Supporting collaborative learning across social media applications. In: Proceedings of CSCL 2009 - Volume 2, International Society of the Learning Sciences (2009)
6. Stahl, G.: Group Cognition: Computer Support for Collaborative Knowledge Building. The MIT Press, Cambridge (2006)
7. Vygotsky, L.: Mind and society: The development of higher mental processes. Harvard University Press, Cambridge (1978)
8. Westerhout, E., Monachesi, P., Markus, T., Posea, V.: Enhancing the learning process: Qualitative validation of an informal learning support system consisting of a knowledge discovery and a social learning component. In: Wolpers, M., Kirschner, P.A., Scheffel, M., Lindstaedt, S., Dimitrova, V. (eds.) EC-TEL 2010. LNCS, vol. 6383, pp. 374–389. Springer, Heidelberg (2010)

Role-Playing Game for the Osteopathic Diagnosis

Pierre Bénech, Valérie Emin, Jana Trgalova, and Eric Sanchez

Institut Français de l'Education - ENS Lyon - EducTice
19 Allée de Fontenay - BP 17424 - 69347 Lyon - cedex 07- France
{pierre.benech,valerie.emin,jana.trgalova,
eric.sanchez}@ens-lyon.fr

Abstract. The aim of this paper is to make a contribution to teaching and learning combining game-based learning and inquiry-based learning. Our main objective is to offer learners conditions enhancing the development of behaviors, attitudes and skills in the osteopathic diagnosis. To this end, we have designed a pedagogical scenario based on an online role-playing game simulating a medical appointment between an osteopath and a patient, using mainly asynchronous forums as the learners are practicing physiotherapists. The game design is based on the analysis of the real activity of an osteopath and of the case studies produced by the school of osteopathy in their previous courses. We present here the design methodology, the conceptual framework and the role-playing game scenario.

Keywords: Role-playing game, Inquiry-based learning, Pedagogical scenarios, Technology enhanced learning, Osteopathic diagnosis.

1 Introduction

For several years researches in the field of Technology Enhanced Learning focus on the design of effective learning situations integrating digital technologies. Socio-constructivist activities and collaborative learning activities have proved their efficiency in the context of problem-based learning for instance. Some of these "innovative" pedagogical methods, such as problem-based, inquiry-based and game-based learning, are particularly relevant in the context of science education and are appropriate for using digital technologies. In this context, designing relevant and high quality learning situations is still a significant concern [1].

In our research, we explore the possibility to apply such innovative pedagogical methods in training trainees-osteopaths to make a diagnosis, in reply to PLP Formation (PLP) request. PLP is a training school of osteopathy and physiotherapy in Lyon, France. It offers an osteopathic training designed for health care practitioners initially trained as physiotherapists. This 5-year training comprises seven 4-5 days of internship per year. The density of the training, combined with their physiotherapist practice, does not allow the trainees to really practise osteopathic diagnosis. PLP thus wishes to design new activities firstly aiming at training in osteopathic diagnosis and secondly aiming at implementing an active pedagogy favourable to the development of professional skills and allowing these activities to be done and monitored at distance.

C. Delgado Kloos et al. (Eds.): EC-TEL 2011, LNCS 6964, pp. 495–500, 2011.

In this paper, we present a Role-Playing Game (RPG) we have designed (section 3), which aims at developing professional skills required in making osteopathic diagnosis. In section 2, we outline the theoretical frameworks and the methodology used before presenting the learning game scenario.

2 Methodology and Design

2.1 Designing Serious Games

The term "serious game" was used long before the introduction of computer and electronic devices into entertainment. Abt [2] gave a useful general definition, which is still considered applicable: "*Reduced to its formal essence, a game is an activity among two or more independent decision-makers seeking to achieve their objectives in some limiting context. A more conventional definition would say that a game is a context with rules among adversaries trying to win objectives. We are concerned with serious games in the sense that these games have an explicit and carefully thought-out educational purpose and are not intended to be played primarily for amusement.*" This definition highlights the key elements of a game: *activity* as a series of actions taking place over time, players or *decision-makers*, *objectives* which are the desired outcomes, and *limiting context* structured by rules. Serious games are thus grounded in the (socio) constructivist learning theory. We add the three registers used in the design of a serious game: pedagogical, didactical and game principles [3], completed by the immersive aspect of the game.

Simulation games and *Role-Playing Games* are of special interest in the field of the acquisition or exercise of skills. They are games in which players assume the roles of characters in a fictional setting. They offer the possibility of replicating an activity in "real life" using a game form. In the case of medical training, future practitioners can practice on virtual patients without risking damages on real ones.

In order to design learning activities supporting the development of professional skills, one needs to analyze a professional activity in a real situation. In the next section we present how we have conducted this analysis.

2.2 Modeling the Activity of Osteopathic Diagnosis

Professional skills are required from an osteopath to make a diagnosis. The approach of professional activity analysis is consistent with the work of professional didactics. Our analysis is based on various data we gathered in order to capture the real osteopath's activity [4]: (1) systems of professional standards elaborated by the osteopaths' trade unions that define the osteopath profession and its principles, (2) videos of the osteopathy diagnosis activity realized for training purposes, by setting up a diagnosis session during which the PLP trainers played the roles of an osteopath and of a patient according to a synopsis based on the professional standards, and (3) interviews with the responsible of PLP aiming at making explicit the process of diagnosis. This data analysis allowed us to model the osteopath's activity and more specifically, the activity of diagnosis, as shown in Fig. 1.

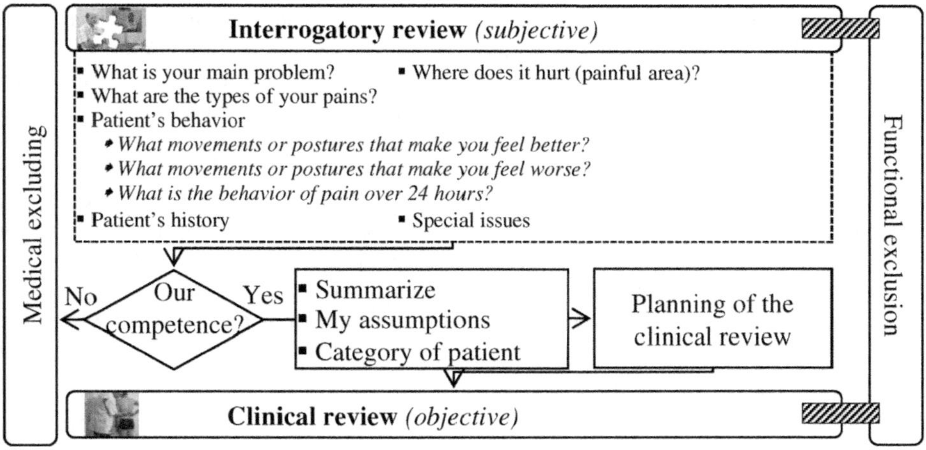

Fig. 1. A model of the osteopathic diagnosis [5]

The activity of osteopathic diagnosis is very much like an inquiry in a sense that the osteopath proceeds by questioning, making hypotheses, carrying out investigations to test them.

To determine the best treatment for the dysfunction observed, the osteopath makes a diagnosis of the patient's problems. It is a process where the practitioner will discuss with the patient's affected structures and propose a treatment. This is why a large part of the osteopathic appointment is devoted to questioning (interrogative review) and physical examination (clinical review). Both reviews require mobilizing medical knowledge in symptomatology and semiology.

The next section presents the conceptual framework our learning scenario design is based on.

2.3 Pedagogical Scenarios

The analysis of teaching practices has been based for decades on several conceptual frameworks: the cognitive psychology and information processing, analysis of representations, ethno-methodology focused on "thinking in action", narratives of experiences, etc. It shows that a teacher as a practitioner has implicit models based on a set of routines [6]: recurrent learning activities, teaching routines, management and enforcement routines. These routines are especially used to orchestrate teaching and learning activities and to choose appropriate strategies with respect to students and contextual elements. At the same time, the interest in design research as a valuable methodology for educational research has been growing. Various theoretical constructs have been proposed to capture abstractions to design learning situations integrating digital technologies and among them *pedagogical design patterns* [7, 8] and *Educational Modelling Languages* (EML) [9]. Design patterns capture recurring features across narratives, encapsulating critical challenges and forces pertaining to a domain of learning design, the interactions between them and possible solution methods. EML aim at providing interoperable descriptions of learning scenarios using XML language. As pointed out by IMS-LD authors themselves [10], an EML, which mainly aims at expressiveness and interoperability, is not intended for a direct manipulation by human users (teachers, designers, engineers…). Specific conceptual models and authoring systems are provided [11] in order to help practitioners to design their own scenarios using patterns and vocabulary nearer to their own practices. ISiS (Intentions, Strategies, and interactional Situations) conceptual framework [12] is one of these and allows structuring the design of learning scenarios by teachers-designers. This framework is based on a goal-oriented approach and offers a specific identification of the intentional, strategic, tactical and operational dimensions of a scenario. ISiS is implemented in ScenEdit authoring environment [12], a graphical tool dedicated to design learning scenarios. It aims to favor sharing and reusing practices by providing patterns for each type of ISiS component.

The next section describes the role-playing game scenario designed by combining these conceptual frameworks: serious game design, activity analysis and pedagogical scenarios design.

3 Role-Playing Game Scenario

When searching for a solution to the PLP request, we opted for a simulation of the ostheopath diagnosis activity. Among possible types of games, a RPG seemed the most suitable as it gives the trainees opportunities to play various roles chosen to represent real roles in the real situation [13]. The RPG we have designed is implemented on a platform using only forums.

We describe in this article only the pedagogical register of the RPG. From the activity analysis (cf. section 2.2), we have identified a teaching pattern corresponding to an inquiry-based learning approach that allows the osteopath to make an osteopathic diagnosis. This pattern enhances the pedagogical register of the RPG prototype that we suggest for PLP. To design the learning game scenario, we have used ISiS model and ScenEdit authoring tool (cf. section 2.3).

The activity analysis allowed us identifying two main pedagogical intentions for this pattern. The first intention is enabling the students to "mobilize knowledge in a real professional situation". The second intention is enabling the students to "develop autonomy in planning the clinical examination" since the trainees-osteopaths are practicing physiotherapists, and therefore they are not used to plan and carry out osteopathy examinations on their own.

To reach these intentions, we define an inquiry-based learning strategy representing the conducting of an osteopathic appointment, detailed in four phases. It has been implemented into the ScenEdit editor in the strategy called "conducting an osteopathic appointment", as shown in Fig. 2.

Phase 1: *Conduct an interview.* The questioning of the patient allows the osteopath to determine precise indications and contraindications of the therapy while building a reflection looking for mechanical dysfunctions of the body.

Phase 2: *Perform clinical examination.* Following the initial summary at the end of the interview, a clinical examination leads the osteopath to detect areas of the patient's body with movement restrictions that may affect her/his health.

Phase 3: *Decide osteopathic care.* The synthesis of these two diagnoses (examination and palpation) enables the osteopath to identify a dysfunction and to determine priorities, the cause or causes of the problem and finally decide osteopathic care to implement.

Phase 4: *Practice osteopathic care*. It is only after this process that the osteopath can start the treatment.

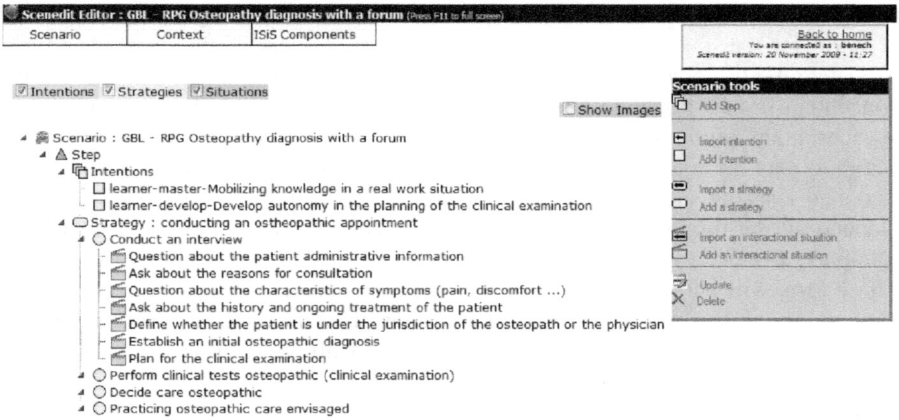

Fig. 2. Implementation into ScenEdit of the osteopathic appointment strategy

The first phase, "Conduct an interview", is declined by seven interactional situations which are collection of interactions and activities "using a specific set of roles, tools, services resources, locations according to the *situational context*" [12]. For instance, the situation "Ask about reasons for the appointment", is the step where three trainees play the role of an osteopath asking questions, via a forum, to a patient played by two trainees who provide answers to the questions. Within both teams, the trainees discuss the questions to ask or the answers to provide via dedicated forums accessible only to the team members (Fig. 3). A description of the patient's case is provided to the patient team which then knows the dysfunction and can respond precisely to the questions of the osteopath team whose aim is to make a diagnosis and identify the dysfunction. The validation of the diagnosis is done by the patient team comparing the diagnosis of the osteopath team with the dysfunction to be discovered.

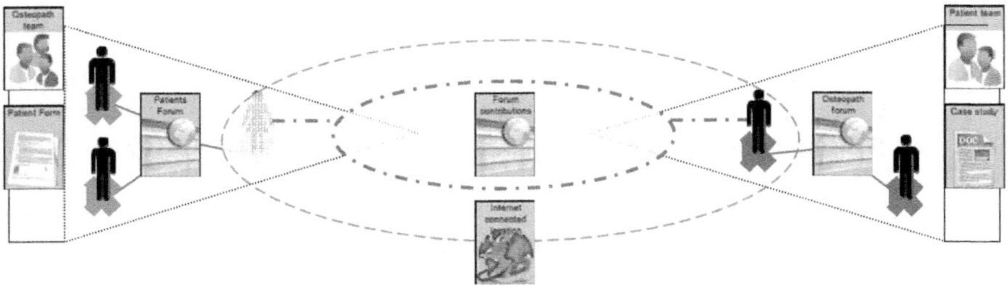

Fig. 3. Schema of the Role-playing game

4 Conclusion

In this paper, we have presented an overview of the design methodology, the pedagogical scenario and the first implementation on a learning platform of a RPG scenario for the training in osteopathic diagnosis. The aim of this work was first to formalize the osteopathic diagnosis using activity analysis and professional didactics conceptual frameworks and then to design a pedagogical scenario using RPG on an asynchronous forum. This learning game scenario requires the involvement of several actors to play the role of the osteopath, but also of the patient. The experiment of the designed learning situation with a group of learners at PLP School of Osteopathy is planned during training sessions in July 2011. This experiment will aim to validate: role-play, interactions between actors, accessibility and empowerment of the game. These tests will allow making necessary adjustments and settings that will improve the quality of the design in order to permanently maintain the delicate balance between the expertise to acquire and the fun intrinsically brought by the game.

References

1. Lockyer, L., Bennett, S., Agostinho, S., Harper, B. (eds.): Handbook of research on learning design and learning objects: Issues, applications, and technologies. Information Science Reference, New York (2009)
2. Abt, C.: Serious Games. The Viking Press, New York (1970)
3. Mariais, C., Michau, F., Pernin, J.P.: The use of Game Principles in the Design of Learning Role-Playing Game Scenarios. In: Proceedings of the 4th European Conference on Game-Based Learning, pp. 462–469. Academic Publishing Ltd. (2010)
4. Leplat, J., Hoc, J.M.: Tâche et activité dans l'analyse psychologique des situations. Cahiers de Psychologie Cognitive 3(1), 49–63 (1983)
5. Bénech, P., Pommerol, P.: Jeu de rôle par forum. Mémoire de master 2 professionnel, CUEEP, Lille 1 (2011)
6. Schank, R.C., Abelson, R.: Scripts, plans, goals and understanding. Erlbaum, Hillsdale (1977)
7. Dimitriadis, Y., Goodyear, P., Retalis, S. (eds.): Using e-learning design patterns to augment learners' experiences. Computers in Human Behavior 25(5) (2009)
8. Emin, V., Pernin, J.-P., Prieur, M., Sanchez, E.: Stratégies d'élaboration, de partage et de réutilisation de scénarios pédagogiques. International Journal of Technologies in Higher Education 4(2), 25–37 (2007)
9. Koper, R., Tattersall, C.: Learning Design: A Handbook on Modelling and Delivering Networked Education and Training. Springer, Heidelberg (2005)
10. Koper, R.: Current Research in Learning Design. Educational Technology & Society 9(1), 13–22 (2006)
11. Botturi, L., Stubbs, S.: Handbook of Visual Languages in Instructional Design: Theories and Pratices. Informing Science Reference, Hershey (2008)
12. Emin, V., Pernin, J.-P., Guéraud, V.: Model and Tool to Clarify Intentions and Strategies in Learning Scenarios Design. In: Cress, U., Dimitrova, V., Specht, M. (eds.) EC-TEL 2009. LNCS, vol. 5794, pp. 462–476. Springer, Heidelberg (2009)
13. Wills, S., Leigh, E., Ip, A.: The Power of Role-based e-Learning: Designing and Moderating Online Role Play. Routledge, London (2010)

M-Learning Manager: An Emerging Profession

Fabrizio Giorgini

eXact learning solutions, Via Portobello,
16039 Sestri Levante, Italy
f.giorgini@exactls.com

Abstract. The m-learning is emerging as a new sector in education and training provision, side by side with face-to-face education, distance education and e-learning. The m-learning introduced new pedagogical and technical paradigms for the design and delivery of m-learning materials. The e-learning manager not necessarily has the knowledge to manage m-learning projects or products thus a new profession is emerging: the m-Learning Manager. This paper focuses on what the mLeMan EC project is doing to define this new profession in terms of needed skill set and related learning materials useful to fill the current professional gap in the m-learning management.

Keywords: m-learning.

1 Introduction

There never was a technology as widely available to citizens as mobile telephony. This technology connects people working at different places and different education and learning paths with opportunities for expert and peer feedback and co-learning. Mobile technology offers unprecedented possibilities for combining the strengths of formal and non-formal education and professional internship. For the first time in the history of the use of technology in education and training, is a technology that will cost the learners nothing, because they own the technology to be used.

The m-learning is emerging as a new sector in education and training provision, side by side with face-to-face education, distance education and e-learning. Just as distance education was recognized and accepted as a field in the 1970s when the great European Open Universities were founded and e-learning was accepted as a field in 1995 when the World Wide Web first became available to educators. We can say that we have been in the process of acceptance of m-learning from the beginning of the 21st century, along with 3G/UMTS and Smartphones.

The new mobile learning arena introduced significant changes in the design of the training programs - the ways they are structured and maintained, the definition of the usability requirements and the assessment methodology (which results only apparently similar to all VET - vocational education and training - systems) [1]. These changes require new specific learning elements that a manager should be able to apply when dealing with m-learning projects and products. In other words, they set the requested pedagogical and technological skills of the manager facing m-learning application development and management.

C. Delgado Kloos et al. (Eds.): EC-TEL 2011, LNCS 6964, pp. 501–505, 2011.

The m-Learning Manager is not a developer, or teacher – she/he has to organize and manage m-learning design, development, evaluation and implementation into her/his, organization. In this light she/he needs to get a picture of development and teaching processes of m-learning. The m-Learning Manager should be aware of the benefits and potential of m-learning, staff needed for m-learning development, resources and organization. The aim of this work is to present what the EC LLP project mLeMan [2] is doing for defining the set of skills an m-Learning Manager should have and how these skills can be developed and assessed.

The project is based on the work of two highly successful EU projects, one of which (EQN - European quality Management [3]) established a Europe-wide certification body based on job role standards, rather than academic content, and the second of which ("ELM - e-learning manager" [4]) developed and mainstreamed a Europe-wide skills recognition and qualification service for e-learning managers.

2 The mLeMan Project Approach

mLeMan project aims at to develop and mainstream Europe wide skills recognition and qualification service for m-Learning Managers, and other individual management level in charge with evaluating sourcing or implementing m-learning solutions. More specifically, the project aims at:

1. to develop, in cooperation with the main European m-learning actors, a skills set for the job role of m-Learning Manager according to the needs of m-learning market and taking into account VET and qualifications systems changes such as the shift to learning outcomes and competence-based systems
2. to develop m-Learning Manager Training Program: learning outcomes, learning objects, multimedia learning resources in the form of task for performance in four languages, in support of the established skills set
3. to develop the DALA System (Development, Assessment, Learning, Assessment) with the functions: set skills card, self assessment, create a learning portfolio, check learning references, attend courses online
4. to configure the new skills set and content in DALA, and pilot the program in all participating member states of the project (at least 100 participants in the pilot) and refine the program based on systematic feedback.

At the present time the project, which started in October 2010, is dealing with point 1, the development of the skills set for the m-Learning Manager job role. A skill set is a group of specific learning elements that a person should be able to apply within a certain job role.

The skill set for the job role of m-Learning Manager has been structured according to skill set standards outlined in the EQN guide [5] and based on the skills definition proposed by the DTI (Department of Trade and Industry) in the UK for NVQ (National Vocational Qualification) standards [6]. It contains the following items:

- **Domain:** An occupational category. E.g. Domain = Process Improvement.
- **Job Role:** A certain profession that covers part of the domain knowledge. E.g. Job role = Yellow Belt, Orange Belt, Green Belt or Black Belt.
- **Unit:** A list of certain activities that have to be carried out in the workplace. It is the top-level skill in the qualification standard hierarchy. Each unit consists of a number of elements.
- **Learning Element:** Description of one distinct aspect of the work performed by a worker, either a specific task that the worker has to do or a specific way of working. Each element consists of a number of performance criteria.
- **Performance criteria:** Description of the minimum level of performance a participant must demonstrate in order to be assessed as competent.
- **Level of cognition:** For each performance criteria there is an intended level of cognition. At the same time this describes the complexity level of the test questions for each performance criteria, according Bloom's Taxonomy – Rev. 2001 [7].

Particular attention has been paid to the differences between the skill set of the e-learning manager defined by the ELM project (see http://www.iscn.com/projects/piconew_skill_portal/DirTree/index.php) and the m-learning manager defined in mLeMan. The following section presents the result of such work and illustrates the identified skill set for the m-Learning Manager.

3 The M-Learning Manager Skill Set

Using the terminology outlined in the skills definition model and including the skills identified during the demand analysis at the beginning of the project, a skills hierarchy for the job role Mobile Learning Manager has been designed. It contains the following Units:

U1. Pedagogical Aspects of M-Learning

The pedagogy of mobile learning straddles two quite different pedagogical systems: the pedagogy of traditional face-to-face education and the pedagogy of distance education. The aim of this unit is to recognize the role of mobile learning in conventional, face-to-face education and in distance education. The objectives are to 1) develop skills in the didactical structuring of mobile learning materials, 2) develop skills in the pedagogy of the use of media in mobile education systems, 3) develop skills in the pedagogy of the development of student support services for students studying by mobile learning and 4) develop skills in the management of the course development process for mobile learning in Textual materials, audio, video and TV materials, location and context sensitive course materials and augmented reality so that the qualified m-Learning Manager is able to manage the production of mobile learning materials in her/his institution.

The Unit is then divided in the following set of sub-units:

U1.E1 Mobile learning characteristics and design principles
U1.E2 Devices and Content
U1.E3 Learning theories & approaches in m-Learning

U2. Mobile Learning Tools and Technologies

Nowadays, there are many different mobile platforms. Each of them has its different standards and methodologies for content display. Often additional tools are needed for the development of educational content. Due to the smaller screen sizes, different way for working and lack of "typical" for computers features, using standard ways for content delivery is not suitable. Therefore new architectures and systems for content development, which are mobile-ready, must be used. In cases, where standard methods like web and/or file downloading cannot be used (like when sensors like camera/gyroscope/GPS must be used), native application development must be done. Unit U2 aims at providing a complete knowledge of existing m-learning tools, their features and the technologies used to create and deliver m-learning contents.

The Unit is then divided in the following set of sub-units:

U2.E1 m-Learning Content Development - main concepts
U2.E2 Technological Layers in m-Learning

U3. M-Learning Management

The aim of the unit is to give a clear vision about the methods of m-learning target group identification, analysis of their needs and constraints such as the operational environment. As well to be aware of the innovation aspects of m-learning management.

The Unit is then divided in the following set of sub-units:

U3.E1 Needs analysis
U3.E2 Innovation and Business Management

For each sub-unit related performance criteria have been also defined. The criteria will be used to assess the level of learning of each learner at the end of the pilot phase. An example of performance criteria follows:

U1.E2.PC1 Knowing which devices should be considered for mobile learning (PDAs/smart phones, digital phones and non-telephony devices including MP3 players, tablets) and their main specifications

U1.E2.PC2 Knowing the type of media elements (text, audio, video, etc.) used in m-Learning content according to the characteristics of the mobile devices

They will be used to assess the learning of the sub-unit U1.E2 Devices and Content.

4 Conclusions

The 2-years Leonardo project mLeMan project started in October 2010 is defining a skill set for the emerging new profession "m-Learning Manager". The skills requirements have been collected with the contribution of several key players and experts of m-learning around Europe.

The definition of the skill set presented in this paper is the first step of the process the will bring us to the certified profession of the m-Learning Manager. The next phase of the project will focus on the development of the learning contents for each identified skill according to the defined performance criteria.

After that, our m-learning manager definition will be submitted to the European Certification and Qualification Association (ECQA, [8]) for the final step of the project: the certification of the new profession of the m-Learning Manager.

The mLeMan project is made in the context of the growth of m-learning into one of the most promising educational technologies of the last years and the paucity of qualifications and programmes at managerial (as opposed to design and development levels) levels in the field.

Acknowledgements. This research has been co-funded by the European Commission within the Leonardo Da Vinci project mLeMan (mobile Learning Manager), grant 509960-LLP-1-2010-1-BG-LEONARDO-LMP.

References

1. Gaskel, A.: Open and distance learning on the move: mobile learning, crossing borders, rethinking roles (Editorial). Open Learning. V. 22(3), 197–200 (2007)
2. MLeMan project, `http://mleman.dipseil.net/`
3. EQN European Quality Network, project information, `http://www.leonardodavinci-projekte.org/prj/2022/project_2022_en.pdf`
4. ELM, e-learning manger project, `http://deis.ie/elm/`
5. EQN guide chapter 2 part 2 – European Qualification Network, `http://www.iscn.com/projects/eucertificates/program/eqn2007-guideline.zip`
6. DTI - Department of Trade and Industry UK, British Standards for Occupational Qualification, National Vocational Qualification Standards and Levels
7. Krathwohl, D.R., Anderson, L.W.: A Taxonomy for Learning, Teaching, and Assessing: A Revision of Bloom's Taxonomy of Educational Objectives (2001)
8. European Certification and Qualification Association (ECQA), `http://www.ecqa.org/`

Self-regulated Learners and Collaboration: How Innovative Tools Can Address the Motivation to Learn at the Workplace?

Teresa Holocher-Ertl[1], Claudia Magdalena Fabian[1], Melody Siadaty[2,3], Jelena Jovanović[4], Kai Pata[5], and Dragan Gasevic[3]

[1] Centre for Social Innovation/Technology and Knowledge, Austria
[2] Simon Fraser University, Canada
[3] Athabasca University, Canada
[4] University of Belgrade, Serbia
[5] Tallinn University, Estonia
holocher@zsi.at, fabian@zsi.at, melody_siadaty@sfu.ca, jeljov@gmail.com, kpata@tlu.ee, dgasevic@sfu.ca

Abstract. This paper describes an innovative motivational approach combined with new tools to support self-regulated learners, stressing the importance of social embeddedness and collaboration for learners at work. We suggest that the social context plays an important role when it comes to defining learning goals, adapting one's strategies to social norms, and collaboratively creating knowledge with colleagues and peers. A first assessment of this motivational approach and early prototypes was conducted within the IntelLEO project by the end-users of three heterogeneous business cases. The results of this evaluation in the biggest business case are presented in this paper.

Keywords: Self-regulated learning, collaboration, motivation, workplace.

1 Introduction

In today's working environments knowledge is reproduced, created, and recombined in fast cycle-times and in problem contexts that are difficult to imitate in educational institutions. Thus, workplace learning is foreseen to be the "next big thing" [1] and social learning in ICT-enabled communities is becoming the dominant source of education in areas where new practical knowledge emerges rapidly and has a short lifetime. Being competent becomes a dynamic and social phenomenon, where individual workers are the managers of their learning paths and the self-determined enhancement of work-relevant knowledge is based on exchanging experience with colleagues. The new ways of learning in the workplace require self-motivated and proactive learners, whom Zimmerman B.J. [2] refers to as self-regulated learners. Self-regulated learning (SRL) activities assume the learner's own direction, and include additional components such as self-monitoring of effectiveness and self-motivation [3], whilst a tutor's direct support and guidance are missing; yet, there are many learning options, as also many opportunities for distraction. Thus, the motivation to get involved and persist in

C. Delgado Kloos et al. (Eds.): EC-TEL 2011, LNCS 6964, pp. 506–511, 2011.

SRL is such a critical factor. In the scope of the IntelLEO project (www.intelleo.eu), the motivational approach and innovative tools aim to address the challenge of motivation for SRL in the workplace.

2 Self-regulated Learners and Motivation

Self-regulated learners engage in self-observation, self-judgment, and self-reactions, which can go simultaneously and are structured in three cyclical phases [4]: 1) *The forethought phase*, where a learner identifies the need to enhance his/her competences, sets his/her learning goals, selects strategies to reach them, and judges the perceived capability and expected outcomes. 2) *The performance phase* contains activities related with self-control and self-observation. The learner instructs (him-)herself, tries to focus his/her attention on the learning activities, works out task strategies and monitors his /her performance. 3) *The self-reflection phase* contains two main processes: self-judgment and self-reaction. According to these phases, at a first glance SRL looks to focus mainly on individual learners and their learning process. However, in our approach we hypothesize that the social context plays an important role especially in the workplace, where employees need to collaboratively enhance their work-relevant knowledge [5]. Taking a closer look at the role of motivation in self-regulation of learning, we also take into consideration the cognitive theories of motivation, highlighting the relevance of achievement goals [6].

The Achievement Goal Theory. In this theory, learners' behavior of actively engaging in learning is influenced by two different sets of goals: learning goals and performance goals. *Learning (mastery)-goal oriented individuals* want to increase their competence and challenges are seen as an opportunity to learn. *Performance goal oriented individuals* want to gain favorable judgments of their competence and challenges are seen as a test to measure competence [7]. It demands the comparison with other individuals (social reference norm), whereas the learning goal orientation seeks for comparison with one's achievements in a specific situation or with existing standards (individual reference norm) [8]. An important contribution to this multi-goal perspective comes from [9], which differentiates between social and academic goals. Social goals reflect desires to achieve a particular social outcome and influence achievement in their own right, as well as together with learning goals. The pursuit of social goals, as being cooperative, compliant, and willing to share, is positively related with successful learning [9]. Thus, research suggests combining orientation towards learning goals and orientation towards social goals to increase learning and performance [6].

Collaboration and the Motivation to Learn. The positive influence of being embedded in a learning community on learners' motivation is also confirmed by other scholars, emphasizing the importance of collaboration for successful learning [10, 11] and suggesting that social factors can be as powerfully motivating as intellectual ones in keeping learners on task. Collaboration enhances the active exchange of ideas within small groups, increases interest among participants, and promotes critical thinking [12]. Empirical work about knowledge sharing in virtual communities implies that community-related outcome expectation and social factors play an important role in knowledge sharing in terms of quantity and quality [13]. A lack of social relationship between the knowledge - seeker and sharer(s) was identified as an import barrier for knowledge sharing [14]. From the viewpoint of internalizing external regulations, [15] argues that external learning goals and regulations will be more successfully harmonized with one's own values, if they are supported by relevant others.

3 IntelLEO's Motivation Concept and Tools

IntelLEO is a research project of the 7th Framework program, which has as one of its primary aims the increase in motivation for learning and knowledge building within and across organizational borders through innovative tools and procedures. To elaborate the motivation concept, we defined two clusters that need to be supported by new tools and procedures: 1) *Support of self-regulated learners* and 2) *Support of collaboration and social embeddedness*. The first cluster requires procedures that support self-regulated planning, setting, observing and evaluating individual learning goals, helping learners to relate personal learning goals and interests to those of the organization, dividing them into intermediate, specific goals. The second cluster emphasizes procedures that strengthen the community as a source of motivation. Pro-social goals, feeling of relatedness, collaborative discussion and enhancement of learning objects, aim to increase the learners' motivation for (non-intrinsically) motivated tasks.

These two clusters are closely interwoven, as the planning, performance and monitoring processes in self-regulated learning are based on resources and activities that are constructed and undertaken collaboratively within and across organizations.

To support the two clusters the IntelLEO project develops generic services for managing collaborative learning activities and contents, as well as an ontological framework for learning context representation. The main tools discussed in this paper are the *Learning Planner (LP)* and *Content and Knowledge Provision* (CKP) tools.

The *LP* allows learners to have ubiquitous access to their personal learning spaces where they can manage and share their personal learning goals, competences, learning paths, activities and resources. In the forethought phase LP provides learners with insight into the organization's expectations via the recommendation of available competences. These recommendations stem from the person's position and his/her corresponding duties/projects/tasks. In addition, the LP shows the "popularity" of any given competence, demonstrated via elements such as comments or the number of times it was added to users' learning goals; thus facilitating the selection of competence(s) to be acquired and supporting the feeling of social embeddedness. To support the performance phase, the LP highlights the latest uploaded knowledge resources related to one's learning goals thus facilitating the access to relevant learning content. It also provides insight into one's social activity stream (e.g. sharing, tagging, adding to a goal, commenting, requesting help, and accessing learning content). This allows learners to pursue their learning activities even after work-related interruptions. The LP supports the self-reflection phase via showing the completeness of each learning goal, competence and activity of the learner. This allows users to have an overall view of their progress in each goal (e.g. with regard to the goal's urgency) and also compare their progress toward the goal with their colleagues; e.g. for a group-based goal.

The *CKP* tool enables users to easily bookmark and annotate online resources, upload different kinds of Knowledge Objects (KOs) into a knowledge repository, and (re-)discover relevant KOs by performing semantic search. It provides *a portfolio of annotations* – annotation of learning resources with tags (plain keywords), ontology concepts, competences, learning goals etc. These annotations are the primary facilitators for later search/retrieval and recommendation in all three phases of SRL.

4 Early Evaluation of the Motivational Concept and Tools

An early validation of the new tools and motivational approach was undertaken, by applying a participatory design approach [16] in the three IntelLEO business cases. The results from this first evaluation in the largest business case, which involved 10 employees of a leading car-manufacturer, are described in this section. The initial evaluation of the tools was conducted with storyboards, combining low-fidelity prototypes with usage scenarios. The evaluation used an open-ended scenario with the aim to collect broad and conceptual answers concerning the prototypes. In this scenario, a fictive employee interacted with the future system to plan his new learning goal, access existing learning resources, search for learning partners and document his own learning experiences. To introduce the storyboard to the participants, demonstration sessions were conducted with end-users in a focus-group setting with 5 participants per focus group. The results from this first evaluation show the relevance of both clusters of the motivational concept (Sect. 3).

In the forethought phase, learners stressed the importance of a *structured access to existing knowledge (cluster 1)* and required services for social embeddedness which helped to enrich the recommendations and suggestions with information coming from the experiences and behavior of colleagues and peers (cluster 2). To increase the motivation for informal workplace learning within and across organizations participants demanded for *intelligent recommendations on WHAT to learn*, according to the learner's level of existing knowledge, the learning activities already done and the competences already acquired by the learner. These recommendations have also to consider the learning experiences and recommendations from co-learners, thus creating a *"thread"* which is based on one's own and the colleagues' learning history.

In the performance phase participants stressed the importance of the LP as a tool to *collaboratively structure and collect information about completely new topics*. In this business case, new learning topics are elaborated in groups and one of the main challenges is to define the most important questions of the new topic, to collaboratively collect and contribute knowledge to answer those questions, and to bring them into a structure, so it can be reused by others at a later stage. Additionally, employees required support in *finding colleagues as co-learners* and the ability to send a request for *help for a competence, a learning activity or a knowledge asset*, if the motivation for informal workplace learning should be increased. In the self-reflection phase the most important aspect discussed was the tracking of one's learning activities and thus demonstrating the completeness of one's learning goals (cluster 1). Allowing individuals to tick off learning activities that were successfully completed, enables learners to monitor what one has already accomplished and what that is still open to successfully achieve a competence. On the one hand, achieving a competence and becoming an "expert" are perceived as motivating, and thus monitoring one's progress is an important driver of motivation. On the other hand, participants pointed out that it might be in return de-motivating when competences change over time and thus one risks to lose this "expert status" again. Another topic discussed in more detail was "time": to understand how much time it needs to acquire a new competence based on the estimation of learning content creators and experiences of previous learners. But also how much time is provided by the organization for employees to increase their competence is seen as relevant. Concerning this issue, performance approach aspects were raised again, for instance what would happen if one cannot finish a learning activity in the given average time frame.

Looking at the goal orientation, the results of this initial evaluation showed strong learning goal orientation of respondents, especially regarding the enhancement of the work-relevant knowledge, which is mainly conducted due to a high intrinsic motivation for learning. When it comes to sharing knowledge with colleagues and partner-organizations, social goals were revealed to play an increasingly important role, like sharing knowledge to help colleagues, supporting teamwork or showing a socially expected behavior. Performance goals turned out to be relevant, but were controversially discussed whenever they came up. The feedback from the participants showed the two sides of the medal: the comparison with a social reference norm is perceived as motivating, if learners feel capable of reaching the expected positive achievement under the given cognitive, social and organizational conditions. If one fails to positively achieve an expected behavior, although effort is put into the activity, then the comparison with others risks de-motivating individuals to get involved in learning and knowledge sharing. A longer evaluation of the IntelLEO tools using full prototypes will provide more detailed inputs on these aspects and how they influence the motivation to learn and build knowledge in an organizational context.

5 Summary

In the IntelLEO project, we took the approach to develop a motivational concept and innovative tools that support the activities of self-regulated learners in the forethought, performance and self-reflection phases, as well as collaboration and social embeddedness in all of these phases. A first evaluation with storyboards of the Learning Planner and the Content and Knowledge Provision tools showed that both, the support of SRL activities and collaboration were relevant. While in the forethought phase being socially embedded in a group of learners supports one's own planning and setting of learning goals, in the performance phase the collaboration with colleagues is key to success to motivate sharing and documenting one's knowledge and to elaborate new organizational and cross-organizational knowledge. In the self-reflection phase, co-learners provide a social reference norm to evaluate the SRL achievements. A detailed evaluation of the motivational approach and related tools for informal learning within and across organizations will follow using full prototypes.

Acknowledgement. This publication was partially supported/co-funded by the European Community/European Union under the Information and Communication Technologies theme of the 7th Framework Program for R&D. This document does not represent the opinion of the European Community, and the European Community is not responsible for any use that might be made of its content.

References

1. Straub, R.: Knowledge work in a connected world: is workplace learning the next big thing? Journal of Applied Research in Workplace E-learning 1(1), 5–11 (2009)
2. Zimmerman, B.J.: A Social Cognitive View of Self-Regulated Academic Learning. Journal of Educational Psychology 81(3) (1989)
3. Rheinberg, F., Vollmeyer, R., Rollett, W.: Motivation and action in self-regulated learning. In: Boekaerts, M., Pintrich, P.P., Zeidner, M. (eds.) Handbook of Self-Regulation, pp. 503–529. Elsevier Academic Press, London (2000)
4. Zimmerman, B.J.: Becoming a self-regulated learner: an overview. Theory Into Practice. FindArticles.com (2002)
5. Jackson, T., Mackenzie, J., et al.: Communal Aspects of Self-Regulation. In: Boekaerts, M., Pintrich, P.P., Zeidner, M. (eds.) Handbook of Self-Regulation, pp. 275–296. Elsevier Academic Press, London (2000)
6. Montalvo, F.T., Gonzalez Torres, M.C.: Self-Regulated Learning: Current and Future Directions. Electronic Journal of Research in Educational Psychology 2(1), 1–34 (2004)
7. Covington, M.V.: Goal Theory, Motivation, and School Achievement: An Integrative Review. Annual Review Psychology 51, 171–200 (2000)
8. Heckhausen, J., Heckhausen, H.: Motivation und Handeln. Springer Medizin Verlag, Heidelberg (2006)
9. Wentzel, K.R.: Social Relationships and Motivation in Middle School: The Role of Parents, Teachers, and Peers. J. of Educational Psychology 90(2), 202–209 (1998)
10. Dillenbourgh, P.: What do you mean by collaborative learning? In: Dillenbourg, P. (ed.) Collaborative-Learning: Cognitive and Computational Approaches, pp. 1–19. Elsevier, Oxford
11. Alvesson, M.: Social identity and the problem of loyalty in knowledge-intensive organizations. Journal of Management Studies 37(8), 1101–1020 (2000)
12. Garrison, D.R.: Will "distance" disappear in "distance studies? A Reaction. J. of Distance Education 14(2) (1999)
13. Chiu, C.M., Hsu, M.H., Wan, E.T.G.: Understanding knowledge sharing in virtual communities: An integration of social capital and social cognitive theories. Decision Support Systems 42, 1872–1888 (2006)
14. Ardichvili, A., Page, V., Wentiling, T.: Motivation and barriers to participation in virtual knowledge-sharing communities of practice. J. of Knowledge Management 7(1) (2003)
15. Ryan, R.M., Deci, E.L.: Self-determination theory and the facilitation of intrinsic motivation, social development, and well-being. American Psychologist 55, 68–78 (2000)
16. Holocher, T., Kieslinger, B., Fabian, C.M.: Participatory methods and their implications in collaborative workplace learning. In: Proc. of World Conference on Educational Multimedia, Hypermedia and Telecommunications 2010, pp. 7–15 (2010)

Active Learning for Technology Enhanced Learning

Artus Krohn-Grimberghe, Andre Busche, Alexandros Nanopoulos, and Lars Schmidt-Thieme

Information Systems and Machine Learning Lab
University of Hildesheim, Germany
{artus,busche,nanopoulos,schmidt-thieme}@ismll.uni-hildesheim.de

Abstract. Suggesting tasks and learning resources of appropriate difficulty to learners is challenging. Neither should they be too difficult and nor too easy. Well-chosen tasks would enable a quick assessment of the learner, well-chosen learning resources would speed up the learning curve most. We connect active learning to classical pedagogical theory and propose the uncertainty sampling framework as a means to the challenge of selecting optimal tasks and learning resources to learners. To assess the efficiency of this strategy, we compared different exercise selection strategies and evaluated their effect on different datasets. We consistently find that uncertainty sampling significantly outperforms several alternative exercise selection approaches and thus leads to a faster convergence to the true assessment. These findings demonstrate that active (machine) learning is consistent with classic learning theory. It is a valuable instrument for choosing appropriate exercises as well as learning resources both from a teacher's and from a learner's perspective.

Keywords: Active learning, recommender systems, student performance prediction, optimal learning, intelligent tutoring systems.

1 Introduction

From a pedagogical point of view, learners should be challenged and guided in an optimal way such that they are neither overwhelmed with too difficult material nor bored with too easy or repetitive material [4]. This approach goes back to Vygotsky's classic "zone of proximal development" (ZPD) as a framework used to describe the corridor of content difficulty in relation to the learner's skill level that is best suited for learning [7]. For an illustration see figure 1.

In machine learning, the area of active learning connects to that challenge. Active learning researches strategies on how a machine learning algorithm can achieve greater accuracy (a better assessment) by carefully eliciting the next training labels (exercise results) to see. One of the most popular active learning frameworks is uncertainty sampling, where learning instances are chosen such that the posterior probability of being positive (pass) is closest to the posterior probability of being negative (fail) [2] .

To the best of our knowledge, we are the first to connect active learning and classical learning theory in general and uncertainty sampling and Vygotsky's Zone of Proximal Development [4,7] in particular.

Our contributions are as follows:

C. Delgado Kloos et al. (Eds.): EC-TEL 2011, LNCS 6964, pp. 512–518, 2011.

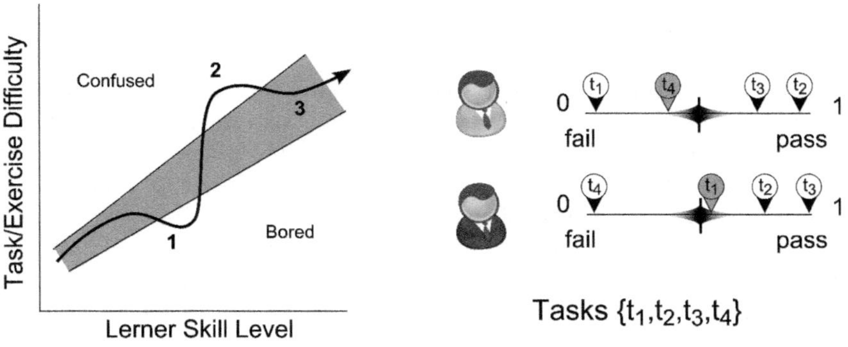

Fig. 1. Left: Vygotsky's Zone of Proximal Development as the area where the tasks/exercises neither bore (1) or overwhelm (2) the learner. Right: Personalized view of different tasks/exercises per learner. For each learner different tasks may be in the Zone of Proximal Development (shaded area on the pass-fail-scale).

- The selection strategy for tasks and learning resources follows a principled approach that directly derives from Vygotsky's Zone of Proximal Development and active (machine) learning.
- The learner needs less interactions with the learning environment for the same rigor in assessment.
- The learner gets assigned exercises that are more appropriate to his current skills.
- The model is allowed to change over time and adapt to the learning curve of the learner, as the systems model is updated with every new observation.
- Due to the collaborative nature of the underlying recommender algorithm, any response elicited from the learner improves the model for the learner and the exercise or resource at hand, helping other learners, too.

2 Related Work

As recommender systems are domain independent, they have been introduced to technology enhanced learning with great success [3]. In e-learning, recommender systems have been used for recommending learning material and exercises to learners. Furthermore, recommender systems have been successfully employed in student performance prediction [6]. In those works, the focus of suggesting and classifying material for learners was on data mining methods and traditional statistical performance metrics. The question of which material is best suitable from a learning theory point of view was usually not addressed.

Measuring the suitability of tasks in the context of educational sciences has been studied, e.g., in [4]. They analyze the Vygotskian constructs of scaffolding and the zone of proximal development within the context of adaptive instructional system design and evaluation. They describe a conceptual framework in which it is tried to maintain a student in an ideal (neither too bored, nor to challenged) state when solving exercises.

Their key approach is based on the idea that students should need less hints to solve a problem set than previously expected by a human teacher (and/or system), aiming at mastering individual steps of a certain problem. In a possible application scenario, they use the students' performance to skip some exercises and/or groups of exercises. Our approach differs as we pro-actively estimate students' performance without the need to manually pre-define sequential learning tasks (ordered by their level of difficulty). Furthermore, we follow a principled machine learning approach in selecting the appropriate tasks. As our approach is collaborative, each learners performance on any task improves our estimation of other learners on those and other tasks, too.

In machine learning active learning refers to the elicitation of new training data for algorithms. The key idea is that the algorithm should be able to determine those instances for training that bear the most additional information and increase the predictor's accuracy most if added to the training data. The crucial part in this setting is the way of choosing "good" additional training samples. In any setting, however, the algorithm obtains labels from an oracle (e.g., a human annotator) in an interactive manner, e.g. through a user interface. Active learning was applied to recommender systems before [5]. However, so far it has not been connected to classical pedagogical theory.

3 Active Learning for Technology Enhanced Learning

In our proposed approach we use active learning as a filtering step on top of a recommender algorithm's predictions. The recommender models the learner's predicted success rate on the exercises or the helpfulness of the learning resources, respectively. Active learning is then leveraged to select the appropriate exercise or resource for presentation to the learner. Finally, the learner's feedback is used for updating the underlying recommender model and the next iteration of recommendations and active learning starts (c.f. algorithm 1). For performance reasons we retrained the algorithm only after obtaining one new training instance for each learner.

Algorithm 1. Training with Active Learning

```
while not converged do
    for all learner l in dataset d do
        predict score on unknown exercise using recommender
        select most informative exercise using active learning
        elicit label for most informative exercise
        add label to training data
    end for
    retrain or update recommender model
end while
```

For our experiments we used the biased matrix factorization as the underlying recommender algorithm, as it is known to be a very strong algorithm in general and to perform well on the specific datasets in particular [6].

The first strategy for selecting the optimal exercise is the random selection strategy. Under the assumption that more data on average improves prediction accuracy, randomly selected exercises are presented to the learner and the obtained results are added to the training partition for the next iteration.

Uncertainty sampling [2] is one of the simplest and most frequently employed frameworks in active learning. In this framework, a tutor presents those task instances to the active learner where the tutor is least certain how the learner would label them. For probabilistic learning models this approach is often straightforward. The biased matrix factorization can make a real-valued prediction for each {learner; exercise}-pair which can be mapped to the 0..1 interval. As such it can be interpreted as a probabilistic model. For our uncertainty sampling implementation we then present each learner that exercise where the prediction is closest to 0.5 and obtain the learner's true score on this exercise and add this item to the training set for the next iteration. A predicted value of 0.5 implies maximum uncertainty about the learners result on the selected task. As such this strategy resembles the Zone of Proximal development closest and consequently should yield the most interesting exercises for the learners.

For an overview of all employed strategies see table 1.

Table 1. Characterization of Selection Strategies for Active Learning Methods employed; each of which based upon Biased Matrix Factorization

Task Selection Strategy	Description
Biased Matrix Factorization	no active learning
Random	randomly select a new exercise per learner
Personalized User Average	predicted score must be closest to learner's average score
Uncertainty Sampling	predicted score is closest to 0.5

Although uncertainty sampling on top of a recommender model is already personalized (as the recommender's predictions already are), we wanted to explore if further personalization is still helpful. Instead of selecting instances near the global 50% uncertainty boundary, we investigated the effect of selecting instances near the learner specific success rate (as sampled from the training data). Thus, those exercises were presented to the learner (later on, added to the training set) that were closest to her average score.

4 Experimental Results

We empirically compare the quality of two active learning strategies against the non-updated model (Biased Matrix Factorization) and the random selection of further training samples (random). For active learning, we choose uncertainty sampling and personalized user average. Recommendation quality is measured by decreasing root mean square error (RMSE) and evaluated on transactional data from adaptive tutoring systems.

We evaluated the proposed methods on the KDD Cup 2010 Educational Data Mining Challenge[1] data. For the KDD Cup 2010 Educational Data Mining Challenge two

[1] https://pslcdatashop.web.cmu.edu/KDDCup/

datasets, "Algebra 2008-2009" and "Bridge to Algebra 2008-2009", were provided. Those datasets capture historical information from computer-aided tutoring systems, each storing several million records about the performance of students on tasks, having a binary indication of success or failure as well as the required knowledge components and skills, among other things like time stamps. Consistent with [6], we extracted the triples {student id; problem group + step name; success} ("Algebra Step") as well as {student id; knowledge component; success} ("Algebra KC") from "Algebra 2008-2009" . From "Bridge to Algebra 2008-2009" we also extracted the triples {student id; knowledge component; success} ("Bridge KC"). For more details see [6].

We splitted all three datasets into a training, an active learning and a holdout partition: 50 per cent of the triples per user came into the training partition, 35 per cent into the active learning partition. The remaining ones went into the holdout partition. For detailed dataset statistics see table 2. The training partition was used to train the biased matrix factorization and optimize the hyper-parameters via grid search (learnrate=0.02, iterations=100, latent dimensions=64, regularization=0.0075 for Algebra Step; learnrate=0.02, iterations=100, latent dimensions=8, regularization=0.0075 for Algebra KC). From the active learning partition the further training examples could be obtained one by one. Evaluation was conducted on the holdout partition which was never touched for training. Prediction quality and performance improvements were measured by improvements in root mean square error (RMSE) on the holdout partition.

Table 2. Dataset statistics

Dataset	Users	Items	Triples
Algebra Step	3,311	877,181	3,848,398
Algebra KC	3,310	2,976	365,776
Bridge KC	6,044	2,944	719,936

Figures 2a,b show the predictive performance of the presented active learning strategies. The continuous yellow line indicates the biased matrix factorization RMSE without any active learning. It can be clearly seen that all strategies improve over this score by obtaining additional training examples during active learning. Randomly selecting further instances for learning already improves accuracy significantly; Personalized user average sampling is generally able to infer even more information on two datasets but fails to improve over random on Algebra KC (results on Bridge KC and the classic Movielens recommender dataset not shown for brevity). As supported by theory, the uncertainty sampling strategy selecting instances with a predicted value closest to global decision boundary outperforms the baseline and the simpler active learning strategies.

The above presented results show that even randomly adding more training data to the model is helpful and over-personalization not backed by theory (personalized user average) does not outperform this in any case. As predicted by active learning literature, uncertainty sampling proved to be the best among the tested approaches. This goes in line with Vygotsky's classical "zone of proximal development" (ZPD), a framework used to describe the corridor of content difficulty in relation to the learner's skill level that is best suited for learning[7]: Learners should be challenged and guided such that they are neither overwhelmed with too difficult material nor bored with to easy or repetitive material [4] (c.f. figure 1a). To be precise, classical pedagogical theory would want challenges of just *slightly above* the decision boundary. As our method by design will pick the exercise (or learning resource) estimated to be closest (but not necessarily exactly at) the decision boundary, this difference should be very subtle though.

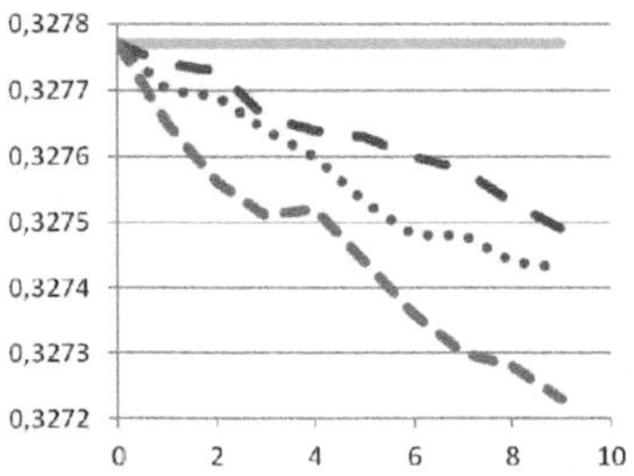

(a) Performance on Steps data

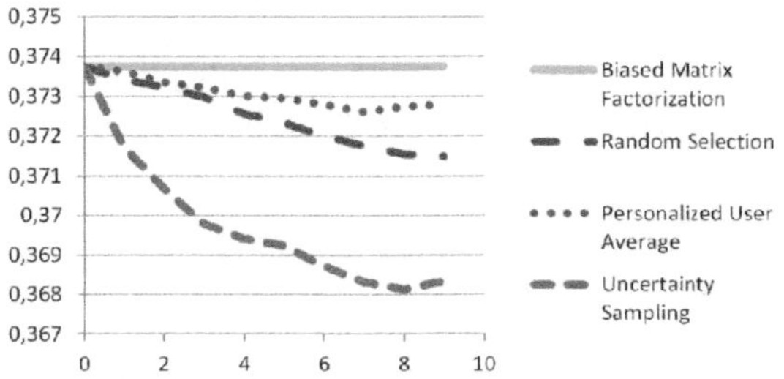

(b) Performance on KnowledgeComponent data

Fig. 2. Prediction error (RMSE) of the various active learning strategies on the Algebra data vs. the number of selected instances from the active learning partition.

A limitation of this approach is the fact that it needs past transactional data from the learners interaction with a tutoring system. But this may even stem from the same exercises but other learners (say, from a previous course), because active learning is also very useful in a cold start situation of a new user or a new exercise [1].

5 Conclusion

Assessment of learners in any learning environment should be swift, thus reducing the burden on teacher and learner. This would safe resources and time better spent on learning relevant lessons or on leisure. The presented active (machine) learning strategy of uncertainty sampling for the personalized selection of exercises and learning materials exactly offers a means for achieving an efficient and prolific learning experience. It connects classical learning theory and machine learning theory as a means to realize Vygotsky's Zone of Proximal Development. Furthermore, we want to emphasis, that not only the speed of assessment is better than with other exercise selection strategies, but also the difficulty of the recommended exercises should exactly match those "[neither] too hard and [nor] too easy" exercises that were stipulated in [6].

For future work, it would be interesting to explore further active learning strategies and evaluate the effects in a user-study. Furthermore, it would be interesting to connect to dynamic user modelling and instructional design of adaptive learning tasks.

References

1. Golbandi, N., Koren, Y., Lempel, R.: Adaptive bootstr. of recommender systems using decision trees. In: Proc. of the 4th ACM Intl. Conf. on Web Search and Data Mining (2011)
2. Lewis, D., Catlett, J.: Heterogeneous uncertainty sampling for supervised learning. In: Proceedings of the International Conference on Machine Learning, ICML (1994)
3. Manouselis, N., Drachsler, H., Vuorikari, R., Hummel, H., Koper, R.: Recommender systems in technology enhanced learning. In: Recommender Systems Handbook: A Complete Guide for Research Scientists & Practitioners (2010)
4. Murray, T., Arroyo, I.: Toward measuring and maintaining the zone of proximal development in adaptive instructional systems. In: Cerri, S.A., Gouardéres, G., Paraguaçu, F. (eds.) ITS 2002. LNCS, vol. 2363, pp. 749–758. Springer, Heidelberg (2002)
5. Rish, I., Tesauro, G.: Active Collaborative Prediction with Maximum Margin Matrix Factorization. In: Inform. Theory and App. Workshop (2007)
6. Thai-Nghe, N., Drumond, L., Horváth, T., Krohn-Grimberghe, A., Nanopoulos, A., Schmidt-Thieme, L.: Factorization techniques for predicting student performance. In: Educational Recommender Systems and Technologies: Practices and Challenges (2011)
7. Vygotsky, L.S.: Mind in Society: The Development of Higher Psychological Processes. Harvard University Press, Cambridge (1978)

A Model for Assessing the Success of Virtual Talent Communities

Daniel Müller[1], Anke Diederichsen[2], Christian Gasper[2], and Stefan Strohmeier[2]

[1] IMC, Altenkesseler Str. 17 D3,
66115 Saarbrücken, Germany
[2] Chair for Management Information Systems, Saarland University,
Post Office Box 151150,66041 Saarbrücken, Germany
daniel.mueller@im-c.de, {a.diederichsen,c.gasper,
s.strohmeier}@mis.uni-saarland.de

Abstract. Design characteristics constitute a promising approach to support researchers and practitioners in developing, implementing and evaluating virtual talent communities (VTC) in order to prevent costly misconceptions in every phase of the software lifecycle. The paper aims at deriving a rigorous and relevant model for assessing the success of VTC, realised by use of the prototype SABINE so that researchers are able to better understand particular system-, information- and service-related VTC success drivers. Practitioners have a valuable means for management interventions, task prioritisations as well as effective and efficient resource allocation at their disposal. The proposed model, validated via an online survey as well as a subsequent partial least squares (PLS) analysis, can be approved successfully. The positive results open up new options for further development and application of the model and VTC.

Keywords: DeLone and McLean IS Success Model, Evaluation Methods for TEL, Design Characteristics, Virtual Talent Communities.

1 Introduction

Facing the ongoing debate on "war for talent" [20] and talent-oriented Human Resource Management (HRM) strategies and systems [26], interactive web-based applications for HR-related purposes are increasingly discussed [10, 19]. Virtual talent communities (VTC), defined as a web-based information systems (IS) for social interaction by different actors (talents, educational institutions, employers, intermediaries) with a shared interest in talent development and employment, are an innovative kind of electronic HRM [29]. Although VTC reflect a promising integrative approach, predominantly partial IS (e.g. learning management systems (LMS), virtual job boards, career websites, social networking sites (SNS)) exist. Rigorous and practically relevant approaches for the development, implementation and evaluation of VTC are needed to be developed and validated. Based on design science [13, 22], it is of importance to investigate the determinants of VTC success systematically in order to assist system designers in building and operating systems that are useful and accepted. Initial research which includes a rigorous foundation to substantiate the VTC concepts

C. Delgado Kloos et al. (Eds.): EC-TEL 2011, LNCS 6964, pp. 519–524, 2011.

confirms the usefulness of the above-mentioned approach and elicits basic design characteristics for VTC [21]. Benefits of VTC (e.g. facilitation of convenient, ubiquitous information and knowledge sharing processes, communication and relationship building, collaboration and transaction [18]) can only be successfully realised if proper development, implementation and (continuous) improvement are assured [9, 16, 22, 27], given the assumption that success is at least manageable to a certain degree [30]. Technically- as well as managerially-oriented literature [14, 31] congruently understand design characteristics as a set of those properties which determine IS success and support decision makers, system developers, and content providers in developing, implementing and (continuously) evaluating VTC based on talent-specific requirements [8, 9, 16, 27]. The purpose of this paper is to derive a rigorous and relevant research model based on earlier research [21] in order to assess the success of VTC from a talent perspective more appropriately [13]. A positive effect of such a research model is its potential to specify the measures for VTC success assessment so that research results could be better compared and findings better validated [13]. Below the research object and the model are specified and a conclusion is given.

2 Specification of the Research Object

The innovative VTC concept as an ideal multi-access integration of actors and functions in regard to HR-relevant processes in general, and talent development and deployment in particular have not been yet fully developed. Thus, the here applied SABINE[1] prototype does not depict an ideal and overall VTC but includes basic search, recommendation and interactivity functions for use in a VTC. As talents participate in trainings in general and e-learning courses in particular to acquire qualifications in order to (better) fit to jobs, the SABINE prototype's purpose is to support the process of acquiring job-fit qualifications. To this, the integrated usage of semantic web technologies (SWT) in the combined areas of e-learning and e-recruiting is evaluated. Unlike other approaches using SWT in either e-learning [2, 12] or e-recruiting [3], SABINE integrates both areas in one platform and aims at recommending bundles of corresponding training and job offers. SWT-usage enables sharing and reusing data across boundaries of applications, enterprises, and also communities [32] provided by standards that allow the computing of the data's underlying meaning (i.e. semantics) [34, p. 15]. For this purpose ontologies, a core SWT, are used. Ontologies "are formalised vocabularies of terms, often covering a specific domain and shared by a community of users. They specify the definitions of terms by describing their relationships with other terms in the ontology" [33]. In the SABINE prototype ontologies are used to model qualifications which characterise a talent's personal profile, learning and job offers. In the backend, publicly available training and job offers are crawled from multiple learning and job exchange platforms using a semantic information access framework[2]. These offers are written in natural language without marking up qualifications. For an individual, it is difficult to identify qualifications needed for

[1] SABINE was developed within the THESEUS-Project supported by the German Federal Ministry of Economics and Technology. http://sabine-demo.de/

[2] http://www.eclipse.org/smila/

a specific job, individual qualification gaps and training offers which fit best to fill these gaps. It is a non-trivial process to comprehend the implicit qualification descriptions in training and job offers and to match them to one's personal qualification profile manually. To solve it, SABINE uses ontologies to analyse the offers and to identify qualifications. At the end of this process, training and job offers are semantically annotated with respective qualifications and stored in a database. The frontend of SABINE is programmed in HTML and JavaScript for a high compatibility to all major browsers. The user is enabled to create a personal profile with standard data (name, date of birth) and personal qualifications. SABINE assists the qualification capturing by providing auto-completion during the input. The suggested qualifications originate from the ontology. Apart from manually creating one's personal profile (in yet another community), application programming interfaces to LMS and SNS (e.g. facebook) provide automatic data import, taking into consideration that the user can decide which data from a specific external source he wants to have imported. The imported data is also analysed with the ontology's assistance. Making the annotated training and job offers as well as the talent's qualification profile available, SABINE is able to recommend offers that provide the best match between the personal qualifications and those inherent in the offers. To enhance ubiquitous access, SABINE functions can be integrated into facebook and iGoogle as a widget [23].

3 Towards a VTC Success Model

As rigorous and relevant research models for measuring VTC success from a talent perspective are missing at present, the subsequent elaboration therefore presents such an instrument which core principles are patterned according to the updated DeLone and McLean IS success model (ISSM) [8]. The ISSM is chosen as it presents a general framework for organising design characteristics relevant to the success of IS/VTC [8]. Basically, the ISSM offers three groups of coarse-granular IS/VTC success predictors, namely, *system quality*, *information quality*, and *service quality* [8]. In order to foster the relevance of the research model as well, based on the ISSM, medium-granular (i.e. operative) system-, information-, and service-related VTC design characteristics were derived by means of a cross-national, talent-focused Delphi study [8, 21, 23 table 1]. In addition to the constructs related to the aforementioned ISSM success dimensions, three further ISSM constructs were considered which are incorporated into the research model due to their relevance to the ISSM, namely talents' behavioural intention to use, satisfaction with a VTC as well as their (perceived) net benefits of using the VTC under consideration [8; for further information regarding the entire set of constructs applied, their definitions and sources included, see 23 table 1]. Based on the associations postulated by the ISSM, a research model is proposed. It assumes that system, information, and service quality as well as the sub-dimensions related to them are linked to talents' behavioural intention to use, respectively their satisfaction with a VTC [8]. Beyond, these success dimensions in turn influence talents' (perceived) net benefits of using a VTC [23, section 4].

The research model is tested pilotly and analysed using partial least squares (PLS), a structural equation modelling technique that is well suited for highly complex predictive models [1, 5]. PLS is the most appropriate given the large number of constructs that result from extending and adapting the ISSM to the VTC context. Also,

PLS requires minimal restrictions in terms of distributional assumptions and sample size [6]. In particular, smartPLS is used for the analysis [25]. All constructs tested are modelled using reflective items as this approach considers items to constitute replaceable reflections of their underlying constructs, which in turn makes the ongoing refinement of the research model highly convenient as only items have to be changed, but not their corresponding construct definition [4]. The final research model includes 64 items representing 20 constructs [23, table 2].

The analysis of each measurement model includes the estimation of the construct reliability as well as the convergent and discriminant validity of its corresponding items.

Cronbach's alpha is almost entirely above the recommended level of 0.70 [17, 24], indicating adequate construct reliability [24]. The two exceptions are *fast* (-0.14) and talents' behavioural intention to use a VTC (0.49), which therefore should be slightly modified in the ongoing research process as low inter-item correlations indicate that not the entire set of items reflects the "appropriate" construct domain systematically, thus producing slight tendencies of unreliability [7]. In the fashion of Cronbach's alpha, the composite reliability of almost all constructs, except *fast* (0.46) and talents' behavioural intention to use a VTC (0.61), demonstrates satisfactory levels of at least 0.70 as well [17, 24], equally indicating adequate construct reliability [24]. The average variance extracted of all constructs shows levels above the recommended limit value of at least 0.50 so that a satisfying construct reliability can be approved entirely [11]. Average variance extracted values confirm the items to exhibit appropriate convergent validity [11]. Discriminant validity of the items deployed is found to be satisfactory as the square root of the average variance extracted of each construct is higher than the correlations of that construct with all other constructs of the research model (Fornell-Larcker criterion) [11]. Discriminant and convergent validity are further approved as almost all items, except *ad-free* (first item) and talents' behavioural intention to use a VTC (third item), demonstrate loadings above 0.50 on their associated constructs [28] and the item loadings within each associated construct are found to be higher than those across constructs. This result is mainly due to the fact that the video-based VTC stimulus applied does not support particular functionalities, respectively specific design characteristics attributed to these functionalities (e.g. *ad-free*). Hence, the current item pool will be revised and further developed in the ongoing validation phase of this research project to ensure universal applicability of the item pool in regard to VTC. A generally recommended minimum of three items per construct is targeted [15]. Based on the current item pool, the significance of the item loadings is analysed using a bootstrapping procedure with 200 samples. Fortunately, all loadings are significant at the 0.05 level, disregarding the following constructs: *Ad-free* (first item), talents' behavioural intention to use a VTC (third item), and *fast* (all items).

4 Conclusions

Within this paper, first steps in the development of a rigorous and relevant research model for assessing the success of VTC from a talent perspective are undertaken. A first validated set of relevant VTC design characteristics can be offered to practice.

The research model may facilitate system evaluators in the assessment of VTC which could lead to improved levels of task prioritization and resource allocation from a VTC decision maker's viewpoint. The research model can help VTC stakeholders to better understand the relative importance of particular design characteristics to VTC success. The proposed research model can serve as a checklist to better assess in how far VTC fulfill the requirements postulated by prior research efforts in this subject domain [21]. However, as mentioned, the research model requires some further development and validation firstly before it can be applied in practice finally.

References

1. Barclay, D., Higgins, C., Thompson, R.: The partial least squares approach to causal modeling: Personal computing adoption and use as an illustration. Technology Studies 2, 285–309 (1995)
2. Bittencourt, I.I., Costa, E., Silva, M., Soares, E.: A computational model for developing semantic web-based educational systems. Knowledge-Based Systems 22, 302–315 (2009)
3. Bizer, C., Heese, R., Mochol, M., Oldakowski, R., Tolksdorf, R., Eckstein, R.: The impact of semantic web technologies on job recruitment processes. Wirtschaftsinformatik 15, 1367–1381 (2005)
4. Bollen, K., Lennox, R.: Conventional wisdom on measurement: A structural equation perspective. Psychological Bulletin 110, 305–314 (1991)
5. Chin, W.W.: Issues and opinions on structural equation modeling. MIS Quarterly 22, 7–16 (1998)
6. Chin, W.W., Marcolin, B.L., Newsted, P.: A partial least squares latent variable modeling approach for measuring interaction effects: Results from a Monte Carlo simulation study and an electronic-mail emotion/adoption study. Information Systems Research 14, 189–217 (2003)
7. Churchill, G.A.: A paradigm for developing better measures of marketing constructs. Journal of Marketing Research 16, 64–73 (1979)
8. DeLone, W.H., McLean, E.R.: The DeLone and McLean model of information systems success: A ten-year update. Journal of Management Information Systems 19, 9–30 (2003)
9. Dennis, A.R., Wixom, B.H., Roth, R.M.: Systems analysis and design, 3rd edn. John Wiley and Sons, Hoboken (2006)
10. Diederichsen, A.: Human resource management-relevant virtual community research: Review and outlook. In: Strohmeier, S., Diederichsen, A. (eds.) Proceedings of the 3rd European Academic Workshop on Electronic Human Resource Management, Germany, Bamberg, pp. 207–228 (2010), http://CEUR-WS.org/Vol-570/paper013.pdf
11. Fornell, C., Larcker, D.F.: Evaluating structural equation models with unobservable variables and measurement error. Journal of Marketing Research 18, 39–50 (1981)
12. Gladun, A., Rogushina, J., García-Sanchez, F., Martínez-Béjar, R., Fernández-Breis, J.T.: An application of intelligent techniques and semantic web technologies in e-learning environments. Expert Systems with Applications 36, 1922–1931 (2009)
13. Hevner, A.R., March, S.T., Park, J., Ram, S.: Design science in information systems research. MIS Quarterly 28, 75–105 (2004)
14. ISO/IEC: ISO/IEC 25000:2005: Software product quality requirements and evaluation (SQuaRE), http://www.iso.org/iso/iso_catalogue/catalogue_tc/catalogue_detail.htm?csnumber=35683

15. Kankanhalli, A., Tan, B.C.Y., Wei, K.-K.: Contributing knowledge to electronic knowledge repositories: An empirical investigation. MIS Quarterly 29, 113–143 (2005)
16. Kavanagh, M.J., Thite, M.: Human resource information systems: Basics, applications, and future directions. Sage, Los Angeles (2009)
17. Lewis, B.R., Templeton, G.F., Byrd, T.A.: A methodology for construct development in MIS research. European Journal of Information Systems 14(4), 388–400 (2005)
18. Lin, H.-F.: Determinants of successful virtual communities: Contributions from system characteristics and social factors. Information & Management 45(8), 522–527 (2008)
19. Martin, G., Reddington, M., Kneafsey, M.B.: Web 2.0 and human resource management: 'Groundswell' or hype? Chartered Institute of Personnel and Development (CIPD), London (2009)
20. Michaels, E., Handfield-Jones, H., Axelrod, B.: The war for talent. Harvard Business School Press, Boston (2001)
21. Mueller, D., Diederichsen, A.: Design characteristics of virtual talent communities: A cross-national, talent-focused, exploratory requirements analysis. In: Proceedings of the 19th European Conference on Information Systems (ECIS) 2011, Finland, Helsinki, pp. 1–12 (2011)
22. Mueller, D., Strohmeier, S., Gasper, C.: HRIS design characteristics: Towards a general research framework. In: Strohmeier, S., Diederichsen, A. (eds.) Proceedings of the 3rd European Academic Workshop on Electronic Human Resource Management, Germany, Bamberg, pp. 250–267 (2010), http://CEUR-WS.org/Vol-570/paper015.pdf
23. Müller, D., Diederichsen, A., Gasper, C., Strohmeier, S.: A model for assessing the success of virtual talent communities (MIS Report 1-2011) (2011), http://www.unisaarland.de/fileadmin/user_upload/Professoren/fr13_ProfStrohmeier/reports/TR-MIS-1-2011.pdf
24. Nunnally, J.C.: Psychometric theory. McGraw-Hill, New York (1978)
25. Ringle, C.M., Wende, S., Will, S.: SmartPLS 2.0, Hamburg (2005), http://www.smartpls.de
26. Schweyer, A.: Talent management systems: Best practices in technology solutions for recruitment, retention, and workforce planning. Wiley, Toronto (2004)
27. Sommerville, I.: Software engineering, 8th edn. Addison-Wesley, Munich (2007)
28. Straub, D.W.: Validating instruments in MIS research. MIS Quarterly 13(2), 147–169 (1989)
29. Strohmeier, S.: Research in e-HRM: Review and implications. Human Resource Management Review 17, 19–37 (2007)
30. Strohmeier, S.: Concepts of e-HRM consequences: A categorisation, review and suggestion. The International Journal of Human Resource Management 20(3), 528–543 (2009)
31. Venkatesh, V., Bala, H.: Technology acceptance model 3 and a research agenda on interventions. Decision Sciences 39(2), 273–315 (2008)
32. W3C semantic web activity, http://www.w3.org/2001/sw/
33. OWL 2 web ontology language: Document overview, http://www.w3.org/TR/2009/REC-owl2-overview-20091027/
34. Yu, L.: A developer's guide to the semantic web. Springer, Berlin (2011)

Towards the Prediction of User Actions on Exercises with Hints Based on Survey Results

Pedro J. Muñoz-Merino, Abelardo Pardo, Mario Muñoz-Organero, and Carlos Delgado Kloos

Universidad Carlos III de Madrid, Department of Telematics Engineering, Avda de la Universidad, 30, E-28911 Leganés (Madrid), Spain
{pedmume,abel,munozm,cdk}@it.uc3m.es

Abstract. The actions a user performs on exercises depending on the different hinting techniques applied, can be used to adapt future exercises. In this paper, we propose a survey for users in order to know their different actions depending on different conditions. The analysis of preliminary results for some questions of the model shows that there is a correlation between some survey questions and the real student actions, but there is a case in which there is not such correlation. For the cases where that correlation exists, this correlation leads to think that some prediction of users actions based on survey results is possible.

Keywords: student behavior, adaptation, exercises, hints, survey.

1 Introduction and Related Work

The generation of adaptive hints for tutoring systems based on exercises has been addressed (e.g. [1], [2]). The paper [1] takes into account the student knowledge level and their mental state so that Bayesian networks can be applied to adapt hints in the Andes system. While article [2] applies the Item Response Theory to calibrate the hints over the SIETTE system based on the student knowledge about the topics. In addition, paper [3] also proposes an adaptive solution for the delivering of hints.

All these works dealt with the adaptation of hint contents but they did not adapt the different hinting techniques of an exercise without changing the knowledge contents. An example of hinting technique adaptation would be that a system can decide if there is some penalty for viewing hints or not depending on the user. The use of different hinting techniques can imply different learning gains in some situations [4].

Monitoring systems can be implemented in order to retrieve these user interactions, and a post-processing of the data must be done in order to have the desired student actions. Different works [3], [5], [6], [7], [8] give excellent ideas about the type of user interactions that can be retrieved into hinting systems, including the number of tries and selected hints, or the incorrect answers. From previous students actions we can have an estimation of their future behavior [1], or the prediction of their test scorings [5], as well as the correlation of different student behaviors with respect to their learning gains [8], or an evaluation of the instructional effectiveness of different exercises [8]. Therefore, user interactions in the hint context are valuable information.

C. Delgado Kloos et al. (Eds.): EC-TEL 2011, LNCS 6964, pp. 525–530, 2011.

In addition, these user interactions can be written into suitable formats called Contextualized Attention Metadata [9] that can bring together different feeds and store different user events. Some tools have been created for monitoring and processing user interactions such as CAMERA [10] with very interesting applications.

Instead of monitoring the different student actions, we propose giving a survey to the users. Although decisions based on the monitoring can be better e.g. for adaptation than those based on survey results, nevertheless sometimes this method can be worth, because the costs of implementation of monitoring and processing systems can be saved, as well as the problem of the initial training for the system is solved. In order to make this approach scalable for many students, a web survey form has been designed so that students can submit their survey answers through the Web.

There are user questionnaires for adaptation purposes such as [11] for feedback taking into account a model about learning styles [12]. There are also other prediction works not applied in the hinting context, such as to try to predict a user model from user behavior in a shopping context [13] or the interests of a museum visitor [14].

2 Student Survey Regarding Exercises with Hints

The survey included three questions to determine students' hint abuse. The questions and the hint abuse concept are in the direction of the work presented in [7]:

1.- If I am quite sure to know the solution of an exercise:
  a) I answer the solution of the problem immediately
  b) I like to view all the hints previously
2.- If I have no idea of the solution of an exercise:
  a) I like to make several attempts before viewing the hints
  b) I like to view the hints previously before making several attempts
3.- If I am not sure to know the answer to an exercise:
  a) I like to make several attempts before viewing the hints
  b) I like to view the hints previously before making several attempts

Next, the survey includes some questions about different hinting techniques without changing the hint contents. Tables 1 and 2 show the rest of the survey questions that include aspects about the importance, fairness, preferences and student behavior. Each question is rated between 1 and 7 by students. Seven categories were selected based on studies that suggest the optimal number of categories (e.g. [15]).

Table 1. Survey questions for rating the importance of different hinting techniques

Question
- Hints that are only available when a user answers a problem incorrectly
- Hint sequence (a hint that is composed by several steps. Each step is a problem)
- Selection of a maximum number of hints out of several that are proposed
- Penalties on the scoring for viewing hints
- Penalties on the scoring for incorrect attempts
- Rewards on the scoring for answering hints correctly
- Penalties on the scoring for answering hints incorrectly

Table 2. Survey questions about the fairness, preferences and the own perceived behavior

Question
- I think it is fair that each incorrect attempt implies penalties on the scoring
- I think it is fair that viewing hints implies penalties on the scoring
- I think it is fair that answering hints incorrectly implies penalties on the scoring
- I think it is fair that answering hints correctly implies rewards on the scoring
- I think it is fair that hints can only be activated as a result of a student incorrect attempt
- When the system shows several hints and the user can select a maximum of them, I think that it is fair such limitation in the number of hints that can be selected
- I prefer hints that are only text instead of hints that are other problems to solve
- I prefer hints that are available directly at the beginning, instead of hints that are only available after an incorrect answer to the initial problem
- I prefer a huge number of hierarchical hint levels (concept of hints about hints)
- I prefer a hint that is divided into a sequence of many problems (hints that are several problems in different steps) instead of a hint that is only a problem
- I prefer questions with score in order to evaluate my knowledge
- The fact of penalizing on the scoring for incorrect attempts lead me not to answer
- The fact of penalizing on the scoring for viewing hints lead me not to select them
- The fact of penalizing on the scoring for answering hints incorrectly lead me not to answer
- The fact of increasing the scoring for answering a hint correctly, lead me to answer

3 Research Question and Description of the Experience

The main research question of this work is if some questions of a specific survey to students are correlated with their future user actions on exercises with hints. There are several factors that can make a difference between the different student survey answers with respect to their real actions, such as the following:

- While some survey questions are directly related to the future students behaviors, there are other questions that do not, e.g. about the user preferences.
- A student can misunderstand a question so that she does not reflect her real thoughts and feelings. Or as the survey is filled in before the user interactions, students cannot understand well the different proposed situations.
- Students can lie or they cannot be aware of their own future actions.

An experience took place in the classroom in 2010 in the context of a Computer Architecture Laboratory Course (http://www.it.uc3m.es/pedmume/LAO/index.html). Students received a survey with most of the questions of Section 2. The survey was filled in by students before their interaction with the system. Next, students performed a pre-test. Next, they interacted with 26 exercises with hints in XTutor during two sessions: the first one of 90 minutes, and the second one of 110 minutes. Each student was provided with different exercises with associated hints, the most relevant for this work were: six exercises with penalties for viewing hints, and other six without any penalties; three exercises with hints directly available, and other three without them. The number of samples was 29, as this is the number of students that filled in the survey correctly and attended both sessions.

4 Preliminary Results

Table 3 shows some students actions information that were used for this research:

Table 3. Retreived information from the tutor regarding the experience

Acronym	Description
T	The pre-test results.The pre-test was composed by 8 questions, and each student obtained an scaled grade between [0,1]
PHS	The percentage of selected hints considering the total number of exercises that a student interacted with. It is considered that a student selected a hint if she selected at least one of the hints before answering the exercise correctly
PW	The percentage of exercises for which a student selected a hint when there were no penalties for doing it
PP	The percentage of exercises for which a student selected a hint when there were penalties for doing it
PD	The percentage of exercises for which a student selected a hint when the hint is directly available
PA	The percentage of exercises for which a student selected a hint when the hint is only available as a result of an incorrect attempt
Diff_HS1	PW – PP: The difference in the percentage of selected hints when there is not any effect on the scoring with respect to having a penalty for selecting hints
Diff_HS2	PD – PA : The difference in the percentage of selected hints when the hint is directly available with respect to not having it directly available

We present some preliminary results that compare three types of student actions with respect to some questions of the survey and the pre-test grades. The selected student actions to be predicted were: *PHS*, *Diff_HS1*, and *Diff_HS2*.The dedicated time for the sessions gave only chance to have enough data for these three actions.

The *Diff_HS1* would be tried to be predicted based on the student answers to the survey question of "The fact of penalizing in the scoring for viewing hints led me not to select many" (*Survey_1*), while the *Diff_HS2* based on the survey question: "I prefer hints that are available directly at the beginning, instead of hints that are only available after an incorrect answer to the initial problem" (*Survey_2*).

Finally, the *PHS* would be tried to be predicted based on the three initial survey questions about hint abuse, but also taking into account the student grade on the pre-test because a student with more knowledge does not need so many hints. For the three initial questions, a student is given a value from [0,3], this is the hint abuse rate (*AR*), depending on the times she prefers to view hints previously. The proposed estimator for trying to predict *PHS* would be the multiplication of *AR* by *T*.

Table 4 shows the Pearson linear correlation coefficient between the commented parameters (where N=29 for all the cases). We can observe that two correlations were statistically significant ($p<0.05$) while one was not. Therefore, students have some perception about their own behavior depending on if there are penalties or not for viewing hints. And their preferences about the hint being available at the beginning or only as a result of an incorrect attempt are according to their real behavior. But in both cases, with the monitored data, a linear model with statistically significant coefficients cannot be provided to predict the user actions based on the survey results.

Table 4. Pearson correlation coefficient between different parameters

Parameters	Correlation	p
AR*T and PHS	0.112	0.282
Survey_1 and Diff_HS1	0.322	0.044
Survey_2 and Diff_HS2	0.365	0.031

The result between *AR*T* and *PHS* can mean that the real student actions about the percentage of hints selected is not according to what they answered in the survey. Indeed, analyzing the data, we can observe that there are three students that answered that they like to view all the hints previously in all the cases but they did not really view many hints. The reasons of this can be the same as commented in section 3.

In addition, we can think that neither *AR*T* is a proper parameter nor *PHS*, and more proper estimators and parameters should be selected to be taken into account in a prediction model. The *AR*T* and *PHS* parameters are high level ones, since *AR*T* considers the hint abuse rate as the result of joining three different survey questions and the total pre-test grade, and *PHS* does not take into account an analysis of each specific question but joins all in a group without a difference between the different initial student knowledge levels for each question. But more specific estimators can be determined, considering for each question the initial student knowledge level in that question and determining if a student really needs a hint or not for it, and if it is according to what she answered in the specific question of the survey.

5 Conclusions

We have proposed a new survey to retrieve the features of a user model regarding different hinting techniques that can be applied to exercises, and about user preferences for viewing hints depending on their knowledge in an exercise.

In one case, there was not a statistically significant correlation between the real student actions and the selected survey questions for that aspect combined with the pre-test grade. On the other hand, there were statistically significant correlations between two student actions and two survey questions. This leads to think that for some cases we can rely on the survey results to try to predict the real user actions.

These are only preliminary results since from this experience we can only analyze three different types of student actions. Furthermore, there are only 29 student samples, and future experiences might complement this study. In addition, other variables might also be included in future studies such as the specific course topics.

Another interesting topic for future research would be to study students' actions to gain the system (e.g. answer the survey questions lying in a way that they know that they will receive more hints) and possible measures to prevent from it.

A challenge is to find the best model that can predict the real student actions based on the different questions of the survey, when this is possible. In order to achieve it, more student interactions with exercises and more student samples would be needed.

Acknowledgments. Work partially funded by the Learn3 project TIN2008-05163/TSI within the Spanish "Plan Nacional de I+D+I", and the Madrid regional community project eMadrid S2009/TIC-1650.

References

1. Gertner, A.S., Conati, C., Vanlehn, K.: Procedural help in Andes: Generating hints using a Bayesian network student model. In: 15th National Conference on Artificial Intelligence, pp. 106–111. AAAI Press, Menlo Park (1998)
2. Conejo, R., Guzmán, E., de-la Cruz, J.-L.P., Millán, E.: An Empirical Study About Calibration of Adaptive Hints in Web-Based Adaptive Testing Environments. In: Wade, V.P., Ashman, H., Smyth, B. (eds.) AH 2006. LNCS, vol. 4018, pp. 71–80. Springer, Heidelberg (2006)
3. Zhoul, Y., Freedman, R., Glass, M., Michael, J.A., Rovick, A., Evens, M.W.: Delivering hints in a dialogue-based intelligent tutoring system. In: 16th National Conference on Artificial intelligence, pp. 128–134. AAII (1999)
4. Muñoz-Merino, P.J., Delgado Kloos, C., Muñoz-Organero, M.: Deciding on Different Hinting Techniques in Assessments for Intelligent Tutoring Systems. International Journal of Innovative Computing Information and Control 7(2), 841–858 (2011)
5. Feng, M., Heffernan, N.T., Koedinger, K.R.: Predicting state test scores better with intelligent tutoring systems: developing metrics to measure assistance required. In: Ikeda, M., Ashley, K.D., Chan, T.-W. (eds.) ITS 2006. LNCS, vol. 4053, pp. 31–40. Springer, Heidelberg (2006)
6. Baker, R., Corbett, A.T., Roll, I., Koedinger, K.R.: Developing a generalizable detector of when students game the system. User Mod. and User-Adapted Int. 18, 287–314 (2008)
7. Aleven, V., Mclaren, B., Roll, I., Koedinger, K.R.: Toward Meta-cognitive Tutoring: A Model of Help Seeking with a Cognitive Tutor. Int. J. Artif. Intell. 16, 101–128 (2006)
8. Feng, M., Heffernan, N.T., Beck, J.: Using learning decomposition to analyze instructional effectiveness in the ASSISTment system. In: 14th International Conference on Artificial Intelligence in Education (AIED 2009), pp. 523–530. IOS Press, Amsterdam (2009)
9. Wolpers, M., Najjar, J., Verbert, K., Duval, E.: Tracking Actual Usage: the Attention Metadata Approach. Educational Technology & Society 10(3), 106–121 (2007)
10. Schmitz, H., Scheffel, M., Friedrich, M., Jahn, M., Niemann, K., Wolpers, M.: CAMera for PLE. In: Cress, U., Dimitrova, V., Specht, M. (eds.) EC-TEL 2009. LNCS, vol. 5794, pp. 507–520. Springer, Heidelberg (2009)
11. Vasilyeva, E., Pechenizkiy, M., De Bra, P.M.E.: Adaptation of Elaborated Feedback in e-Learning. In: Nejdl, W., Kay, J., Pu, P., Herder, E. (eds.) AH 2008. LNCS, vol. 5149, pp. 235–244. Springer, Heidelberg (2008)
12. Felder, R.M., Silverman, L.K.: Learning and Teaching Styles in Engineering Education. Journal of Engineering Education 78, 674–681 (1988)
13. Mori, J., Matsuo, Y., Koshiba, H., Aihara, K., Takeda, H.: Predicting Customer Models Using Behavior-Based Features in Shops. In: Houben, G.-J., McCalla, G., Pianesi, F., Zancanaro, M. (eds.) UMAP 2009. LNCS, vol. 5535, pp. 126–137. Springer, Heidelberg (2009)
14. Bohnert, F., Zukerman, I., Berkovsky, S., Baldwin, T., Sonenberg, L.: Using Interest and Transition Models to Predict Visitor Locations in Museums. AI Communications 21 (AICom) – Special Issue on Recommender Systems, 195–202 (2008)
15. Malhotra, N.K.: Questionnaire design and scale development. In: The Handbook of Marketing Research Uses Misuses and Future Advances, pp. 83–94. Sage Publications, Inc., Thousand Oaks (2006)

Analyzing 5 Years of EC-TEL Proceedings

Wolfgang Reinhardt[1], Christian Meier[1], Hendrik Drachsler[2], and Peter Sloep[2]

[1] University of Paderborn
Department of Computer Science, Computer Science Education Group
Fuerstenallee 11, 33102 Paderborn, Germany
{wolle,chrmeier}@upb.de
[2] Open University of the Netherlands
Centre for Learning Sciences and Technologies
6401 DL Heerlen, The Netherlands
{Firstname.Lastname}@ou.nl

Abstract. Over the past five years, EC-TEL has established itself as a major conference on learning with and through technology. Nearly 600 researchers have contributed about 230 full and short papers on various topics in the domain of Technology Enhanced Learning. In this paper we analyze the contributions of five years of EC-TEL and identify prolific authors, successful co-author-networks and most cited publications. The analysis reveals a very fragmented EC-TEL community that is strongly influenced by a relatively small set of authors.

Keywords: research 2.0, bibliometrics, awareness, visualization.

1 Introduction

Scholarly practice focuses on the improvement of the society and therefore conducts research on pressing problems. The daily work process of researchers can be roughly distinguished into reviewing and reflecting new publications, preparing experiments, and publishing the results and the data on scientific conferences or in international journals. Over centuries researchers have gathered around recent publications that served as boundary objects between different communities of practice and reciprocally influenced each other's work. With today's unmanageable amount of published work it becomes increasingly difficult to monitor one's own research domain and to find relevant publications that concur with one's own interests and research projects. In the past years research has explored collaboration of scientists by means of co-authorships of publications. In the TEL community, such endeavors were undertaken for example in [2,5,7]. We argue that structured analyses of scientific publications and visualizations synthesizing the results can help all interested stakeholders in the scientific process to be more aware about content and connections and thus may serve as decision support.

The EC-TEL started as a European conference on Technology Enhanced Learning in 2006. The aim of the founders was to provide a unique forum for

C. Delgado Kloos et al. (Eds.): EC-TEL 2011, LNCS 6964, pp. 531–536, 2011.

research related to Technology Enhanced Learning in Europe and world-wide [6]. The topics of the conference are dealing with e-learning, knowledge management, and workplace learning. With an acceptance rate of about 20% EC-TEL has established itself as one of the main conferences in the domain of Technology Enhanced and E-Learning.

In this paper we present a bibliometric and scientometric analysis of the published full and short papers of the 5 years of EC-TEL. The remainder of the paper is structured as follows: In Section 2 we present the research question that guided our work. In Section 3 we discuss how the data was obtained and processed. In Section 4 we present the results of our analyses and thereafter in Section 5 we report some limitations of the conducted study. Finally in Section 6, we discuss implications and opportunities for future research and draw some requirements for a future scientific event management system.

2 Research Question

The analyses in this article are carried out in order to understand how the EC-TEL conference series can be described using bibliometric and scientometric measures. Moreover, we aim at raising awareness about peculiarities in the analyzed data with the goal to show the fissured EC-TEL community. The overall research question of this paper can thus be stated as "*Which awareness support functions can be derived from the analysis of the EC-TEL publications to improve the support of conference organizers and attendees?*"

3 Data Collection and Processing

The papers analyzed in this study have been collected from the Digital Library of Springer [10]. All full and short papers of the 5 years of EC-TEL were downloaded in PDF format and used for the following analyses. Poster papers, invited papers and keynote abstracts have not been considered in the study. Altogether we analyzed 229 papers from 574 unique authors. On average there were 46 papers in each year of the conference with 148 unique authors per year. Each year the authors referenced 751 unique publications in their accepted papers.

3.1 Processing and Analysis Preparation

The extraction and cleaning of bibliographic data was one of the most complex and time-consuming steps in the processing of the data. First, we used an open source command-line tool called *pdftotext* which is available for many Linux distributions [3]. Pdftotext allows to extract the plaintext from PDF files, which is the prerequisite for the analysis in the next step: the analysis using ParsCit [1]. ParsCit extracts metadata about the given publication itself and about the contained references. As the output of ParsCit is not always correct and some metadata might be extracted wrongly, a manual cleaning of the metadata was required in order to provide high quality input for the analysis. After extracting and cleaning the metadata we needed to persistently store it in a database which provided access to this data for the next analysis step. For the research presented here, we used the model of Artefact-Actor-Networks (AANs; [8]) as data storage and data access framework. The latest reference implementation of AANs is described in [9]. Finally, we used the visualization component of the AAN which provides files in the GEXF file format. The graph export served as input for the visual analysis using Gephi [4]. With Gephi we created visual representations of networks that emerged from the published papers and their content-based analysis.

4 Analyses

In this section we report about the results of the paper analyses conducted. All EC-TEL authors whose names are used within this publication have agreed on their appellation.

4.1 Authorship Analysis

We started our analysis with the exploration of the authors of EC-TEL. Therefore, we counted the number of papers that had been published by each individual author between 2006 and 2010 and analyzed the distribution of publications over individual authors. The analysis shows that 77.4% (444 authors) of all authors have published only one paper in EC-TEL between 2006 and 2010. It is very rare to find authors with more than 3 papers (less than 5% (24 authors) of all authors). The 13 authors that contributed 5 or more papers to the conference (2.3% of all authors) account for 56 unique papers (24.5% of all papers). From an individual point of view, Pierre Dillenbourg contributed 4 papers to the 2008 edition of EC-TEL and thus accounted for 7.7% of all papers in this year. The 9 authors that published 6 or more papers in EC-TEL are: 1) Ralf Steinmetz (Germany, 8), 2) Christoph Rensing (Germany, 8), 3) Marcus Specht (Netherlands, 7), 4) Rob Koper (Netherlands, 6), 5) Erik Duval (Belgium, 6), 6) Pierre Dillenbourg (Switzerland, 6), 7) Stefanie Lindstaedt (Austria, 6), 8) Jelena Jovanović (Serbia, 6) and 9) Dragan Gašević (Canada, 6).

4.2 Co-authorship Analysis

Scientific publications always contain information about the authors who contributed to the publication. The smallest *co-authorship* relation consists of two authors who contributed to one paper. By analyzing co-authorship information on a larger corpus of scientific publications, it is possible to identify groups of persons who work closely together. Figure 1 visualizes the overall co-authorship network of the EC-TEL conference series.

From the set of 229 papers, 11 papers (4.8%) were created by a single author and thus not taken into consideration for the co-authorship analysis. More than 50% of all papers (123) were written by two or three authors and still more than 11% of the papers (27) were co-authored by 6 or more authors. The largest co-authorship network for a single paper consisted of 11 authors. In the co-authorship network we can identify 103 weakly connected subnetworks, where the largest co-authorship cluster consists of 75 authors that contributed 33 papers (14.4% of the overall papers).

The co-authorship analysis reveals a very fragmented community that is only weakly connected (overall density of only 0.007). The existing larger co-authorship clusters mostly connect authors from one subdomain of TEL only.

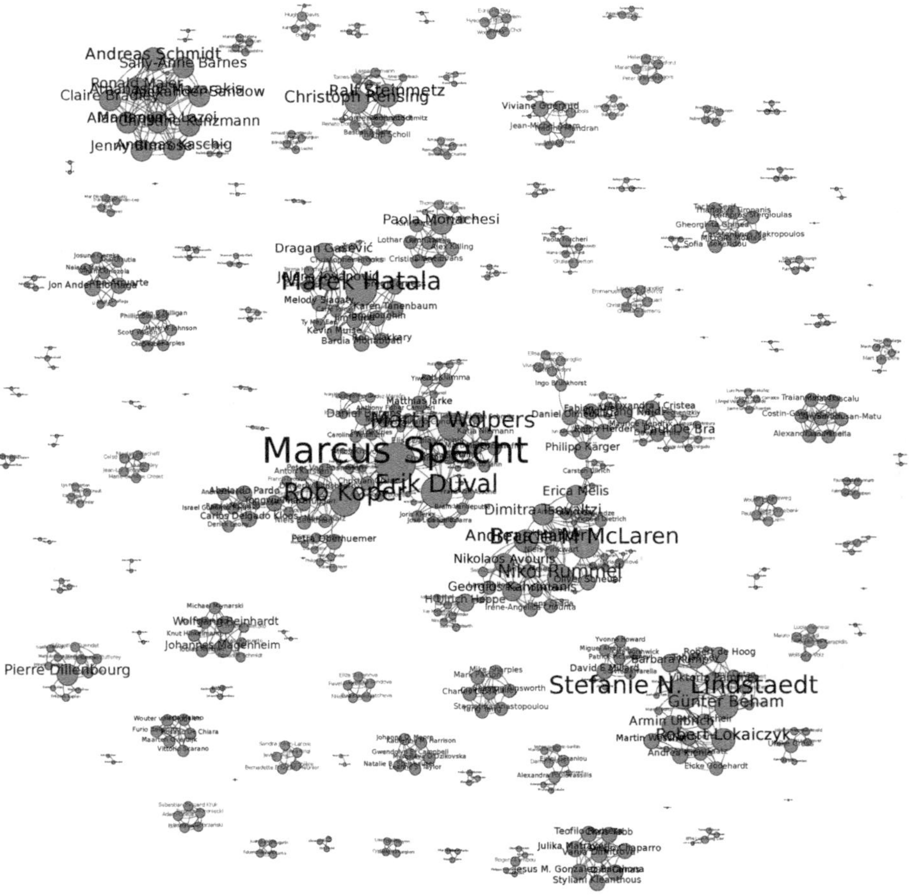

Fig. 1. Co-authorship network for EC-TEL 2006-2010

4.3 Citation Analysis

Within the EC-TEL corpus we find 3.919 references of which 3.401 are unique. On average, each published paper has 17 references with only 1 reference as minimum and 70 references as maximum. The most cited paper that has been published within EC-TEL is on *MACE*, received 3 citations has been published by Stefaner et al. in 2007 [11]. In total, 43 papers from the first 4 years of EC-TEL have received 52 references. The five most cited authors within the EC-TEL corpus are Erik Duval (87 citations), Peter Brusilovsky (70), Rob Koper (54), Alexandra Cristea (44), and Wolfgang Nejdl (39).

5 Limitations

The results presented in this paper are limited to the publication activities within the EC-TEL conference and do not claim a general validity for the whole TEL community. The explorative data analysis in our research was based on a relatively small set of publications and did not take into consideration poster publications or invited publications from EC-TEL 2006 - 2010. Despite the fact that EC-TEL is a young and relatively small conference, it represents an important section of the whole TEL domain.

6 Conclusions and Further Research Opportunities

In this paper we analyzed 229 papers from 574 authors that have been published in the EC-TEL conference between 2006 and 2010 and presented the results of this exploratory study. The analysis shows a very fragmented community that is dominated by some prolific authors, who account for a large percentage of all publications. This observation underscores the need formulated by the STELLAR Network of Excellence to "*overcome this fragmentation and reach a real multi- and trans-disciplinary approach that TEL research needs*" [12]. This conclusion is further strengthened by other results of ours, which show that authors with five or more publications in EC-TEL 2006 - 2010 (2.3% of the authors) account for 24.5% of all papers. Moreover, the co-authorship analysis shows that the community of EC-TEL authors is not strongly coherent; the analysis shows 103 weakly connected clusters and an overall density of only 0.007.

Future conference management systems could should pick up the analyses presented here in order to provide conference organizers and attendees with easy-to-understand data and visualizations of the connections between papers and authors of a conference. The information that can be extracted from scientific papers and their metadata (e.g. co-citation or bibliographic coupling of papers) can be used to make authors aware of and recommend fellow researchers that they might not yet know but that would be worth cooperating with.

For future research it would be interesting to create a representative paper corpus of the whole TEL field to gain an authentic overview of the TEL community. With such an overview, evidence could be found for the impact of the community activities of the STELLAR Network of Excellence as well as for the international visibility of European TEL by the analysis of large corpora of citations.

References

1. Councill, I.G., Giles, C.L., Kan, M.-Y.: ParsCit: an Open-source CRF Reference String Parsing Package. In: LREC (2008)
2. Fisichella, M., Herder, E., Marenzi, I., Nejdl, W.: Who are you working with? - visualizing tel research communities. In: ED-MEDIA 2010 - World Conference on Educational Multimedia, Hypermedia & Telecommunications (2010)
3. Foolabs. Xpdf (2011), http://www.foolabs.com/xpdf/ (accessed April 15, 2011)
4. Gephi NGO. Gephi - the open graph viz platform (December 2011), http://gephi.org (accessed April 15, 2011)
5. Henry, N., Goodell, H., Elmqvist, N., Fekete, J.-D.: 20 Years of Four HCI Conferences: A Visual Exploration. Intl. Journal of Human-Computer Interaction 23(3), 239–285 (2007)
6. Nejdl, W., Tochtermann, K. (eds.): Innovative Approaches for Learning and Knowledge Sharing. Springer, Heidelberg (2006)
7. Ochoa, X., Méndez, G., Duval, E.: Who we are: Analysis of 10 years of the ED-MEDIA Conference. In: Siemens, G., Fulford, C. (eds.) Proceedings of World Conference on Educational Multimedia, Hypermedia and Telecommunications (ED-MEDIA) 2009, Chesapeake, VA, pp. 189–200 (2009), AACE
8. Reinhardt, W., Moi, M., Varlemann, T.: Artefact-Actor-Networks as tie between social networks and artefact networks. In: Proceedings of the 5th International ICST Conference on Collaborative Computing: Networking, Applications and Worksharing (CollaborateCom 2009) (November 2009)
9. Reinhardt, W., Varlemann, T., Moi, M., Wilke, A.: Modeling, obtaining and storing data from social media tools with Artefact-Actor-Networks. In: Proceedings of the 18th Intl. Workshop on Personalization and Recommendation on the Web and Beyond (2010)
10. Springer. Springerlink (2011), http://www.springerlink.com/ (accessed April 15, 2011)
11. Stefaner, M., Dalla Vecchia, E., Condotta, M., Wolpers, M., Specht, M., Apelt, S., Duval, E.: MACE – enriching architectural learning objects for experience multiplication. In: Duval, E., Klamma, R., Wolpers, M. (eds.) EC-TEL 2007. LNCS, vol. 4753, pp. 322–336. Springer, Heidelberg (2007), http://dx.doi.org/10.1007/978-3-540-75195-3_23
12. STELLAR. Concept and project objectives (2011), http://www.stellarnet.eu/about/ (accessed April 15, 2011)

Growing beyond Innovators – ICT-Based School Collaboration in eTwinning

Riina Vuorikari, Anne Gilleran, and Santi Scimeca

European Schoolnet, Rue de Treves, 61
1050 Brussels, Belgium
{Riina.Vuorikari,Anne.Gilleran,Santi.Scimeca}@eun.org

Abstract. We investigate how eTwinning, the community for schools in Europe that promotes teacher and school collaboration through the use of Information and Communication Technologies (ICT), has grown in its first five years. eTwinning currently attracts over 130'000 teachers in Europe who can be identified as "innovators" in using ICTs in cross-border school collaboration projects. Using the indicator called eTwinning reach, we find that on average, 2.64% of European teachers already participate in eTwinning. In this paper, we investigate the synergies between eTwinning and professional development. We argue that in order to grow beyond innovators and to attract early adopters in innovative use of ICTs in education, eTwinning should be given a more prominent and official status among national teachers' professional development activities. This, in turn, would greatly enhance both formal and informal teachers' professional development schemes in the countries that participate in the eTwinning action.

Keywords: school collaboration, information and communication technologies, teachers, professional development, spread of innovation.

1 Introduction

eTwinning (www.etwinning.net) is defined as "the community for schools in Europe". It promotes teacher and school collaboration through the use of Information and Communication Technologies (ICT). Currently, thirty-two countries in Europe participate in eTwinning. Since 2005, eTwinning has been one of the most successful actions of the school education programme (Comenius) under the European Union's Lifelong Learning Programme.

From a survey conducted in late 2008, we know that one third of the respondents said they signed up for eTwinning to improve their teaching skills (Crawley et *al.*, 2009). Moreover, more than 75% of the respondent teachers stated that their eTwinning project had had an impact, or even a high impact, on improving their ICT skills, communication skills, teaching skills and interdisciplinary working skills, as well as learning about new teaching methods. Additionally, and somewhat unexpectedly, when describing their eTwinning project, over 90% described their eTwinning project as, "it was fun". On the other hand, OECD's report (OECD, 2009)

C. Delgado Kloos et al. (Eds.): EC-TEL 2011, LNCS 6964, pp. 537–542, 2011.

finds that "informal dialogue to improve teaching" is mentioned by teachers as the most common activity for professional development with a participation rate of over 90% in most countries. Interestingly, 87% of teachers also report that participation in "informal dialogue to improve teachers" has a moderate or high level of impact on their professional development. Similarly, involvement in a "professional development network", which eTwinning can also be considered to be, ranks high (80%) as regards to the perceived impact on teachers' development, although it is not among the high participation rated activities (40%). This can indicate that eTwinning could have a high potential for satisfying some of the professional development needs of teachers in Europe.

The main question examined in the paper is: What is the relationship between eTwinning and professional development, and how eTwinning can cross the gap between innovators and early adaptors? We focus on presenting a new parameter called *eTwinning reach* with the help of which we can study the spread of eTwinning applying the idea of Diffusion of Innovation (Rogers, 1962). The starting point of the analysis is by looking at eTwinning statistics in Section 2. Based on that, synergies between eTwinning and teachers' professional development are elaborated (Section 3). Section 4 elaborates on future directions for studying eTwinning using Social Network Analysis and visualisation (Tellnet, 2011).

2 Monitoring eTwinning Activities: Core Statistics and *eTwinning Reach*

eTwinning is regarded as the community for schools in Europe. Teachers from all participating countries can register and use the eTwinning online tools to find each other, meet virtually, exchange ideas and practice examples. The eTwinning online tools are provided by the Central Support Service (CSS), the coordinating body of eTwinning run by European Schoolnet. Additionally, each country involved has a National Support Service (NSS) that represents and promotes the eTwinning action.

Since its inception, eTwinning statistics have been gathered on the eTwinning Portal (www.eTwinning.net) using different methods such as user registration; the interaction of users with the different tools and website analytics to monitor visitors on the site (Crawley *et al.,* 2009). We now introduce a new parameter called *eTwinning reach*, which better allows gauging eTwinning's place within the country. *eTwinning reach* is rooted in the popular idea of Diffusion of Innovations (Rogers, 1962), "a theory of how, why, and at what rate new ideas and technology spread through cultures."

The following method is used: *eTwinning reach* = the registered eTwinning users of a country / teacher population within this country. To calculate it, the OECD data[1] from 2007 (OECD statistics of educational personnel) and the eTwinning data from May 2010 to May 2011 is used. In Table 1 we also show website statistics for one year (Oct 2009 to Oct 2010). The latter number of visitors contains returning visitors and first visits on the eTwinning Portal; the latter is around 32% throughout the countries.

---

[1] http://stats.oecd.org/

Table 1. eTwinning reach

Country	Teachers (OECD 2007)	Visitors on eTwinning portal/year	Registered users on eTwinning in May	eTwinning reach May 2010	eTwinning reach May 2011
Austria	100,984	43,610	914	0.90%	1.06%
Belgium	189,930	124,153	1,382	0.70%	0.90%
Czech Republic	105,818	137,322	2,935	2.80%	**3.73%**
Estonia	17,423	30,589	1,320	7.60%	**9.74%**
Finland	68,442	46,840	1,472	2.20%	**2.83%**
France	707,609	401,956	9,298	1.30%	1.86%
Germany	835,980	275,667	4,606	0.60%	0.75%
Greece	148,627	195,768	3,225	2.20%	**2.65%**
Hungary	135,030	48,746	1,003	0.70%	1.01%
Iceland	6,218	9,810	274	4.40%	**7.16%**
Ireland	60,718	22,049	671	1.10%	1.43%
Italy	723,870	454,420	7,365	1.00%	1.36%
Luxembourg	6,973	3,036	103	1.50%	1.74%
Netherlands	245,876	65,108	1,684	0.70%	0.93%
Norway	89,480	26,944	1,042	1.20%	1.52%
Poland	521,037	523,791	9,895	1.90%	2.57%
Portugal	157,239	83,339	2,239	1.40%	1.90%
Slovakia	63,184	88,161	2,111	3.30%	**4.46%**
Slovenia	22,290	34,693	564	2.50%	**3.70%**
Spain	484,289	477,897	7,966	1.60%	2.23%
Sweden	140,326	61,170	1,992	1.40%	1.98%
Turkey	590,494	496,357	5,941	1.00%	**3.62%**
United Kingdom	788,575	184,044	8,549	1.10%	1.52%
Average:	-	-	-	**1.90%**	**2.64%**

On the two right-hand columns of Table 1, the data for eTwinning reach are presented. On average, in May 2010, 1.9% of the teacher population in countries indicated in Table 1 had registered on the eTwinning Portal. In May 2011, the average had increased to 2,64% of the teachers' population in Europe. In general, rather small countries show higher percentages: Estonia, Iceland, Slovakia, Slovenia and Finland, but also the Czech Republic have reached beyond 2.5% of their teacher population. Similar indicators can be calculated not only for teachers, but also for schools. With some early indicative data collected from NSS, it is estimated that the eTwinning reach for schools in around an average of 25%.

Interesting observations can be derived from this information:

(1) On average, the eTwinning action concerns 2,64% of the potential teaching population within the participating countries. According to Rogers' model of diffusion of innovation (1962), eTwinning in most countries still remains limited to

teachers who are "innovators" in using ICT for cross-country school collaboration. Some countries have passed the 2.5% milestone of "innovators" (e.g., Estonia, Iceland, Slovakia, the Czech Republic, Greece, Slovakia, Slovenia, Finland and Turkey) and are currently targeting the segment of "early adopters" within their teacher population. Several of these countries have a relatively small population, although the Czech Republic has over 10 million and Turkey over 70 million inhabitants.

(2) The indicator *eTwinning reach* can be used to monitor growth of eTwinning over a period of time.

(3) If the *eTwinning reach* indicator is interpreted alone without the context of general interest on the eTwinning portal (e.g., by using web analytics), it can give an impression that the eTwinning action is hardly well known in the countries that implement it. However, if one examines by country the number of visitors to the eTwinning Portal compared to the number of registered users, it shows that, in general, there are many more visits on the Portal from participating countries than there are registered users (in Table 1, compare figures "visitors on eTwinning Portal" with "registered users"). This may be interpreted as a level of interest in eTwinning and in particular in Norway, Iceland, Estonia, Greece, Slovenia, Slovakia, Poland and the Czech Republic, the number of visitors on the Portal indicates a strong interest in eTwinning.

(4) Based on calculations regarding the number of schools present in each country (note, not part of Table 1), it can be seen that the ration between available schools and registered schools is quite high, up to 25%. This demonstrates that the coverage of schools is much higher than the coverage of teachers. This in turn could potentially provide an opportunity to increase the eTwinning teachers' reach through local dissemination and growth (see Scimeca *and al.*, 2009 for a typology of growth in school networks).

3 Finding Synergies between eTwinning and Local Professional Development

From Vuorikari (2010), it can be seen that there is a strong link between eTwinning and professional development. Out of twenty-eight countries for which we were able to gather the information (Vuorikari, 2010), only in seven countries (Estonia, Hungary, Lithuania, Poland, Portugal, Slovenia and Spain) eTwinning activities can be fully taken into account for formal professional development. In eleven countries, some synergies were found, whereas in nine countries, the situation was the opposite: there was no link between eTwinning and formal professional development. Therefore, we can conclude that in 58% of the twenty-eight countries, eTwinning can be used at least to some extent to support the goals of professional development programmes.

There are many similar features, one of which is the status of eTwinning as part of formal professional development and career advancement programmes. We have seen examples of recognition and accreditation of eTwinning activities where they are built-in elements of formal professional development opportunities, e.g., participation in an eTwinning project and resources produced within a project can be used to gain career credits; eTwinning online training courses or workshops count for professional development; some ambassador-type activities can be encouraged with monetary incentives.

Apart from synergies found in professional development and eTwinning, several framework conditions exist in countries that are successful in eTwinning. For example, teachers' ICT maturity and skills count as underlying conditions; these can be related to the wider scale of professional development opportunities in the given country. According to TALIS "Index of development need", there is a demand for professional development in many eTwinning countries. Also, concerning infrastructure and availability of computers in schools, there is more variation between the countries studied. All these framework conditions most likely influence eTwinning and teachers' participation in it in one way or another, but precise understanding of the dynamics is difficult to achieve from this type of study.

A clearer picture can be drawn of how eTwinning interacts with both *formal* and *informal* professional development opportunities. In this case, by *formal* professional development we mean that the needs and goals of a national and/or local professional development programme strongly interact with eTwinning. This results, for example, in an eTwinning online course that is offered to teachers as part of the other general professional development offerings. By *informal* professional development we mean that teacher's involvement in an eTwinning project, for example, is recognised to support the goals of a national and/or local professional development programme, and therefore teacher's participation in such a project can count, in a way or another, towards formal recognition, e.g., career credits or advancement in the teacher's career.

Apart from formal recognition, the value of informal recognition that teachers get from being part of eTwinning is also notable. This informal recognition manifests itself in different ways and is hard to measure. Most importantly, it seems that teachers participating in eTwinning activities can gain positive status within their own work environment (e.g., school), and also outside (e.g., parents' interest in their project). Informal recognition can also be personal self-fulfilment, for example, the satisfaction of learning new skills and perspectives through eTwinning. Moreover, informal recognition can also be translated into behaviour in the work environment. For example, it can be very easy for a teacher to participate in eTwinning because of a well-developed and active professional development culture in the school. Lastly, it can be the collegial attitude of a school head or a colleague in agreeing to arrange replacement teaching while the teacher participates in eTwinning activities outside of the school. All the eTwinning stories (see in Vuorikari, 2010), which were gathered from NSS interviews as "typical" cases, also confirm this finding.

The current research supports these types of informal practices: the kind of professional development a teacher participates in is more important than the amount of time invested. The net effects of days of professional development are small and only significant in a few countries, whereas indicators of participation in networks and mentoring (and in some countries also in workshops and/ or courses) have significant and stronger net associations with teaching practices in a majority of countries" (OECD, 2009, 117).

4 Growing beyond Innovators, Attracting Early Adaptors

This paper suggests that eTwinning has great potential to make a significant contribution to teachers' continuous professional development and lifelong learning at European level as well as at national and local level. It gathers evidence on how eTwinning creates a wider framework for ICT-based school collaboration that offers both formal and informal professional development opportunities also giving teachers either formal and/or informal recognition for their input and effort. Having emphasised eTwinning as a professional development network for teachers, it becomes interesting and important to gain deeper understanding of how such a network can actually support and foster teachers' development also in the future, and how it can engage a wider scale of practitioners and not only innovators in the area of pedagogical use of ICTs for school collaboration.

The Tellnet project (2011) will focus on studying eTwinning through visualisation techniques, Social Network Analysis (SNA) and prospective scenario building exercises. Current policies and research acknowledge teachers' role in developing innovation in education and training practices. There is particularly a need to harness the transformative potential of ICT for innovative and effective teaching and learning approaches. The assumption is that a Learning Network is a suitable support for competence development (Koper et *al.,* 2005). Such networks can have a key role in supporting in-service teachers in their changing role, encouraging their professional development and sharing good practices. Therefore, it is important to understand the dynamic and multidirectional flow of social influence within such networks.

References

1. Crawley, C., Gilleran, A., Scimeca, S., Vuorikari, R., Wastiau, P.: Beyond School Projects, A report on eTwinning 2008-2009. Central Support Service for eTwinning (CSS), European Schoolnet (2009)
2. OECD, Creating Effective Teaching and Learning Environments: First results from TALIS (2009)
3. Rogers, E.M.: Diffusion of innovations. Simon and Schuster, New York (1962)
4. Scimeca, S., Dumitru, P., Durando, M., Gilleran, A., Joyce, A., Vuorikari, R.: European Schoolnet: enabling school networking. European Journal of Education 44(4), 475–492 (2009), doi:10.1111/j.1465-3435.2009.01407.x
5. Tellnet, Teachers Lifelong Learning Network (2011), `http://www.tellnet.eun.org`
6. Koper, R., Rusman, E., Sloep, P.: Effective Learning Networks. Lifelong learning in Europe 1, 18–27 (2005)
7. Vuorikari, R.: eTwinning Report 2010: Teachers' professional development: an overview of current practice. European Schoolnet (2010)

Author Index